S0-ARD-757

PEARSON BACCALAUREATE

Mathematical Studies

2nd Edition

ROGER BROWN • RON CARRELL • DAVID WEES

Supporting every learner across the IB continuum

Published by Pearson Education Limited, Edinburgh Gate, Harlow, Essex, CM20 2JE.

www.pearsonglobalschools.com

Text © Pearson Education Limited 2013
Edited by Krysia Winska, Maggie Rumble and Gwen Burns
Designed by Tony Richardson
Typeset by Techtype
Original illustrations © Pearson Education Ltd 2013
Cover photo © **Science Photo Library Ltd:** Ted Kinsman.

First published 2008

This edition published 2013

18 17 16 15 14 13
IMP 10 9 8 7 6 5 4 3 2 1

British Library Cataloguing in Publication Data
A catalogue record for this book is available from the British Library

ISBN 978 1 447 93847 7

Printed in Spain by Graficas Estella

Acknowledgements
The publisher would like to thank the following for their kind permission to reproduce their photographs:

(Key: b-bottom; c-centre; l-left; r-right; t-top)

Corbis: 473bl, Bettmann 469tl, 474bl, Richard T. Nowitz 468tr, Robert Llewellyn 471r, Ron Chapple 178b, Rudy Sulgan 184b; **Fotolia.com:** B. and E. Dudzinscy 103tr, B. and E. Dudzinscy 103tr, Benshot 103br, bradcalkins 197t, Eric Isselée 92cl, sanderstock 159br; **Getty Images:** SMC Images 178cr; **Science Photo Library Ltd:** 469br, Genral Electric Research and Development Center / Emilo Segre Visual Archives / American Institute of Physics 479br; **Shutterstock.com:** 155cr, 184t, 476-477bc, Anthony Hall 472cl, Ivonne Wierink 179tr, Lucie Zapletalova 478tl, Rafael Ramirez Lee 472tr, Svetlana Privezentseva 173br, Tomasz Trojanowski 468bl

All other images © Pearson Education

Every effort has been made to contact copyright holders of material reproduced in this book. Any omissions will be rectified in subsequent printings if notice is given to the publishers.

Websites
There are links to relevant websites in this book. In order to ensure that the links are up to date and that the links work we have made the links available on our website at www.pearsonhotlinks.co.uk. Search for this title **Pearson Baccalaureate Mathematical Studies 2nd edition for the IB Diploma** or ISBN 9781447938477.

Dedication

I would like to dedicate this textbook to my wife Vasilia and my son Thanasi. Without their support and understanding this project would not have been possible.

David Wees

I would like to dedicate this book to my muse and wife Sharlene. Her immutable support enabled me to accomplish my best writing.

Ron Carrell

I would like to dedicate this textbook to my children Mark and Kylie, who as students taught me a lot about teaching and learning. As well, I dedicate this book to my wife and partner Birgit, who has always supported me as I have explored the many facets of mathematics education.

Roger Brown

Contents

Contents _____

Introduction

This book has been written to help you prepare for and pass the Mathematical Studies (MS) exam. In particular, four chapters have been included to help with certain aspects of the course.

The first is Chapter 1: Prior Learning Topics. This chapter reviews all of the Prior Learning Topics (PLT) content on pages 14 and 15 in the Mathematical Studies SL Guide (MSSLG). As stated in the guide, students are not required to be familiar with all of PLT topics before they start the MS course. However, students have very different levels of knowledge when they enter the course, so we thought it would be helpful to provide a review. In doing so, you will find that there will be some duplication of material in the later chapters. This should not concern you. The more problems you solve, the better prepared you will be for the exams. Further, your instructor may be able to 'fine tune' which topics you need and which you do not. Our advice to you is to understand the nature of the PLT chapter and make appropriate use of it.

The second is Chapter 14: Review Questions. This chapter contains 123 questions (often with multiple parts) that review every Syllabus Content Detail (SCD). It can be of great value if you solve each problem and strive for understanding as you work through each Syllabus Content (SC) area. You are advised not to wait until the 'last minute' to begin preparation for your examination. Chapter 14 alone will take some time to work through. Keep track of your progress and identify those areas in which you are successful as well as those areas that you need to review.

The third is Chapter 15: Practice Papers. This chapter includes two complete examinations: two Paper 1's and two Paper 2's. The papers have been constructed with the most appropriate questions that could be found for our current curriculum. Understanding the concepts behind each problem and being able to work through each paper in the allotted amount of time will increase your chances of success in your IB exams.

The fourth is Chapter 16: The Project. The project comprises 20% of your entire IB Exam grade. It is well worth your while to spend some time completing this task to the best of your ability. This chapter provides the criteria used for grading, a checklist to help keep you focused, and a sample project for reference.

At the beginning of each chapter, you will see a heading 'Assessment Statements'. These refer to the Syllabus Content Details (SCD). The 'Assessment Statements' are included to help you remember which SCD you are studying.

The questions in the written exercises are generally grouped according to type. The first third, or so, of each set of exercises allows you to review your comprehension of the basic concepts and ideas. No conscious effort was made to simulate the IB style questions. The middle third are similar to Paper 1 style questions. In the last third we made every attempt to simulate, with language, style, and notation, Paper 2 questions, and again you will find several questions designated as such.
Good luck!

Roger Brown
Ron Carrell
David Wees

Information boxes

Throughout the book you will see a number of coloured boxes interspersed through each chapter. Each of these boxes provides different information and stimulus as follows.

> **Assessment statements**
> 3.1 Basic concepts of symbolic logic: definition of a proposition; symbolic notation of propositions.
> 3.2 Compound statements: implication, equivalence, negation, conjunction, disjunction, exclusive disjunction. Translation between verbal statements and symbolic form.

You will find a box like this at the start of each section in each chapter. They are the numbered objectives for the section you are about to read and they set out what content and aspects of learning are covered in that section.

What are some of the ways that mankind has dealt with the concept of *infinite*? Exactly what does it mean to be infinitely large?

In addition to the Theory of Knowledge chapter, there are TOK boxes throughout the book. These boxes are there to stimulate thought and consideration of any TOK issues as they arise and in context. Often they will just contain a question to stimulate your own thoughts and discussion.

Some books on number theory include proofs that show there are infinitely many prime numbers, that there are infinitely many levels of infinite, that the square root of any prime number is not a rational number and that π is irrational.

These boxes contain interesting information which will add to your wider knowledge and help you see how mathematics connects with other subjects.

Use of the Graphing Display Calculator. Throughout the text examples and solutions have been completed on the Texas Instruments TI84, with the operating system as shown.

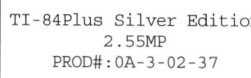

These facts are drawn out of the main text and are highlighted. This makes them useful for quick reference and they also enable you to identify the core learning points within a section.

The International System of Units is abbreviated to SI. This is from the French phrase Le Système International d`unités.

These boxes indicate examples of internationalism within the area of study. The information in these boxes gives you the chance to think about how mathematics fits into the global landscape. They also cover environmental and political issues raised by your subject.

These boxes can be found alongside questions, exercises and worked examples and they provide insight into how to answer a question in order to achieve the highest marks in an examination. They also identify common pitfalls when answering such questions and suggest approaches that examiners like to see.

● **Examiner's hint:** It is commonly accepted to only use the unit of measure, cm, or cm^2, with the final answer.

These boxes direct you to the Pearson Hotlinks website, which in turn will take you to the relevant website(s). On the web pages you will find background information to support the topic, video simulations and the like.

 For the proof that $\sqrt{2}$ is irrational, visit www.pearsonhotlinks.co.uk, enter the ISBN for this book and click on weblink 1.1.

Now you are ready to start. Good luck with your studies!

Prior Learning Topics

Overview

The purpose of this chapter is to review topics that you are expected to be familiar with. These topics include: numbers, fractions, algebra, geometry, trigonometry, financial mathematics and statistics. Questions from the IB examinations will assume knowledge of these topics. Therefore, you should refer to these topics, concepts and solutions as needed, as you work through the text.

Each section will provide the basic concepts and building blocks necessary for mastery of the forthcoming chapters.

1.1 Numbers

Classifying numbers

The idea of number is older than recorded history. Mankind's ability to count probably began with the necessity for recording quantities, using tally bones. The oldest example may be the piece of baboon leg bone showing 29 notches or tally marks. This was found in Swaziland and is believed to date back to 35 000 BCE

There have been many number systems used throughout history. A few of these are: Egyptian, Babylonian, Greek, Roman, Chinese-Japanese, Mayan, and Hebrew systems. Today we use the Hindu-Arabic number system. This system is named after the Hindus, who invented it, and the Arabs who introduced it to western civilization.

The numbers in the Hindu-Arabic system can be classified according to the properties that each has. Listed below are the classifications (sets) you will need to know.

Natural numbers $= \mathbb{N} = \{0, 1, 2, 3, \ldots\}$
Integer numbers $= \mathbb{Z} = \{\ldots, -3, -2, -1, 0, 1, 2, 3, \ldots\}$
Positive integer numbers $= \mathbb{Z}^+ = \{1, 2, 3, \ldots\}$
Rational numbers $= \mathbb{Q} = $ numbers that can be expressed as a ratio of two integers $\left(\dfrac{p}{q} : q \neq 0\right)$.

$$Q = \left\{ \begin{array}{l} \ldots, -\frac{1}{3}, -\frac{1}{2}, -\frac{1}{1}, 0, \frac{1}{1}, \frac{1}{2}, \frac{1}{3}, \ldots \\ \ldots, -\frac{2}{3}, -\frac{2}{2}, -\frac{2}{1}, \quad \frac{2}{1}, \frac{2}{2}, \frac{2}{3}, \ldots \\ \ldots, -\frac{3}{3}, -\frac{3}{2}, -\frac{3}{1}, \quad \frac{3}{1}, \frac{3}{2}, \frac{3}{3}, \ldots \end{array} \right\}$$

Positive rational numbers $= \mathbb{Q}^+ = \{x \mid x \in \mathbb{Q}, x > 0\}$
Irrational numbers $= \mathbb{Q}' = \{$real numbers that are not rational$\}$
Real numbers $= \mathbb{Q} \cup \mathbb{Q}' = \{$all numbers on the number line$\}$
Positive real numbers $= \mathbb{R}^+ = \{x \mid x \in \mathbb{R}, x > 0\}$

$\{x \mid x \in \mathbb{Q}, x > 0\}$
This notation is read: 'the set of all x, such that x is an element of the set of rational numbers and x is greater than zero.'

A Venn diagram is often helpful in visualizing the relationships between sets of numbers.

Real numbers

Rational numbers	Irrational numbers
$-2, \frac{3}{5}, 0.68\dot{6}\dot{8}, -0.5, \sqrt{25}$ Integer numbers $..., -2, -1, 0, 1, 2...$ Natural numbers $0, 1, 2, 3, ...$	$\sqrt{2}$ $-\sqrt{17}$ $\log 23$ π e $0.121221222\ ...$ $\sin 13°$

Figure 1.1 Venn diagram for the set of real numbers.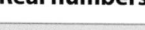

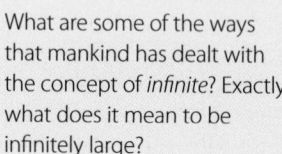
There are exactly the same number of numbers in the set of natural numbers as there are in the set of integer numbers or even the set of rational numbers! However, there are more real numbers than there are natural numbers! There is more than one level of infinity!

What are some of the ways that mankind has dealt with the concept of *infinite*? Exactly what does it mean to be infinitely large?

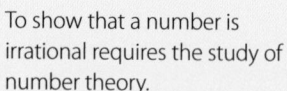
To show that a number is irrational requires the study of number theory.

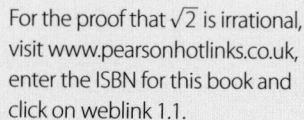
For the proof that $\sqrt{2}$ is irrational, visit www.pearsonhotlinks.co.uk, enter the ISBN for this book and click on weblink 1.1.

Example 1.1

Classify each of the following numbers as $\mathbb{N}, \mathbb{Z}, \mathbb{Z}^+, \mathbb{Q}, \mathbb{Q}^+, \mathbb{Q}', \mathbb{R}, \mathbb{R}^+$.

a) 3 b) 0.4 c) $\sqrt{2}$ d) $-\frac{37}{5}$

Solution

a) Since $3 = \frac{3}{1}$, 3 is a member of $\mathbb{N}, \mathbb{Z}, \mathbb{Z}^+, \mathbb{Q}, \mathbb{Q}^+, \mathbb{R}$, and $\mathbb{R}^+$.

b) Since $0.4 = \frac{4}{10}$, 0.4 is a member of $\mathbb{Q}, \mathbb{Q}^+, \mathbb{R}$, and $\mathbb{R}^+$.

c) Since $\sqrt{2}$ cannot be expressed as the ratio of two integers, $\sqrt{2}$ is a member of $\mathbb{Q}', \mathbb{R}$, and $\mathbb{R}^+$.

d) $-\frac{37}{5}$ is located in the 37th column to the left of 0 and down to the 5th row of the chart on page 1. Therefore, $-\frac{37}{5}$ is a member of $\mathbb{Q}$ and $\mathbb{R}$.

Order of operations

In order to avoid confusion when performing a series of arithmetic operations, we follow a standard order:

Step 1: Eliminate all **P**arentheses.

Step 2: Simplify all **E**xponents.

Step 3: Perform **M**ultiplication and **D**ivision as you come to them, reading from left to right.

Step 4: Perform **A**ddition and **S**ubtraction as you come to them, reading from left to right.

An easy way to remember these steps is by using the mnemonic:
Please **E**xcuse **M**y **D**ear **A**unt **S**ally (**PEMDAS**)

Example 1.2

Simplify each of the following.

a) $3 \cdot 5 - 7 + 8 \div 2$

b) $2(9 - 5)^2 + -1 \times 10 \div 2$

c) $\sqrt{4^2 + 3^2}$

Solution

a) $3 \cdot 5 - 7 + 8 \div 2 = 15 - 7 + 4 = 8 + 4 = 12$

b) $2(9 - 5)^2 + -1 \times 10 \div 2 = 2(4)^2 + -10 \div 2 = 2 \cdot 16 + -5$

   $= 32 + -5 = 27$

c) $\sqrt{4^2 + 3^2} = \sqrt{16 + 9} = \sqrt{25} = 5$

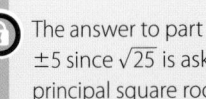 The answer to part c) is not ± 5 since $\sqrt{25}$ is asking for the principal square root, which, by definition, is always positive.

Number theory

Number theory is the study of integers. This review section will cover the basics: prime numbers, factors and multiples.

 Some books on number theory include proofs that show there are infinitely many prime numbers, that there are infinitely many levels of infinite, that the square root of any prime number is not a rational number and that π is irrational.

Prime numbers

A prime number is defined as a natural number greater than 1 whose only positive divisors are 1 and itself.

For example, 5 is a prime number since the only natural numbers (other than 0) that divide into it (without a remainder) are 1 and 5.

As a counter-example, 6 is not a prime number, since both 2 and 3 divide into 6.

All other natural numbers (0 and 1 excluded) that are not prime are called **composite**.

By definition, the first prime number is 2.

Below is a partial list of prime numbers.

Prime numbers $= \{2, 3, 5, 7, 11, 13, 17, 19, 23, \ldots\}$

That there are infinitely many primes is not a foregone conclusion since, as you think of larger and larger numbers, there are also more and more numbers that have a chance of dividing into that number, thus making that large number not prime!

Example 1.3

Determine if 137 is a prime number.

Solution

Start dividing 137 by natural numbers to see if any divide into 137 without a remainder. $\dfrac{137}{2} = 68.5$

$\dfrac{137}{3} = 45.\overline{6}$ and so on.

Do this procedure using all of the natural numbers that are smaller than or equal to the $\sqrt{137}$ plus one (i.e. $11.7 + 1 = 12$).

$\dfrac{137}{12} = 11.41\overline{6}$

Since none of those natural numbers divide 137 evenly, 137 must be prime.

 You only have to test the numbers up to the approximate value of $\sqrt{137}$ since 12×12 is greater than 137.

Factors

Factors are numbers that are multiplied together.

In this book, factors are considered to be natural numbers other than 0.

For example, 2 and 5 are called factors of 10 since $2 \cdot 5 = 10$.

A commonly worded question is to ask for a natural number to be written as a product of primes, often called the **prime factorization** of that number.

Example 1.4

Write 48 as a product of primes.

Solution

$$48 = 2 \cdot 24$$
$$= 2 \cdot 2 \cdot 12$$
$$= 2 \cdot 2 \cdot 2 \cdot 6$$
$$= 2 \cdot 2 \cdot 2 \cdot 2 \cdot 3$$
$$= 2^4 \cdot 3$$

Another popular technique used is called the **factor tree**.

Figure 1.2 Writing the prime factorization of 48 using a factor tree.

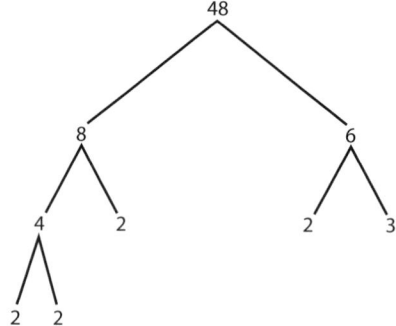

Example 1.5

List all of the natural number factors of 45.

Solution

The factors of 45 are 1, 3, 5, 9, 15, 45 since those are the only natural numbers to divide into 45 (without a remainder).

The greatest common factor (GCF) of two or more natural numbers is the greatest number that will divide into all of the given numbers.

Example 1.6

Find the greatest common factor of 24 and 36.

A notation that can be used to express the idea of the greatest common factor is 'GCF(24, 36)'.

Solution

The factors of 24 are 1, 2, 3, 4, 6, 8, **12**, 24.

The factors of 36 are 1, 2, 3, 4, 6, 9, **12**, 18, 36.

As you can see, there are five common factors for 24 and 36, but there is only one that is the largest, the greatest. Therefore, the GCF(24, 36) = 12.

Multiples

A **multiple** of a natural number is the product of that number and another natural number.

There are infinitely many multiples of natural numbers (except 0).

For example, 30 is a multiple of 3 since $3 \cdot 10 = 30$.

Example 1.7

List the first five multiples of each of the following.
a) 3
b) 17
c) 100

Using the TI Calculator, repeated addition can be completed as shown.

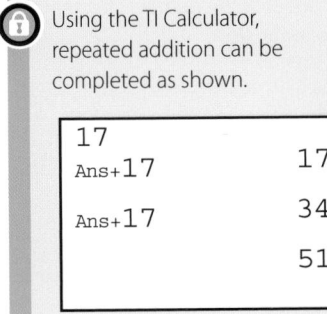

Solution
a) $3 \cdot 1 = 3, 3 \cdot 2 = 6, 3 \cdot 3 = 9, 3 \cdot 4 = 12, 3 \cdot 5 = 15$
b) Another method involves adding 17 to each preceding number:
    $17, 17 + 17, 34 + 17, 51 + 17, 68 + 17 \Rightarrow 17, 34, 51, 68, 85$
c) The first five multiples of 100 are 100, 200, 300, 400, 500.

The least common multiple (LCM) of two or more numbers is the smallest number that the given numbers will divide into.

Example 1.8

Find the least common multiple of 12 and 18.

Using the TI Calculator, the LCM can be found as shown.

Solution
The multiples of 12 are: 12, 24, **36**, 48, 60, **72**, 84, 96, **108**, ...

The multiples of 18 are: 18, **36**, 54, **72**, 90, **108**, ...
As you can see, there are infinitely many common multiples, but there is only one that is the smallest, the least. Therefore, the LCM(12, 18) = 36.

Ratios, percentages and proportions

A **ratio** is a quotient (a fraction) of two numbers.

A **percentage** is a part of 100.

A **proportion** is an equation involving two ratios.

An example of a ratio is $\frac{250}{30}$. You can think of this ratio as 250 for every 30. We can apply ratios in our everyday lives. This ratio could represent 250 kilometres for every 30 litres or it could represent 250 apples for every 30 horses. Some ratios can be simplified in order to make large numbers less cumbersome. $\frac{250}{30}$ can be simplified by dividing both the numerator (the top number) and the denominator (the bottom number) by the GCF(30, 250).

Since 10 is the GCF, then $\frac{250}{30} = \frac{250 \div 10}{30 \div 10} = \frac{25}{3}$.

The International System of Units is abbreviated to SI. This is from the French phrase Le Système International d'unités.

Can you see the number 100 disguised as the '%' symbol?

An example of a percentage would be 20 per cent or 20%.

20% can be written as the ratio $\frac{20}{100}$.

An example of a proportion would be $\frac{20}{100} = \frac{1}{5}$. Sometimes a proportion is written in this style: 20:100 = 1:5 where 20 and 5 are called the **extremes** and 100 and 1 are called the **means**. You should be able to see that $20 \times 5 = 100 \times 1$. Putting the concept of proportions into words we say, 'the product of the extremes is the product of the means'. This idea can be written as a general rule.

The proof, or verification, of this rule involves multiplying both sides of the proportion by the product of the numbers in the denominators.

$$\text{If } \frac{a}{b} = \frac{c}{d} \text{ then } a \cdot d = c \cdot b, \text{ where } b, d \neq 0.$$

Example 1.9

Consider the fact that 5280 feet = 1.609 kilometres.
a) Write a ratio for feet to kilometres.
b) Write a proportion using 10 560 feet and 3.218 kilometres.
c) Write a proportion describing 30% of 5280 feet.

Solution

a) $\frac{5280}{1.609}$ b) $\frac{5280}{1.609} = \frac{10\,560}{3.218}$ c) $\frac{30}{100} = \frac{x}{5280}$

Exercise 1.1

1. List the set of integers between -5 and 3 inclusive.

2. Using the chart on page 1, describe where the rational number $\frac{16}{27}$ can be located.

3. Using the chart on page 1, write the fourth row for the set of rational numbers.

4. What is the first integer number to the right of 0 on a number line?

5. What is the first rational number to the right of 0 on a number line?

6. What is the first real number to the right of 0 on the number line?

7. List the set of natural numbers that are multiples of 2.

8. List the set of natural numbers that are one more than the multiples of 2.

9. List the first twenty prime numbers.

10. Classify each of the following as $\mathbb{N}, \mathbb{Z}, \mathbb{Z}^+, \mathbb{Q}, \mathbb{Q}^+, \mathbb{Q}', \mathbb{R}, \mathbb{R}^+$, prime.
    a) 2   b) -5 c) $\frac{2}{3}$ d) $\sqrt{5}$ e) 10^{50}
    f) π g) $\frac{22}{7}$ h) 0   i) 1   j) -1.23×10^{-3}

11. List the positive integer factors of:
    a) 18   b) 45   c) 72   d) 100

12. List the first five positive multiples for:
    a) 3   b) 13   c) 19

13. Write each of the following as a product of primes:
    a) 108   b) 75   c) 5733

14. Determine if each of the following is prime:
    a) 253   b) 257   c) 391   d) 421

15. Find the GCF and LCM of each of the following:
    a) 36, 48   b) 18, 24, 100   c) 9, 12, 33

16. Simplify each of the following as an exact answer or as indicated:

a) $(-2)^2 \cdot 3 + 5 \div 4$

b) $-2^2 \times 4^{-1} + 6 \times 2 - 1$

c) $4000\left(1 + \frac{0.056}{12}\right)^{12 \cdot 10}$, correct to 2 decimal places

d) $\dfrac{1}{\sin 23° - \sin 37°}$, correct to 2 decimal places

e) $7 + 6 \div 2 - 3 \times 4 + 4 \div 2 + 6$

f) $2 + 3^{-1} - \dfrac{\log 13}{\log 2}$, correct to 2 decimal places

g) $\sqrt{15^2 - 5^2}$

h) $4 \div 2 \times 3$

17. Consider the fact that 720 feet = 120 fathoms.

a) Write a ratio for feet to fathoms.

b) Write a proportion using 2160 feet and 360 fathoms.

c) Write a proportion describing 90% of 360 fathoms.

18. Consider the fact that 12 months = 365.25 days.

a) Write a ratio for months to days.

b) Express that ratio as the simplified quotient of two integers.

c) Write a proportion using 48 months and correct number of days.

d) Express that ratio as the simplified quotient of two integers.

e) Write a proportion describing 25% of 48 months.

1.2 Fractions

In general, a fraction is the quotient of any two quantities. Another name for a fraction is a ratio. For example, $\frac{\text{miles}}{\text{hour}}$ is the fraction (a ratio) expressing a distance in miles and time in hours. In arithmetic, a simple fraction is usually considered as the quotient of two integer numbers such as $\frac{2}{3}$. There are many types of fractions: complex, similar, rational, proper, improper, terminating, non-terminating, continued, partial and repeating, to name a few. Fractions can be added, subtracted, multiplied, divided, simplified, changed to decimals, and so on. This section will review the important properties of simple fractions.

● **Examiner's hint:** When taking the IB exams you may leave the fraction in unsimplified form, e.g. $\frac{6}{8}$, and still receive full credit. However, you should always try to simplify to lowest terms.

Simplifying fractions

The 'top' number of a fraction is the numerator. The 'bottom' number is called the denominator.

A **proper fraction** has a numerator that is less than its denominator.

An **improper fraction** has a numerator that is larger than its denominator.

A simple fraction is said to be simplified if the GCF (numerator, denominator) = 1. For example, $\frac{4}{7}$ is said to be simplified since GCF (4,7) = 1.

When the greatest common factor of two numbers, such as 4 and 7, is equal to one, then those numbers are said to be relatively prime.

Example 1.10

Simplify $\frac{12}{18}$

Solution 1

Since the GCF$(12, 18) = 6$, then $\frac{12}{18} = \frac{2 \cdot 6}{3 \cdot 6} = \frac{2}{3}$.

Solution 2

$\frac{12}{18} = \frac{12 \div 6}{18 \div 6} = \frac{2}{3}$

Solution 3

Press the following keys on your calculator:

ALPHA, Y = , 1, 12, >, 18, ENTER

The simplified fraction is shown.

The keystrokes ALPHA, Y = , 1, 12, >, 18, ENTER will simplify the unsimplified fraction.

Multiplying fractions

When multiplying one fraction by another, multiply the numerators and multiply the denominators.

Example 1.11

Find the product of $\frac{5}{9} \cdot \frac{4}{7}$

Solution 1

$\frac{5}{9} \cdot \frac{4}{7} = \frac{5 \cdot 4}{9 \cdot 7} = \frac{20}{63}$

Solution 2

Press the following keys on your calculator:

ALPHA, Y = , 1, 5, >, 9, >, ×, ALPHA, Y = , 1, 4, > 7, ENTER

The answer is as shown.

When multiplying a fraction by an integer, rewrite the integer as a fraction whose numerator is the integer and whose denominator is 1 and then follow the above rules.

For example, $5 = \frac{5}{1}$

Example 1.12

Find the product of $\frac{5}{8} \cdot 3$, leaving your answer as an improper fraction.

Solution 1

$$\frac{5}{8} \cdot 3 = \frac{5}{8} \cdot \frac{3}{1} = \frac{5 \cdot 3}{8 \cdot 1} = \frac{15}{8}$$

Solution 2

Press the following keys on your calculator:

ALPHA, Y = , 1, 5, >, 8, >, ×, 3, ENTER

The answer is as shown.

$$\frac{5}{8} * 3 \qquad\qquad \frac{15}{8}$$

Dividing fractions

When dividing two fractions, multiply the first by the reciprocal of the second.

Example 1.13

Find the quotient of $\frac{3}{7} \div \frac{11}{5}$

Solution 1

$$\frac{3}{7} \div \frac{11}{5} = \frac{3}{7} \cdot \frac{5}{11} = \frac{3 \cdot 5}{7 \cdot 11} = \frac{15}{77}$$

Solution 2

Press the following keys on your calculator:

ALPHA, Y = , 1, 3, >, 7, >, ÷, ALPHA, Y = , 1, 11, >, 5, ENTER

The answer is as shown.

$$\frac{3}{7} / \frac{11}{5} \qquad\qquad \frac{15}{77}$$

A **complex fraction** is a fraction in which both the numerator and denominator are simple fractions.

For example, $\dfrac{\frac{4}{5}}{\frac{7}{9}}$

When simplifying a complex fraction, treat it as the division of two fractions: the numerator times the reciprocal of the denominator.

Example 1.14

Simplify the complex fraction $\dfrac{4/5}{7/9}$

Solution

$$\frac{4/5}{7/9} = \frac{4}{5} \cdot \frac{9}{7} = \frac{4 \cdot 9}{5 \cdot 7} = \frac{36}{35}$$

Adding and subtracting fractions

It is important to know the process for adding and subtracting fractions for algebra, as in Exercise 1.2.

Fractions that are similar have common or like denominators.

Finding the sum or difference of two fractions requires that the fractions be similar.

When adding fractions that are similar, add the numerators and keep the same denominator.

When subtracting two similar fractions, subtract the numerators and keep the same denominator.

Example 1.15

Find the sum of $\frac{2}{3} + \frac{4}{5}$, leaving the answer as an improper fraction.

Solution 1

Multiplying $\frac{2}{3}$ by $\frac{5}{5}$ did not change its value. It just 'unsimplified' it.

$$\frac{2}{3} + \frac{4}{5} = \frac{2}{3} \cdot \frac{5}{5} + \frac{4}{5} \cdot \frac{3}{3} = \frac{10}{15} + \frac{12}{15} = \frac{10 + 12}{15} = \frac{22}{5}$$

Solution 2

Press the following keys on your calculator.

ALPHA, Y = , 1, 2, 3, >, +, ALPHA, Y = , 1, 4, >, 5, ENTER

The answer is as shown.

$$\frac{2}{3} + \frac{4}{5}$$
$$\frac{22}{15}$$

Example 1.16

Find $\dfrac{1}{3} - \dfrac{7}{10}$.

Solution 1

$$\frac{1}{3} - \frac{7}{10} = \frac{1}{3} \cdot \frac{10}{10} - \frac{7}{10} \cdot \frac{3}{3} = \frac{1 \cdot 10}{3 \cdot 10} - \frac{7 \cdot 3}{10 \cdot 3} = \frac{10}{30} - \frac{21}{30} = \frac{10 - 21}{30} = \frac{-11}{30}$$

These steps will be skipped in Solution 2, the 'cross-multiply' method.

Solution 2

$$\frac{1}{3} - \frac{7}{10} = \frac{1 \cdot 10 - 3 \cdot 7}{\underbrace{3 \cdot 10}} = \frac{10 - 21}{30} = \frac{-11}{30}$$

Even this step can be skipped!

Exercise 1.2

1. Simplify each of the following:

a) $\frac{24}{36}$ b) $\frac{75}{125}$ c) $\frac{512}{128}$ d) $\frac{255}{153}$

2. Find each product or quotient. Leave your answer as a proper or improper fraction in simplified form.

a) $\frac{3}{4} \cdot \frac{6}{7}$ b) $\frac{3}{4} \div \frac{6}{7}$ c) $\frac{64}{200} \cdot -30$ d) $16 \cdot \frac{5}{8}$ e) $\frac{-12/28}{3/7}$

f) $\frac{72}{96} \div \frac{21}{12}$ g) $\frac{a}{b} \cdot \frac{c}{d}$ h) $\frac{e}{f} \div \frac{g}{h}$ i) $2 \div \frac{1}{2}$ j) $\frac{5}{8} \div 4$

3. Find each sum or difference. Express your answer as a proper or improper fraction in simplified form.

a) $\frac{1}{2} + \frac{3}{5}$ b) $\frac{7}{11} - \frac{3}{4}$ c) $6 + \frac{4}{5}$ d) $-\frac{3}{10} - 3$ e) $2\frac{3}{5} + 4\frac{7}{10}$

f) $12\frac{1}{3} - 15\frac{7}{9}$ g) $\frac{a}{b} + \frac{c}{d}$ h) $\frac{x}{y} - \frac{y}{x}$ i) $z - \frac{1}{y}$ j) $1 - \frac{1}{x}$

1.3 Algebra

Expanding and factorizing

An axiom is a fundamental statement we assume is true without proof. We **must** accept some statements as true or other statements, called theorems, would not be possible to prove. One such statement is called the **distributive axiom**. It is one of the eleven fundamental statements called the **field axioms**.

The distributive axiom states:

$$a(b + c) = ab + ac, \text{ for all } a, b, c \in \mathbb{R}.$$

You should recall that when reading a mathematics equation you must read from left to right as well as from right to left.

For example, reading and applying the axiom from left to right we have: $3(4 + 5) = 3 \cdot 4 + 3 \cdot 5$. Reading and applying the axiom in this manner is often called expanding.

Conversely, reading and applying the axiom from right to left we have: $7 \cdot 9 + 7 \cdot 13 = 7(9 + 13)$. Reading and applying the axiom in this manner is often called factorizing.

Since a variable, such as x, is merely a symbol used to represent a real number, the distributive axiom continues to hold true when they are used. For example, $5(x + 7) = 5x + 5 \cdot 7 = 5x + 35$.

 For a historical perspective of algebra, visit www.pearsonhotlinks.co.uk, enter the ISBN for this book and click on weblink 1.2.

 For an entire list of the field axioms, visit www.pearsonhotlinks.co.uk, ISBN for this book and click on weblink 1.3.

Expanding

Terms are expressions that are being added or subtracted.

Factors are expressions that are being multiplied.

You can think of **expanding** as 'going' from one term to many terms.

An expression such as $(x + 2)(x + 3)$ is considered as one term with two factors.

Example 1.17

Expand each of the following using the distributive axiom. Do not simplify your answer.

a) $6(4 + 8)$ b) $-4(x + 3)$ c) $\sqrt{2}(y - 7)$

Solution

a) $6(4 + 8) = 6 \cdot 4 + 6 \cdot 8$

b) $-4(x + 3) = -4x + -4 \cdot 3$

c) $\sqrt{2}(y - 7) = \sqrt{2}(y + -7) = \sqrt{2}y + \sqrt{2} \cdot (-7)$

Polynomials are expressions having more than one term. For example, $x^2 + 5x + 6$ is a polynomial with three terms. When expanding the product of two binomials (a polynomial with two terms), the distributive axiom may still be used.

Example 1.18

Expand and simplify the product $(x + 2)(x + 3)$.

Solution

Think of the $(x + 2)$ as the 'a' in the axiom and x and 3 as 'b' and 'c' respectively. Therefore,

$$
\begin{aligned}
(x + 2)(x + 3) &= (x + 2) \cdot x + (x + 2) \cdot 3 \\
&= x \cdot (x + 2) + 3 \cdot (x + 2) \\
&= x \cdot x + x \cdot 2 + 3 \cdot x + 3 \cdot 2 \\
&= x^2 + 2x + 3x + 6 \\
&= x^2 + 3x + 2x + 6 \\
&= x^2 + 5x + 6
\end{aligned}
$$

The above solution suggests a shorter method for finding the product. The mnemonic is **FOIL**. The letters stand for: **First, Outside, Inside, Last.**

In other words, multiply the **First** terms x and x, and then the **Outside** terms x and 3, and then the **Inside** terms 2 and x, and finally the **Last** terms 2 and 3.

Example 1.19

Expand and simplify $(x + 4)(x + 7)$ using the FOIL method.

Solution

$$
\begin{aligned}
(x + 4)(x + 7) &= x \cdot x + x \cdot 7 + 4 \cdot x + 4 \cdot 7 \\
&= x^2 + 7x + 4x + 28 \\
&= x^2 + 11x + 28
\end{aligned}
$$

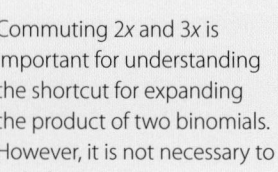

See the commutative axiom for multiplication in the field axiom list.

Commuting $2x$ and $3x$ is important for understanding the shortcut for expanding the product of two binomials. However, it is not necessary to include this step.

The first step is usually not written, except as an explanation.

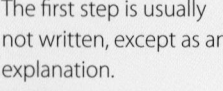

Example 1.20

Expand and simplify $(y + 5)^2$.

Solution

Rewrite $(y + 5)^2$ as $(y + 5)(y + 5)$.
$$(y + 5)(y + 5) = y^2 + 5y + 5y + 25 = y^2 + 10y + 25$$

As with most mathematics, a concept is thought of, a rule made, and a shortcut developed. One shortcut that can be developed (and hence eliminate the middle step of the solution in the example above) is:

$$(a + b)^2 = a^2 + 2ab + b^2$$

Even though the FOIL mnemonic is specific to multiplying two binomials, the idea can be used to expand the product of binomials and trinomials.

Example 1.21

Find and simplify the product $(x + 2)(x^2 + 3x + 5)$.

Solution
$$(x + 2)(x^2 + 3x + 5) = x^3 + 3x^2 + 5x + 2x^2 + 6x + 10$$
$$= x^3 + 5x^2 + 11x + 10$$

Are you able to see the '**FOIL**' method at work here?

Factorizing

A polynomial has been factorized when the answer is in the form of one term. The one term may possibly contain many (two or more) factors.

For example, $x^2 + 7x + 12$ has three terms, but when factorized as $(x + 3)(x + 4)$ it is expressed as only one term (with two factors).

We know that the trinomial has been 'factorized' correctly since when it is expanded', the trinomial, $x^2 + 7x + 12$, reappears.

The concept of factorizing involves expanding. To be able to factorize simple polynomials correctly, you must always expand your (factorized) answer in order to see if the original polynomial has reappeared.

Example 1.22

Factorize $x^2 + 10x + 24$.

Solution

Start by writing two sets of open parentheses: $(\_ + \_)(\_ + \_)$.

Next, fill in the blanks with your best guess so that when you expand the answer you will get $x^2 + 10x + 24$ back again.

Try this guess: $(x + 3)(x + 8)$. This is a good guess since $3 \cdot 8 = 24$.

Now test the guess by expanding:
$$(x + 3)(x + 8) = x^2 + 8x + 3x + 24 = x^2 + 11x + 24$$

Since this is not $x^2 + 10x + 24$, the factorized form of $(x + 3)(x + 8)$ was not correct. In other words, 3 and 8 were not the correct choices.

Therefore, try another combination of numbers. Try 4 and 6.

Test the new guess by expanding:
$$(x+4)(x+6) = x^2 + 6x + 4x + 24 = x^2 + 10x + 24$$

Since this is the original polynomial, $x^2 + 10x + 24$ factors correctly as $(x+4)(x+6)$.

Rearranging formulae

Rearranging formulae involves the use of axioms and theorems to transform equations into different but equivalent forms.

For example, the equation $y = 2x + 1$ can be transformed into an equivalent form such as $y - 1 = 2x$. These forms are not equal since the left sides are different, but they are equivalent since they both have the same solution set. In this case the solution set is a set of ordered pairs. A few such ordered pairs would be $(-1, -1)$, $(0, 1)$, and $(1, 3)$.

The strict method for transforming equations requires the use of axioms, definitions and theorems. Although very interesting, those concepts are beyond the scope of this course. There are two practical methods that are often used. The first is used in the beginning of transforming equations and the second is used after proficiency has been gained in the first method.

Method I: Transform $y = 3x - 2$ into the form $ax + by + c = 0$, where $a, b, c \in \mathbb{Z}$.

   Step 1: Write the equation: $y = 3x - 2$

   Step 2: Thinking of the order of operations in reverse, add 2 to both sides:

$$\begin{array}{r} y = 3x - 2 \\ +2 \qquad\quad 2 \\ \hline \end{array}$$

   Step 3: Simplify the result $y + 2 = 3x$

   Step 4: Add $-3x$ to both sides: $y + 2 = 3x$

$$\begin{array}{r} +-3x \qquad 3x \\ \hline \end{array}$$

   Step 5: Simplify the result: $-3x + y + 2 = 0$

Method II: This method involves thinking of the steps in Method 1, not writing them all down, and just simplifying the results.

   Step 1: Write the equation: $y = 3x - 2$

   Step 2: Add 2 to both sides: $y + 2 = 3x$

   Step 3: Add $-3x$ to both sides $-3x + y + 2 = 0$

● **Examiner's hint:** When you are unsure, or the pressure of a test (i.e. classroom, IB, AP, PSAT, SAT, ACT) is intense, Method I is almost foolproof, it is just that it takes so much longer.

You can clearly see that Method II is faster since there is less work involved. However, Method II does take a little practice as there are more opportunities for careless mistakes.

The phrase 'y' is in terms of 'x' means that on the left side of the equation is the variable y and on the right side are terms that involve the variable x. In the example $y = 7x + 5$, x does not appear to be present with the constant 5. However, you can think about 5 as $5 \cdot x^0$ and thus 'see' x in all of the terms.

Example 1.23

Write, in words, which variable is in terms of the other(s).

a) $A = \pi r^2$

b) $P = 2l + 2w$

Solution

a) A is in terms of r.

b) P is in terms of l and w.

Example 1.24

Given the formula for simple interest, $I = \dfrac{Crn}{100}$, solve for r in terms of C, n and I.

Solution

$$I = \frac{Crn}{100}$$

$$100\,I = Crn \text{ (both sides multiplied by 100)}$$

$$\frac{100\,I}{Cn} = r \text{ (both sides divided by } Cn)$$

$$\therefore \quad r = \frac{100I}{Cn}$$

 The symbol '$\therefore$' is read as 'therefore'.

Evaluating expressions

To evaluate an expression is to find a number value for the expression.

To evaluate a polynomial expression means to substitute the given value(s) for the variable(s) and then write a simplified answer.

From algebra, it is known that $a^{-b} = \dfrac{1}{a^b}$, when $b > 0$. For example, $3^{-2} = \dfrac{1}{3^2}$.

Example 1.25

Evaluate the following:

a) $5 - 2 \cdot 4$

b) $x^2 + 3x + 1$, for $x = 4$

c) 5^{-2}

Solution

a) $5 - 2 \cdot 4 = 5 - 8 = -3$

b) $x^2 + 3x + 1 = 4^2 + 3 \cdot 4 + 1 = 16 + 12 + 1 = 29$

c) $5^{-2} = \dfrac{1}{5^2} = \dfrac{1}{25}$

Solving linear equations in one variable

To solve an equation means to find an answer that will satisfy the equation.

A linear equation is an equation in which the variable's exponent is 1 and the graph of the related function is a straight line.

 See Chapter 4 for a more detailed explanation of linear functions.

Example 1.26

Solve for x.

a) $2x + 3 = 13$ b) $15x - 2(x + 7) = 10x + 19$

Solution

The following solutions will make use of Method II on page 14.

a) $2x + 3 = 13$

$$2x = 13 - 3 = 10$$
$$\therefore \quad x = \frac{10}{2} = 5$$

You should check that 5 is the solution by substituting it back into the original equation and verifying that 5 satisfies the equation.

Does $2 \cdot 5 + 3 = 13$?

Yes, $10 + 3 = 13$.

Therefore, since 5 satisfies the original equation, 5 is the solution.

b) $15x - 2(x + 7) = 10x + 19$

$$15x - 2x - 14 = 10x + 19$$
$$13x - 14 = 10x + 19$$
$$13x - 10x = 19 + 14$$
$$3x = 33$$
$$\therefore \quad x = \frac{33}{3} = 11$$

> As you become more proficient at solving equations, steps such as the 3rd and 4th ones can be skipped.

Example 1.27

Solve the equation $\frac{x}{2} = \frac{3}{5}$.

Solution

Method I: $10 \cdot \frac{x}{2} = 10 \cdot \frac{3}{5}$ (multiply both sides by 10, the LCM)

$$5x = 2 \cdot 3 \quad \text{(simplify)}$$
$$5x = 6$$
$$\therefore \quad x = \frac{6}{5}$$

Method II: $5 \cdot x = 2 \cdot 3$ (cross-multiply)

$$5x = 6 \quad \text{(simplify)}$$
$$\therefore \quad x = \frac{6}{5}$$

> Method II can often save at least one step.

Exercise 1.3

1. Expand each of the following. Leave your answer as a simplified polynomial.

   a) $5(x + 3)$ b) $-3(y - 7)$ c) $x(x + y)$

   d) $z(w - t)$ e) $(x + 9)(x + 2)$ f) $(r + 1)(r - 7)$

   g) $(2y + 3)(3y - 2)$ h) $(x + 4)(x - 4)$ i) $(2a + 3)^2$

   j) $(3z - 1)(3z + 1)$ k) $(x + 3)(x^2 + 2x + 4)$ l) $(g - 5)(g^2 - 7g - 1)$

2. Factorize each polynomial completely over the set of integers.

   a) $5x - 5$ b) $3y + 6$ c) $2x^2 + 6x + 8$

   d) $5z^2 - 15z + 45$ e) $x^2 + 5x + 6$ f) $y^2 + 8y + 15$

g) $z^2 - z - 2$　　　h) $w^2 + 4w - 21$　　　i) $x^2 - 16$

j) $r^2 - 25$　　　k) $2x^2 - 11x - 21$　　　l) $3m^2 + 10m + 3$

3. Solve for the underlined variable in terms of the other variable(s).

a) $\underline{y} - x = 5$　　　b) $2x + \underline{y} - 7 = 0$　　　c) $2\underline{z} + 4w = 7$

d) $3\underline{r} - 5s - 6 = 1$　　　e) $\underline{r}t = d$　　　f) $p = \dfrac{360}{\underline{b}}$

g) $u = a + (\underline{n} - 1)d$　　　h) $u = a + (n - 1)\underline{d}$　　　i) $x = \dfrac{-b}{2\underline{a}}$

j) $x = \dfrac{-b}{2\underline{a}}$　　　k) $A = \dfrac{\underline{h}(a + b)}{2}$　　　l) $r = \dfrac{S_{xy}}{S_x \cdot \underline{S_y}}$

4. Evaluate each expression for the given value of the variable.

a) πr^2; $r = 4$　　　b) $\dfrac{4}{3}\pi r^3$; $r = 3$

c) $x^2 + 5x + 1$; $x = 2$　　　d) $3y^2 - 4y + 5$; $y = -1$

e) $\dfrac{2}{x - 1}$; $x = 1.1$　　　f) $\dfrac{2}{x - 1}$; $x = 1.01$

g) $\dfrac{2}{x - 1}$; $x = 1.001$　　　h) $\dfrac{2}{x - 1}$; $x = 1$

i) $u + (n - 1)d$; $u = 3, n = 25, d = 4$

j) $\dfrac{u(1 - r^n)}{1 - r}$; $u = 2, r = 0.5, n = 10$

k) $C\left(1 + \dfrac{r}{100}\right)^n - C$; $C = 1000, r = 6, n = 30$

l) $C\left(1 + \dfrac{r}{100t}\right)^{nt} - C$; $C = 1000, r = 6, n = 30, t = 12$

m) 7^{-3}　　　n) 2^{-5}　　　o) $\left(\dfrac{1}{2}\right)^{-1}$　　　p) $\left(\dfrac{3}{4}\right)^{-2}$

5. Solve each equation.

a) $2x + 1 = 5$　　　b) $3y - 7 = 8$

c) $5(z + 1) = 15$　　　d) $-3(2x - 3) = -9$

e) $2r + 7 = -3r + 27$　　　f) $-4t - 1 = 2t + 5$

g) $2x + 2 = -(x + 1)$　　　h) $-(7w - 1) = 7w + 13$

i) $\dfrac{2}{3}y - 1 = 8$　　　j) $\dfrac{-4}{7}t + \dfrac{3}{14} = \dfrac{5}{7}$　(Hint: Use Method I)

k) $\dfrac{x}{5} = \dfrac{3}{8}$　　　l) $\dfrac{-3}{7} = \dfrac{y}{4}$

m) $\dfrac{4}{r} = \dfrac{9}{11}$　　　n) $\dfrac{2x + 1}{3} = \dfrac{5}{7}$

o) $\dfrac{-3z - 2}{5} = \dfrac{-5z + 2}{3} + 2$　(Hint: Use Method I)

p) $\dfrac{-3x + 13}{2} + 1 = \dfrac{2}{3}$　(Hint: Use Method I)

1.4　Algebra extended

Rewriting linear equations in two variables

Linear equations can be expressed in many forms. Some of these are:

- gradient-intercept
- standard
- point-gradient. (This form is not required at this level.)

Any particular equation can be rewritten in any one of the forms.

Each form is useful in its own way. (See Chapter 3.)

This section will concentrate on rewriting a given linear equation into one of the above forms.

$y = mx + c$ is called the gradient-intercept form of a linear equation.

Example 1.28

Rewrite $2x - 7y = 5$ in the form $y = mx + c$.

Solution

$$2x - 7y = 5$$
$$-7y = -2x + 5$$
$$y = \frac{-2x + 5}{-7}$$
$$y = \frac{-2}{-7}x + \frac{5}{-7}$$
$$\therefore \quad y = \frac{2}{7}x + \frac{-5}{7}$$

Why is division by zero not allowed? For an explanation, visit www.pearsonhotlinks. co.uk, enter the ISBN for this book and click on weblink 1.4.

You can see that m is $\frac{2}{7}$ and that c is $\frac{-5}{7}$.

An equally good answer would be:

$$y = \frac{2}{7}x - \frac{5}{7}$$

$ax + by + d = 0$ is called the standard form of a linear equation.

Example 1.29

Rewrite $y = \frac{2}{3}x - 5$ in the form $ax + by + d = 0$, where $a, b, d \in \mathbb{Z}$.

Solution

$$3 \cdot \left(y = \frac{2}{3}x - 5\right)$$
$$3y = 2x - 15$$
$$3y - 2x + 15 = 0$$
$$\therefore \quad -2x + 3y + 15 = 0$$

You can see that $a = -2$, $b = 3$, and $d = 15$.

An equally good answer would be:

$$2x - 3y - 15 = 0$$

Both sides of the equation $-2x + 3y + 15 = 0$ were multiplied by -1.

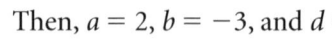

Then, $a = 2$, $b = -3$, and $d = -15$.

Solving a system of linear equations in two variables

A system of linear equations contains at least two equations.

Straight line graphs are associated with the equations.

The graphs are to be considered coplanar (lying in the same plane).

The graphs of the equations could:

- intersect at one point
- intersect at all points (both lines would be the same)
- not intersect at all (the lines would be parallel).

Solving a system of equations means finding the point, if one exists, where the graphs intersect.

This review will concentrate on finding the solution for a system of two equations each having the same two variables. This discussion will also assume that the graphs intersect thus producing one unique solution. This solution will be in the form of an ordered pair.

Example 1.30

Solve the system: $2x + 3y = 2$

$\qquad\qquad\qquad 5x + 2y = -6$

Solution
The Linear Combination Method

Step 1: Multiply both sides of the equation by a number so that, when you add the equations together, the sum of one of the terms will be 0.
- In this case, both sides of the first equation are multiplied by 5, and both sides of the second equation are multiplied by -2.
- The notation $5 \cdot (2x + 3y = 2)$ is really a short form of $5 \cdot (2x + 3y) = 5 \cdot 2$.
- A good hint is to always add instead of subtract. This will reduce careless mistakes.
- The symbol $\Rightarrow$ is read as 'implies'. (See Chapter 9.)

Step 2: Solve the equation that results from the addition.

$$2x + 3y = 2 \quad \Rightarrow \quad 5 \cdot (2x + 3y = 2) \quad \Rightarrow \quad 10x + 15y = 10$$
$$5x + 2y = -6 \Rightarrow -2 \cdot (5x + 2y = -6) \Rightarrow \underline{-10x - 4y = 12}$$
$$0 + 11y = 22$$
$$\therefore \quad y = 2$$

Step 3: Now substitute $y = 2$ back into any one of the above equations in order to find the x-value. In this case we chose $2x + 3y = 2$, since it looked the easiest.

$$2x + 3 \cdot 2 = 2$$
$$2x = -4$$
$$\therefore \qquad x = -2$$

- Therefore, the solution to the system of equations is the ordered pair $(-2, 2)$. This means that the graphs of the equations will intersect at the ordered pair $(-2, 2)$.

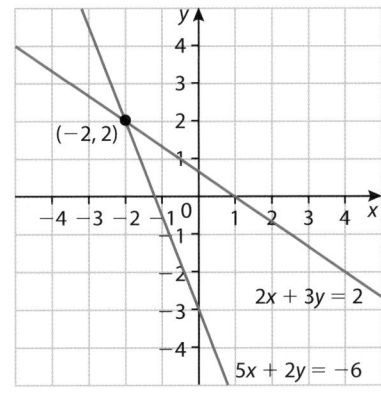

Step 4: It is also a good idea to check this solution by substituting it into the other equation.

$$5 \cdot 2 + 2 \cdot -2 \stackrel{?}{=} -6$$
$$10 + -4 \stackrel{?}{=} -6$$
$$-6 \stackrel{\checkmark}{=} -6$$

Example 1.31

Solve the system:
$$y = 2x + 1$$
$$3x - 5y = 2$$

Solution

The Substitution Method

Step 1: Substitute the right side of the first equation into the y-value in the second equation.
- This method is very useful when one of the equations has one variable in terms of the other variable.
$$3x - 5 \cdot (2x + 1) = 2$$

Step 2: Solve for x:
$$3x - 5 \cdot (2x + 1) = 2$$
$$3x - 10x - 5 = 2$$
$$-7x = 7$$
$$\therefore \quad x = -1$$

Step 3: Back-substitute to find the y-value.
- Although either equation can be used, the first one will be the most efficient since y is already in terms of x.
$$y = 2x + 1$$
$$y = 2 \cdot -1 + 1$$
$$\therefore \quad y = -1$$

Step 4: Check to see if this solution satisfies the other equation.
$$3(-1) - 5(-1) \stackrel{?}{=} 2$$
$$2 \stackrel{\checkmark}{=} 2$$
- Since $2 = 2$, we know that the solution $(-1, -1)$ satisfies both equations and therefore represents the solution of the system.
- As before, this solution represents the ordered pair where the lines intersect.

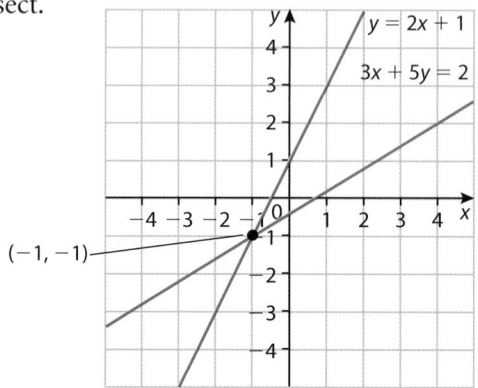

Order relations

Order relations involve the use of:

- $<$ read as 'less than'
- $\leqslant$ read as 'less than or equal to'
- $>$ read as 'greater than'
- $\geqslant$ read as 'greater than or equal to'.

These relations are often called **inequalities**.

It is advisable, although not necessary, to keep the variable on the left side of the equation.

- For example, $x < 5$ represents the same set of numbers as $5 > x$, but is usually easier for you to graph on a number line when written as $x < 5$.

When graphing on a number line, make the graph extend in the same direction as the inequality symbol is pointing.

- For example, the graph of $x < 5$ would point like this: $\longleftarrow\!\!\circ$
- The graph of $x \geqslant 7$ would point like this: $\bullet\!\!\longrightarrow$
- When graphing using $<$ or $>$ indicate the end of the graph with an open circle. This tells the reader that you are not including the endpoint.
- When using $\leqslant$ or $\geqslant$ indicate the end of the graph with a closed circle. This tells the reader that you are including the endpoint.

The rules for the use of order relations when solving inequalities are:

- the same for addition and subtraction, and you will *keep the same* inequality symbol when performing those operations.
- the same for multiplication and division *except* that you will *reverse* the inequality symbol you are using when performing those operations with *negative* numbers.

Example 1.32

Solve and graph the solution on a number line: $2x + 1 > 5$

Solution

$$2x + 1 > 5$$
$$2x > 5 - 1$$
$$2x > 4$$
$$x > \frac{4}{2}$$
$$\therefore \quad x > 2$$

 Notice that the inequality symbol stayed the same when subtracting.

 Notice that the inequality symbol stayed the same when dividing by 2.

Example 1.33

Solve and graph on a number line: $3x + 7 \geqslant 8x + 27$

Solution

$$3x + 7 \geqslant 8x + 27$$
$$3x - 8x \geqslant 27 - 7 \quad \text{Hint: Notice that the inequality symbol stayed the same when we subtracted.}$$
$$-5x \geqslant 20 \qquad \text{Hint: The inequality symbol is \underline{still} the same.}$$

$$x \leqslant \frac{20}{-5}$$ Hint: When we divided by -5, the inequality symbol reversed!

$$\therefore \quad x \leqslant -4$$

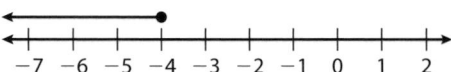

Intervals on a real number line

The real number line is completely filled up with the rational ($\mathbb{Q}$) and irrational ($\mathbb{Q}'$) numbers.

There are no gaps on the real number line. Every point has one and only one real number assigned to it and conversely, every real number has one and only one point assigned to it.

There are several ways to express a group of real numbers you wish to discuss. For example, if you wish to discuss the group of real numbers greater than 5, you could write that idea in any one of the following ways:

- $x > 5$
- $\{x \mid x > 5, x \in \mathbb{R}\}$
- $\{x \mid x > 5\}$
- $(5, \infty)$

The last notation is called **interval notation** and is explained below.

- A parenthesis, (or), is an indication not to include the real number next to it. In $(5, \infty)$, the parenthesis next to the 5 would mean that the group of numbers being discussed would not include 5, but would include any real number to the right of 5 on the number line, i.e. 5.1, 5.01, 5.001, 5.0001, etc.
- The infinity symbol, ∞, will always have a parenthesis next to it as it is assumed that you can never reach infinity.
- A bracket, [or], would indicate to the reader that you mean to include the number next to it in the group of numbers you wish to discuss. For example, $[5, 7)$, would be the group (the set) of real numbers from 5 to 7. The number 5 would be included in the set, but 7 would not be. The graph would look like:

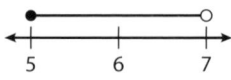

Example 1.34

Write each of the following inequalities in interval notation.

a) $x > 7$ b) $-3 \leqslant x < 9$

Solution

a) $(7, \infty)$ b) $[-3, 9)$

$\{x \mid x > 5, x \in \mathbb{R}\}$ is read as, 'the set of all x such that x is greater than 5 and x is an element of the real numbers.'

The set of real numbers would be implied when writing $\{x \mid x > 5\}$.

A word of caution – the interval $(2, 8)$ can be confused with the ordered pair $(2, 8)$. It is therefore important to keep the context of the question in mind at all times.

1. Write each of the following in the form $y = mx + c$, where $m, c \in \mathbb{Q}$.

   a) $2x + y = 9$ b) $3x - 2y = 7$ c) $\frac{3}{2}x + \frac{5}{2}y = 4$

   d) $6x - \frac{4}{7}y = 1$ e) $0.2x + 2.8y = 0.5$

2. Write each of the following in the form $ax + by + d = 0$, where $a, b, d \in \mathbb{Z}$.

   a) $\frac{2}{3}x - \frac{5}{6}y = 1$ b) $y = 4x - 5$ c) $0.5x + 1.2y = 0.7$ d) $y = \frac{-3}{7}x + \frac{2}{7}$

3. Write each of the following in the form $ax + by + d = 0$, where $a, b, d \in \mathbb{Z}, a > 0$.

   a) $y = \frac{1}{2}x - \frac{3}{4}$ b) $y = 0.3x + 0.5$

4. Solve each system using the Linear Combination Method. Leave your answer in exact form.

   a) $2x + 3y = 12$
   $5x - 2y = 11$

   b) $4x - 7y = 3$
   $3x - 5y = 1$

   c) $5x - 6y = 2$
   $-4x + 2y = -3$

   d) $-9x - 7y = 8$
   $2x + 6y = 11$

5. Solve each system using the Substitution Method. Leave your answer in exact form.

   a) $y = 2x + 1$
   $3x + y = 11$

   b) $x + y = 2$
   $3x + 2y = 1$

   c) $y = 5x - 1$
   $y = -6x + 7$

   d) $y = \frac{4}{5}x + 1$
   $y = \frac{2}{3}x - 6$

6. Solve each inequality. Graph the solution set on a number line.

   a) $2x - 1 > 9$

   b) $-3z - 2 \geqslant 19$

   c) $5(t - 3) + 1 < 2t + 4$

   d) $7 \leqslant -(r - 1)$

   e) $\frac{-6}{7}m + 1 \leqslant \frac{3}{14}m + 2$

   f) $8 > 2w - 10$

7. Write the following inequalities in interval notation.

   a) $x > 7$ b) $y \leqslant 4$ c) $-5 \leqslant z \leqslant 6$

   d) $13 < t \leqslant 25$ e) $r \geqslant -3$ f) $8 < x < 12$

8. Write the following interval notations in inequality form.

   a) $[2, \infty)$ b) $(-\infty, 9)$ c) $[-2, 8)$

   d) $(3, 10)$ e) $(-\infty, \infty)$ f) $(-6, -1]$

   g) $[3, 4]$ h) $(-5, 0)$

1.5 Geometry

Basic concepts

In any mathematical system there must be a starting place or beginning position. This place is called the **undefined terms**. Undefined terms are those 'things' we believe exist but we just cannot write a definition for. For example, 'addition' $(+)$ is an undefined term until we give meaning to it with the addition tables. In geometry there are three undefined terms.

- Point: it has no length, width, or depth. It has no dimensions. For example, think of a pencil dot on a piece of paper.

 How do these terms relate to the ideas expressed in Euclid's *Elements*?

- Line: it has no width or depth. It has only length. A line extends infinitely far in one dimension. For example, think of a red laser beam shot through a smoke-filled room.
- Plane: it has no depth. It has length and width only. A plane extends infinitely far in two dimensions. For example, think of the smooth surface of a lake.

The above examples are only useful to the extent that they are trying to convey what we all already believe is true. The three undefined terms are only concepts we can think of mentally. It does seem, however, that we intuitively know what they mean.

Once the three undefined terms have been established, then the study of geometry can proceed. The next ideas that need to be discussed are **definitions**. An example of a geometric definition is **space**.

Space is defined as the collection of all points. Space has length, width and depth. The best example of space is the universe.

Once the undefined terms and some definitions have been established, then **postulates** may be introduced.

A postulate is a statement of fact in which we have complete faith, but are unable to prove. For example, the statement 'every two points will contain one and only one line' is considered a postulate since it agrees fundamentally with our common and intuitive senses. However, we cannot prove it. We simply accept it as fact. We believe it is true. We have faith it is true.

Finally, after the undefined terms, some defined terms, and some postulates have been introduced, then **theorems** can be hypothesized.

A theorem is a statement of fact that can be proven true based upon the undefined terms, defined terms and the postulates.

- One of the best known theorems is Pythagoras' theorem.
- Another one is: The sum of the measures of the angles of a triangle is 180 degrees.

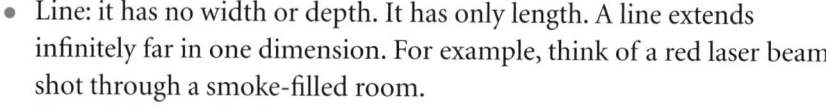

How is it that we can discuss something that we cannot see or touch? If we can think about something, then must that something exist?

For some of the proofs of Pythagoras' theorem, visit www.pearsonhotlinks.co.uk, enter the ISBN for this book and click on weblinks 1.5 and 1.6.

For a more complete discussion on the basic concepts of geometry, visit www.pearsonhotlinks.co.uk, enter the ISBN for this book and click on weblink 1.7.

Exercise 1.5A

1. List three examples you might use to describe a point.
2. List three examples you might use to describe a line (or line segment).
3. List three examples you might use to describe a plane (or piece of a plane).
4. How does 'circular reasoning' differ from 'logical reasoning'?
5. Why is subtraction not an undefined term?
6. List three examples of geometry definitions.
7. How do postulates differ from undefined terms?
8. List three examples of geometry postulates.
9. How do theorems differ from postulates?
10. List three examples of geometry theorems.

Perimeter and areas of two-dimensional shapes

The **perimeter** of a closed shape is defined as the distance around that shape.

If the closed shape is a polygon, the perimeter is the sum of the lengths of its sides.

If the closed shape is a circle, the perimeter is called the circumference.

Below are some of the formulae used to compute perimeter.
- Perimeter of a rectangle: $P = 2l + 2w$, where l is the length and w is the width.
- Perimeter of a square: $P = 4s$, where s is the length of a side.
- Perimeter of a triangle: $P = a + b + c$, where a, b and c are the lengths of the sides.
- Perimeter of a circle: $C = 2\pi r$, where C is the circumference and r is the radius.

The **area** of a closed shape is defined as the number of square units it contains.

In layman's terms, area is the size of the surface of a closed figure.

In some geometry books, an area postulate is used prior to defining area.

Below are some of the many formulae used to compute areas.
- Area of a rectangle: $A = (l \times w)$, where l is the length and w is the width.
- Area of a parallelogram: $A = (b \times h)$, where b is the base and h is the height.
- Area of a triangle: $A = \frac{1}{2}(b \times h)$, where b is the base and h is the height.
- Area of a trapezium: $A = \frac{1}{2}(a + b)h$, where a and b are the parallel sides and h is the height.
- Area of a circle: $A = \pi r^2$, where r is the radius.

 See Chapter 2 for a more detailed explanation of SI units.

In most countries the SI (Le Système International) units for length and area are used. The unit for length is metre (m) and area is square metre (m^2). The metre is made up of smaller sub units according to a decimal system.

1 m = 100 centimetres (cm)
1 cm = 10 millimetres (mm)

When calculating areas we need to be a little careful.

1 square metre (m^2) = 1 m $\times$ 1 m = 100 cm $\times$ 100 cm = 10 000 cm^2
1 square centimetre (cm^2) = 10 mm $\times$ 10 mm = 100 mm^2

Example 1.35

Find the perimeter and area of a rectangle whose dimensions are 45 cm long and 30 cm wide.

Solution

Perimeter: $P = 2(l + w) = 2(45\,\text{cm} + 30\,\text{cm}) = 2(75\,\text{cm}) = 150\,\text{cm}$

Area: $A = l \times w = 45\,\text{cm} \times 30\,\text{cm} = 1350$ square cm $= 1350\,\text{cm}^2$

Example 1.36

Find the perimeter and area for the given circle
a) exactly   b) to 3 significant figures.

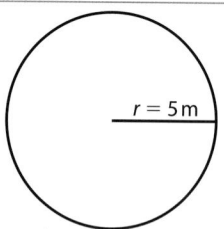

Solution

Perimeter: $C = 2\pi r = 2\pi \times 5 = 10\pi$ m. This is the exact answer.

$C = 2\pi r = 2 \times 3.14159 \times 5 = 31.4159 = 31.4$ m to 3 s.f.

This is the approximate answer.

Area: $A = \pi r^2 = \pi \times 5^2 = \pi \times 25 = 25\pi$ square metres $= 25\pi\,\text{m}^2$.

This is the exact answer.

$A = \pi r^2 = 3.14159 \times 5^2 = 78.53975 = 78.5\,\text{m}^2$ to 3 s.f.

This is the approximate answer.

● **Examiner's hint:** It is commonly accepted only to use the unit of measure, cm, or cm², with the final answer.

Exercise 1.5B

1. For each of the following diagrams find the perimeter. Express the answer exactly.

   a) Parallelogram:

   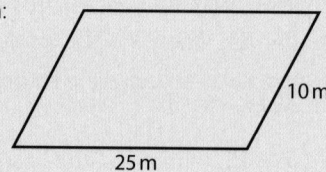

   b) Triangle:

   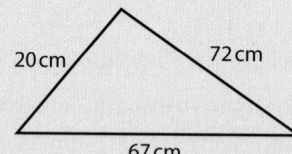

   c) Trapezium:

   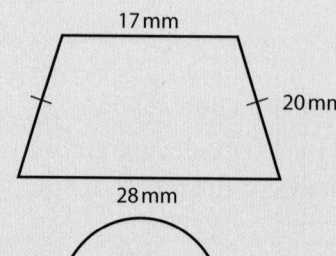

   d) Circle:

e) Compound shape:

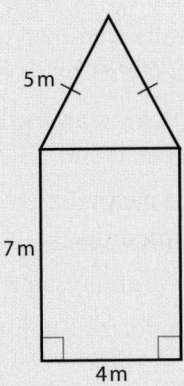

2. Find each of the following:
   a) The exact area of a rectangle whose dimensions are 100 m by 50 m.
   b) The exact area of a triangle whose base is 22.5 cm and height is 10.4 cm.
   c) The area of a circle whose radius is 17 feet, giving your answer correct to 3 s.f.
   d) The area of the trapezium below, giving your answer correct to 1 decimal place.

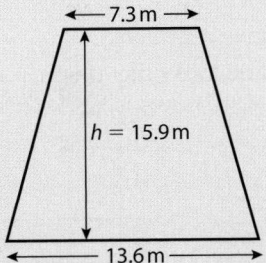

e) The exact area of the compound shape below.

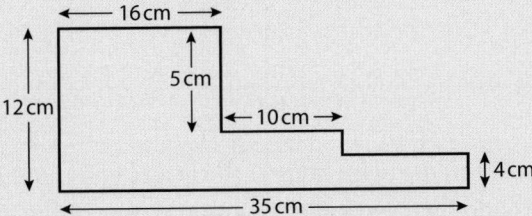

f) The approximate area of the compound shape below correct to 1 decimal place.

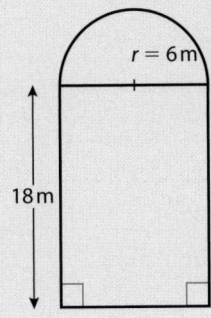

How is it possible to think of a line, which is an undefined term, in terms of points which have no dimensions?

Are there more points on the line or on the plane? How does this help us understand the universe in which we live?

Plotting on the *x*-, *y*-coordinate plane

Think of a line as infinitely many points crammed together in a straight row so that there is no space between them.

Think of a plane as infinitely many points crammed together on a flat surface so that there is no space between them.

The word 'coordinate' is derived from the prefix 'co' meaning 'together with' and 'ordinal' meaning 'of a certain order'.

Therefore, 'coordinate' means 'a set of numbers in a specified order'.

Hence, a coordinate system is a way to locate a point either on a line, in a plane or in space.

In a coordinate plane only two numbers are required to locate a point.
- The first number is called the **abscissa**, often referred to as *x*. The second is called the **ordinate**, often referred to as *y*.
- Such a system is often referred to as an *x*-, *y*-coordinate plane.
- The axes are often labelled *x* for the horizontal movement and *y* for the vertical movement. They meet at the origin $(0, 0)$.
- To the left and below $(0, 0)$ are the negative numbers and to the right and above are the positive numbers.
- An *x*-, *y*-coordinate plane looks like this:

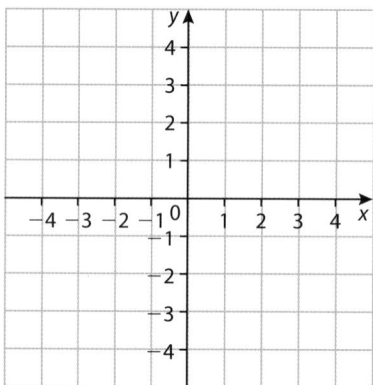

- To locate a point in the plane start at the origin and move horizontally (right or left) and then vertically (up or down) according to the ordered pair given.

Example 1.37

Locate and plot $(2, 3)$ on the *x*-, *y*-coordinate plane.

Solution
Start at the origin.
Move 2 units to the right.
Move up 3 units parallel to the *y*-axis.

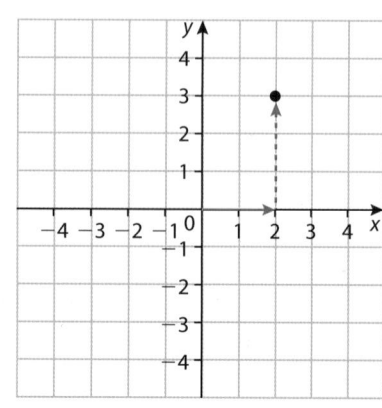

Exercise 1.5C

On the same x-, y-coordinate axes locate and plot each coordinate.

1. $(3, 4)$ **2.** $(-3, 2)$ **3.** $(-4, -5)$ **4.** $(1, 1)$ **5.** $(0, 3)$

6. $(-2, 0)$ **7.** $(3, -3)$ **8.** $(-1, 5)$ **9.** $(-2, 1)$ **10.** $(5, -1)$

Midpoint of a line segment

The midpoint of a line segment is defined to be the point, expressed in x and y coordinates, where a line segment is broken into two equal pieces. Simply put, it is the point where the middle of a line segment occurs.

To find this point, given the coordinates of the endpoints of the line segments, we find the mean of the x component of each coordinate and the mean of the y component of each coordinate. Therefore, the formula for the midpoint of a line is:

$$M(x_M, y_M) = \left(\frac{x_1 + x_2}{2}, \frac{y_1 + y_2}{2}\right)$$

 For a helpful applet that shows the midpoint of a line, visit www.pearsonhotlinks.co.uk, enter the ISBN for this book and click on weblink 1.8

Example 1.38

Given $A(2, 3)$ and $B(-6, 0)$, find the midpoint of A and B.

Solution

We know that $M(x, y) = \left(\frac{x_1 + x_2}{2}, \frac{y_1 + y_2}{2}\right)$, therefore:

$$M(x, y) = \left(\frac{2 + -6}{2}, \frac{3 + 0}{2}\right)$$
$$= \left(\frac{-4}{2}, \frac{3}{2}\right)$$
$$= (-2, 1.5)$$

● **Examiner's hint:** Always write the formula you intend to use to solve a problem. Some of the marks for a problem come from your method, which you can demonstrate with the use of the proper formula.

Occasionally we will be required to work out one of the endpoints of the line segments, given the midpoint of the line and one of the other endpoints. To do this, we use some relatively straightforward algebra.

Example 1.39

Find the coordinates of B, given that A is $(4, 5)$ and the midpoint of $[AB]$ is $M(1, 3)$.

Solution

First we write down the formula, $M(x_M, y_M) = \left(\frac{x_1 + x_2}{2}, \frac{y_1 + y_2}{2}\right)$, and substitute the values from the question.

$$M(1, 3) = \left(\frac{4 + x_2}{2}, \frac{5 + y_2}{2}\right)$$

This leads to a pair of two-step algebra problems which we need to solve.

$$1 = \frac{4 + x_2}{2} \quad \text{and} \quad 3 = \frac{5 + y_2}{2}$$

We solve each of these equations in a similar way. We multiply by 2, and isolate the variable using subtraction.

$$2 = 4 + x_2 \quad \text{and} \quad 6 = 5 + y_2$$
$$-2 = x_2 \quad \text{and} \quad 1 = y_2$$

 Why do you think the formula for the midpoint of a line segment involves finding means?

 1. Midpoint formula:

$$M(x_M, y_M) = \left(\frac{x_1 + x_2}{2}, \frac{y_1 + y_2}{2}\right)$$

2. The midpoint of a line segment is the centre of the line segment.

From the solutions to these two equations, we see that the coordinates of B are $(-2, 1)$. We can use the procedure shown in the first example to verify this fact.

---

1. Find the midpoint of each line segment.
   a) $A(2, 3)$ and $B(5, 6)$
   b) $A(-4, -4)$ and $B(4, 4)$

2. Find the midpoint of each side of triangle ABC.
   $A(-3, -2)$, $B(3, 3)$, and $C(6, -2)$

3. Draw the triangle using the coordinates $A(-4, 2)$, $B(4, 4)$, and $C(0, -5)$. Find the midpoint of each side of the triangle, then join the midpoints together to form another triangle.

4. Given A and M, the midpoint of $[AB]$, find B.
   $A(-2, -2)$ and $M(-2, 3)$

5. A has coordinates $(6, -4)$ and B has coordinates $(-4, 2)$.
   a) Find the coordinates of M, the midpoint of $[AB]$.
   b) Find the coordinates of the midpoint of $[BM]$.

6. Show that the midpoint of $[AC]$ is the same as the midpoint of $[DB]$ in $ABCD$.

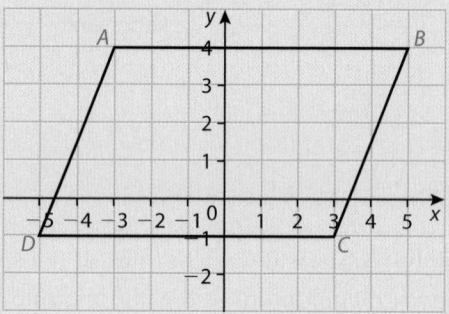

7. Redha throws a ball from the corner of a field. When the ball is halfway to where he estimates it will land, he notices that the ball is exactly 40 m north and 30 m west of where he stands. Where will the ball land, relative to Redha?

Right triangles and Pythagoras' theorem

Right triangles

A right triangle is a triangle that contains a 90 degree angle. See Figure 1.3 on page 31.

A, B, C are called the vertices of the triangle.

The angles of a triangle can be named in several different ways. For example, Angle A can be named as $\angle A$, or $\measuredangle A$, or $\sphericalangle A$, or $\angle BAC$, or $\measuredangle CAB$, or $B\hat{A}C$, or $C\hat{A}B$.

The sum of the measures of all three angles of a triangle is $180°$.

Angle C is called the right angle.

● **Examiner's hint:** $B\hat{A}C$ and $C\hat{A}B$ are often used by the IBO examiners.

A right angle has a measure of 90 degrees (90°).

Side AB can be named as $\overline{AB}$ or $[AB]$.

The length of $[AB]$ can be named as c or AB.

The hypotenuse is the side opposite the right angle. In this triangle $[AB]$ is the hypotenuse.

The sides that form the right angle are called the legs of the triangle. In this triangle, the legs are $[BC]$ and $[AC]$ while their lengths are a and b respectively.

● **Examiner's hint:** $[AB]$ is the common usage for the IBO examiners.

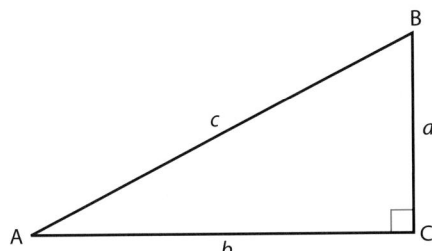

◀ **Figure 1.3** Right triangle.

Pythagoras' theorem

Pythagoras, who was born around 575 BC, is credited with discovering and proving arguably the most important theorem in mathematics: Pythagoras' theorem.

For a more complete history of Pythagoras, visit www. pearsonhotlinks.co.uk, enter the ISBN for this book and click on weblink 1.9.

The theorem states that if $\triangle ABC$ is a right triangle, then $a^2 + b^2 = c^2$, where a and b are the lengths of the legs and c is the length of the hypotenuse. (See diagram above.)

For example, if $\triangle ABC$ is a right triangle then the sides could have length $a = 3$, $b = 4$, and $c = 5$, since $3^2 + 4^2 = 5^2$.

3, 4, 5 is called a Pythagorean triple.

Other popular Pythagorean triples are: $(6, 8, 10)$, $(5, 12, 13)$, $(9, 12, 15)$, $(8, 15, 17)$.

The converse (see Chapter 9) of Pythagoras also holds true: if a, b, and c are the lengths of the sides of the triangle and $a^2 + b^2 = c^2$, then the triangle is a right triangle.

Not all converses hold true. Chapter 9 explains this further.

Example 1.40

If $\triangle DEF$ is a right triangle, which of the following sets could be the lengths of the sides?

a) 7, 24, 26   b) 9, 40, 41   c) 12, 36, 35

d) 66, 63, 16   e) 28, 53, 45

Solution

a) $7^2 + 24^2 = 625$; $26^2 = 676$.   ∴ 7, 24, 26 cannot be the lengths of the sides of $\triangle DEF$.

b) $9^2 + 40^2 = 1681$; $41^2 = 1681$. $\therefore$ 9, 40, 41 *could* be the lengths of the sides of $\triangle DEF$.

c) $12^2 + 35^2 = 1369$; $36^2 = 1296$. $\therefore$ 12, 36, 35 cannot be the lengths of the sides of $\triangle DEF$.

d) $16^2 + 63^2 = 4225$; $66^2 = 4356$. $\therefore$ 66, 63, 16 cannot be the lengths of the sides of $\triangle DEF$.

e) $28^2 + 45^2 = 2809$; $53^2 = 2809$. $\therefore$ 28, 53, 45 *could* be the lengths of the sides of $\triangle DEF$.

Example 1.41

If $\triangle ABC$ is a right triangle with C at the vertex of the right angle and $a = 3$, $b = 7$, solve for c.

Solution

Since $\triangle ABC$ is a right triangle with C at the vertex of the right angle, then we know that $a^2 + b^2 = c^2$. Therefore, by substitution:

$$3^2 + 7^2 = c^2$$
$$9 + 49 = c^2$$
$$58 = c^2$$
$$\therefore \qquad c = \sqrt{58}$$

Exercise 1.5E

1. Given $\triangle RST$ below, name each of the following:

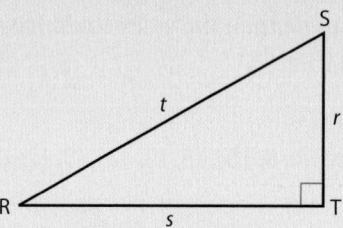

a) The hypotenuse
b) The right angle (use three different letter designations)
c) The two legs
d) $\angle R + \angle S$
e) Two expressions for the length of [RS]
f) Two expressions for the length of [ST]
g) Two expressions for the length of [RT]

2. Which of the following could be the lengths of the sides of a right-angled triangle?
   a) 65, 72, 97   b) 21, 20, 27   c) 1, $\sqrt{2}$, 1   d) 1, 1, 1.414
   e) 1, 2, $\sqrt{3}$ f) 6.5, 42, 42.5   g) $\dfrac{39}{7}, \dfrac{89}{7}, \dfrac{80}{7}$

3. List all of the following that are Pythagorean triples.
   a) 12, 35, 37   b) $\dfrac{1}{2}, 1, \dfrac{\sqrt{3}}{2}$ c) 10, 10, 20   d) 41, 9, 40
   e) 8.100, 15.200, 17.223   f) 1.5, 2, 2.5   g) 55, 48, 73

4. Using △RST above, solve each of the following as required.

a) If $r = 6$ and $s = 8$, find t (exactly).

b) If $r = 7$ and $s = 11$, find t (exactly).

c) If $r = 8$ and $s = 12$, find t (correct to 1 decimal place).

d) If $r = 23$ and $s = 35$, find t (correct to 1 decimal place).

e) If $r = 10$ and $t = 26$, find s (exactly).

f) If $s = 17$ and $t = 30$, find r (exactly).

g) If $s = \frac{13}{5}$ and $t = \frac{65}{7}$, find r (correct to 1 decimal place).

 Pythagoras created a school in Crete for mathematicians to exchange ideas. However, when he disagreed with a particularly controversial result of one of his disciples Pythagoras had him drowned!

Distance between two points

The distance between two points A and B is equal to the length of the line segment joining A to B (see Figure 1.4).

To find the distance between two points on the coordinate plane, there are two widely accepted methods. One is to use the distance formula $\left(d = \sqrt{(x_2 - x_1)^2 + (y_2 - y_1)^2}\right)$ which is given in the formula sheet. The other method is to draw a right-angled triangle using the two coordinates, and apply Pythagoras' theorem. Both of these methods involve the same calculations, and both are acceptable ways of finding the distance between two points.

To use the distance formula for the points $A(x_1, y_1)$ and $B(x_2, y_2)$, simply substitute the values of x_1, y_1, x_2 and y_2 into the formula, and then evaluate the formula to find the value of d.

Example 1.42

Find the distance between $A(1, 2)$ and $B(4, -5)$ to the nearest tenth.

Solution

Use the distance formula $d = \sqrt{(x_2 - x_1)^2 + (y_2 - y_1)^2}$ and substitute.

$$d = \sqrt{(4 - 1)^2 + (-5 - 2)^2}$$
$$= \sqrt{(3)^2 + (-7)^2}$$
$$= \sqrt{9 + 49}$$
$$= \sqrt{58}$$
$$d \approx 7.6$$

The other method is more useful if the problem given has the two points shown in a plot, or if the problem provides a space to plot the two points carefully (like square paper). Once we have the two points plotted, we can create a right-angled triangle, as shown in Figure 1.5. With this right-angled triangle, we can count how many units long the two sides adjacent to the right angle are, and use these measurements in Pythagoras' theorem.

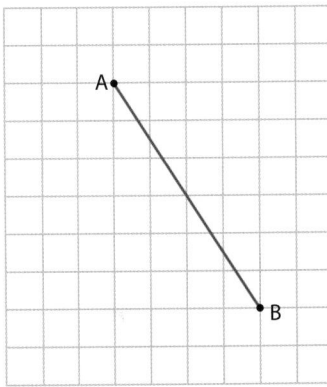

Figure 1.4 The line segment *AB*.

 Can you prove that the Pythagorean theorem and the distance formula are equivalent methods for finding the distance between two points?

 For a helpful applet that demonstrates the effect of changing the location of the points when using the distance formula, visit www.pearsonhotlinks.co.uk, enter the ISBN for this book and click on weblink 1.10.

 Despite the fact that we attribute the theorem that bears his name to Pythagoras, the same theorem was in use in what is now modern day Iran and in China at least 500 years before he started using it.

 How do we measure the distance between two points on Earth?

Example 1.43

In Figure 1.5, how long is $[AB]$?

Figure 1.5 Creating a right-angled triangle from two points. ▶

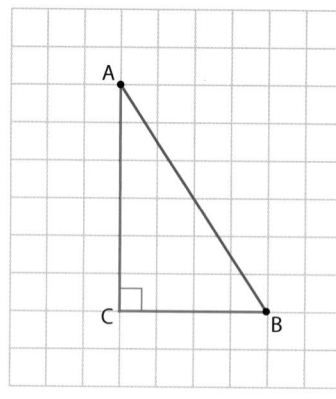

Distance formula:

$$d = \sqrt{(x_2 - x_1)^2 + (y_2 - y_1)^2}$$

The distance formula is equivalent to the Pythagorean theorem.

Solution

The height of the triangle shown is 6 units and the base of the triangle is 4 units.

Use Pythagoras' theorem, $a^2 + b^2 = c^2$, to solve.

$$6^2 + 4^2 = c^2$$
$$36 + 24 = c^2$$
$$60 = c^2$$
$$\sqrt{60} = c \text{ or } c \approx 7.75$$

Exercise 1.5F

1. Find the distance between each pair of points.
    a) $A(1, 3)$ and $B(1, 8)$
    b) $A(0, 0)$ and $B(3, 3)$
    c) $A(-1, -1)$ and $B(-4, -5)$

2. Find the lengths of the line segments in the diagram below.

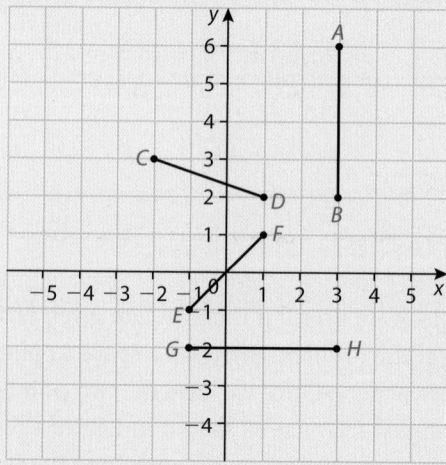

3. Use the distance formula to confirm that *ABCD* is a parallelogram.

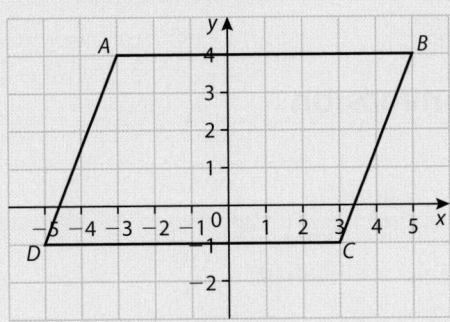

4. A taxi driver is 4 blocks north and 3 blocks east of Union Square in Manhattan, NY.
His next customer is located 2 blocks south and 1 block west of Union Square.
   a) Calculate the straight-line distance the taxi driver is from his next customer.
   b) If the taxi driver can only travel south and east, how many blocks does he need to travel to pick up his customer?
   c) How much distance would he save if he could travel straight to his customer?

5. a) Find the length of each side of the triangle *ABC*.
   b) Round each side length to the nearest whole number.

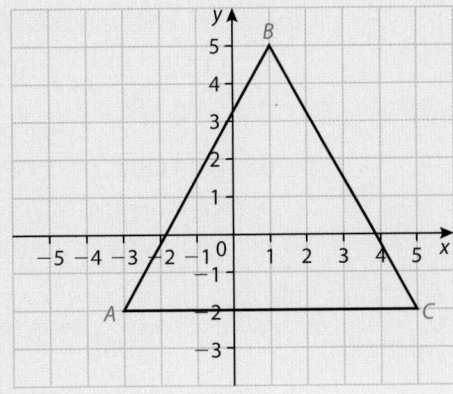

Financial mathematics

This section will review the names and abbreviations for common world currencies. We will also review two ways to convert between currencies.

World currencies

The following are examples of some currencies and their abbreviations:

Country	Currency	Abbreviation
United Kingdom	pound	GBP or GB£
United States	dollar	USD or US$
France	euro	EUR or €
South Africa	rand	ZAR
Japan	yen	JPY or ¥
Australia	dollar	AUD or AU$
Spain	euro	EUR or €

 There are a total of about 190 countries, and 167 currencies in use in the world today. For a complete list of all of the world currencies, visit www.pearsonhotlinks.co.uk, enter the ISBN for this book and click on weblink 1.11.

 To access a table listing many of the current exchange rates between currencies, visit www.pearsonhotlinks.co.uk, enter the ISBN for this book and click on weblink 1.12.

When currencies are exchanged (euro for Japanese yen, for example), there is a fee called a commission for doing the exchange. This concept will be discussed in Chapter 2.

Currency conversion

Example 1.44

If 1 USD = 0.501 013 GBP, find
a) the number of GBP for 1250 USD
b) the number of USD for 750 GBP.

Solution

a) Method I:

$$1\text{ USD} = 0.501\,013\text{ GBP}$$
$$1250 \cdot (1\text{ USD}) = 1250 \cdot (0.501\,013\text{ GBP})$$
$$\therefore\quad 1250\text{ USD} = 626.266\,25\text{ GBP} = 626.27\text{ GBP to 2 decimal places.}$$

Method II: When using this method, it is important to think of the equation 1 USD = 0.501 013 GBP as the proportion

$$\frac{1\text{ USD}}{0.501\,013\text{ GBP}} \quad\text{or as}\quad \frac{0.501\,013\text{ GBP}}{1\text{ USD}}.$$

$$\frac{1\text{ USD}}{0.501\,013\text{ GBP}} = \frac{1250\text{ USD}}{x\text{ GPB}}$$

$$(1\text{ USD})(x\text{ GBP}) = (0.501\,013\text{ GBP})(1250\text{ USD})$$

$$\therefore 1 \cdot x = 0.501\,013 \cdot 1250 = 626.266\,25 = 626.27\text{ to 2 decimal places.}$$

b) Method I:

$$1\text{ USD} = 0.501\,013\text{ GBP}$$
$$\frac{1\text{ USD}}{0.501\,013} = \frac{0.501\,013\text{ GBP}}{0.501\,013}$$
$$\therefore\qquad 1\text{ GBP} = 1.995\,956\text{ USD to 7 s.f.}$$

Hence, $750(1\text{ GBP}) = 750(1.995\,956\text{ USD})$

and $\qquad 750\,\text{GBP} = 1496.97\text{ USD to 2 decimal places.}$

Method II:

$$\frac{x\,\text{USD}}{750\text{ GBP}} = \frac{1\text{ USD}}{0.501\,013\text{ GBP}}$$

$$(x\,\text{USD})(0.501\,013\text{ GBP}) = (1\text{ USD})(750\text{ GBP})$$

$$0.501\,013\, x = 750$$

$$x = \frac{750}{0.501\,013}$$

$$\therefore\qquad x = 1496.97\text{ to 2 d.p.}$$

 Make sure you use words when setting up proportion so that you can see 'like' words are in the numerators and 'like' words are in the denominators.

Think of USD and GBP as variables and 'cancel' them out.

Exercise 1.6

1. Using a website, find the currency and ISO abbreviation for each of the countries.

a) Austria   b) Denmark   c) China   d) Israel

e) Mexico   f) Saudi Arabia   g) Taiwan

2. Using the website link 1.12, find the currency and the most recent exchange rate (per United States dollar) for each country.

   a) Belgium   b) Botswana   c) Hong Kong   d) United Kingdom
   e) Norway   f) Russia   g) Switzerland

3. If 1 USD = 10.810 65 MXN, then using Method I, convert:

   a) 225 United States dollars to Mexican pesos correct to 2 decimal places
   b) 350 Mexican pesos to United States dollars correct to 2 decimal places.

4. If 1 USD = 1.300 540 NZD, then using Method II, convert:

   a) 575 United States dollars to New Zealand dollars correct to 2 decimal places
   b) 2000 New Zealand dollars to United States dollars correct to 2 decimal places.

5. If 1 NOK = 20.819 041 JPY, then using Method I, convert:

   a) 300 Norway krone to Japanese yen correct to the nearest yen.
   b) 16 450 Japanese yen to Norway krone correct to the nearest krone.

6. If 1 EUR = 35.9852 RUB, then using Method II, convert:

   a) 1000 Russian rubles to euros correct to the nearest euro.
   b) 1000 euros to Russian rubles correct to the nearest ruble.

1.7 Statistics

Statistics is largely concerned with the collection of data and the subsequent analysis of that data. A more thorough discussion will be presented in Chapters 11 and 12. This section will merely review some of the very basic concepts: data collection and visual data representation in the forms of bar charts, pie charts and pictograms.

Data collection

Data is information that is used for calculating or measuring. For example:
- A baseball manager might record how many throws a pitcher has made.
- A vehicle manufacturer might record how many hours it takes to make a car.
- A student might record all of her exam grades.

Data can be recorded in several ways. For example:
- Using 'tally marks': e.g. ⅢⅡ // might indicate seven shots taken on goal during a football game.
- Using the actual data: e.g. 12.1, 12.7, 13.2, 12.5 might indicate the number of seconds that four people ran the 100 metre dash.

There are three commonly used ways that data can be classified.
- Nominal: no order or ranking can be used with this type of data. For example:
  - political affiliation
  - religious preference.

- Ordinal: an order or ranking can be used, but specific differences between the rankings cannot be determined. For example:
  - letter grades A, B, C, D, F
  - 1st, 2nd, or 3rd place.
- Interval: an order or ranking can be used and specific differences can be determined. For example:
  - test scores
  - height.

Bar charts

Bar charts have the following characteristics:

- The frequency is displayed on the vertical axis.
- The description of each item of data is given below the horizontal axis.
- The bars may not touch.
- The bars have uniform width.
- The bars are often coloured or shaded for visual appeal.

● **Examiner's hint:** Visual displays, such as bar charts, are encouraged when writing the Mathematical Studies project.

Example 1.45

Data has been collected from 120 teenage girls on the type of music they enjoy. The data is shown in the table below. Construct a bar chart to represent the data.

Rock	Country	Blues	Hip-hop
35	25	20	40

Solution

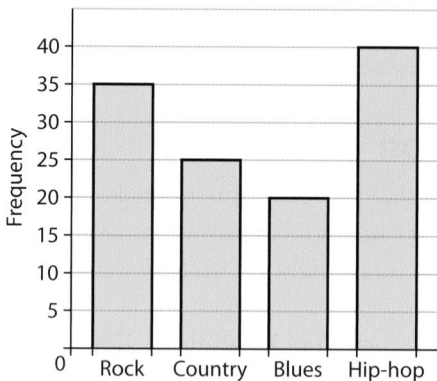

Pie charts

A pie chart is a graph in the shape of a circle. It has the following characteristics:

- It is divided into pie-shaped or wedge-shaped sections.
- Each section represents a percentage of the total amount of data collected.
- The size of each wedge is in terms of degrees and is a percentage of 360°.

- The formula for determining each wedge size is:

Wedge size (in degrees) $= \dfrac{\text{number of data pieces per category}}{\text{total amount of data collected}} \times 360°$.

Example 1.46

Determine each wedge size for the data given in Example 1.45.

Solution

Rock wedge: degrees $= \dfrac{35}{120} \cdot 360° = 105°$

Country wedge: degrees $= \dfrac{25}{120} \cdot 360° = 75°$

Blues wedge: degrees $= \dfrac{20}{120} \cdot 360° = 60°$

Hip-hop wedge: degrees $= \dfrac{40}{120} \cdot 360° = 120°$

Example 1.47

Use the data in Example 1.45 and the calculated degrees in Example 1.46 to construct a pie chart.

Solution

Follow the steps listed below:

Step 1: Draw a large circle.

Step 2: Draw a radius (it is usually drawn horizontally).

Step 3: Measure out 105° with a protractor (usually counterclockwise) and make a mark on the circle.

Step 4: Connect the centre of the circle to that point.

Step 5: Using that radius, measure 75° and make a mark on the circle.

Step 6: Connect the centre to that mark and continue until all four angles have been drawn.

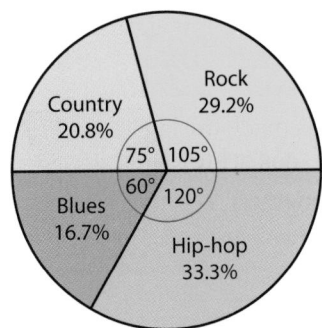

The number of degrees should add to 360°.

The percentages should add to 100%. (The total might miss a little due to rounding.)

You do not have to label the wedges with the degrees.

It is customary to label the wedges with a description and with the percentages.

It is also common to shade or colour the wedges for a better visual presentation.

Pictograms

A pictograph (or pictogram), as the name suggests, is a graph that uses pictures to describe or show the data that has been collected.

A general type of picture or symbol is used that best represents the data. For example, basketballs might be used for the number of points scored in a game.

Pictographs have the following characteristics:
- Part, a half or a quarter, of the picture is sometimes used to approximate less than a full amount. For example:
  - If one car represented 1000 accidents, a half of a car would represent 500 accidents.
- A legend is necessary to tell the reader how much of the data the picture represents.
- The scale is usually on the left vertical axis or the bottom horizontal axis.
- Pictographs are easy and fun to look at.

Example 1.48

The following data represents the number of cellphone calls that were made during the first three months of 2008 by a randomly selected household. Construct a pictograph, using a suitable symbol.
1 symbol = 100 calls.

Month	Number of calls
January	400
February	300
March	500

Solution

Number of cellphone calls made during the first three months of 2008. This is an example of a vertical pictograph.

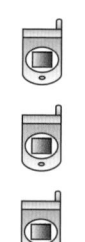

January   February   March

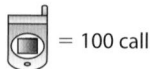 = 100 calls

1. Classify each as nominal, ordinal, or interval data.

a) Hair colour   b) Class in high school   c) Weight

d) IQ score   e) Race (Korean, African-American, Hispanic, etc.)

f) Gender   g) Opinion about one's class schedule

h) Music ratings   i) Movie genres   j) Age

2. Construct a bar chart for each set of data below.

a) Last year car Company A sold 600 vehicles, Company B sold 400 vehicles and Company C sold 800 vehicles.

b) Last month Read-a-Lot book store sold the following number of books:

Type	Number sold
Action	350
Non-fiction	200
Romance	400
Science fiction	250

3. Construct a pie chart for each set of data below.

a) It was found that in an elementary school, there were 60 students with blond hair, 130 with brown hair, 80 with black hair and 30 with red hair.

b) A golf store keeps track, on a weekly basis, of the number (in dozens) of inexpensive, moderately priced or expensive golf balls it sells. The data is recorded in the chart below from a randomly selected week.

Price	Number of dozens
Inexpensive	70
Moderately priced	100
Expensive	50

4. Construct a pictograph for each set of data below.

a) The number of buses that a school district used over the last four years is recorded below.
Let 1 bus picture = 20 buses. Construct this as a horizontal graph.

Year	Number of buses
2006	100
2005	80
2004	60
2003	40

b) The All-Sports sporting goods store kept track of how many basketballs it sold during the NBA season. The data from four months is recorded below.
Let one ball picture = 10 basketballs. Construct this as a vertical graph.

Month	Number of basketballs sold
March	40
April	45
May	60
June	75

Number and Algebra

2

Assessment statements

1.1 The sets of natural numbers, ℕ; integers, ℤ; rational numbers, ℚ; and real numbers, ℝ.

1.2 Approximation: decimal places; significant figures. Percentage errors. Estimation.

1.3 Expressing numbers in the form $a \times 10^k$ where $1 \leqslant a < 10$ and k is an integer.
Operations with numbers in this form.

1.4 SI (*Système International*) and other basic units of measurement: for example, kilogram (kg), metre (m), second (s), litre (l), metre per second (ms^{-1}), and Celsius scale.

1.5 Currency conversions

1.6 Use of GDC to solve: pairs of linear equations in two variables, and quadratic equations.

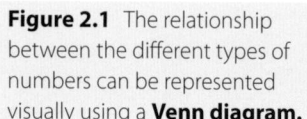

Artefacts, dating from 30 000 BCE, can be found which show notches cut into them similar to today's tally marks. Obviously these notches were used for counting, and are an example of an early form of written language. How has the development of number systems affected civilization?

For a complete definition of real numbers, visit www.pearsonhotlinks.co.uk, enter the ISBN for this book and click on weblink 2.1.

Figure 2.1 The relationship between the different types of numbers can be represented visually using a **Venn diagram.**

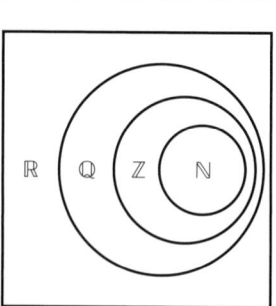

Overview

By the end of this chapter you will be able to:
- explain number concepts, including different number systems
- show how the differences between these number systems affect approximations and calculations
- express numbers in scientific notation or standard form
- understand and know how to use SI (*Système International*) units
- convert one type of currency to another, and calculate commission charges on such transactions
- solve systems of linear and quadratic equations.

2.1 Organization of numbers

1.1 The sets of natural numbers, ℕ; integers, ℤ; rational numbers, ℚ; and real numbers, ℝ.

The four main groups of numbers are the natural numbers (ℕ), the integers (ℤ), the rational numbers (ℚ) and the real numbers (ℝ). The natural numbers are those numbers you use for counting, such as 0, 1, 2, 3, etc. The integers include the natural numbers, as well as the negative counting numbers, such as -2, -1, 0, 1, 2, etc. The rational numbers include the integers (we will see why later) and all numbers that can be written as a fraction or, more properly, as a ratio of two integers. The real numbers are the rational and the irrational numbers.

Rational numbers

The rational numbers include any number that can be written in the form $\frac{a}{b}$ where a and b are integers and b is not equal to 0. Some examples of the rational numbers include $\frac{1}{2}$, 0.23, $\frac{-5}{3}$, and 16. The number 0.23 is included in the list because it is equal to $\frac{23}{100}$, and 16 is included because it is equal to $\frac{16}{1}$. Note that $\frac{-5}{3}$ is equal to $-1.6\dot{6}$. All integers are therefore also included in the rational numbers, as are all decimal numbers where the decimal either terminates or repeats.

Real numbers

The real numbers include all numbers that can be represented by a decimal, including π, $\sqrt{5}$, e and many more. If you look at the decimal expansions of these numbers, they never repeat, and they never terminate. When you write down 3.14 to approximate π, you are rounding it off to 2 decimal places. In general, any square root of a number that is a prime number will be irrational. Numbers that belong in $\mathbb{R}$ but do not belong in $\mathbb{Q}$ are called irrational numbers (the symbol for which is $\mathbb{Q}'$).

Example 2.1

Which number groups ($\mathbb{N}$, $\mathbb{Z}$, $\mathbb{Q}$ or $\mathbb{R}$) include the following numbers?
0.454 454 445..., $\frac{\pi}{2}$, $\sqrt{16}$, 0.000 000 002

Solution

For 0.454 454 445..., although it is clear we have a pattern in the digits, no group of the digits repeats itself exactly, so this number is not a rational number. It belongs in $\mathbb{R}$ but not in $\mathbb{Q}$. For $\frac{\pi}{2}$, although this is a ratio of two numbers, one of the two numbers is not an integer and there is no way to simplify the fraction so that it will be an integer. This number would also be grouped in $\mathbb{R}$. $\sqrt{16}$ is equal to 4, so this number would be most appropriately grouped in $\mathbb{N}$ and hence $\mathbb{Z}$, $\mathbb{Q}$, and $\mathbb{R}$ as well.
0.000 000 002 is a rational number as it can be written as $\frac{2}{1\,000\,000\,000}$; therefore it should be grouped in $\mathbb{Q}$ and hence it belongs to $\mathbb{R}$ as well.

If you know that the German word for numbers is Zahlen, it is clear why we use the symbol $\mathbb{Z}$ for integers.

For more about integers, visit www. pearsonhotlinks.co.uk, enter the ISBN for this book and click on weblink 2.2.

For the first 100 digits of π, visit www.pearsonhotlinks.co.uk, enter the ISBN for this book and click on weblink 2.3.

Exercise 2.1

1. List four examples of a number from each of:
   a) natural numbers
   b) rational numbers
   c) integers
   d) real numbers

2. Find two numbers a and b such that $\frac{a}{b} = 1.9$ which shows that 1.9 is rational.

3. Explain why the number 5.17 is rational.

4. List all of the natural numbers less than 10.

5. Give an example of a rational number between 3.1 and 3.2. Express your solution in the form $\frac{a}{b}$ where a and b are integers, $b \neq 0$.

6. Using Figure 2.1 as an example, draw a Venn diagram to represent the categories of each of the following numbers: -4, 4.5, $\sqrt{2.25}$, 0, $0.\dot{1}$

7. Look at the following list of numbers:

$-2.7, 35, 300\,000, \frac{\pi}{3}, \sqrt{17}, 0$

a) Which numbers are rational?

b) Which numbers are integers?

c) Which numbers are irrational?

8. To which sets ($\mathbb{N}, \mathbb{Z}, \mathbb{Q}, \mathbb{Q}', \mathbb{R}$) does 2^{10} belong?

9. Indicate which of the following numbers are irrational:

$0.\dot{1}, \sqrt{19}, \sqrt{64}, \frac{\pi}{3}, 0.121\,121\,112\,111\,12..., \frac{\sqrt{3}}{2}$

10. A is defined to be the set of numbers $\{4, -5, 6.3, 5.\dot{5}, \frac{3}{2}, \sqrt{3}, 2\pi, 0\}$.

a) List the elements in A but not in $\mathbb{Q}$.

b) List the elements in A but not in $\mathbb{Z}$.

c) List the elements in A but not in $\mathbb{N}$.

d) List the elements in A and in $\mathbb{Q}'$.

e) Give three examples of numbers in $\mathbb{N}, \mathbb{Z},$ and $\mathbb{Q}$.

11. A is defined to be all of the prime numbers less than 20.
B is defined to be all integers less than 20.
C is defined to be all numbers less than 10.

a) Give an example of a number which is in A, B, and C.

b) List all of the elements of A.

c) Copy Figure 2.1 and then draw and label circles onto it to represent the sets A, B, and C.

At the time of writing, the world record for the number of digits of π was 1 241 100 000 000 digits, set in 2002 by Yasumasa Kanada of Tokyo University. See www.pearsonhotlinks.co.uk, enter the ISBN for this book and click on weblink 2.4.

2.2 Numbers in calculations

1.2 Approximation: decimal places; significant figures. Percentage errors. Estimation.

Estimation

Advances in computer technology have allowed us a high degree of accuracy in carrying out calculations, but this would not have been possible without some important mathematical developments earlier in our history.

Decimal notation allows us to write numbers to a chosen degree of accuracy. If we want to know the result of a calculation to within one thousandth, we write down our number with 3 decimal places. For example, 3.237 is written to 3 decimal places (since there are three digits to the right of the decimal point) or to the nearest one thousandth, since the 7 represents the one thousandths position.

Sometimes we want to limit the accuracy to which we write our answer, because of limitations with our original data. For instance, if we measure a table length as 185 centimetres, any calculations we use that number in are

hundreds
tens
units
tenths
hundredths
thousandths

203.667

one decimal place
two decimal places
three decimal places

▲

Figure 2.2 This chart allows us to see the relationship between the number of decimal places of accuracy we want, and its position.

limited to three digits, or significant figures, of accuracy. If we had instead measured the table to the nearest metre and found it to be 2 metres long, we could have only one significant figure in any answers to further calculations involving this length.

Example measurement	Number of significant figures
230.4 to nearest 0.1 cm	4
230 to nearest 1 cm	3
230 to nearest 10 cm	2
200 to nearest 100 cm	1

Table 2.1 This table shows the relationship between measurements and significant figures.

If we want to find the error in a measurement or calculation, we use the formula percent error $= \frac{v_E - v_A}{v_E} \times 100$, where v_E is the exact value of the measurement and v_A is the approximate value of the measurement. Note that although percent error can be positive or negative, **absolute value** of the error is equal to the size of the error.

Example 2.2

If the height of a door is measured to be 220 cm, to the nearest centimetre, find:

a) the minimum possible height of the door

b) the maximum possible height of the door

c) the maximum error in this measurement.

Solution

a) Since we know the height of the door to the nearest centimetre, the most we could be wrong is 0.5 cm. Therefore, the minimum height is 219.5 cm.

b) Using a similar argument as part a), the maximum height is 220.5 cm.

c) Percent error $= \frac{v_E - v_A}{v_E} \times 100$ or percent error $= \frac{v_E - v_A}{v_E} \times 100$

$$= \frac{220.5 - 220}{220.5} \times 100 \qquad = \frac{219.5 - 220}{219.5} \times 100$$

$$= 0.227\% \qquad = -0.228\%$$

Since $|0.228\%|$ is the larger of the two possible errors, it is the maximum error.

If we want to find the error in a calculation, we have to take the individual measurements in the calculation. Find a lower and an upper limit on each measured value used in the calculation. Use these values to place a lower and upper limit on the result of the calculation. Finally we can use our percent error formula, as in the example above.

Example 2.3

If Majed measures the radius, r, of a cylindrical drum to be 11 cm and the height, h, of a cylinder to be 35 cm, both of which we can assume to be accurate to at least the nearest centimetre, find the possible error in his calculation of the volume of the cylinder. (Volume of a cylinder $V = \pi r^2 h$)

Solution

First, Majed's calculated volume would be approximately 13304.6 cm³. We also know that $10.5 \leqslant r \leqslant 11.5$ and that $34.5 \leqslant h \leqslant 35.5$ from the previous example. Hence:

$$V = \pi(10.5)^2(34.5) \approx 11949.4 \text{ cm}^3, \text{ or}$$

$$V = \pi(10.5)^2(35.5) \approx 12295.8 \text{ cm}^3, \text{ or}$$

$$V = \pi(11.5)^2(34.5) \approx 14333.9 \text{ cm}^3, \text{ or}$$

$$V = \pi(11.5)^2(35.5) \approx 14749.4 \text{ cm}^3.$$

We do all four of these calculations to find the minimum and maximum possible values for our actual volume. Our possible percent errors are therefore:

$$\text{percent error} = \frac{v_A - v_E}{v_E} \times 100 = \frac{13304.6 - 11949.4}{11949.4} \times 100 \approx 11.3\%$$

$$\text{percent error} = \frac{v_A - v_E}{v_E} \times 100 = \frac{13304.6 - 14749.4}{14749.4} \times 100 \approx -9.80\%$$

Since $|11.3\%| > |-9.80\%|$, our solution is that the largest percent error in his calculation is 11.3% (to 3 significant figures).

When performing calculations, it is important to check that the final answer makes sense. To do this, use nice round numbers for calculations to first estimate the result.

Example 2.4

Estimate the surface area of a cylinder with a radius of 13.35 m and a height of 25.8 m.

Solution

The surface area of a cylinder is $A = 2\pi rh + 2\pi r^2$. Let $r = 13$ m, $h = 26$ m and $\pi = 3.14$.

$$A = 2(3.14)(13)(26) + 2(3.14)(13)^2$$

$$= 2122.64 \text{ m}^2 + 1061.32 \text{ m}^2$$

$$= 3183.96 \text{ m}^2$$

$$A \approx 3180 \text{ m}^2$$

The other way to do estimation is to consider the reasonableness of the solution. If you calculate a solution that seems unreasonable, especially given the context of the question, then you may need to recalculate your answer.

1. Write 412.4563
   a) to the nearest tenth
   b) to the nearest one hundreth
   c) to 3 decimal places
   d) to 3 significant figures
   e) to the nearest integer
   f) to the one hundreds place.

2. For each of the following numbers, write to 3 significant figures:
   a) 345.678
   b) 34 567.8
   c) 0.000 345 678
   d) 1010.1
   e) 1200.02
   f) 19

3. Find the percent error for each of the following, to 3 significant figures:
   a) a measurement of 200 cm, to the nearest centimetre
   b) a measured volume of a vase of 2.3 litres (l), when the actual volume is 2.5 l
   c) the length of a sports field is 100 m, measured to the nearest centimetre
   d) π as displayed on your calculator and $\frac{22}{7}$

4. The age of the universe, according to the 'Big Bang theory', used to be quoted as 15 000 000 000 years. With recent advances in technology, this has been revised to 13.6 billion years. Assuming the revised figure is accurate, what is the percent error between the original estimated age and the more current age of the universe?

5. Estimate the volume of a cuboid ($V = l \times w \times h$) with dimensions of 2.1 m, 5.9 m and 6.4 m respectively. Find the percent error in your estimation and the exact volume of the cuboid.

6. The length, width and height of a cardboard box are measured at 44 cm, 32 cm, and 25 cm respectively.
   a) Using the formula $V = l \times w \times h$, find the volume of the box.
   b) The manufacturer indicates that the actual dimensions of the box are 44.3 cm, 32.4 cm, and 24.6 cm. Find the exact volume of the box using these dimensions.
   c) Find the percent error between the measured volume and the actual volume.

7. The area of a quilt is 2.35 m^2.
   a) Round this area to the nearest whole m^2.
   b) Round this area to the nearest tenth.
   c) Find the percent error between your solutions to the previous parts and the actual area of the quilt.

8. A problem has an **exact** answer of $x = 2.125$.
   a) Write down the **exact** value of x^2.
   b) State the value of x given correct to **2** significant figures.
   c) Calculate the percentage error if x^2 is given correct to **2** significant figures.

9. a) Given the equation $p = r^2 + 2qr$, calculate the exact value of p when $q = 3.6$ and $r = 24$.
   b) Write your answer to 2 significant figures.
   c) Calculate the percent error between the exact answer from a) and the approximate answer from b).

10. The total length of 1000 identical hinges is 3064.2 cm. Calculate the length of one hinge, in metres.
    a) Give your answer exactly.
    b) Give your answer correct to three significant figures.
    c) Find the percent error between your answers to part a) and part b).

11. In the expression, $x = \dfrac{2p^3 + 3p}{q + r}$, p has a value of 2.03, q has a value of 4.51 and r has a value of 3.92.
  a) Find the value of x.
  b) If p, q, and r are rounded to **1 decimal place**, what is x?
  c) If p, q, and r are rounded to **1 significant figure**, what is x?
  d) Find the percent error in your answer to part b).
  e) Find the percent error in your answer to part c).

12. a) Find the value of $\sqrt{a^2 + b^2}$ when $a = 2.78$ and $b = 3.06$.
  b) Round your answer to part a) to:
     (i) **2 significant figures**
     (ii) **2 decimal places**.
  c) Round a and b to **1 decimal place**, and find the value of $\sqrt{a^2 + b^2}$ to:
     (i) **2 significant figures**
     (ii) **2 decimal places**.
  d) Find the percent error in your calculation for c) (ii).

(2.3) **Standard form**

> **1.3** Expressing numbers in the form $a \times 10^k$ where $1 \leqslant a < 10$ and k is an integer. Operations with numbers in this form.

Standard form is just another way of representing numbers, and it is generally used to simplify calculations with either very large or very small numbers. Standard form is sometimes referred to as scientific notation.

A number written in standard form will be in the form $a \times 10^k$, where $1 \leqslant a < 10$ and k is an integer. For example, we would write 34 530 000 as 3.453×10^7, and 0.000 453 as 4.53×10^{-4}. How we find these numbers is by realizing that each time we multiply by 10 we move the decimal point once right, and each time we divide by 10 we move the decimal point once left. Positive powers of 10 correspond to multiplication by 10, negative powers of 10 correspond to division by 10. So, 1.65×10^5 means multiply 1.65 by 10 five times, which would mean move the decimal point five places to the right, and hence this is 165 000.

Table 2.2 Changing a number to standard form.

1. Find the leftmost non-zero digit.	00.00453
2. Count the number of decimal places from just right of this digit to the current decimal point.	0.000453
3. If we counted left, our k will be negative; if we counted right, our k will be positive.	$k = -4$
4. Write down our number in standard form, placing the decimal point where we counted from before.	4.53×10^{-4}

The real advantage to representing numbers in standard form is that we can do calculations with them much more easily. For example, 3×10^5 times 4×10^6 can be rewritten as $3 \times 4 \times 10^5 \times 10^6 = 12 \times 10^{11} = 1.2 \times 10^{12}$. Here we had to change 12×10^{11} back into standard form by recognizing that $12 = 1.2 \times 10^1$.

Calculations with numbers in standard form are relatively straightforward when we are doing multiplication or division, but when we do addition or subtraction, we have to change each number so that their powers of 10 are the same. For example, $4.5 \times 10^4 + 3.6 \times 10^5$ becomes $4.5 \times 10^4 + 36 \times 10^4 = 40.5 \times 10^4$, and then finally we convert the result back into standard form as 4.05×10^5.

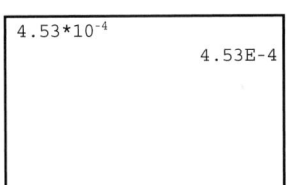

Figure 2.3 Here we can see that our calculators do not write $\times 10^{-4}$, they use E instead.

Example 2.5

The total mass of 256 identical pencils is 4.24 kg. Calculate the mass of one pencil, in kg.

a) Give your answer exactly.
b) Give your answer correct to 3 significant figures.
c) Write your answer to part b) in the form $a \times 10^k$ where $1 \leqslant a < 10$ and k is an integer, also written as $k \in \mathbb{Z}$.

Solution

a) $\dfrac{4.24}{256}$ kg $= 0.0165625$ kg

b) 0.0166 kg

c) 1.66×10^{-2} kg

 The largest number that we have a name for is a 'Googolplex' which is 1 followed by 10^{100} zeros, which is much larger than the number of particles in our universe. See www.pearsonhotlinks.co.uk, enter the ISBN for this book and click on weblink 2.5.

Exercise 2.3

1. Write the following numbers in the form $a \times 10^k$ where $1 \leqslant |a| < 10$ and k is an integer.
   a) 435 000
   b) 1900
   c) 0.0087
   d) −12 005
   e) 0.2
   f) 323
   g) −0.004 56
   h) −19.6
   i) 19 000
   j) 0.157

2. Write the following numbers without using exponents.
   a) 2.006×10^7
   b) 8.8×10^{-4}
   c) -2×10^3
   d) 5.06×10^5
   e) 3.39×10^{-5}
   f) -7.5×10^1
   g) 9×10^{-2}
   h) -3.3×10^{-5}
   i) 4.3×10^0
   j) -5.4×10^6

3. Evaluate, leaving your answer in the form $a \times 10^k$ where $1 \leqslant |a| < 10$ and k is an integer.
   a) $(4 \times 10^5) \times (2 \times 10^3)$
   b) $(-3 \times 10^4) \times (5 \times 10^6)$
   c) $(1.2 \times 10^6) \times (8.5 \times 10^2)$
   d) $\dfrac{4.5 \times 10^8}{1.5 \times 10^5}$
   e) $\dfrac{2.4 \times 10^4}{1.2 \times 10^8}$
   f) $\dfrac{1.6 \times 10^{-4}}{3.2 \times 10^3}$

4. Evaluate, leaving your answer in the form $a \times 10^k$ where $1 \leqslant |a| < 10$ and k is an integer.

   a) $3.4 \times 10^3 + 5.4 \times 10^6$

   b) $2.7 \times 10^4 + 3.04 \times 10^5$

   c) $9.8 \times 10^{-3} + 2.4 \times 10^{-4}$

   d) $5.06 \times 10^7 - 5 \times 10^5$

   e) $1.4 \times 10^{-3} - 1.4 \times 10^{-2}$

   f) $1.7 \times 10^2 - 8.6 \times 10^{-1}$

5. Evaluate, leaving your answer in the form $a \times 10^k$ where $1 \leqslant |a| < 10$ and k is an integer.

   a) $\dfrac{3.4 \times 10^4 + 5.4 \times 10^4}{1.1 \times 10^3}$

   b) $\dfrac{1.8 \times 10^6 + 8.4 \times 10^5}{6.6 \times 10^3}$

   c) $\dfrac{1.03 \times 10^{-4} + 1.7 \times 10^{-5}}{5 \times 10^2 + 7 \times 10^2}$

6. A large rectangular field has a length of 3.4×10^4 m and a width of 2.7×10^4 m.

   a) Find the area of the field in m².

   b) Find the area of the field in km².

   c) Round your answer to part b) to **the nearest hundred**.

   d) Calculate the percent error between your exact answer to b) and your answer to c).

7. a) If $x = \dfrac{2a + b}{c}$ and $a = 2.5 \times 10^5$, $b = 5 \times 10^5$ and $c = 2 \times 10^4$, find x.

   b) Write your answer to part a) in the form $a \times 10^k$ where $1 \leqslant |a| < 10$ and k is an integer.

8. The mean radius of the Earth is approximately 6.373×10^3 km.

   a) Using the formula $V = \dfrac{4}{3}\pi r^3$, find the volume of the Earth in km³, writing your answer in the form $a \times 10^k$ where $1 \leqslant |a| < 10$ and $k \in \mathbb{Z}$.

   b) Using the same formula, but a radius of 6.3×10^3 km, find the volume of the Earth.

   c) Find the percent error between your answers to a) and b), assuming a) to be the more accurate result.

9. The population of the Earth is approximately 6.6×10^9 people. If this increases by 2% each year, about how many people will inhabit the Earth in five years? Give your answer in the form $a \times 10^k$ where $1 \leqslant |a| < 10$ and k is an integer.

10. Every year, a small music company sells 2.3×10^7 CDs. They have 1.2×10^3 different musicians.

    a) On average, how many CDs for each musician does the company sell?

    b) If sales increase by 20% and the number of musicians drops by 30%, how many CDs for each musician does the company sell? Round your answer **to the nearest thousand**.

11. A typical star is composed of about 1.2×10^{57} hydrogen atoms. There are about 400 000 000 000 stars in the typical galaxy and about 8.0×10^{10} galaxies in the universe.

    a) About how many hydrogen atoms are there in the universe?

    b) Given that the mass of 6.02×10^{23} hydrogen atoms is 1 gram, find the approximate mass of all of the hydrogen atoms in the universe.

12. The mass of the Earth is approximately 5.9742×10^{24} kg. The mass of the Sun is approximately $1.988\,92 \times 10^{30}$ kg.

    a) About how many times heavier is the Sun than the Earth?

    b) If a typical person weighs 65 kg, how many times heavier is the Earth than the typical person?

    c) If every time a rocket is launched from the Earth 2000 kg of material is lost from the Earth, how many launches would it take before the Earth is entirely depleted?

    d) There are approximately 40 launches per year. Use your answer to part c) to calculate how many years such a depletion would take.

Why do different cultures have different systems of measurement?

2.4 International units of measure

| 1.4 | SI (*Système International*) and other basic units of measurement: for example, kilogram (kg), metre (m), second (s), litre (l), metre per second (m s⁻¹), Celsius scale. |

For more information about SI units, visit www.pearsonhotlinks.co.uk, enter the ISBN for this book and click on weblinks 2.6 and 2.7.

This system of measurement was developed in France around the time of the revolution in the 1790s. The main reason why it has been implemented so broadly worldwide is that calculations and conversions involving metric units are easier (because they are in powers of 10) than those using the older imperial units.

There are three countries in the world still officially using non-metric systems of measure: Liberia, Myanmar, and the United States.

In the SI system, each type of unit, for example the metre or the gram, is used as a base. To change the unit of measure, you just need to use the system of prefixes given below.

Prefix	tera	giga	mega	kilo	hecto	deca	deci	centi	milli	micro	nano
Symbol	T	G	M	k	h	da	d	c	m	μ	n
Conversion factor	10^{12}	10^{9}	10^{6}	10^{3}	10^{2}	10^{1}	10^{-1}	10^{-2}	10^{-3}	10^{-6}	10^{-9}

◀ **Table 2.3** Common SI prefixes.

To convert from one unit of measure to another, you divide the starting unit factor by the end unit factor, and multiply this by the quantity of the unit. For example, to convert 34 km into centimetres:

$$\frac{10^3}{10^{-2}} \times 34\,\text{km} = 3\,400\,000\,\text{cm}$$

Note that if either of the starting units has no prefix, then you would use a factor of 1 in your calculations.

Remember that when converting a unit of area or volume, each unit needs to be converted. For instance, if converting a volume in m³ to cm³, you have to remember this is really m × m × m being converted to cm × cm × cm. In other words, a volume in m³ is converted to cm³ by multiplying by $10^2 \times 10^2 \times 10^2$ (or 10^6). You must consider this whenever an SI unit is raised to a power.

Example 2.6

Find the volume in m³ of the rectangular prism shown right.

Solution

The volume of a rectangular prism is $V = l \times w \times h$. Substituting the values shown from the diagram, we arrive at $V = (20\,\text{cm})(10\,\text{cm})(5\,\text{cm}) = 1000\,\text{cm}^3$. Next we convert our answer from cm³ to m³. Each cm needs to be converted into m; therefore we need to multiply by 1, and divide by 10^2 three times:

$$1000\,\text{cm}^3 = 1000 \times \frac{1}{100} \times \frac{1}{100} \times \frac{1}{100}\,\text{m}^3 = 0.001\,\text{m}^3.$$

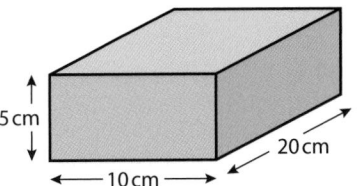

1 m³ ≠ 100 cm³, but if we mistakenly used this conversion, we would calculate the volume to be 10 m³, out by a factor of 10 000!

Example 2.7

a) The volume of a vase is 1570 ml. Find the volume of the vase in litres.

b) The SI unit for force is the newton. Given that:
$$F = ma$$
where F is the force on an object, m is the object's mass in kg and a is the acceleration of the object in m s^{-2}, find the units of a newton in terms of kg, m and s.

c) The density (ρ)of an object is equal to the mass of the object divided by the volume of the object. If the dimensions of a 1 kg cube of metal are each 10 cm, find the density of the cube in kg m^{-3}.

Solution

a) We divide 1570 ml by 1000 to convert to litres, hence $\dfrac{1570 \text{ ml}}{1000} = 1.570 \text{ l}$.

b) Substituting the units of mass and the units of acceleration into the formula we arrive at:
$$F = (\text{kg})(\text{m s}^{-2})$$
$$F = \text{kg m s}^{-2}$$

c) First we convert 10 cm to 0.1 m. The volume of the cube is then $(0.1 \text{ m})^3$. To find the density we do the following:
$$\rho = \frac{1 \text{ kg}}{0.001 \text{ m}^3}$$
$$\rho = 1000 \text{ kg m}^{-3}$$

Exercise 2.4

1. Convert each measurement into metres.

  a) 2600 cm   b) 16 km   c) 85 μm   d) 250 dm   e) 1.2 Gm

2. Convert each area into cm^2.

  a) 2 m^2 b) 190 mm^2 c) 36 km^2 d) 92 hm^2 e) 5.6 $\times$ 10^4 nm^2

3. Convert each volume into litres (remembering that 1000 cm^3 is 1 litre).

  a) 3 m^3 b) 250 cm^3 c) 345 ml   d) 25 kl   e) 6.7 dl

4. Convert each mass into kilograms.

  a) 200 mg   b) 34 g   c) 59 Mg   d) 393 μm   e) 6.25 dag

5. For each prefix, determine what number you would need to **multiply** by in order to convert to the deci prefix.

  a) centi   b) nano   c) deka   d) kilo   e) micro

6. The area of a square piece of land is given as 1.35 km^2 to the nearest km^2.

  a) What is the area of the land in m^2?

  b) What is the maximum error in the measurement of the land in km^2?

  c) What is the percent error of the area of the land in km^2?

  d) What is the percent error of the area of the land in m^2?

7. A 20 Mg boat has a speed of 13 m s^{-1}. Find its energy using the formula $E = \frac{1}{2} mv^2$, where m is the mass in kg and v is the speed in m s^{-1}. Write your answer in the form $a \times 10^k$ where $1 \leqslant a < 10$ and $k \in \mathbb{Z}$, using the appropriate units.

8. The length of a building is measured as 100 m to the nearest metre.
   a) Write down the maximum error in this measurement.
   b) Calculate the percent error of this measurement.
   c) The building is 10 023 cm long when measured to the nearest cm. What is the percent error in this measurement?

9. A painter needs to buy enough paint to cover an entire house with two coats of paint. If each bucket of paint contains 2 l of paint, each litre of paint can cover an area of $250 \, m^2$, and the surface area of the house is $140 \, dam^2$, find the minimum number of whole buckets of paint the painter needs to purchase.

10. The width of the Milky Way galaxy is about 150 million light years (a light year is defined to be the distance light travels in one year, and the speed of light is $3 \times 10^8 \, m \, s^{-1}$).
    a) Find the distance across the Milky Way in km.
    b) Using your answer to part a), calculate the amount of time a rocket travelling at $2.5 \times 10^6 \, m \, s^{-1}$ would need to cross the galaxy.

11. A hectare is defined to be the area covered by a square one hectometre on each side.
    a) How many square metres is one hectare?
    b) How many times larger is a square kilometre than a hectare?
    c) Each year in the United States about $17 \, 000 \, km^2$ of forest is burned in wildfires. How many hectares is this?
    d) Use the formula $S_A = 4\pi r^2$, where S_A is the surface area and r is the radius of the Earth (about 6378.1 km), to find the surface area of the Earth in km^2.
    e) Use your answer to d) to calculate how many years it will take for the area of forest burned in the USA to exceed the surface area of the Earth.

12. One newton is the force required to give a mass of 1 **kilogram** an acceleration of 1 **metre** per **second** per **second**.
    a) According to this definition, what are the SI units for a newton?
    b) How many newtons of force must be applied to accelerate a 10 kg mass object by $9.8 \, m \, s^{-2}$?
    c) 1000 newtons of force are applied to a 65 kg object. By how much is this object accelerated?

2.5 Currency Conversions

| 1.5 | Currency conversions. |

Different countries developed their own currencies, so now each currency has its own value. Since each country has to import and export goods, exchange rates between different currencies are calculated. These exchange rates are often displayed using a table.

When two currencies are compared, the value of one currency is given in the units of the other. Sometimes this exchange rate is given between only two countries. Alternatively, exchange rates are given for multiple countries, such as in the table below. We can see the exchange rates given between US dollars, GB pounds sterling, and Japanese yen.

 For a website with an easy to use currency converter, visit www.pearsonhotlinks.co.uk, enter the ISBN for this book and click on weblink 2.8

Table 2.4 Currency exchange rates for three countries.

	US $	GB £	JAPAN ¥
US $	1	0.51	117
GB £	1.96	1	229.3
JAPAN ¥	0.008 55	0.004 36	1

In this table we can see that £1 is worth US$1.96, and ¥229.3. To find the value of any other amount of £ in the other currencies, we multiply the number of £ by the exchange rate. So £800 is equal to US$1568, and ¥183 440.

Commission

Another type of conversion table is used when currency is bought and sold by a currency broker. A currency broker earns money by selling currency for more than they pay for it. They also often charge an additional commission fee, which is taken out of the money their customers want to convert. An example table is shown below.

Table 2.5 Buying and selling prices for three currencies into GB£.

	Buying price (£)	Selling price (£)
US $1	0.48	0.49
AUS$1	0.43	0.44
CAN$1	0.49	0.5

From this chart we can see that if we have US$1, the currency broker will buy our dollar from us for £0.48, and sell us our dollar back for £0.49. In practice, a currency broker will buy currency from one person and sell it to another, making money in the process.

Why does currency have worth?

When a broker charges a commission, they charge it in the currency that they buy and sell. In the example above, this would be GB£. A typical commission is 2.5% of the total amount of currency converted.

Example 2.8

How do different cultures handle currency?

Arnold has £450 and on his return to the United States goes to a currency broker to convert his money. If the currency broker buys £ at a rate of $1 equals £0.51, how many US$ will Arnold receive?

Solution

Each currency has its own value, and an exchange rate with all of the other currencies. We either multiply or divide by the exchange rate depending on whether we are converting to or from the given currency.

Since $1 equals £0.51, £450 = $882.35. We divide £450 by 0.51 to arrive at this answer, since we expect a larger number of US$ (since each $1 has less value than £1).

Exercise 2.5

1. If the exchange rate from £ to CAN$ is £1 equals CAN$2, convert the following into CAN$.
   a) £1000   b) £10 000   c) £930   d) £3.45

2. If the exchange rate from US$ to Thai baht is US$1 equals 35 baht, convert the following into Thai baht.
   a) $35   b) $800   c) $250   d) $120

3. If the currency exchange rate between ¥ (Japanese yen) and US$ is $1 equals ¥120, convert the following into US$.

a) ¥360   b) ¥24   c) ¥23 000   d) ¥1000

4. If the currency exchange rate between Martian yoldas and Venusian anthmas is 1 yolda equals 13.2 anthmas, convert each of the following either into yoldas or anthmas.

a) 132 anthmas   b) 24 yoldas   c) 4.57 anthmas   d) 740 yoldas

5. Use the table to convert between the following currencies.

	€ (euros)	$	£
€	1	1.5	0.75
$	0.67	1	0.5
£	1.33	2	1

a) $500 US to €   b) € 320 to £   c) € 320 to US$ d) £120 to US$

6. Samantha travels to Myanmar from the United States where the currency is worth 6 kyat per US$.

a) She converts $1400 into Myanmar kyat and is charged a 2% commission. Find out how many Myanmar kyat she has.

b) At the end of her trip, she is again charged a 2% commission when she converts her remaining Myanmar kyat back into $. If she is left with exactly $50, how many Myanmar kyat did she spend?

7. Rafael travels around Europe visiting Greece, Turkey, and Switzerland. Use the table below for the following questions.

	€	Turkish lira	Swiss franc
€	1	1.73	1.65
Turkish lira	0.58	1	0.95
Swiss franc	0.60	1.03	1

a) Rafael travels from Greece to Turkey. How many Turkish lira will he have, if he converts all of his €3000?

b) In the airport, on the way to Switzerland, he converts his remaining 2000 Turkish lira into Swiss francs. If he is charged a 3% commission, how many Swiss francs can he spend?

8. The following shows the exchange rate between one British pound (GBP) and other currencies.

EXCHANGE RATES		
	Bank buys foreign currency	**Bank sells foreign currency**
Denmark (KR)	11.38	10.78
Finland (MKK)	7.00	6.60
France (FFR)	10.05	9.45
Germany (DM)	2.854	2.798
Greece (DR)	292	266
NO COMMISSION CHARGED		

Geraldine eats a meal while on holiday in Greece. The meal costs 4256 drachma (DR).

a) Calculate the cost of the meal in British pounds.

The Williams family go to Germany. Before leaving, they change GBP 600 into German marks.

  b) Calculate the number of German marks they receive for GBP 600, giving your answer correct to **two** decimal places.

They spend DM 824 in Germany, and on returning to the United Kingdom, they change their remaining German marks into British pounds.

  c) Calculate the number of British pounds they receive, correct to **two** decimal places.

9. The exchange rate from US dollars (USD) to French francs (FFR) is given by 1 USD = 7.5 FFR. Give the answers to the following correct to **two** decimal places.
  a) Convert 115 US dollars to French francs.
  b) Roger receives 600 Australian dollars (AUD) for 2430 FFR. Calculate the value of the US dollar in Australian dollars.

10. Zog from the planet Mars wants to change some Martian Dollars (MD) into US dollars (USD). The exchange rate is 1 MD = 0.412 USD. The bank charges 2 % commission.
  a) How many US dollars will Zog receive if she pays 3500 MD?
   Zog meets Zania from Venus where the currency is Venusian Rupees (VR). They want to exchange money and avoid bank charges. The exchange rate is 1 MD = 1.63 VR.
  b) How many Martian dollars, to the nearest dollar, will Zania receive if she gives Zog 2100 VR?

11. A bank in Canada offers the following exchange rate between Canadian dollars (CAD) and euros (EUR). The bank sells 1 CAD for 1.5485 EUR and buys 1 CAD for 1.5162 EUR. A customer wishes to exchange 800 Canadian dollars for euros.
  a) Find how many euros the customer will receive.
  b) The customer has to cancel his trip and changes his money back later when the rates are 'sells 1 CAD = 1.5546 EUR, buys 1 CAD = 1.5284 EUR'. Use the 'we sell' information to find how many Canadian dollars he receives.
  c) How many Canadian dollars has he lost on the transaction?
  d) If he then exchanges his remaining Canadian dollars for US dollars at a rate of 1 CAD equals 1.02 USD, how many Canadian dollars will he make, if the exchange rate later changes to 1 CAD equals $0.97 USD?

12. Veronika decides to take a trip to Jamaica. At the bank she converts 50 000 Russian rubles into Jamaican dollars at an exchange rate of 1 JMD = 0.3476 RUB.
  a) Calculate how many Jamaican dollars she receives.
  b) If she is charged a 3% commission, how much is the commission in Russian rubles?
   Veronika decides to invest in the Jamaican dollars she bought instead of going on her trip. In 5 years, the exchange rate between rubles and dollars has changed to 1 JMD = 0.35 RUB.
  c) How many Russian rubles will Veronika receive if she exchanges her Jamaican dollars?
  d) Veronika needs to save US$5000 in preparation for her emigration to the United States. If the exchange rate between US$ and rubles is US$1 = 24.67 RUB, calculate how many Russian rubles she still needs to save.

2.6 Simultaneous equations

1.6 Use of GDC to solve: pairs of linear equations in two variables, and quadratic equations.

Simultaneous equations are two or more linear equations that include two or more different variables. We will cover solving these equations using the GDC.

$$3x + 5y = 16$$
$$3x - 3y = -2$$

Figure 2.4 An example of a pair of linear equations.

Elimination

Suppose we have the following two equations where the x and y variables are both on the same side of each equation:

$$4x + 5y = 28$$
$$3x - 2y = -2$$

We need to multiply each of the equations so that we end up with one of the variables in each equation having the same coefficient:

$$3 \times (4x + 5y = 28)$$
$$4 \times (3x - 2y = -2)$$

We end up with:

$$12x + 15y = 84$$
$$12x - 8y = -8$$

To continue solving, we can cancel the x variable out of the equation by subtracting one equation from the other:

$$12x + 15y = 84$$
$$\underline{- \ (12x - 8y = -8)}$$
$$23y = 92$$

Dividing both sides of the equation by the coefficient of y gives us the value of y:

$$\frac{23y}{23} = \frac{92}{23}$$
$$\therefore \quad y = 4$$

We substitute this value of y in one of the original two equations to find the value of x:

$$4x + 5(4) = 28$$
$$4x + 20 = 28$$
$$4x = 8$$
$$\text{Hence, } x = 2$$

The last thing we should do is confirm our solution is correct by substituting both values for x and y into the other original equation:

$$3(2) - 2(4) = -2$$
$$6 - 8 = -2$$
$$-2 = -2$$

Having confirmed that the right-hand side of the equation equals the left-hand side of the equation, we can confirm that and $x = 2$ and $y = 4$.

Substitution

If our original equations are in a different form, it may be more efficient to use the substitution method:

$$2x - 5y = 25$$
$$4y = 3x - 27$$

In the substitution method, the objective is to substitute an expression in one of the variables, for the other variable. Since y is the only variable on the left-hand side of the second equation in this example, we will begin by isolating it by dividing both sides of the equation by the coefficient of y, in this case 4:

$$4y = 3x - 27$$
$$y = \frac{3}{4}x - \frac{27}{4}$$

We then take this expression for y, and substitute in for the y variable in the other equation:

$$2x - 5\left(\frac{3}{4}x - \frac{27}{4}\right) = 25$$

First we simplify this expression and then we solve it for x:

$$2x - \frac{15}{4}x + \frac{135}{4} = 25$$

The easiest way to solve an equation that involves fractions is to multiply the entire equation by the common denominators of those fractions first.

$$4 \times \left(2x - \frac{15}{4}x + \frac{135}{4}\right) = 25$$
$$8x - 15x + 135 = 100$$

Now we collect like terms, and finally solve for x:

$$-7x = -35$$
$$\therefore \quad x = 5$$

We then continue as we did for the elimination method, substituting this value in for x in the other equation, and solving for y:

$$4y = 3x - 27$$
$$4y = 3(5) - 27$$
$$4y = 15 - 27$$
$$4y = -12$$
$$\therefore y = -3$$

We should then, as a precaution against making an error, check that this solution for x and y works in the other original equation:

$$2x - 5y = 25$$
$$2(5) - 5(-3) = 25$$
$$10 - (-15) = 25$$
$$25 = 25 \quad \checkmark$$

So our solution is $x = 5$ and $y = -3$.

Using your GDC

There are a few ways of solving these problems using your graphics display calculator, all of which are fairly straightforward.

The program **PlySmlt2** is a great program for helping you solve pairs of simultaneous equations (and we will also see for quadratic equations in the next section). Once you have downloaded it and installed the software, try and run it. You should see the following screen:

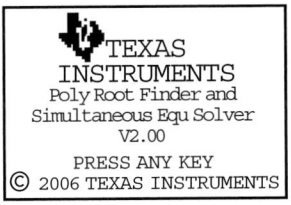

Press any key, and you see:

We want to press 2, and we see the following:

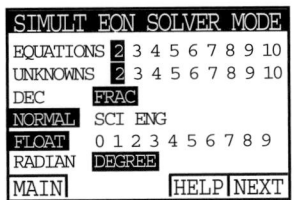

We press GRAPH (below the NEXT tab) and we get the following:

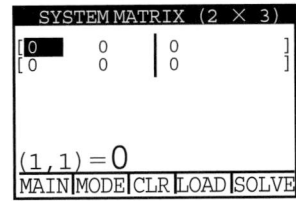

One of the programs you are allowed to use for your exams on the TI calculators is **PlySmlt2**. To download it, go to www.pearsonhotlinks.co.uk, enter the ISBN for this book and click on weblink 2.9. For instructions on installation of the program, click on weblink 2.10, and then click *downloads*.

◀ **Figure 2.5** The opening screen of PlySmlt2.

◀ **Figure 2.6** The main menu of PlySmlt2.

◀ **Figure 2.7** The first screen of the simultaneous equation solver. Set up the screen of the solver by pressing 2, ENTER, 2.

◀ **Figure 2.8** Entering the coefficients.

Here we enter in the coefficients of the equation on the left, and the numbers from the right-hand side of the equation on the right. So if our equations were:

$$14x + 17y = 45$$
$$13x - 12y = 14$$

We would enter:

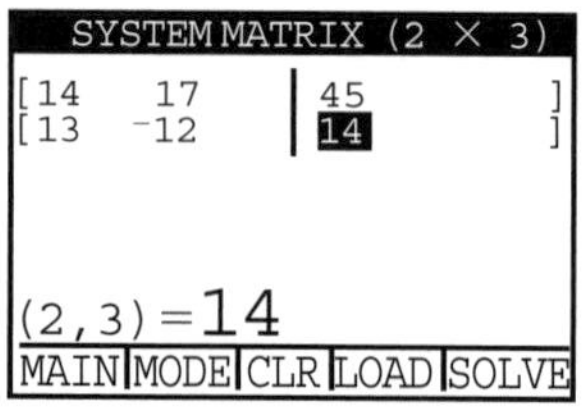

Figure 2.9 Entering the coefficients.

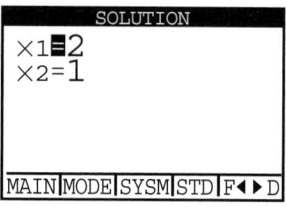

Next, we press the GRAPH key, which is below SOLVE on the screen, and the calculator almost instantly solves the equations for us, giving us the solution of:

Figure 2.10 Solution to the equations.

X2 is really the *y*-value.

Exercise 2.6

1. Solve for *x* and *y*.

a) $2x + 5y = 14$
$\; 3x - 2y = 2$

b) $10x + 7y = 234$
$\; 3x + 8y = 141$

c) $8x + 7y = 10.45$
$\; 5x + 2y = 4.75$

d) $17x + 10y = 60.1$
$\; 15x + 3y = 40.8$

e) $4x + 5y = 27.5$
$\; 5x + 6y = 33.5$

f) $x + y = 23$
$\; x - y = 7$

2. Solve for *x* and *y*.

a) $4x + y = 25$
$\; 3x = y + 10$

b) $y = 4x - 3$
$\; 2x + 3y = 117$

c) $y = \frac{2}{3}x + 1$
$\; 3x - 2y = 8$

d) $\frac{4}{5}x - 3 = 2y$
$\; 3x - 5y = -\frac{1}{2}$

e) $2x + 3y = 17$
$\; 4x = -2y - 2$

f) $0.75x + 0.5y = 11$
$\; 0.5x = y - 6$

3. Solve for *x* and *y*.

a) $5x - 6y = 13.2$
$\; 4x + 2y = 16$

b) $9.3x - 2.7y = 11.1$
$\; 8.4x + 1.9y = 13.1$

c) $0.73x + 0.45y = 14$
$\; 0.89x + 0.53y = 13$

d) $4.01x + 6.03y = 505.6$
$\; 3.26x + 5.98y = 498.7$

e) $4x - 6y = 0.07$
$\; 13x + 2y = 0.65$

f) $-0.95x + 1.4y = 3.45$
$\; 0.73x + 1.6y = 5.4$

4. Solve for *x* and *y*.

a) $y = 4x + 2$
$\; y = 3x + 5$

b) $y = -3x + 4$
$\; y = 7x + 1$

c) $y = \frac{2}{3}x + 4$
$\; y = -5x + 1$

d) $y = \frac{4}{5}x - 8$
$\; y = \frac{-4}{5}x + 3$

e) $2y = 5x + 3$
$\; 3y = 6x - 4$

f) $\frac{5}{6}y = \frac{2}{5}x + 7$
$\; 3y = \frac{3}{4}x + 8$

5. Solve for *x* and *y*.

a) $-4x - 5y = 8$
$\; 2x + 4y = -9$

b) $y = 3x + 2$
$\; 2y = 4x - 1$

c) $\frac{2}{3}y = 5x - 7$
$\; y = 2x + 1$

d) $3x - 4y = -17$ e) $250x + 160y = 53.5$ f) $y = 6x + 3$

     $8x = 5y$ $150x + 180y = 40.5$ $y = \frac{4}{3}x - 2$

6. Molly goes to a local store on Tuesday and buys 3 apples and 5 bananas for £2.55. The next day she returns to the same store and, at the same prices, buys 5 apples and 4 bananas for £2.69.
   a) Write a pair of linear equations to represent this problem.
   b) Find the price of an apple and the price of a banana at this store.

7. A particle travels along the path $s = at^2 + bt + 5$, where t is the time in seconds and s is the distance from the particle's starting position in metres. After 3 seconds the particle is 47 m from its starting position, and after 4 seconds the particle is 69 m from its starting position.
   a) Write two equations to represent the information given in this problem.
   b) Solve these two equations to find the values of a and b.

8. Jacques can buy six CDs and three DVDs for $163.17, or he can buy nine CDs and two DVDs for $200.53.
   a) Express the above information using two equations relating the price of CDs and the price of DVDs.
   b) Find the price of one DVD.
   c) If Jacques has $180 to spend, find the exact amount of change he will receive if he buys nine CDs.

9. The length of a rectangle is known to be two times longer than its width. The perimeter of the rectangle is $P = 2L + 2W$, where L and W are the length and width of the rectangle respectively.
   a) Write down two equations that represent the information given in the problem.
   b) If the perimeter of the rectangle is 60 cm, find the length and width of the rectangle.

10. A number is equal to twice another number increased by 7. The first number is also equal to 3 times the second number increased by 6.
   a) Write a pair of equations to represent the information given in the problem.
   b) Using your equations, or otherwise, find the value of the numbers.

11. At the local market, Leila buys 2 mangoes and 3 kiwi fruit for $5.10. Later that day, at the same market, she pays $12.19 for 5 more mangoes and 7 more kiwi fruit for her friends.
   a) Write two equations to represent the information given in the problem.
   b) Solve these two equations to find the price of a mango, and the price of a kiwi fruit.
   c) Leila estimates that 6 mangoes and 10 kiwi fruit will cost her $16. Find the percent error in her estimation.

12. It is known that:

   $4p + 8q = 35.8$

   $7p + 5q = 30.25$
   a) Find p and q.
   b) Use your answers from a) to find the value of $\dfrac{2p + q}{p^2}$.
   c) Round your answer from b) to:
        (i) **the nearest whole number**
        (ii) **2 significant digits**.

(2.7) Quadratic equations

| **1.6** | Use of GDC to solve: pairs of linear equations in two variables, and quadratic equations. |

A quadratic equation is any equation that can be transformed into the form $ax^2 + bx + c = 0$, where a, b and c are any real number, except $a \neq 0$.

In general, we will not be able to use simple algebraic manipulation to solve these equations – we need to use some special techniques.

Solving by factorizing when $a = 1$

In the last chapter, we reviewed how to factorize expressions of the form $x^2 + bx + c$, which is a quadratic equation with $a = 1$. To actually solve an equation of the form $x^2 + bx + c = 0$, we need to set each factor of the equation equal to zero, and solve the two linear equations that result, and these two solutions (which could end up being the same number!) are our solutions to the equation.

Example 2.9

Solve the equation $x^2 - 8x + 12 = 0$.

Solution

First we factorize the expression, as reviewed in Chapter 1, and get $(x - 6)(x - 2) = 0$. Our next step is to realize that if we have two numbers (here $x - 6$ and $x - 2$) whose product is zero, then one or both of those numbers must be zero. Hence, either $x - 6 = 0$ or $x - 2 = 0$, and therefore $x = 6$ or $x = 2$.

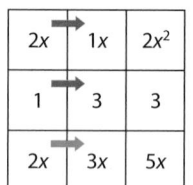

Figure 2.11 We multiply across the red arrows, and add across the blue arrow.

Figure 2.12 Now our solution works!

Solving by factorizing when $a = 2$

Since the equations we are solving are of the form $ax^2 + bx + c = 0$, factoring them is more difficult, when $a \neq 1$. To factorize these quadratic equations, use the following method:
1. List all of the factors of c.
2. List all of the factors of a.
3. Try each combination of factors in the expression $(a_1x + c_1)(a_2x + c_2)$ for a_1, a_2, c_1, and c_2. For each combination, multiply out the binomials until you find out which combination is equal to the original equation.

A useful way of organizing this 'trial and error' is to use a table. Suppose we are trying to solve $2x^2 + 7x + 3 = 0$. Try $(2x + 3)(x + 1)$ as in Figure 2.11.

We notice that this does not give us the correct solution, so we have to try again (Figure 2.12).

Our equation factorized is therefore $(2x + 1)(x + 3) = 0$. This leads to two new equations, $2x + 1 = 0$ and $x + 3 = 0$, which lead to the solutions $x = -\frac{1}{2}$ or $x = -3$.

Solving using the quadratic formula

If our equation is in the form $ax^2 + bx + c = 0$, the two solutions are

$$x = \frac{-b + \sqrt{b^2 - 4ac}}{2a} \text{ or } x = \frac{-b - \sqrt{b^2 - 4ac}}{2a}.$$

We write this as $x = \frac{-b \pm \sqrt{b^2 - 4ac}}{2a}$

where $\pm$ means that there are actually two possibilities. To solve any quadratic equation, we simply have to identify the values of a, b, and c, and substitute them into the quadratic formula given above.

Example 2.10

Solve $3x^2 + 10x + 3 = 0$.

Solution

From the equation, we see that $a = 3$, $b = 10$, and $c = 3$. Therefore:

$$x = \frac{-(10) \pm \sqrt{(10)^2 - 4(3)(3)}}{2(3)}$$

$$= \frac{-10 \pm \sqrt{100 - 36}}{6}$$

$$= \frac{-10 \pm \sqrt{64}}{6}$$

$$x = \frac{-10 + 8}{6} \text{ or } x = \frac{-10 - 8}{6}$$

$$x = \frac{-1}{3} \text{ or } x = -3$$

Note that the quadratic formula also allows us to solve quadratic equations when the equation does not factorize, and in these cases leads to irrational solutions.

Solving using the GDC

From a previous section in this chapter, we learned about a program we can install on our TI calculators called 'PlySmlt2'. This program can also be used to solve quadratic equations.

Example 2.11

Find the solution(s) to the equation $x^2 - 9x + 8 = 0$.

Solution

1. Start the Polysmlt2 application, and press ENTER.	TEXAS INSTRUMENTS Poly Root Finder and Simultaneous Equ Solver V2.00 PRESS ANY KEY © 2006 TEXAS INSTRUMENTS
2. Press 1 to choose 'Poly Root Finder'.	MAIN MENU **1:**Poly Root Finder 2:Simult Eqn Solver 3:About 4:Poly Help 5:Simult Help 6:Quit Poly Smlt

 There is a song to remember the quadratic formula, which you can download by going to www.pearsonhotlinks.co.uk, entering the ISBN for this book and clicking on weblink 2.11.

3. As this is a quadratic the Order is 2. Press 2 and then press ENTER.

```
POLY ROOT FINDER MODE
ORDER    1 2 3 4 5 6 7 8 9 10
REAL     a+bi  re^(0i)
DEC      FRAC
NORMAL   SCI ENG
FLOAT    1 2 3 4 5 6 7 8 9 10
RADIAN   DEGREE
MAIN                 HELP NEXT
```

4. We press GRAPH (below the NEXT tab) and enter the coefficients of our quadratic equation as a2, a1, and a0.

```
a2×2+a1×+a0=0
  a2=1
  a1=-9
  a0=8

MAIN MODE CLR LOAD SOLVE
```

5. We press GRAPH (below the SOLVE tab).

```
a2×2+a1×+a1=0
  ×1▉8
  ×2▉1

MAIN MODE CDEF STD F◄►D
```

Exercise 2.7

1. Solve for x.

a) $x^2 + 6x + 8 = 0$ b) $x^2 - 7x + 10 = 0$

c) $x^2 + 10x + 25 = 0$ d) $x^2 + 4x + 4 = 0$

e) $x^2 + 5x - 6 = 0$ f) $x^2 - 9x + 20 = 0$

2. Solve for x.

a) $2x^2 - 7x + 3 = 0$ b) $4x^2 + 24x + 35 = 0$

c) $3x^2 + 16x + 5 = 0$ d) $5x^2 - 21x + 4 = 0$

e) $6x^2 + 28x + 16 = 0$ f) $8x^2 - 38x + 45 = 0$

3. Solve for x.

a) $x^2 + 8x = -16$ b) $x^2 = 9x - 20$

c) $5x + 4 = -x^2$ d) $x^2 = 100$

e) $x^2 - 36 = 0$ f) $x^2 = 9x$

4. Solve for x.

a) $2x^2 + 14x + 20 = 0$ b) $3x^2 + 15x + 18 = 0$

c) $4x^2 + 28x + 24 = 0$ d) $3x^2 - 12x + 12 = 0$

e) $5x^2 + 10x - 240 = 0$ f) $10x^2 - 30x - 400 = 0$

5. Solve for x.

a) $0.25x^2 + 1.75x = -1.5$ b) $x^2 + \frac{10}{3}x + 1 = 0$

c) $10x^2 - 1000 = 0$ d) $4.5x^2 + 15x + 12 = 0$

e) $\frac{1}{3}x^2 + \frac{1}{2}x + \frac{1}{6} = 0$ f) $0.01x^2 - 0.08x + 0.15 = 0$

6. A cannonball has an equation of motion modelled by $h = -2.5t^2 + 25t$, where t is the time in seconds and h is the height of the ball in metres.
   a) Find the times when the cannonball is on the ground.
   b) What does the smallest of these two times mean?

7. The equation $x^2 + ax + b = 0$ has solutions $x = 2$ and $x = 5$. Find a and b.

8. The amount of profit earned in a month in the XYZ widget factory can be modelled by the equation $P = 5x^2 - 850x$, where P is the profit in dollars and x is the number of widgets created in a month.
   a) Find the 'break-even' amounts of widgets (when the profit is $0).
   b) What does the smaller of the two solutions mean?

9. a) Factorize the expression $x^2 - 25$.
   b) Factorize the expression $x^2 - 3x - 4$.
   c) Using your answer to part b), or otherwise, solve the equation $x^2 - 3x - 4 = 0$.

10. The area of the shaded region below is given by $A = (x - 6)(x - 5)$. When the area is $12\,\text{m}^2$, find the value of x.

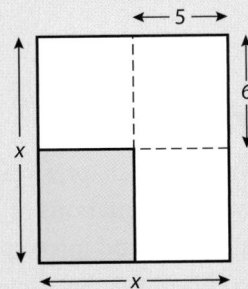

11. A rectangle has dimensions $(5 + 2x)$ metres and $(7 - 2x)$ metres.
   a) Show that the area, A, of the rectangle can be written as $A = 35 + 4x - 4x^2$.
   b) The following is the table of values for the function $A = 35 + 4x - 4x^2$.

x	-3	-2	-1	0	1	2	3	4
A	-13	p	27	35	q	r	11	s

   (i) Calculate the values of p, q and r.
   (ii) On graph paper, using a scale of 1 cm for 1 unit on the x-axis and 1 cm for 5 units on the y-axis, plot the points from your table and join them up to form a smooth curve.

   c) Answer the following, using your graph or otherwise.
      (i) Write down the equation of the axis of symmetry of the curve.
      (ii) Find one value of x for a rectangle whose area is $27\,\text{m}^2$.
      (iii) Using this value of x, write down the dimensions of the rectangle.

   d) (i) One the same graph, draw the line with equation $A = 5x + 30$.
      (ii) Hence, or otherwise, solve the equation $4x^2 + x - 5 = 0$.

12. All of the measurements in the figure shown right are in metres.
   a) Write an expression for the area of the shaded region.
   b) When, according to this expression, is the area of the shaded region equal to zero?
   c) Explain whether the solutions you found in part b) make sense.

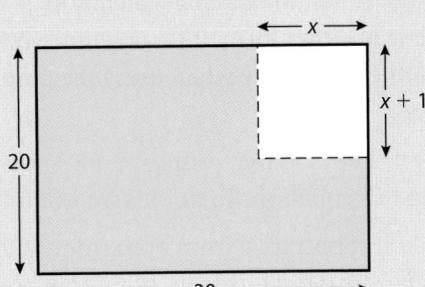

Geometry

Assessment statements

5.1   Equation of a line in two dimensions: the forms
       $y = mx + c$ and $ax + by + d = 0$.
       Gradient; intercepts. Points of intersection of lines. Lines with gradients
       m_1 and m_2. Parallel lines $m_1 = m_2$; perpendicular lines $m_1 \times m_2 = -1$.

Overview

By the end of this chapter you will be able to:
- represent the equations of lines in two dimensions
- find gradients and intercepts of lines
- find points of intersection of lines
- explain how to tell if lines are parallel or perpendicular, by examining their gradients.

3.1 The equation of a line

Rene Descartes, a French philosopher and mathematician, created much of the framework for analytical geometry. Hence, many of our variable names in coordinate geometry are from French words.

Why do you think we have more than one form of the equation of a line?

We need to be able to convert between the $y = mx + c$ and the $ax + by = d$ forms of a line.

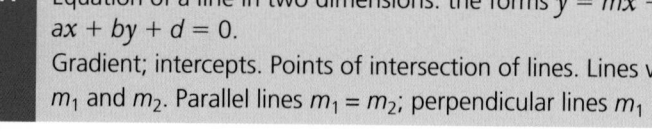
5.1   Equation of a line in two dimensions: the forms $y = mx + c$ and
       $ax + by + d = 0$.
       Gradient; intercepts. Points of intersection of lines. Lines with gradients
       m_1 and m_2. Parallel lines $m_1 = m_2$; perpendicular lines $m_1 \times m_2 = -1$.

Linear equations are so named because when graphed they always form a straight line. These linear equations are often represented with an equation of the form $y = mx + c$, which is known as the **gradient intercept form**. However, there is an alternate representation known as the **standard form** of the equation of a line which is $ax + by + d = 0$. Given the equation of a line in either form, it is sometimes necessary to convert one form into another, especially when using the graphing display calculator to draw a graph.

If our line is in the form $ax + by + d = 0$, we need to use algebra to isolate y in this equation. To do this we will follow these steps:

Step 1: Subtract d from both sides of the equation. $ax + by = -d$
Step 2: Subtract ax from both sides of the equation. $by = -ax - d$
Step 3: Divide by b (if $b = 0$ then we cannot change forms). $y = -\dfrac{a}{b}x - \dfrac{d}{b}$

Now we have our line in the correct form, because we can let $m = -\dfrac{a}{b}$
and $c = -\dfrac{d}{b}$.

To convert from the form $y = mx + c$ to the form $ax + by + d = 0$, where

a and b are integers, we would like to express m in the form $\frac{a}{b}$. So assuming we have written m as a rational number we can do the following:

Step 1: Subtract c from both sides of the equation. $y - c = \frac{a}{b}x$

Step 2: Multiply both sides of the equation by b. $by - cb = ax$

Step 3: Subtract ax from both sides of the equation. $-ax + by - cb = 0$

We now have our line in the correct form.

Exercise 3.1

1. Convert the following into $y = mx + c$ form.
  a) $y + 2 = 3x$
  b) $y - 5 = \frac{1}{3}x$
  c) $2y = 4x$
  d) $y + 5 = x$
  e) $3y = 4x + 6$
  f) $y + 1 + x = 0$

2. Convert the following into $ax + by + d = 0$ form, where a, b, and c are integers.
  a) $y = 3x + 1$
  b) $y = -2x - 2$
  c) $y = \frac{1}{2}x - 1$
  d) $y = -3x + 2$
  e) $y - 2 = 5x$
  f) $y + 3x = 7$

3. Convert the following into $y = mx + c$ form.
  a) $\frac{1}{2}y - 2 = x$
  b) $2y - 3x = 0$
  c) $4x + 3y - 3 = 0$
  d) $5x + 10y = 5$
  e) $6 = y - x$
  f) $2x + 3y = 6$

4. Convert the following into $ax + by + d = 0$ form, where a, b, and c are integers.
  a) $\frac{1}{2}y = \frac{1}{3}x - 2$
  b) $y = -\frac{3}{4}x - \frac{1}{2}$
  c) $y = \frac{1}{3}x + \frac{1}{5}$
  d) $2y = \frac{2}{5} - 3x$
  e) $5x - \frac{1}{2}y = \frac{2}{7}$
  f) $y - \frac{1}{6} = \frac{2}{3}x$

5. Convert each of the following into $y = mx + c$ form and $ax + by + d = 0$ form, where a, b, and c are integers.
  a) $y + 2x = 3$
  b) $2x + 1 = 3y$
  c) $\frac{1}{4}x - 3y = 2$
  d) $2y - 3x = 6$
  e) $x + y = 4$
  f) $3x + 4 = \frac{1}{3}y$

6. Given $x - d = y$:
  a) Write an expression for x in terms of y.
  b) For what value of d does this line pass through $(10, 22)$?

7. Given $\frac{2}{3}x + by = 30$:
  a) Write an expression for y in terms of x.
  b) For what value of b does this line pass through $(15, 8)$?

8. The height of a student in the XYZ school is related to the length of their arm using the formula $5a - 2h + 20 = 0$, where a is the length of the student's arm in cm and h is their height in cm.
  a) Rearrange this formula for h in terms of a.
  b) How tall is a student with an arm length of 50 cm?
  c) If a student is 160 cm tall, how long is their arm?

9. The amount of time spent doing homework by a student, in minutes per night, is related to the student's years of schooling using the formula $m - 20y - 10 = 0$, where m is the number of minutes and y is the number of years.
  a) Rearrange this formula for m in terms of y.
  b) Find the number of minutes a night a student should be doing homework in their 7th year of school.
  c) In which school year should a student be doing 250 minutes of homework?

10. The relationship between the total profit in dollars, P, and the number of months, x, since a company was formed follows the formula $3P - 60\,000x = 0$.

a) Express the profit in terms of the number of months.
b) If 3 years have passed since the company was formed, find the total profit earned.
c) Find the increase in profit each month.
d) Find out what month the company earned more than $250 000.

11. The length of a rectangle with a width of 20 is related to the perimeter by the expression:

$$2L - P + 40 = 0$$

a) Rewrite the expression for L in terms of P.
b) Find the length of a rectangle with a perimeter of 200 and a width of 20.
c) Rewrite the expression for P in terms of L.
d) Find the perimeter of a rectangle with a width of 20 and a length of 25.
e) Write an expression for the relationship between the width and length of a rectangle with a perimeter of 200.

3.2 The gradient of a line

> **5.1** Equation of a line in two dimensions: the forms $y = mx + c$ and $ax + by + d = 0$.
> Gradient; intercepts. Points of intersection of lines. Lines with gradients m_1 and m_2. Parallel lines $m_1 = m_2$; perpendicular lines $m_1 \times m_2 = -1$.

Where else in mathematics do we use this concept of one change in a quantity divided by another change?

Modern-day differential calculus is based on this simple concept of slope, but is capable of describing amazingly complex concepts.

The gradient of a line is the change in the dependent variable divided by the change in the independent variable. Using the symbol m to mean the gradient, this would be $m = \dfrac{\text{change in } y}{\text{change in } x}$ for lines drawn on the coordinate plane. To find the change in a quantity, we subtract the initial value from the final value of the quantity. So this definition becomes $m = \dfrac{y_2 - y_1}{x_2 - x_1}$, where (x_1, y_1) and (x_2, y_2) are any two points on the line.

Conveniently, if we are given the equation of a line in the form $y = mx + c$, the m in this equation is in fact the gradient of the line. So if we are given the equation of a line in this form, finding the gradient is just a matter of reading the information from the equation.

If we are instead given the graph of a line, as shown in Figure 3.1, then we have to use the definition of the gradient to find the equation.

Example 3.1

Find the gradient of the line passing through A and B as shown in Figure 3.1.

Solution

From Figure 3.1, we see that the coordinates of A are $(-3, -3)$ and that the coordinates of B are $(0, -1)$. Use the definition of the gradient:

$$m = \frac{y_2 - y_1}{x_2 - x_1} = \frac{(-1) - (-3)}{(0) - (-3)} = \frac{2}{3}.$$ We can leave m in fraction form as $\dfrac{2}{3}$.

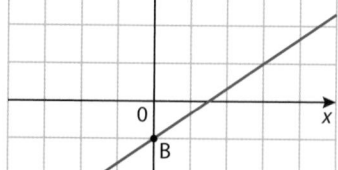

Figure 3.1 The graph of a line on the coordinate plane.

If we are given two points on a line, without the graph of the line or the equation, we can immediately substitute the coordinates of these two points into the definition of the gradient.

Example 3.2

Find the gradient of the line passing through the points $C(-2, 5)$ and $D(6, -1)$.

Solution

Use the definition of the gradient: $m = \dfrac{y_2 - y_1}{x_2 - x_1} = \dfrac{(5) - (-1)}{(-2) - (6)} = \dfrac{6}{-8} = -\dfrac{3}{4}$.

We can leave m as the reduced fraction $-\dfrac{3}{4}$.

 For an applet that demonstrates the effect of changing the location of the points when calculating the gradient of the line, visit www.pearsonhotlinks.co.uk, enter the ISBN for this book and click on weblink 3.1.

 The gradient of a line can be found using the formula
$$m = \dfrac{y_2 - y_1}{x_2 - x_1}.$$

Exercise 3.2

1. Find the gradient of the line that contains each pair of points.
   a) $A(1, 1)$ and $B(4, 5)$
   b) $A(0, 0)$ and $B(-3, 2)$
   c) $A(-2, -2)$ and $B(1, 1)$
   d) $A(2, -3)$ and $B(5, 6)$

2. Find the gradient of each line segment in the diagram right.

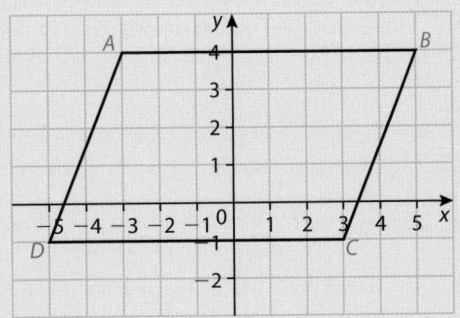

3. Write down the gradients of the following lines.
   a) $y = \frac{3}{4}x + 2$
   b) $y = \frac{1}{3}x + \frac{1}{2}$
   c) $y = -\frac{2}{5}x + 2$
   d) $y = \frac{1}{3} + \frac{4}{5}x$
   e) $y = 3 - \frac{1}{2}x$
   f) $y = 3x + \frac{1}{2}$
   g) $y = 4x$
   h) $y = 3$
   i) $y = x$

4. Find the gradients of the following lines.
   a) $2x + 4y = 3$
   b) $5x - 2y = 0$
   c) $x + y = 6$
   d) $4x - y = 8$
   e) $3x + 5y + 15 = 0$
   f) $y - 3 = \frac{2}{3}x$

5. Find the gradients of all of the
line segments in the diagram
on the right.

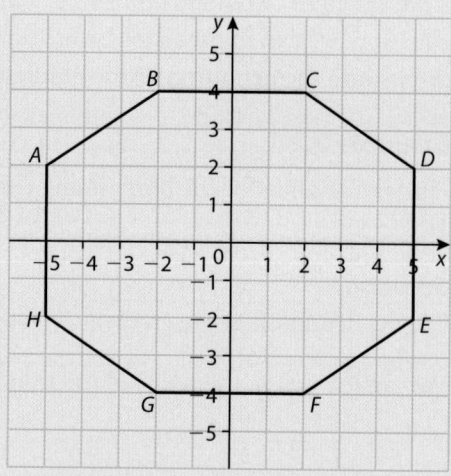

6. If $D = 20t$ represents the distance D, in metres, travelled by a motorcyclist in
time t seconds, find:
a) the distance travelled by the motorcyclist in 20 seconds
b) the speed of the motorcyclist
c) how long it will take the motorcyclist to travel 3000 m.

7. The rate at which a cylindrical water basin is filled with water is 30 litres per second.
a) If the basin starts with 60 l, write an expression for the amount of water in the
basin at time t using V to represent how much water is currently in the basin.
b) Find how long it takes the basin to fill to its maximum volume of 240 l.

8. The rate at which Ayumi can paint a building is 40 m² per hour.
a) Write an expression for the total area, A, Ayumi can paint in t hours.
b) What total area can she paint in 50 hours?
c) How long does it take her to paint 400 m²?

9. John can run up 60 steps in 2 minutes, and 360 steps in 12 minutes.
a) Assuming the rate at which John runs up steps is constant, write an
expression to represent the number of steps, S, John can run up in t minutes.
b) How long will it take John to run up 550 steps?
c) How many steps will John have completed in 30 seconds?

10. Carl is playing pool. He notices that his cue and his cue ball are lined up as in
the diagram below.

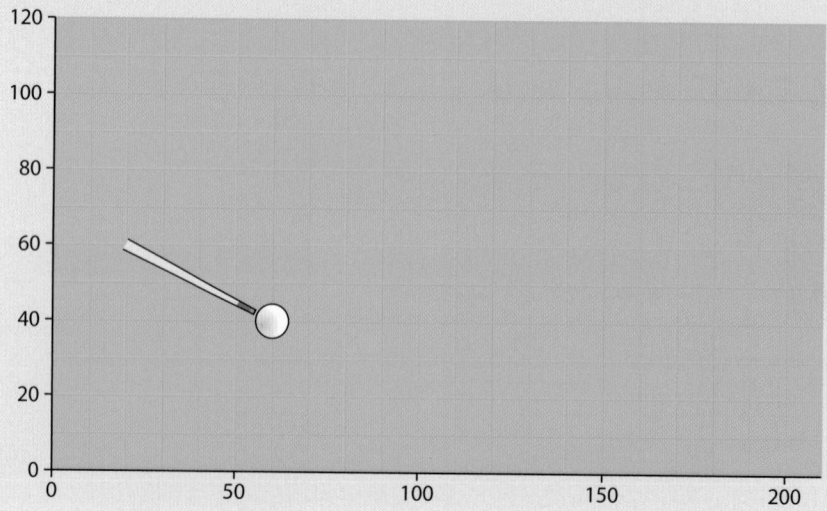

a) Find the gradient of the line his cue makes with the cue ball.
b) Extend the line drawn above so that it meets the bottom edge of the pool table.
c) Estimate the coordinates of the point at which the ball will hit the edge of the pool table (ignore the diameter of the cue ball).
d) Assuming the cue ball reflects off the edge of the pool table at the same angle at which it hit the rail, find the gradient of the line that represents the ball immediately after the ball hits the edge of the table.
e) Estimate the coordinates of the point at which the cue ball will hit the edge of the pool table on its second bounce (ignore the diameter of the cue ball).

11. A circle is drawn with [OP] as a radius of the circle. Another line is drawn such that it touches the circle exactly once at P and makes a 90° angle with [OP].

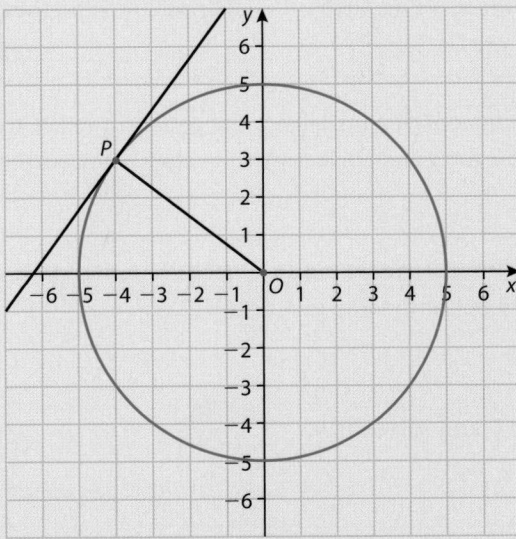

a) Find the gradient of the line through (OP).
b) Find the gradient of the line that is at a 90° angle to (OP).
c) Find the length of [OP].
d) Find the midpoint of [OP].

(3.3) The intercepts of a line

5.1 Equation of a line in two dimensions: the forms $y = mx + c$ and $ax + by + d = 0$.
Gradient; intercepts. Points of intersection of lines. Lines with gradients m_1 and m_2. Parallel lines $m_1 = m_2$; perpendicular lines $m_1 \times m_2 = -1$.

The intercepts of a line are defined to be the points where the graph of a line intersects with the x- and y-axes. These will always exist, unless the line is parallel to either the x- or y-axis.

To find the x-intercept or the y-intercept from the graph of a line, examine the graph and determine the coordinates where it crosses the

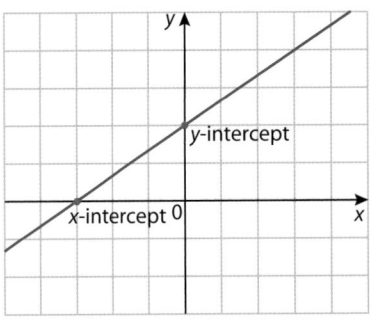

Figure 3.2 We can see here the x- and y-intercepts of a line.

appropriate axis. Looking at Figure 3.2 again, we can see the y-intercept is $(0, 2)$ and the x-intercept is $(-3, 0)$.

To find the y-intercept of a line from the equation of a line given in the form $y = mx + c$, substitute the value of 0 for x since anywhere on the y-axis the value of x is 0. The coordinates of the y-intercept will then be $(0, y)$.

Example 3.3

Find the y-intercept of the line $y = -\frac{3}{4}x + 2$.

Solution

Substituting 0 for x, we see that $y = -\frac{3}{4}(0) + 2 = 2$. So the coordinates of the y-intercept are $(0, 2)$.

To find the x-intercept of a line, given the equation in the form $y = mx + c$, we substitute 0 for y and solve for the x value. The coordinates will then be $(x, 0)$.

Example 3.4

Find the x-intercept of the line $y = -\frac{3}{4}x + 2$.

Solution

Substitute 0 for y and solve for x.

$$0 = -\frac{3}{4}x + 2$$
$$-2 = -\frac{3}{4}x$$
$$-8 = -3x$$
$$\frac{-8}{-3} = x$$
$$\frac{8}{3} = x$$

Therefore, the coordinates of the x-intercept are $\left(\frac{8}{3}, 0\right)$.

> The x- and y-intercepts are the locations where the line crosses the x- and y-axes respectively. We find these locations using algebraic manipulation after substituting the value of 0 for y when finding the x-intercept or for x when finding the y-intercept.

Exercise 3.3

1. Find the y component of the y-intercept of each line.
   a) $y = 3x + 1$
   b) $y = \frac{2}{3}x - 3$
   c) $y = 4x + 5$
   d) $y = -\frac{1}{2}x$
   e) $y = 3x + \frac{1}{2}$
   f) $y = x$

2. Find the x component of the x-intercept of each line.
   a) $y = 4x + 8$
   b) $y = \frac{2}{5}x - 4$
   c) $y = x + 3$
   d) $y = 3x$
   e) $y = -2x + 6$
   f) $y = x - 6$

3. Find the coordinates of the y-intercept of each line.
   a) $2x + 3y - 6 = 0$
   b) $4x + y = 6$
   c) $3x - 2y + 5 = 0$
   d) $x + y = 10$
   e) $2x - y + 5 = 0$
   f) $x - 2y = 3$

4. Find the coordinates of the *y*-intercept for each line drawn.

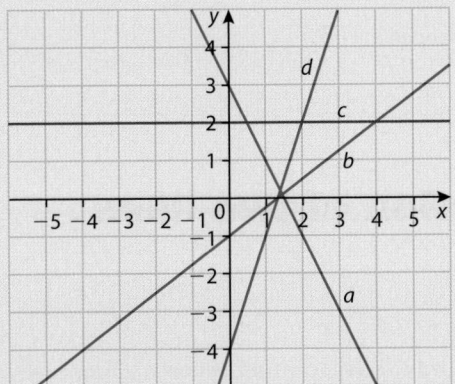

5. Find the coordinates of the *x*-intercept of each line.

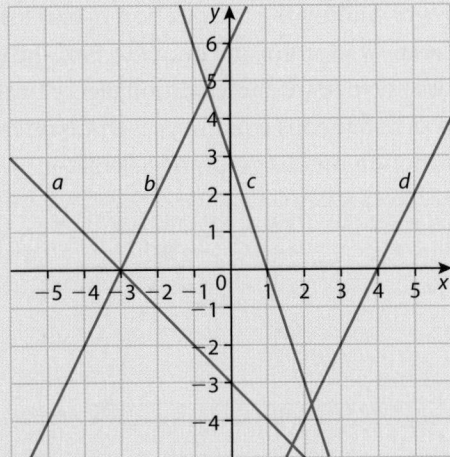

6. Two points are given as *A*(1, 5) and *B*(4, 2) and a line is drawn through them.
   a) Find the gradient of the line (*AB*).
   b) Find the equation of the line.
   c) Find the coordinates of the *y*-intercept of the line.
   d) Find the coordinates of the *x*-intercept of the line.

7. A wasp flies through the air along the path $y = -\frac{4}{5}x + 8$, where *x* is its distance across the ground, and *y* is its height in the air, both in feet.
   a) Find the height of the wasp when $x = 0$.
   b) For what value of *x* does the wasp reach the ground?

8. The equation of a line is $5x + 2y = 20$.
   a) Find the coordinates of the *y*-intercept, *A*.
   b) Find the coordinates of the *x*-intercept, *B*.
   c) Find the length of [*AB*].

9. The *y*-intercept of a line segment is (0, 10) and the midpoint of the same line segment is (4, 7).
   a) Find the gradient of the line.
   b) Find the equation of the line. Write your answer in the form $y = mx + c$.
   c) Find the *x*-intercept of the line.

10. The cost *c*, in Australian dollars (AUD), of renting a bungalow for *n* weeks is given by the linear relationship $c = nr + s$, where *s* is the security deposit and *r* is the amount of rent per week.

Ana rented the bungalow for 12 weeks and paid a total of 2925 AUD.

Raquel rented the same bungalow for 20 weeks and paid a total of 4525 AUD.

Find the value of
a) r, the rent per week;
b) s, the security deposit.

3.4 Intersection of two lines

> **5.1** Equation of a line in two dimensions: the forms $y = mx + c$ and $ax + by + d = 0$.
> Gradient; intercepts. Points of intersection of lines. Lines with gradients m_1 and m_2. Parallel lines $m_1 = m_2$; perpendicular lines $m_1 \times m_2 = -1$.

There are multiple ways of finding the intersection of two lines. We can graph both lines, and visually examine the graph to find the point of intersection. We can also solve for the intersection algebraically, treating the pair of lines as two simultaneous equations as in Chapter 2. We can also use the intersect function on our graphing calculator.

Figure 3.3 Finding the intersection of two lines visually is not always this easy.

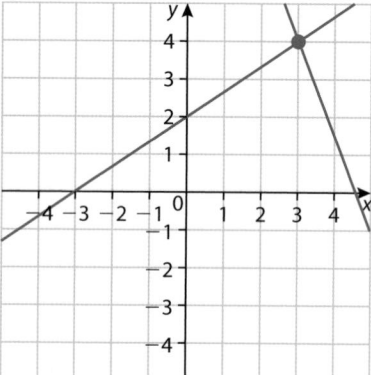

When two lines intersect at integer coordinates, as in Figure 3.3, finding the coordinates of the point of intersection, in this case $(3, 4)$, is easy. However, for more complicated examples, we quite often have to estimate the coordinates of the point of intersection, especially if the scale of the graph is unusual, or the lines do not meet at integer coordinates.

To find the intersection of two lines algebraically, we treat the pair of lines as a pair of simultaneous equations, and solve for x and y. This will give us the coordinates of the point of intersection.

Example 3.5

Find the intersection of the lines $y = 3x + 4$ and $y = 2x + 5$ algebraically.

Solution

Since we have y isolated in both equations, the easiest way to solve these two equations is to use substitution.

$$3x + 4 = 2x + 5$$
$$x + 4 = 5$$

$x = 1$

$y = 2(1) + 5 = 7$

Check: $7 = 3(1) + 4$

Therefore, the point of intersection is $(1, 7)$.

To use a TI calculator to solve this problem, see the instructions below.

Step 1: Turn on the calculator and press Y =

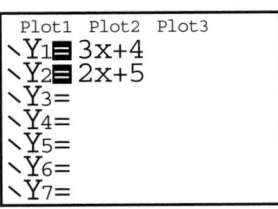

Step 2: Enter the equations of the lines onto separate lines.

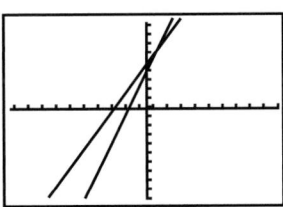

Step 3: Press GRAPH

Step 4: Press 2ND and then press TRACE

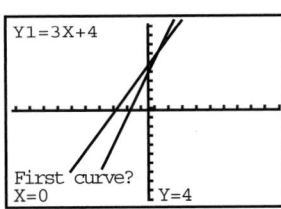

Step 5: Press 5

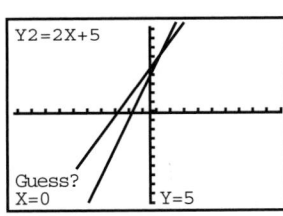

Step 6: Press ENTER twice

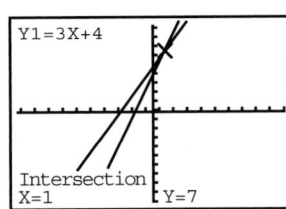

Step 7: Use the < and > keys to move the cursor closer to the point of intersection, then press ENTER

The intersection of two lines can be found in multiple ways, either by visual inspection, using algebraic methods, or using a GDC.

So we can see that the point of intersection is $x = 1$ and $y = 7$, or $(1, 7)$ written in coordinate form.

Exercise 3.4

1. Find the intersection of the lines below.

a)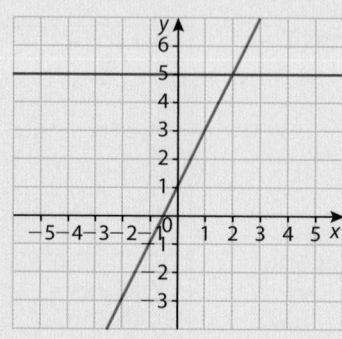

b)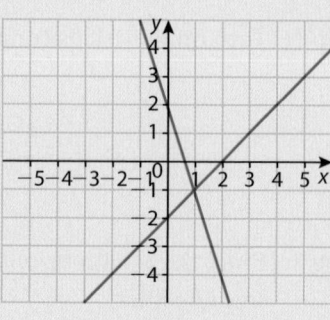

c)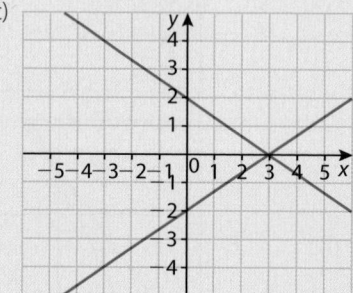

d)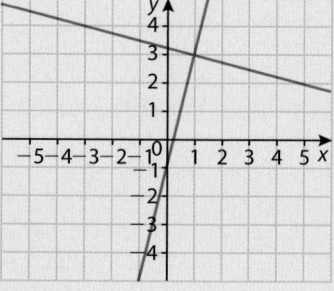

2. Find the intersection of the lines below.

a) $y = 3x + 1$
   $y = -3x + 1$

b) $y = -\frac{1}{2}x + 3$
   $y = \frac{1}{2}x - 1$

c) $y = 2x - 3$
   $y = x - 1$

d) $y = x - 3$
   $y = -x + 3$

3. Find the intersection of the following pairs of lines.

a) $3x + 2y = 0$
   $x = 5y$

b) $-x + 2y = 10$
   $y = 2x + 2$

c) $y + 1 = 4x$
   $y = \frac{1}{3}x - 1$

d) $5y = -4x + 10$
   $x + y = 3$

4. Find the points of intersection of the following lines.

a) $2x + 5y - 7 = 0$
   $3x - 4y + 1 = 0$

b) $7x - 8y = 0$
   $\frac{2}{3}x - 9y = 0$

c) $y = 3$
   $x = -5$

d) $y = -4$
   $x - 6y = -7$

5. Find the intersections of the line segments in the diagram below.

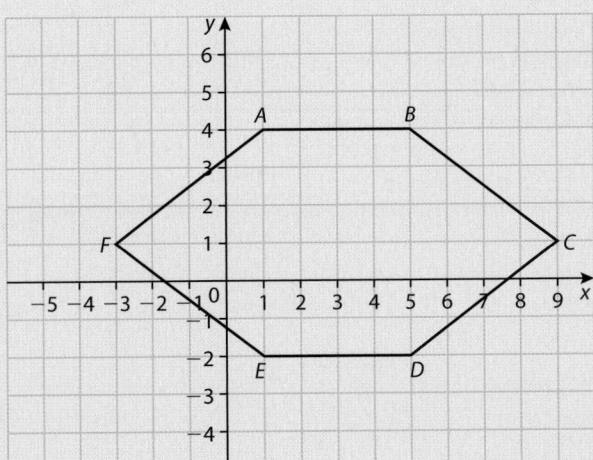

6. The cost to rent videos at video store A is $y = 2x + 10$, where x is the number of videos rented and y is the total cost in US$. For video store B, the cost to rent videos is given by the formula $y = 3x + 4$ in US$.
a) What is the initial membership costs for both video stores?
b) For what number of videos do both video stores cost the same?

7. Ship A travels along the path $2x - 3y = 3$ and ship B travels along the path $y = \frac{2}{5}x + 3$.
a) For ship A's path, find y in terms of x.
b) Find the location where the path of the two ships cross.

8. The data below represents the distance y north and the distance x east from the starting position of an aeroplane.

x (km)	20	40	60	80
y (km)	12	24	36	48

a) Plot this information on a graph.
b) Find the gradient of the resulting line.
c) Write down the equation of this line.
d) If another plane flies along the path $y = \frac{1}{2}x + 10$, where will their paths cross?

9. Two young trees, 6 feet and 9 feet tall, are planted in a garden. The smaller tree grows at a rate of 2 feet per year, and the taller tree grows at a rate of 1.5 feet per year.
a) Write down two equations to represent the growth of each tree.
b) Which year will the smaller tree first pass the taller tree in height?

10. Akira deposits 1000 Yen in a bank account, which earns 100 Yen interest each year. Yuki deposits 1500 Yen in a bank account, but only earns 50 Yen per year.
a) Write expressions for the amount of money Akira and Yuki have in their accounts over time.
b) How many years will it take for Akira to have more money in his account?

11. a) Find the four intersections of the four lines in the diagram.
b) Use these intersections to calculate the lengths of the four sides of the quadrilateral, in the diagram, to confirm that this quadrilateral is a rhombus.

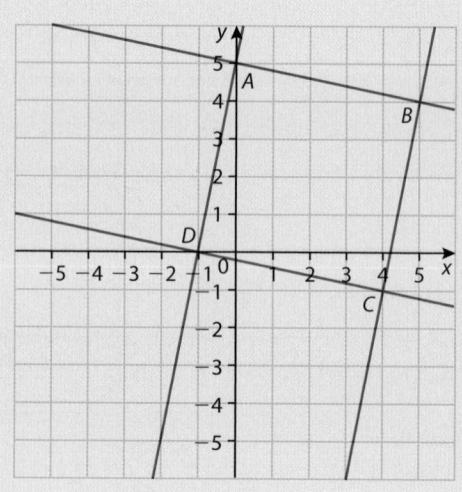

3.5 Parallel and perpendicular lines

5.1 Equation of a line in two dimensions: the forms $y = mx + c$ and $ax + by + d = 0$.
Gradient; intercepts. Points of intersection of lines. Lines with gradients m_1 and m_2. Parallel lines $m_1 = m_2$; perpendicular lines $m_1 \times m_2 = -1$.

Two coplanar lines are parallel if and only if they have the same gradient. This is equivalent to saying that the two lines have the same direction. If the lines are both written in the form $y = mx + c$, we just read both of the values of m and confirm they are the same. If one or both lines are written in a different form, we use algebra to change the form of the lines, as in Section 3.1. If we are examining the graphs of the lines, we should calculate the gradient of both lines using their graphs, and compare gradients. It is possible that two lines can be *almost* parallel, and a quick visual inspection will not spot this, so we must be careful.

Example 3.6

Determine whether the lines $y = 4x + 1$ and $2x - 8y + 3 = 0$ are parallel.

Solution
First change $2x - 8y + 3 = 0$ into $y = mx + c$ form.

$$2x - 8y = -3$$
$$-8y = -2x - 3$$
$$y = \frac{-2}{-8}x - \frac{3}{-8}$$
$$y = \frac{1}{4}x + \frac{3}{8}$$

From this we can see that $4 \neq \frac{1}{4}$, and since the gradients of the lines are not equal, the two lines are not parallel.

Two lines are perpendicular if the graphs of their equations form a 90° angle at their point of intersection. It turns out that if we know the two gradients of the lines, m_1 and m_2, we can also determine whether the two lines are perpendicular. If $m_1 = -\dfrac{1}{m_2}$, the two lines are perpendicular. We should always make sure to check this formula to confirm two lines are perpendicular, since a visual confirmation of a 90° angle is not sufficient.

Example 3.7

Determine whether the lines $y = \frac{1}{3}x + 4$ and $y = -3x - 2$ are perpendicular.

Solution
Use the equation $m_1 = -\dfrac{1}{m_2}$. Here $m_1 = \dfrac{1}{3}$ and $m_2 = -3$.

$$\frac{1}{3} = -\frac{1}{-3}$$

Since a negative times a negative is a positive, the right-hand side of the equation equals the left-hand side of the equation, which means that the two lines must be perpendicular.

 Parallel lines have the same gradient.

The product of the gradients of two perpendicular lines is -1.

1. Determine which of the following pairs of lines are parallel.
   - a) (i) $y = 3x + 2$
     - (ii) $y = 3x + 7$
   - b) (i) $2x + 3y = 6$
     - (ii) $4x - 6y = 12$
   - c) (i) $y = -2x + 1$
     - (ii) $x + 2y = 3$
   - d) (i) $y - 3 = 2x$
     - (ii) $2y - 5 = 3x$
   - e) (i) $y = \frac{1}{2}x - 1$
     - (ii) $2y = x + 2$
   - f) (i) $y = \frac{3}{4}x - 1$
     - (ii) $y = -\frac{4}{3}x + 2$

2. Determine which of the following pairs of lines are perpendicular.
   - a) (i) $y = \frac{1}{2}x + 1$
     - (ii) $y = -2x - 3$
   - b) (i) $y = 3x$
     - (ii) $y = \frac{1}{3}x$
   - c) (i) $y - 3x = 4$
     - (ii) $2y - 5x = 7$
   - d) (i) $y + 5x = 0$
     - (ii) $10x + 2y = 3$
   - e) (i) $8x + 5y = 9$
     - (ii) $y = \frac{5}{8}x - 1$
   - f) (i) $y = 7x$
     - (ii) $y = -\frac{1}{6}x + 2$

3. Determine which of the following pairs of line are perpendicular.
   - a)

   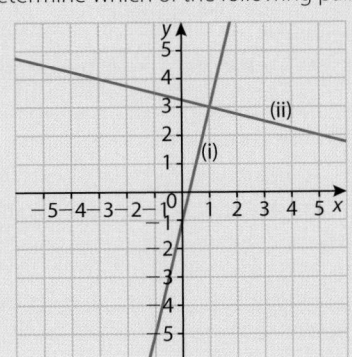

   - b)

   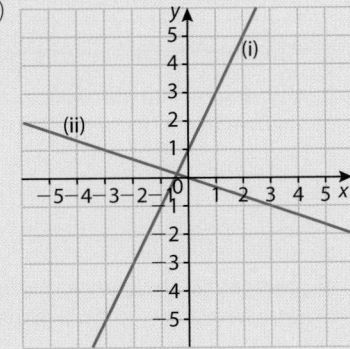

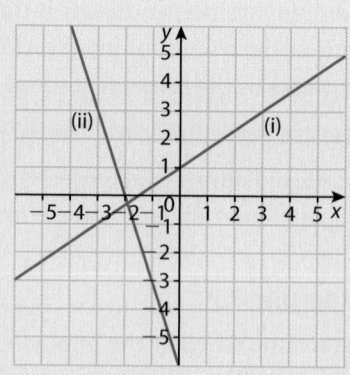

4. Determine which of the following pairs of line are parallel.

a)

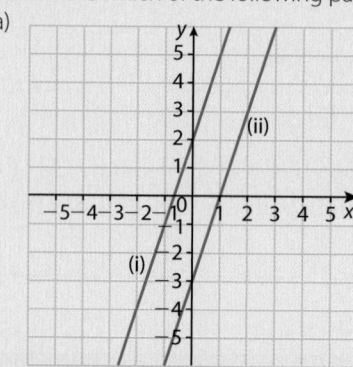

b)

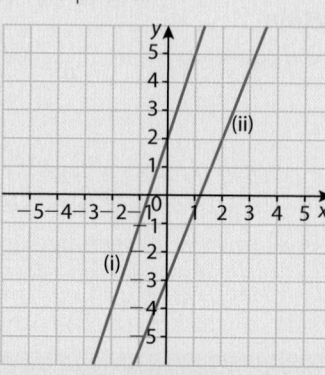

c)

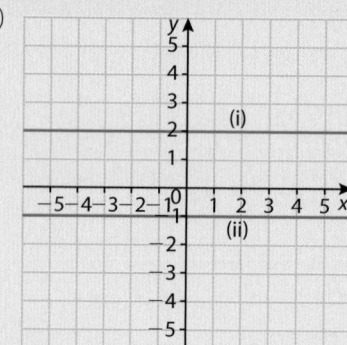

d)

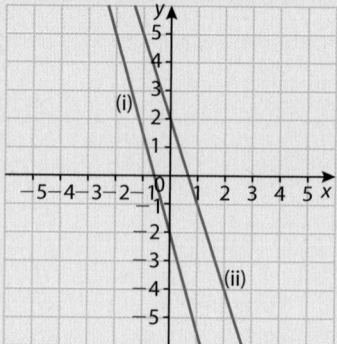

5. Write down the equation of a line that is:
   a) parallel to $y = 3x + 2$ and which passes through $(1, 2)$
   b) perpendicular to $y = x$ and which passes through $(3, 6)$
   c) parallel to $y = -\frac{1}{2}x - 1$ and which passes through $(4, -5)$
   d) perpendicular to $y = \frac{4}{5}x + 7$ and which passes through $(5, -1)$.

6. For $A(1, 1)$, $B(4, 5)$, $C(8, 2)$, $D(5, -2)$ define the quadrilateral $ABCD$.
   a) Find the gradient of $[AB]$, $[BC]$, $[CD]$, and $[DA]$.
   b) Confirm that opposite sides of $ABCD$ are parallel, and explain your reasoning.

7. Use the information from Question 6 for this problem.
   a) Confirm that all four vertices of $ABCD$ form right angles.
   b) Find the length of all four sides of $ABCD$, and hence show that it is a square.

8. Eileen travels 4 km due north and then 3 km due east. From the same initial position, Henry travels 4 km due east and 3 km due south.
   a) Plot their positions on a coordinate graph, using a scale of 1 cm = 1 km, with due north as the positive y-axis.
   b) Find the gradient of each of their resultant journeys.
   c) Show that the angle between their journeys is 90°.

● **Examiner's hint:** When given problems involving the four cardinal directions (north, east, south, west) it is helpful to draw the directions on your picture.

9. The graph right shows the cost per day to rent a car from two rival car companies.

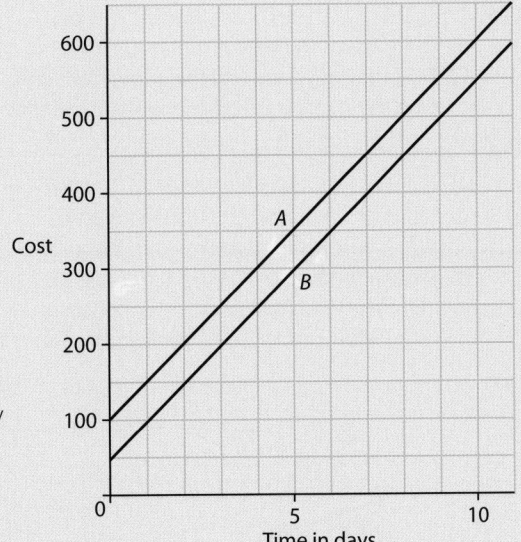

a) Find the gradients of lines *A* and *B*, and hence the cost per day to rent a car.

b) Find the *y*-intercepts for lines *A* and *B*, and hence the initial cost to rent a car.

c) Confirm that the cost to rent a car from company *A* will always exceed the cost to rent a car from company *B*.

10. Insect A travels along the path $3x + 4y = 12$ and insect B travels along the path $y = \frac{4}{3}x - 2$.

a) Write the path for insect A in the form $y = mx + c$.

b) Draw a diagram to represent this information.

c) Confirm that their paths meet at a 90° angle.

11. Three points are given: $A(0, 4)$, $B(6, 0)$, and $C(8, 3)$.

a) Calculate the gradient (slope) of line (AB).

b) Find the coordinates of the midpoint, M, of the line (AC).

c) Calculate the length of $[AC]$.

d) Find the equation of (BM) giving your answer in the form $ax + by + d = 0$, where a, b and $d \in \mathbb{Z}$.

e) State whether the line (AB) is perpendicular to the line (BC), showing clearly your working and reasoning.

12. Three points $A(1, 3)$, $B(4, 10)$, and $C(6, -1)$ are joined to form a triangle. The midpoint of $[AB]$ is D and the midpoint of $[AC]$ is E.

a) Find the coordinates of D and E.

b) Plot the points A, B, C, D, and E.

13. Given $ax + y = d$:

a) Write an expression for *y* in terms of *x*.

b) Find the value of *y* if $x = 20$, $a = 3$, and $d = -5$.

14. a) Copy the grid right, and draw a straight line with a gradient of -3 that passes through the point $(2, 0)$.

b) Find the equation of this line.

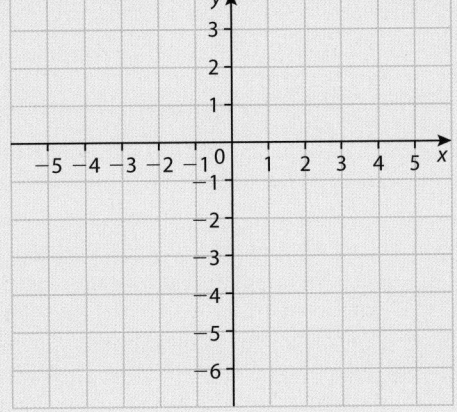

15. Consider the line $h: 2x + y + 4 = 0$.
a) Write down the gradient of h.
b) Find the coordinates of the y-intercept of h.
c) Find the coordinates of the x-intercept of h.

16. Given A(0, 0), B(0, 6), and C(3, $3\sqrt{3}$):
a) (i) Find the distance between A and B.
(ii) Find the distance between A and C.
(iii) Find the distance between B and C.
b) Write a sentence to describe triangle ABC.
c) (i) Find the gradient of the line through A and C.
(ii) Find an equation of the line through A and C.

17. The vertices of quadrilateral $ABCD$, as shown in the diagram, are, $A(-8, 8)$, $B(8, 3)$, $C(7, -1)$, and $D(-4, 1)$.

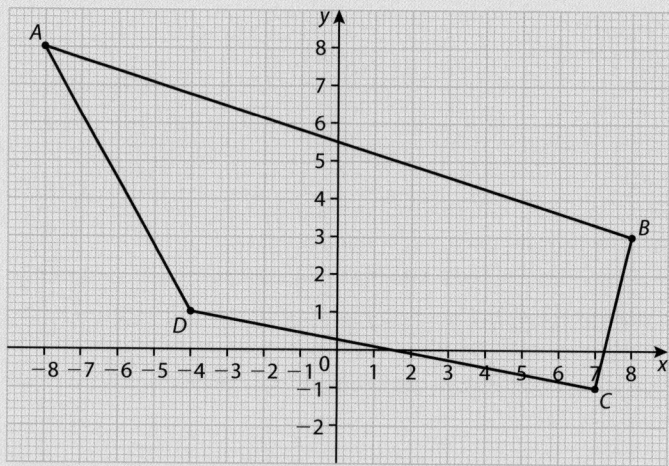

The equation of the line through A and C is $3x + 5y = 16$.
a) Find the equation of the line through B and D, expressing your answer in the form $ax + by = c$, where a, b and $c \in \mathbb{Z}$.
The lines (AC) and (BD) intersect at point T.
b) Calculate the coordinates of T.
The gradient of the line (AB) is $-\frac{5}{16}$.
c) Calculate the gradient of the line (DC).
d) State whether or not (DC) is parallel to (AB) and give a reason for your answer.

4 Mathematical Models

Assessment statements

6.1 Concept of a function, domain, range, and graph. Function notation, eg $f(x)$, $v(t)$, $C(n)$. Concept of a function as a mathematical model.

6.2 Linear models. Linear functions and their graphs, $f(x) = mx + c$.

6.3 Quadratic models. Quadratic functions and their graphs (parabolas): $f(x) = ax^2 + bx + c$; $a \neq 0$
Properties of a parabola: symmetry; vertex; intercepts on the x-axis and y-axis. Equation of the axis of symmetry, $x = -\dfrac{b}{2a}$

6.4 Exponential models. Exponential functions and their graphs:
$f(x) = ka^x + c$; $a \in \mathbb{Q}^+$, $a \neq 1$, $k \neq 0$.
$f(x) = ka^{-x} + c$; $a \in \mathbb{Q}^+$, $a \neq 1$, $k \neq 0$
Concept and equation of a horizontal asymptote.

6.5 Models using functions of the form $f(x) = ax^m + bx^n + \ldots$; $m, n \in \mathbb{Z}$.
Functions of this type and their graphs. The y-axis as a vertical asymptote.

6.6 Drawing accurate graphs. Creating a sketch from information given. Transferring a graph from GDC to paper. Reading, interpreting, and making predictions using graphs. Included all the functions above and additions and subtractions.

6.7 Use of a GDC to solve equations involving combinations of the functions above.

Overview

By the end of this chapter, you will be able to:
- understand the concepts of relation and function
- find the domain and range
- graph linear functions
- graph quadratic functions by using properties of symmetry
- simplify expressions of the form a^b, $b \in \mathbb{Q}$
- understand the concept of asymptotic behavior and find equations of lines of asymptote
- graph exponential functions
- graph polynomial functions
- graph linear, quadratic, and exponential functions using a GDC
- solve real-world problems involving linear, quadratic and exponential models.

 For a brief look at a history of functions, visit www. pearsonhotlinks.co.uk, enter the ISBN for this book and click on weblink 4.1.

Mathematicians and scientists often develop tools as ways of describing, and making predictions about the world around them. Such tools are known as mathematical models. For example, if you want to go travelling to another country you will want to know the cost of items in that country and whether it is cheaper to buy the items in the country you are visiting rather than in your home country. Let's consider a person who is travelling

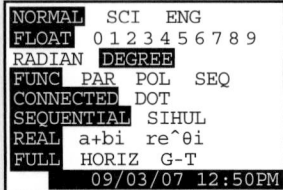

When using your TI calculator, press MODE and make sure that your calculator is set up as in the diagram below:

```
NORMAL  SCI  ENG
FLOAT  0 1 2 3 4 5 6 7 8 9
RADIAN  DEGREE
FUNC  PAR  POL  SEQ
CONNECTED  DOT
SEQUENTIAL  SIHUL
REAL  a+bi  re^θi
FULL  HORIZ  G-T
        09/03/07 12:50PM
```

Look in the fourth row:
     FUNC = Function
     PAR = Parametric
     POL = Polar
     SEQ = Sequence
For this chapter make sure that you set your calculator in FUNC mode.

from the USA to South Africa. Currently the exchange rate between US Dollars (USD) and South African Rand (ZAR) is 1 USD = 8.5 (ZAR). This equivalence can be used to form a mathematical model, or equation.

$$USD = \frac{ZAR}{8.5}$$

i.e. An item costing 25 ZAR will cost $\frac{25}{8.5} = 2.94$ USD. We can also write an equation to convert ZAR into USD;

$$ZAR = 8.5 \; USD$$

This chapter will look at different functions that can be used to model a range of phenomena and physical objects, such as bridges, or people in motion as shown below.

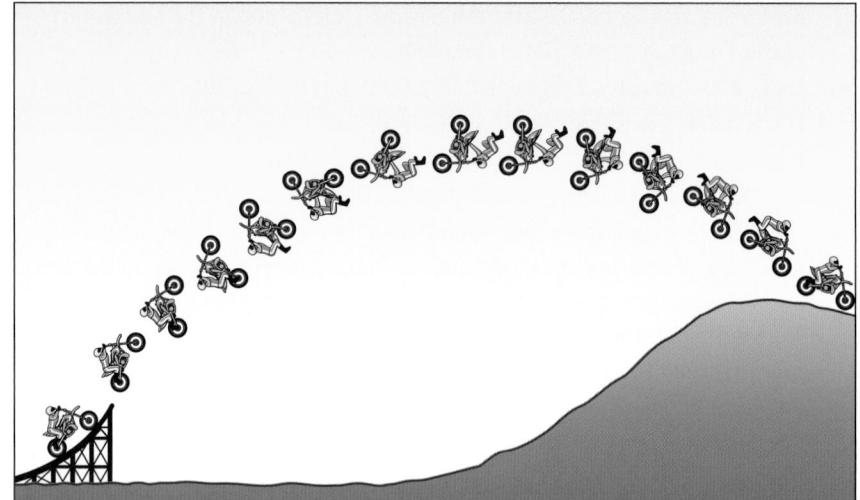

However, before we begin we need to consider what a function is.

4.1   Relations and functions

6.1   Concept of a function, domain, range, and graph. Function notation, eg $f(x)$, $v(t)$, $C(n)$. Concept of a function as a mathematical model.

In our world there are many relationships in which one thing depends upon another thing. Our height, at least for a while, depends upon how old we are. The price of a house depends on the economic idea of supply and demand. Weight loss depends on reducing calorie intake.

Another popular way to express these ideas is to use the word **function**. A good golf score is a function of long hours of practice. A healthy relationship is a function of compromise.

Mathematically, we define relations and functions as follows:
A relation is any set of ordered pairs.

For example:
A = {(1, 2), (2, 3)} is a relation since it is a set of two ordered pairs.
B = {(1, 2), (1, −2), (5, 7)} is also a relation consisting of three ordered pairs.

Now consider sets of ordered pairs in graphic form.

The graphs of $x - 3 = (y - 2)^2$ (Figure 4.1) and $y = 2^x$ (Figure 4.2) are also relations, since graphs really consist of infinitely many ordered pairs (see Chapter 1).

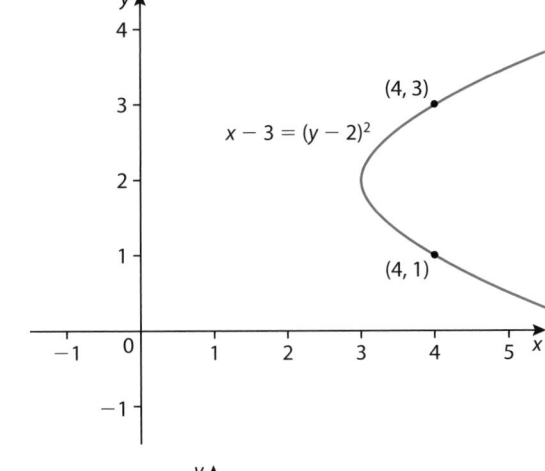

Figure 4.1 The graph of $x - 3 = (y - 2)^2$.

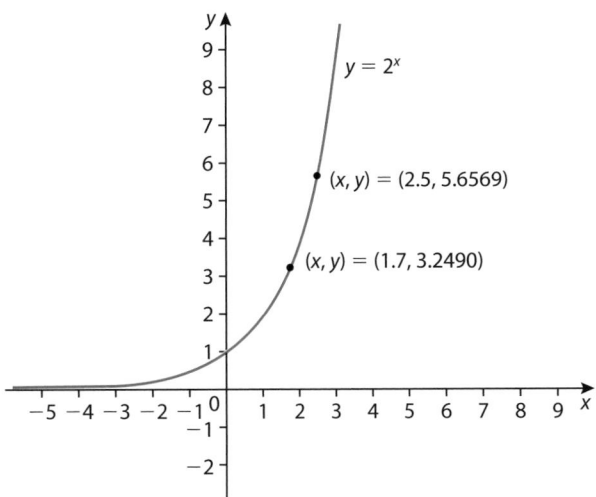

Figure 4.2 Can you see how this set of ordered pairs ($y = 2^x$) differs from the set in Figure 4.1?

A function is defined as a set of ordered pairs (a relation) such that no two ordered pairs have the same first element. In other words, for a set of ordered pairs to be considered a function, all ordered pairs must have different first elements.

Hence, set A above is a function, but set B is not, since '1' is a first element in two of the ordered pairs.

The graph, in Figure 4.1, of $x - 3 = (y - 2)^2$ does not represent a function, since if $x = 4$, then y will equal both 1 and 3. In other words, (4, 1) and (4, 3) both exist on the graph and therefore, by definition, the set of ordered pairs represented by the graph is **not** considered a function.

The answer to the question posed in Figure 4.2 is that the set of ordered pairs represented by the graph is a function, since every x-value will be paired with one and only one y-value.

The domain of a function (or relation) is the set of independent values that are allowed.

The range of a function (or relation) is the set of dependent values that are allowed.

 To determine whether or not a graph can be considered a function, use the vertical line test. Draw a vertical line on the graph. If the line intersects the graph in two or more places, the graph is **not** a function. This is because two ordered pairs would have the same first elements.

The independent values are often thought of as the 'x-values' or as the 'first elements' and the dependent values as the 'y-values' or the 'second elements'.

A mapping diagram is a drawing depicting the pairing of the independent and dependent values.

There are different types of mappings. For example, there is a mapping that maps real numbers to ordered pairs:

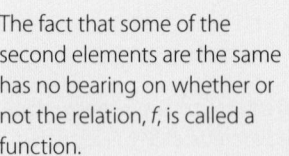

Example 4.1

Given the mapping named as f,
a) list the elements in the domain of f
b) list the elements in the range of f
c) list the ordered pairs that describe the relation f
d) is f a function?

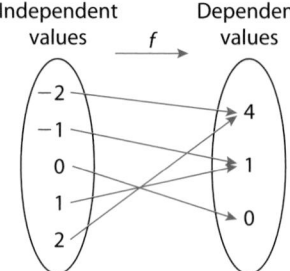

The fact that some of the second elements are the same has no bearing on whether or not the relation, f, is called a function.

● **Examiner's hint:** Do not list the elements of the range more than once.

Solution
a) Domain $= \{-2, -1, 0, 1, 2\}$
b) Range $= \{0, 1, 4\}$
c) $\{(-2, 4), (-1, 1), (0, 0), (1, 1), (2, 4)\}$
d) This mapping diagram does describe a **functional** relation since all of the first elements are different.

Exercise 4.1

1. Explain the mathematical concept of a relation.
2. Explain the mathematical concept of a function.
3. Using the word 'function', describe a relationship between daily calorie intake and weight.
4. Describe a 'non-mathematical' relationship between two 'things'.
5. List a set of five ordered pairs that describe a relation that is not a function.
6. List a set of five ordered pairs that describe a function.
7. Draw a mapping diagram for the relation $y = 2^x$, where $x \in \{-2, -1, 0, 1, 2\}$.
   a) Is this relation a function? Why?
   b) List the elements in the domain.
   c) List the elements in the range.
8. Draw a mapping diagram for the relation $y = \pm\sqrt{x}$, where $x \in \{0, 1, 4, 9\}$.
   a) Is this relation a function? Why?
   b) List the elements in the domain.
   c) List the elements in the range.
9. Determine if each of the following sets of ordered pairs is a function. Give a reason for each answer.
   a) $\{(1, 2), (2, 3), (4, 5)\}$
   b) $\{(-1, 1), (1, 1) (2, 8), (-2, 8), (3, 27), (-3, -27)\}$
   c) $\{(-2, 5), (-1, 5), (0, 5), (1, 5), (2, 5)\}$
   d) $\{(7, 2), (7, 1), (7, 0), (7, -1), (7, -2)\}$

e)

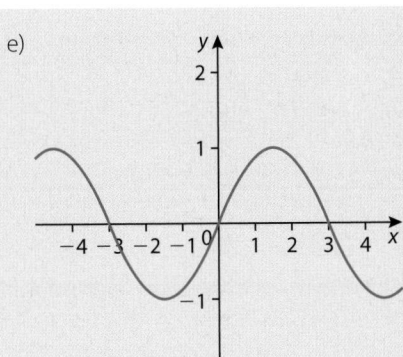

f)

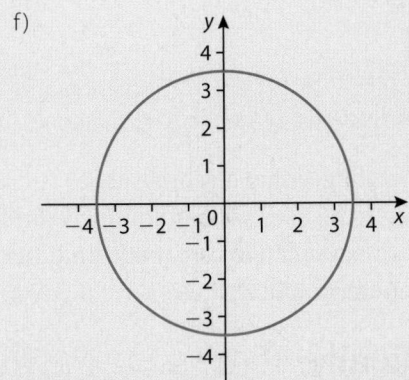

g)

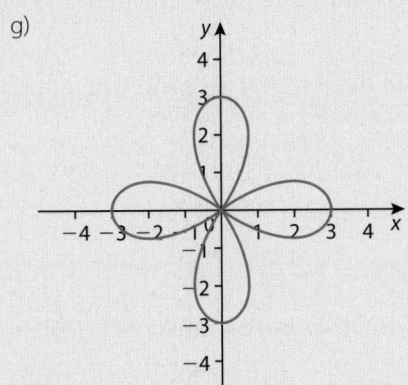

h)

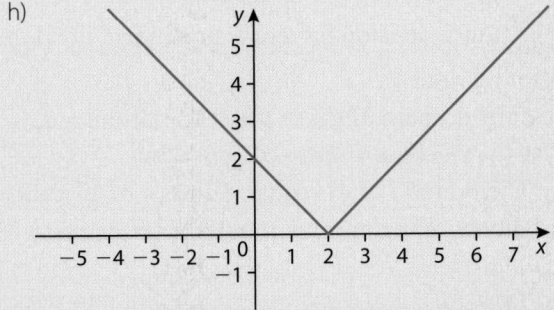

10. Sketch the graph of a non-functional relation. Explain, in words and by drawing lines on the sketch, why the graph is not a function.

11. Sketch the graph of a functional-relation. Explain, in words and by drawing lines on the sketch, why the graph is a function.

12. Say whether each of the following relations is a function or not. Explain why.

a) $y = x^2$

b) $x = y^2$

c) $y = 2x + 3$

d) $x = \frac{1}{2}y - 3$

e) $y = 2^x$

f) $y = \sqrt{4 - x^2}$

g) $y = \pm\sqrt{x^2 - 16}$

h) $x = |y|$

 ## Domain and range

| 6.1 | Concept of a function, domain, range, and graph. Function notation, e.g. $f(x)$, $v(t)$, $C(n)$. Concept of a function as a mathematical model. |

The concept of domain and range is finding, either algebraically or graphically, which x-values are allowed to be used when finding y-values, and which y-values will result after those calculations are made. In other words, what ordered pairs will exist on the graph.

Finding domain algebraically

The domain of a function is the set of allowable values the independent variable may take on.

 The independent variable's generic name is 'abscissa'.

There are two situations that you need to be concerned with when finding the domain of a function:

1. Division by zero.

2. Taking the square-root of a negative number.

Would early humans have had a need for the number 0? How long has the number 0 been used?

For example, consider the function: $g(x) = \dfrac{1}{x - 2}$

If it helps you, replace $g(x)$ with y. In other words, think: $y = \dfrac{1}{x - 2}$.

Ask yourself what value of x will produce a division by zero. The domain will then be restricted from that value.

If $x = 2$, $y = \dfrac{1}{2 - 2} = \dfrac{1}{0}$. Since division by zero is undefined, the domain must be restricted from the value of 2.

Another way to express the domain of g is to use set notation: Domain of $g = D(g) = \{x : x \in \mathbb{R}, x \neq 2\}$.

This notation is read: 'The domain of g is equal to the set of all x such that x is an element of the set of real numbers and x is not equal to 2.'

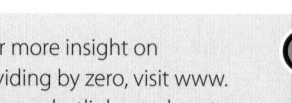 For more insight on dividing by zero, visit www.pearsonhotlinks.co.uk, enter the ISBN for this book and click on weblink 4.2.

Example 4.2

Let $f(x) = \sqrt{x - 1}$. Determine the domain of f.

Solution

Since the square root of a negative number is not an element of the set of real numbers, $x - 1 \geqslant 0$. Therefore, $x \geqslant 1$.

$$\text{Hence, } D(f) = \{x : x \in \mathbb{R}, x \geqslant 1\}$$

 When the domain is the set of real numbers, the '$x \in \mathbb{R}$' notation may be (and usually is) omitted.

Finding domain and range graphically

The domain and range of a function (or relation) can always be found by examining the graph.

Step 1: Draw an accurate graph.

Step 2: To find the domain, draw vertical lines on the graph paper to the x-axis.
- If the vertical line intersects the graph, the x-value on the x-axis is part of the domain.
- If the vertical line does not intersect the graph, the x-value on the x-axis is not part of the domain.

Step 3: To find the range, draw horizontal lines on the graph paper to the y-axis.
- If the horizontal line intersects the graph, the y-value on the y-axis is part of the range.
- If the horizontal line does not intersect the graph, the y-value on the y-axis is not part of the range.

 Finding the range of a function algebraically is beyond the scope of this course. However, below are the basic steps used to find the range of simple functions.
- Solve for x in terms of y.
- Apply the methods used in examples 4.2 and/or 4.3.

• **Examiner's hint:** 'Draw' means draw **accurately** on graph paper. 'Sketch' means give a general shape of the graph.

Example 4.3

Find the domain and range $y = f(x)$ shown in the graph below.

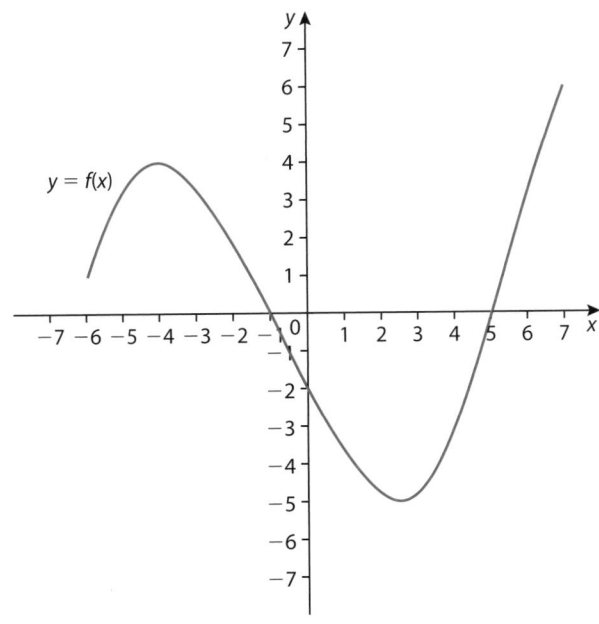

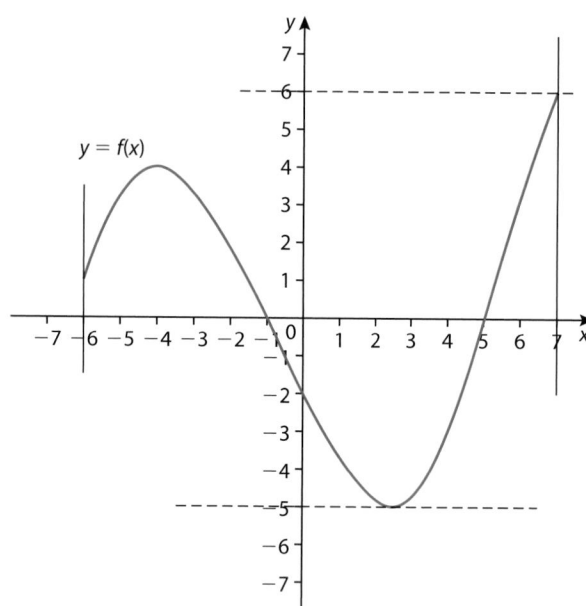

$y = f(x)$

Solution

By drawing vertical lines (see the solid lines here) you can see that the leftmost x-value is -6 and the rightmost x-value is 7. All vertical lines drawn between those x-values will intersect the graph and the x-axis. Therefore, the domain is:

$$D(f) = \{x: -6 \leqslant x \leqslant 7, x \in \mathbb{R}\}.$$

By drawing horizontal lines (see dashed lines) you can see that the lowest y-value is -5 and the highest y-value is 6. All horizontal lines drawn between those y-values will intersect the graph and the y-axis. Therefore, the range is:

$$R(f) = \{y: -5 \leqslant y \leqslant 6\}.$$

Exercise 4.2

For each of the sets A to D:
a) state the domain
b) state the range
c) state if the relation is a function and explain your answer.

1. $A = \{(0, 1), (1, 2), (2, 4), (3, 8), (4, 16)\}$

2. $B = \left\{(1, 0), \left(30, \frac{1}{2}\right), \left(45, \frac{\sqrt{2}}{2}\right), \left(60, \frac{\sqrt{3}}{2}\right), (90, 0), \left(120, \frac{\sqrt{3}}{2}\right), \left(135, \frac{\sqrt{2}}{2}\right), \left(150, \frac{1}{2}\right), (180, 0)\right\}$

3. $C = \{(4, -2), (1, -1), (0, 0), (1, 1), (4, 2)\}$

4. $D = \{(-2, 4), (-1, 1), (0, 0), (1, 1), (2, 4)\}$

Determine the domain for each of the following functions.

5. $y = 2x + 1$

6. $f(x) = x^2 + 7x + 12$

7. $y = \frac{1}{x}$

8. $g(x) = \frac{2}{x - 1}$

9. $f(x) = \sqrt{x}$

10. $g(x) = \sqrt{2x - 3}$

For each graph 11–14, determine the domain and range.

11.

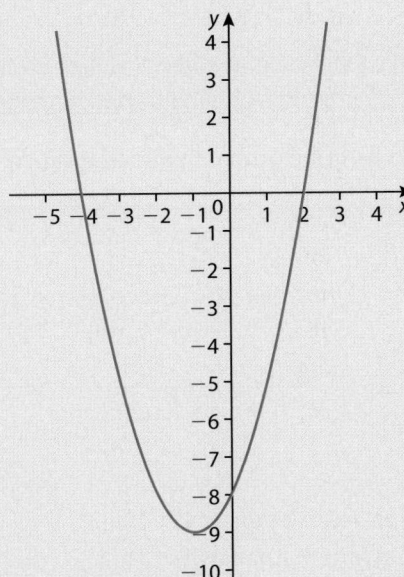

12.

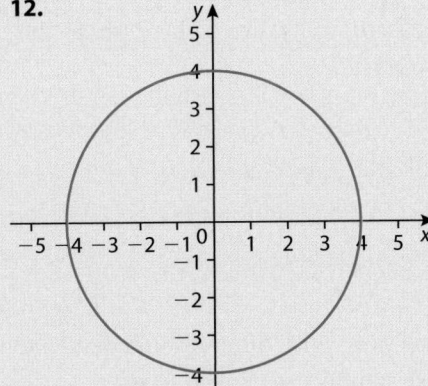

13.

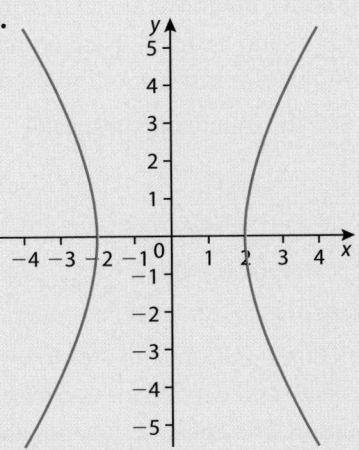

14.

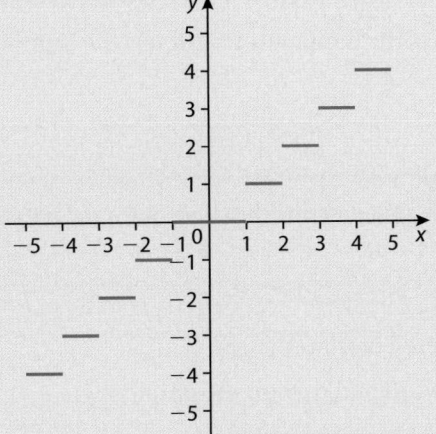

15. If $f(t) = 1000(1 + 0.07)^t$, where t represents elapsed time in years and $f(t)$ represents an amount of money after t years, find the domain and range of f.

16. If the perimeter of a regular polygon is given by $P(x) = 25 + 2x$, where x is the length of a side, find the domain and range of P.

4.3 Function notations

6.1	Concept of a function, domain, range, and graph. Function notation, e.g. $f(x)$, $v(t)$, $C(n)$. Concept of a function as a mathematical model.

There are many notations used to describe ordered pairs that satisfy a given function. For example:

1. A table of values:

x	y
-2	4
-1	1

2. In words: If $x = -2$, then $y = 4$.
   If $x = -1$, then $y = 1$.

3. In function notation:

$f(-2) = 4.$ This can be read as 'f at -2 equals 4'.
$f(-1) = 1.$ This can be read as 'f of -1 equals 1'.

The above notations are all describing the set of ordered pairs: $\{(-2, 4)$ and $(-1, 1)\}$. Each notation is useful in its own right. Perhaps the most popular is the notation used in number 3.

This is the independent variable

$$f(x)$$

This is the dependent variable.
This is the y–value.

Therefore we can write, $y = f(x)$.

Function notation allows us to write functions with different independent variables and dependent variables whose symbols are linked to the question. For example, it is possible to determine the air temperature, Celsius (°C), by counting the number of chirps of a male cricket using the function $T(n) = \dfrac{n}{7.2} + 4$, where T is the temperature and n is the number of chirps counted in a minute.

Example 4.4

Write each of the following in words and describe the concept for the notation.
a) $f(3) = 5$ b) $g(-4) = 2$ c) $h(x) = 2x + 1$

Solution

a) f at 3 equals 5. $(3, 5)$ is an ordered pair on the graph of f.

b) g of -4 equals 2. $(-4, 2)$ is an ordered pair on the graph of g.

c) h of x equals $2x + 1$. $(x, 2x + 1)$ is an ordered pair on the graph of h.

When evaluating a function for a given value, simply substitute the value for the variable.

Can 'Mathematics' be considered a language unto itself? Why are symbols used instead of writing the ideas as words?

One reason that the function notation is so powerful is that it can be embedded. A function of a function can then be expressed as: $f(g(x))$.
For example, if $f(x) = 2x$ and $g(x) = \sqrt{x}$, $f(g(x)) = f(\sqrt{x}) = 2\sqrt{x}$.

Example 4.5

Given that $f(x) = x^2 - 2$, evaluate each of the following:

a) $f(3)$ b) $f(-1)$ c) $f(k)$ d) $f(2 + h)$

• **Examiner's hint:** In the IB examinations, students are expected to be able to read the IB notation, but may write answers with the maths language and notations they are comfortable with.

Solution

a) $f(3) = 3^2 - 2 = 9 - 2 = 7$

b) $f(-1) = (-1)^2 - 2 = 1 - 2 = -1$

c) $f(k) = k^2 - 2$

d) $f(2 + h) = (2 + h)^2 - 2 = (2^2 + 2 \cdot 2h + h^2) - 2 = 4 + 4h + h^2 - 2$
   $= 2 + 4h + h^2$

Example 4.6

Students counted the number of chirps of crickets in 3 locations around the city. Given that the temperature can be calculated using the function $T(n) = \dfrac{n}{7.2} + 4$, determine the temperature in each of the following cases.

a) The students counted 62 chirps in a minute

b) $n = 225$

c) $T(144)$

Solution

a) 62 chirps in a minute $\therefore n = 62$, $T(62) = \dfrac{62}{7.2} + 4 = 12.6$ (3 *s.f*)

b) $n = 225$, $T(225) = \dfrac{225}{7.2} + 4 = 35.3$ (3 *s.f*)

c) $T(144) = \dfrac{144}{7.2} + 4 = 24$

Exercise 4.3

1. Write each of the following in words:
  a) $f(1) = 5$ b) $r(-4) = -2$ c) $g(a) = b$

2. Describe the concept for each notation:
  a) $f(2) = 9$ b) $g(-3) = 7$ c) $v(c) = d$

3. Given that $f(x) = 2x^2 - 3$, evaluate each of the following:
  a) $f(1)$ b) $f(0)$ c) $f(a)$
  d) $f(2 + h)$ e) $f(\pi)$

4. Given that $g(t) = \sqrt{t^2 - 4}$, evaluate each of the following:
  a) $g(3)$ b) $g(-2)$ c) $g(1)$
  d) $g(r)$ e) $g(1 + h)$

5. If $f(x) = x^2 + 1$, find $f(x + h)$.

6. If $g(x) = x^2 - x$, find $g(x + h) - g(x)$.

7. If $f(x) = 2000\left(1 + \dfrac{0.08}{12}\right)^{12x}$, find $f(30)$.

8. If $A(r) = 4\pi r^2$, find $A(10)$ correct to 3 significant figures.

Note
$f(g(x))$ is called a composite function. It is a function of a function. It is read as 'f at g at x'. It is a notation that requires a double substitution. The first requires finding $g(x)$ and the second requires finding f at the $g(x)$ value.

For Questions 9–14, let $f(x) = 2x + 1$, $g(x) = x^2$, and $h(x) = \sqrt{x - 1}$, and find each the following:

9. $f(g(2))$ **10.** $f(h(5))$ **11.** $g(f(-3))$

12. $h(g(4))$ **13.** $f(g(h(10)))$ **14.** $h(f(g(1)))$

$f \circ g$ is another notation used to denote a composite function. It is read as 'f operat g' or, simply, 'f op g'. $(f \circ g)(3)$ can be thought of as $f(g(3))$.

For Questions 15–20, use the same function definitions as in 9 – 14 above and find each of the following:

15. $(g \circ f)(-1)$ **16.** $(h \circ g)(\sqrt{5})$ **17.** $(f \circ f)(2)$

18. $(h \circ f)(0.5)$ **19.** $(h \circ g)(1)$ **20.** $(g \circ h)(4)$

4.4 Linear functions

> **6.2** Linear models.
> Linear functions and their graphs, $f(x) = mx + c$.

As the name suggests, a linear function is a relation in which all of the ordered pairs form a straight line. Recall from Chapter 1 that a line is an undefined term in geometry, but we think of it as being made up of infinitely many points joined so closely together that there is no space between them.

Linear functions can be written in several ways. Examples are given below.

- $y = 2x + 1$
- $f(x) = 2x + 1$

Each notation expresses the idea of ordered pairs that lie on the graph. From geometry, we know that we can draw a unique line through any two points. Therefore, in order to draw the graph of a linear function, at least two ordered pairs must be produced and the points plotted.

There are several ways to produce a set of at least two ordered pairs:

- producing a table of values
- finding function values
- finding the x- and y-intercepts
- using one ordered pair and the gradient.

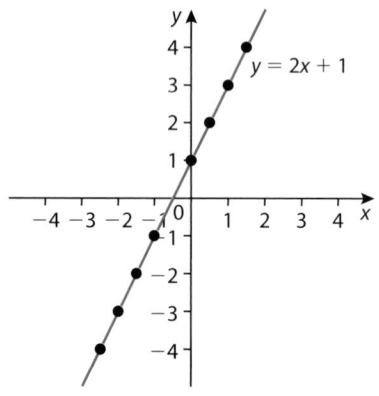

Figure 4.3 Here you see many ordered pairs that satisfy $y = 2x + 1$. Between each pair of them are infinitely many more. This set of points is called a *line*, and the equation is called a *linear function*.

When you use the everyday expression 'varies directly as', you are describing a linear relationship or a linear function through the origin. For example: simple interest varies directly as the number of years, or the circumference of a circle varies directly as the radius.

Producing a table of values

Example 4.7

Produce a table of values, without a calculator, for $y = 2x + 1$.

Solution

Let $x = -1$, then $y = 2 \cdot -1 + 1 = -1$.

Let $x = 0$, then $y = 2 \cdot 0 + 1 = 1$.

Let $x = 1$, then $y = 2 \cdot 1 + 1 = 3$.

x	y	(x, y)
-1	-1	$(-1, -1)$
0	1	$(0, 1)$
1	3	$(1, 3)$

Example 4.8

Produce a table of values with the TI calculator for $y = 2x + 1$.

Solution

Method I: Type the following keystrokes:

  Y=, 2x+1, 2ND WINDOW, TblStart= −1, ΔTbl=1,

  Indpnt: **Auto**, Depend: **Auto**, 2ND GRAPH

- You will now see a table of values.
- Each ordered pair satisfies $y = 2x + 1$.
- Use the ∧ or ∨ to see more ordered pairs.

Method II: Type the following keystrokes:

  Y=, 2x+1, 2ND WINDOW, TblStart= −1, ΔTbl=**1**,

  Indpnt: **Ask**, Depend: **Auto**, 2ND GRAPH

- You should now a see table with no values in it.
- If there are values present, use the DEL key to delete them.

Type −1, press ENTER and the calculator will return −1.

Type 0, press ENTER and the calculator will return 1.

Type 1, press ENTER and the calculator will return 3.

Continue the above process to produce more ordered pairs.

When the table fills up only the last entry will change.

The 2ND key is coloured blue and therefore activates all of the 'blue' function keys. For example, typing the keys 2ND, WINDOW accesses the function TBLSET.

Finding functional values

Example 4.9

Find ordered pairs for $x \in \{-1, 0, 1\}$ where $f(x) = 2x + 1$.

$(-1, -1) \in f$ means that $(-1, -1)$ lies on the graph of f.

Solution

$f(-1) = 2 \cdot -1 + 1 = -1 \Rightarrow (-1, -1) \in f.$

$f(0) = 2 \cdot 0 + 1 = 1 \Rightarrow (0, 1) \in f.$

$f(1) = 2 \cdot 1 + 1 = 3 \Rightarrow (1, 3) \in f.$

Finding x- and y-intercepts

- The x-intercept is the point at which the graph crosses the x-axis.
- All x-intercepts have the form of $(x, 0)$.
- To find the x-intercept, let $y = 0$ and then solve for x.
- The y-intercept is the point at which the graph crosses the y-axis.
- All y-intercepts have the form of $(0, y)$.
- To find the y-intercept, let $x = 0$ and then solve for y.

Example 4.10

Find the x- and y-intercepts for the linear function $f(x) = 2x + 1$.

Solution

Think of $f(x) = 2x + 1$ as $y = 2x + 1$.
If $y = 0$, then $0 = 2x + 1$.
$$-2x = 1$$
$$\therefore \quad x = \frac{-1}{2}. \text{ Hence, the } x\text{-intercept is } \left(\frac{-1}{2}, 0\right).$$
If $x = 0$, then $y = 2 \cdot 0 + 1 = 1$. $\therefore$ the y-intercept is $(0, 1)$.

Using one ordered pair and the gradient

The gradient of a linear function is the same as the slope of the line.
There are several notations for the gradient of a line:

- m
- $\dfrac{\Delta y}{\Delta x}$, where '$\Delta$' means 'the change in'.
- $\dfrac{y_2 - y_1}{x_2 - x_1}$
- $\dfrac{\text{rise}}{\text{run}}$ (from one point to another)

$$\therefore \quad \text{gradient} = \text{slope} = m = \frac{\Delta y}{\Delta x} = \frac{\text{change in } y}{\text{change in } x} = \frac{y_2 - y_1}{x_2 - x_1} = \frac{\text{rise}}{\text{run}}$$

Example 4.11

Describe the meaning of a gradient of $\dfrac{-2}{3}$.

Solution

A gradient $\dfrac{-2}{3}$ means that the slope of a line is constant from any one point to another point on the line. It means that, from point A, the next point could be found by 'rising' -2 units (2 units down) and 'running' 3 units (3 units to the right).

Example 4.12

If a linear function has a gradient of $\dfrac{-2}{3}$ and point A has coordinates $(4, 6)$, find two more points that lie on the line.

Solution

From $(4, 6)$: rise -2 units from $y = 6$ $(6 + -2 = 4)$ and then run 3 units from $x = 4$ $(4 + 3 = 7)$.
$\therefore$ another point on the line would be $(7, 4)$.

From $(4, 6)$: rise 2 units from $y = 6$ $(6 + 2 = 8)$ and then run -3 units from $x = 4$ $(4 - 3 = 1)$.
$\therefore$ another point on the line would be $(1, 8)$.

See the diagram below.

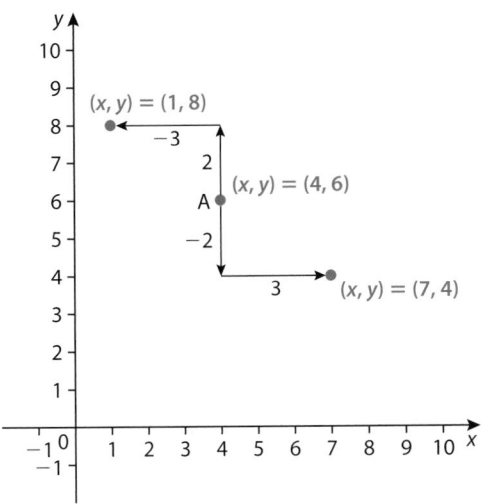

General form of a linear function

The general form of a linear function is defined as:

$y = mx + c$, where

- m is the gradient
- c is the y-intercept.

Example 4.13

Given $f(x) = \frac{3}{5}x + 2$, find the gradient and the y-intercept.

Solution

Using the above definition, the gradient $= \frac{3}{5}$ and the y-intercept $= 2$.

Graphing linear functions

To draw the graph of a linear function:

Step 1: Rewrite the function as an equation in terms of x.

Step 2: Use one of the methods previously outlined to plot at least two ordered pairs (two points).

Step 3: Connect the points.

Step 4: Pay attention to the domain if it is required.

Step 5: Draw the graph using IB 2-mm graph paper and a straight edge.

Example 4.14

Draw the graph of $f(x) = -3x + 2$.

Solution

The following steps will result in an accurately drawn graph.

Step 1: Rewrite the function notation as $y = -3x + 2$.

Step 2: Use 2-mm graph paper.

Step 3: Let 1 cm = 1 unit.

● **Examiner's hint:** Make sure you number the interval marks on the axes of your graph and label the x- and y-axes.

Step 4: Label the x- and y-axes.

Step 5: Use one of the previous methods to produce and plot at least two ordered pairs.

Step 6: Connect the points with a straight edge.

Step 7: Label the graph as:
- $f(x) = -3x + 2$, or
- $y = -3x + 2$.

(Note: Both intercepts and an extra (table) value were found.)

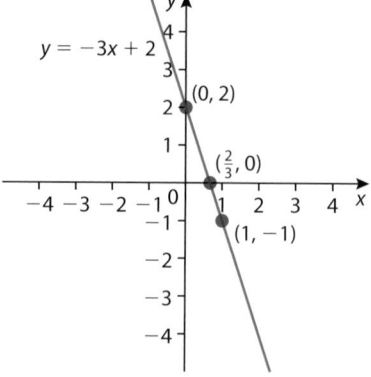

Writing linear functions

You need to know how to write a linear function
- when you know the gradient and one ordered pair
- when you know two ordered pairs.

Example 4.15

Write a linear function in the form $f(x) = mx + c$ when $m = \frac{3}{5}$ and the ordered pair $(-2, 4)$ lies on the graph of f.

Solution

Think of $f(x) = mx + c$ as $y = mx + c$. Substitute $\frac{3}{5}$ for m and $(-2, 4)$ for (x, y).

Solve for c and back-substitute to write the equation.

$$\therefore \quad 4 = \frac{3}{5} \cdot -2 + c$$

$$4 + \frac{6}{5} = c$$

Hence, $c = \frac{26}{5}$.

Therefore, $y = \frac{3}{5}x + \frac{26}{5}$, or $f(x) = \frac{3}{5}x + \frac{26}{5}$.

Example 4.16

Given that $(-3, 4)$ and $(5, 2)$ are ordered pairs that satisfy a linear function, express that function as $f(x) = mx + c$.

Solution

Since the ordered pairs lie on a line, write $y = mx + c$

Calculate $m = \dfrac{y_2 - y_1}{x_2 - x_1} = \dfrac{2 - 4}{5 - (-3)} = \dfrac{-2}{8} = \dfrac{-1}{4}$.

Hence, by way of substitution, $y = \frac{-1}{4}x + c$.

In order to find c, the y-intercept, substitute either of the points that lie on the line for x and y.

Using (5, 2) we have $2 = \frac{-1}{4} \cdot 5 + c \Rightarrow 2 + \frac{5}{4} = c \Rightarrow \frac{13}{4} = c$.

$\therefore$ by substitution, $y = \frac{-1}{4}x + \frac{13}{4} \Rightarrow f(x) = \frac{-1}{4}x + \frac{13}{4}$.

Exercise 4.4

1. For any two points that lie on a line, how many points are between them?

2. List two different ways that a linear function can be written.

3. List four methods for finding a set of ordered pairs.

4. If $y = 3x - 2$, produce a table of at least two values under the condition that $x \in \mathbb{Z}$ and $-2 \leqslant x \leqslant 2$.

5. If $y = -5x + 1$ produce a table of values, using your GDC, under the following conditions: TblStart = 0, ΔTbl = 0.5, and $0 \leqslant x \leqslant 3$.

6. If $y = 2.56x - 3.47$, produce a table of values, using your GDC, under the following conditions: Indpnt: Ask, $0 \leqslant x \leqslant 1.4$, and $\Delta x = 0.2$.

7. If $y = -3 - 3x$, produce a table of values when: Indpnt: Ask, $-2 \leqslant x \leqslant 2$, and $\Delta x = 0.4$.

8. If $f(x) = 4x + 2$, find: $f(-1)$, $f(0)$, $f(1)$.

9. If $g(x) = \frac{1}{2}x - 1$, find: $g(-2)$, $g(2)$, $g(6)$.

10. A linear function has a gradient of $\frac{3}{4}$ and a point (2, 3) that lies on its graph. Find two ordered pairs, one on each side of (2, 3), that also lie on the graph.

11. A linear function has a slope of $\frac{-2}{5}$ and the ordered pair $(-4, 6)$ lies on its graph. Find two other ordered pairs, one on each side of $(-4, 6)$, that also lie on the graph.

12. Find the x- and y-intercepts for each of the following linear functions.
    a) $y = -2x - 1$
    b) $f(x) = 4x - 2$
    c) $f(x) = dx + h$

13. a) If 5 is the x-intercept, what ordered pair is associated with 5 and on what axis does that point exist?
    b) If -4 is the y-intercept, what ordered pair is associated with -4 and on what axis does that point exist?

14. a) Write an example of a linear function that does not have an x-intercept.
    b) What is the equation of the of the x-axis?
    c) Why doesn't the y-axis represent a function?

15. Draw the graph of each of the following under the following conditions:
    - Use 2-mm graph paper (if possible).
    - Let 1 cm = 1 unit.
    - Label the x- and y-axes.
    - Plot at least two ordered pairs.
    - Connect the points with a straight edge.
    - Label the graph.

a) $y = 2x + 2$

b) $y = -4x - 3$

c) $f(x) = -x + 1$

d) $g(x) = 1.5x - 2.5$

16. Write a linear function in the form $f(x) = mx + c$ for each of the following, under the given set of conditions.

a) $m = \frac{2}{3}, c = \frac{7}{3}$.

b) $m = \frac{-1}{4}$, $(0, 3)$ lies on the graph of f.

c) $m = -3, (1, 2)$ lies on the graph of f.

d) $m = \frac{4}{5}, (-2, -1) \in f$.

e) Both $(-1, 3)$ and $(4, -7)$ lie on the graph of f.

f) Both $(2, 2)$ and $(-5, 8)$ lie on the graph of f.

g) The x-intercept is 3 and the y-intercept is 4.

h) The x-intercept is -5 and the y-intercept is 1.

i) Both the x- and y-intercept are 0. $(1, 1) \in f$.

 ## Linear models

There are many examples of linear models in the world around us, for example the conversion of money from one currency to another is a linear function as is the conversion of temperatures from Fahrenheit to Celsius or vice-versa.

Example 4.17

It is known that water freezes at $32°$ F (Fahrenheit) and $0°$ C (Celsius). It is also known that water boils at $212°$ F and $100°$ C.

a) Write a linear function to convert Fahrenheit degrees to Celsius degrees.

b) Find the Celsius temperature that corresponds to a temperature of $75°$ F.

Solution

a) The general ordered pair is in the form (F, C). Therefore we know that $(32, 0)$ and $(212, 100)$ satisfies the function that relates F° and C°. The general form of a linear function is $f(x) = mx + c$. In this application the function will take the form:

$$f(\text{F}) = m \cdot \text{F} + c, \text{ where C} = f(\text{F}).$$

$$\text{Hence, } m = \frac{100 - 0}{212 - 32} = \frac{100}{180} = \frac{5}{9}.$$

$$0 = \frac{5}{9} \cdot 32 + c \Rightarrow c = \frac{-160}{9}$$

$$\therefore f(\text{F}) = \frac{5}{9} \cdot \text{F} - \frac{160}{9} \quad \text{or} \quad \text{C} = \frac{5}{9} \cdot \text{F} - \frac{160}{9}.$$

b) $f(75) = \frac{5}{9} \cdot 75 - \left(\frac{160}{9}\right) = 23.9°$ C to 3 significant figures.

There are also instances where data can be collected and used to make predictions about other related phenomena or objects. However, one must be careful as not all such occurrences are linear and care should be taken when determining whether phenomena or a set of objects does indeed match a linear model.

Example 4.18

In supermarkets you will often find shopping baskets stacked inside each other, that is, the baskets nest inside each other. The measurements for a typical basket such as the one shown, are length = 42 cm, width = 20 cm, height from bottom to top = 22 cm, and thickness of the ridge = 3 cm.

a) Complete the table for the heights of the baskets when they are nested inside each other.

Number of baskets	1	2	3	4	5	10	20
Height (cm)	22	25	28	–	–	49	79

b) Determine the gradient for the linear function, using the ordered pairs $(2, 25)$ and $(10, 49)$.

c) What do you notice about the gradient?

d) Write down the y-intercept for the equation.

e) Write down the equation of the line which describes the number of shopping baskets and the resulting height of the stack.

f) Draw a graph of the function.

g) Using your equation, or graph, determine the number of baskets required for a stack 3 metres high.

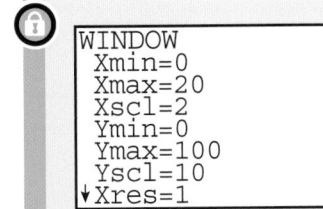

WINDOW
Xmin=0
Xmax=20
Xscl=2
Ymin=0
Ymax=100
Yscl=10
▼Xres=1

To draw the graph on the TI Calculator enter the equation Y1 = 3X + 19 and then set the window as shown above. Once you have entered the numbers for Xmin, Xmax, Xscl, Ymin, Ymax, and Yscl press the GRAPH button.

3 cm

22 cm

20 cm

42 cm

Solution

a) Complete the table for the heights of the baskets when they are nested inside each other.

Number of baskets	1	2	3	4	5	10	20
Height (cm)	22	25	28	31	34	49	79

Note that the height increases by 3 cm each time, that is, by the height of the thickness of the top edge of the basket.

b) Previously we have seen that the gradient can be calculated using two ordered pairs using the formula gradient $m = \dfrac{y_2 - y_1}{x_2 - x_1}$ where (x_1, y_1) is the first ordered pair and (x_2, y_2) the second ordered pair,

   Gradient of the line for our given example is $m = \dfrac{49 - 25}{10 - 2} = \dfrac{24}{8} = 3$

c) The table increases by 3 each time, which is the thickness of the top edge of the basket. This is equal to the gradient, or the change in the height as the number of baskets increases.

d) The y-intercept occurs when x is zero. It is not practical to have zero baskets but we can determine the solution for zero baskets by subtracting 3 from the height of one basket. i.e.

   $22 - 3 = 19$ cm.

   Therefore our y-intercept is 19.

e) The gradient $m = 3$ and the y-intercept is 19.

The equation for the height of the stack of baskets as a function of the number of baskets is given by

$h(n) = mn + c$ where $h(n)$ is the height of n baskets and $n \in \mathbb{Z}^+$, $h(n) = 3n + 19$

f) To draw a graph we use the GDC.

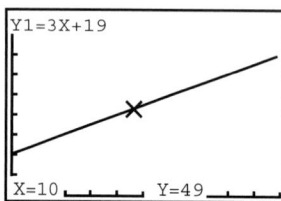

g) To find how many baskets are required for a height of 3 metres we can either read off the graph, or use the equation we found previously.

Remember that the heights were calculated in centimetres. We therefore need to convert 3 m to 300 cm.

Our equation is $h(n) = 3n + 19$, where $h(n) = 300$. We need to solve for n.

$h(n) = 3n + 19 = 300$

$3n = 300 - 19 = 281$

$3n = 281$

$n = \dfrac{281}{3} = 93.667$

As we cannot have a part shopping basket then the number of shopping baskets required for a stack of 3 metres is 94.

Exercise 4.5

1. The cost c, in Australian dollars (AUD), of renting a bungalow for n weeks is given by the linear relationship $c = nr + s$, where s is the security deposit and r is the amount of rent per week.

   Freja rented the bungalow for 12 weeks and paid a total of 6250 AUD, which included a security deposit of 850 AUD.

   a) Write down the linear model which represents her total costs for the bungalow.

   b) Determine the cost of one week's rent.

   Sasha, a very good friend of Freja, rented a bungalow across the road from a different company. Given that Sasha paid the same rent as Freja,

   c) Write down the linear model which represents her total costs for the bungalow.

   Sasha, paid a total of 10 700 AUD for 20 weeks' rent including security deposit.

   d) Find the value of Sasha's security deposit.

   e) Freja suggested that Sasha should have rented an apartment from Beachside Apartments where the security deposit was 850 AUD and the rent was 495 AUD per week. Would this option have cost less for Sasha?

2. Snow has been falling for the past 10 hours. The snow depth was measured after 5 hours and was found to be 8 cm. The depth of snow after 10 hours was 12 cm. Assuming that the rate of increase in the depth of snow can be modelled by a linear function and there is no change to the rate at which the snow is falling, determine the depth of snow after 24 hours.

3. The Flatwhite travelling coffee shop sells cups of coffee for $0.75. The daily cost of producing the cups of coffee using imported ingredients along with shop rental and local council charges can be described by the linear function $C(x) = 0.45x + 150$, where x is the number of cups of coffee.

a) Write down the fixed cost.

b) Write down the equation for the profit.

c) Determine the profit (or loss) when 100 cups of coffee are sold in a day.

d) What is the minimum number of cups of coffee that need to be sold during the day to make a profit?

4. The cost of repairing a motor vehicle at Garage A is 36 GBP per hour with a fixed cost of 50 GBP.

a) Show that the cost function $C(t)$ can be written as $C(t) = 50 + 36t$, where t is the time in hours.

b) If it takes 4 hours to repair a particular car, find the total cost of the repairs at Garage A.

c) Jane paid 365 GBP to have her car repaired at Garage A, and she delivered the vehicle at 7.30 a.m. What time did she collect her car, assuming the car mechanic had a 30-minute lunch break?

Jane's friend Sam had his car repaired at Garage B, where the fixed cost is 25 GBP and the hourly rate is 41 GBP.

d) Write down the cost function for Garage B.

e) Plot graphs for each function on the same grid, and determine the number of hours when the cost to repair a vehicle at the two garages is the same.

5. A cottage is being constructed in an isolated location in the forest and the builder must use a narrow trail to carry all goods to the construction site. The builder is trying to determine how many hours it will take to shift 1000 bricks and he has constructed a table to calculate the length of time needed.

Hour	0	1	2	3	4	...
Number of bricks left to be moved	1000	988	976	964	952	...

a) Determine the gradient of the linear model.

b) Write down the linear model for the number of bricks left to be moved.

c) How many hours will it take for the 1000 bricks to be moved, assuming the builder continues at the same rate?

After working for 12 hours the builder stops for the day.

d) How many bricks had the builder shifted at that time?

On the following day, a friend arrives to help the builder and they will use a small wheelbarrow to move 45 bricks at a time. They can move two loads an hour, i.e. 90 bricks.

e) Write down the new linear model for the number of bricks left to be moved.

f) How many hours on the second day will it take to shift all the remaining bricks?

4.6 Quadratic functions

Apollonius of Perga (262 BC – 190 BC) was known as 'The Great Geometer'. In his book, *Conics*, he introduced the term 'parabola'. To find out more, visit www.pearsonhotlinks.co.uk, enter the ISBN for this title and click on weblinks 4.3, 4.4, and 4.5.

6.3   Quadratic functions and their graphs, parabolas: $f(x) = ax^2 + bx + c$; $a \neq 0$
Properties of a parabola: symmetry; vertex; intercepts on the x-axis and y-axis. Equation of the axis of symmetry, $x = -\dfrac{b}{2a}$

The quadratic function that we will study in this section is called a **parabola**. It is one of the four conic sections.

Like all functions, a quadratic function is a set of ordered pairs. The first word, quadratic, defines that set of ordered pairs as a second degree polynomial function of the form: $f(x) = ax^2 + bx + c$, where $a, b, c \in \mathbb{R}$, $a \neq 0$.

All quadratic functions have similar characteristics.
- They are ∪-shaped up or ∩-shaped down.
- They have a minimum value or a maximum value.
- The point at which the maximum or minimum value occurs is called the vertex or the turning point.
- They have symmetry with respect to a vertical line called the axis of symmetry.
- At least three points, but preferably five points, are required to draw the graph.
- There is always a y-intercept.
- Sometimes there are no x-intercepts.
- The graph is called a parabola.

In this course, a, b, and c will be limited to the set of rational numbers.

Example 4.19

Draw the graph of $f(x) = 2x^2 - 5x - 3$. Describe the characteristics of the graph.

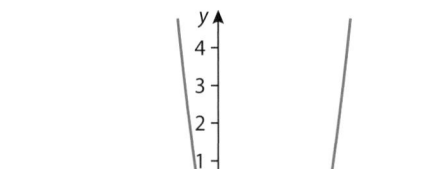

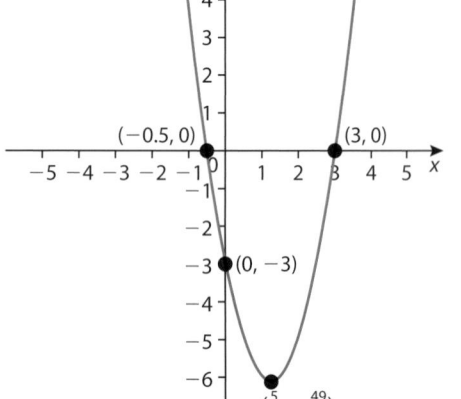

Solution

a) It has a minimum value: $-\dfrac{49}{8}$.

b) The axis of symmetry is $x = \dfrac{5}{4}$.

c) At least three points have been plotted.

d) The y-intercept is -3.

e) The x-intercepts are -0.5 and 3.

This section will examine how to draw the graph of a quadratic function and find the common characteristics listed above.

There are several methods by which the graph of a quadratic function may be drawn:
- point-plotting using a table of values
- plotting the x- and y-intercepts and the vertex
- using the axis of symmetry and its properties.

Point-plotting using a table of values

Since a quadratic function is a set of ordered pairs and since we already know its general shape, $\cup$ or $\cap$, simply plot a sufficient number of points so that the graph 'reveals' itself.

Example 4.20

Draw the graph of $f(x) = x^2 + x - 3$ using the point-plotting method.

Solution

The easiest method to produce a set of ordered pairs is to make a table of values using the method in Section 4.4.

```
TABLE SETUP
 TblStart=-3■
 ΔTbl=0.5
Indpnt: Auto Ask
Depend: Auto Ask
```

X	Y1	
-3	3	
-2.5	.75	
-2	-1	
-1.5	-2.25	
-1	-3	
-.5	-3.25	
0	-3	

X=-3

X	Y1	
0	-3	
.5	-2.25	
1	-1	
1.5	.75	
2	3	
2.5	5.75	
3	9	

X=3

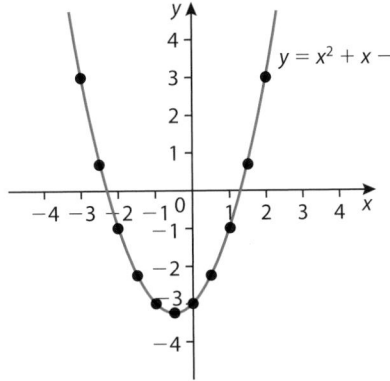

$y = x^2 + x - 3$

Plotting the x- and y-intercepts and the vertex

Once you have knowledge of the shape of a parabola, you will be able to draw a reasonable graph using only three points. One of those points will need to be the vertex, or turning point, and two other very good choices will be the x-intercepts. An easy fourth point to use would be the y-intercept.

There are several ways to find the x-intercepts, if they exist. There will always be a y-intercept and only one way to find to find it. There are several ways to find the vertex.

In Chapter 2 you learned to solve quadratic equations by factorizing, by using the quadratic formula, and by using your GDC.

For example, you learned that the solutions for the equation $x^2 - 2x - 3 = 0$ are $x = -1$ or $x = 3$.

Think of the above equation as $0 = x^2 - 2x - 3$. Now, replace 0 with y and write: $y = x^2 - 2x - 3$. Finally, let $y = f(x)$. Therefore, $f(x) = x^2 - 2x - 3$. As you can see, there is a clear connection between the equation and the associated function.

The solutions, $x = -1$ or $x = 3$, are called:
- answers to the equation
- roots to the equation.

They are also called 'zeros' of the function. The reason that they are called 'zeros' is that, when they are substituted for x in $f(x) = x^2 - 2x - 3$, $f(x)$ will be 0.

In other words, $f(-1) = (-1)^2 - 2(-1) - 3 = 0$. This implies that $(-1, 0) \in f$.

And, $f(3) = 3^2 - 2(3) - 2 = 0$. This implies that $(3, 0) \in f$.

The ordered pairs, $(-1, 0)$ and $(3, 0)$ are called the x-intercepts.

To find the x-intercepts of a quadratic function, let $f(x) = 0$ and solve the resultant equation by
- factorizing
- using the quadratic formula
- using your GDC.

To find the y-intercept of a quadratic function, let $x = 0$ and solve for y.

Example 4.21

Given the quadratic function $f(x) = x^2 - 4x - 5$, find the x- and y-intercepts.

Examiner's hint: The formula $x = \dfrac{-b}{2a}$ is in the IB Formula Booklet. See 'Equation of axis of symmetry'.

Solution

Let $f(x) = 0$ and solve for x. $0 = x^2 - 4x - 5$
$$(x - 5)(x + 1) = 0$$
$$x - 5 = 0 \text{ or } x + 1 = 0$$
$$\therefore x = 5 \text{ or } x = -1$$

Hence, the x-intercepts are $(5, 0)$ and $(-1, 0)$.

Let $x = 0$ and solve for $f(x)$.
$f(x) = 0^2 - 4(0) - 5 = -5$.
$\therefore$ the y-intercept is $(0, -5)$.

The vertex or turning point occurs at the highest or lowest point of the parabola.

There are several ways to find the vertex:
1. Find the mean average of the x-intercepts to find the abscissa and then substitute that value into the function to find the ordinate.
2. Use $\left(\dfrac{-b}{2a}, f\left(\dfrac{-b}{2a}\right)\right)$, where $f(x) = ax^2 + bx + c$.
3. Type the following keystrokes on your TI calculator:
   - Y=
   - $x^2 + x - 3$ (for example)
   - ZOOM 6 (to see if the vertex is the highest or lowest point)
   - 2ND, TRACE (which accesses CALC)
   - 3:minimum (if the vertex is the lowest point) or 4:maximum (if the vertex is the highest point)
   - $<, <,$ or $>, >,$ etc. (until the cursor is to the left of the vertex)
   - ENTER
   - $>, >,$ or $<, <,$ etc. (until the cursor is to the right of the vertex)
   - ENTER
   - $<,$ or $>$ (until the cursor is between the left- and right-bound pointers)
   - ENTER
   $\therefore$ the vertex is $(-\frac{1}{2}, -3\frac{1}{4})$.

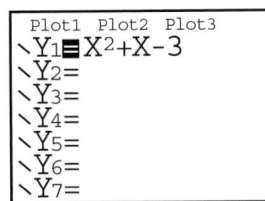

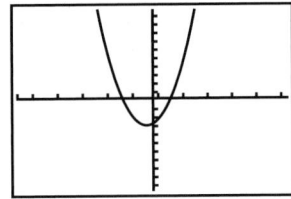

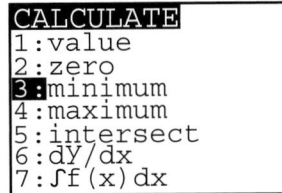

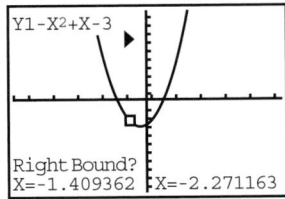

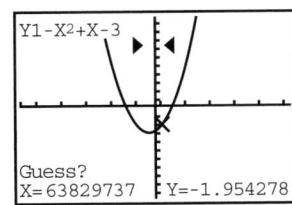

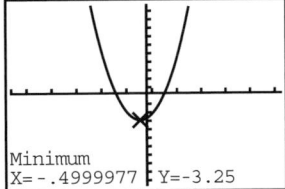

4. Use a calculus method. (See Chapter 13.)
5. Completing the square. (This method is beyond the scope of this course.)

Example 4.22

Find the vertex of the parabola $f(x) = x^2 - 4x - 5$.

Solution

The abscissa is : $x = \dfrac{-b}{2a} = \dfrac{-(-4)}{2 \cdot 1} = 2.$

The ordinate is: $y = f\left(\dfrac{-b}{2a}\right) = f(2) = 2^2 - 4(2) - 5 = -9.$

Therefore, the vertex is $(2, -9)$.

Example 4.23

Draw the graph of $f(x) = x^2 - 4x - 5$ by plotting the x- and y-intercepts and the vertex.

Solution

Solution: See above for the required points.

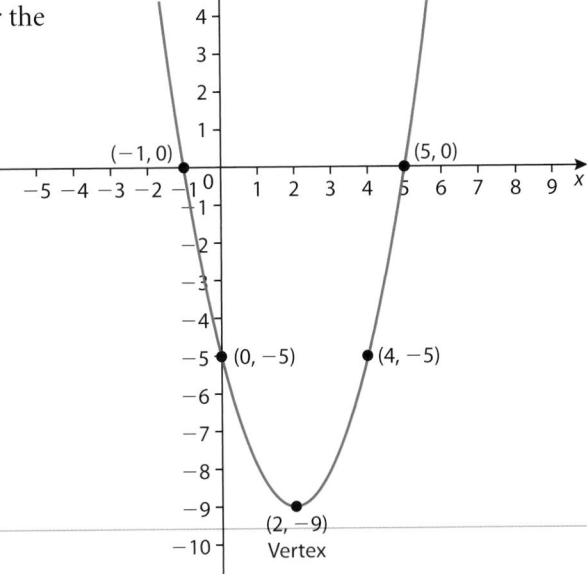

> A function is said to be **even** if $f(-x) = f(x)$, for all x in the domain of f. A function is said to be **odd** if $f(-x) = -f(x)$, for all x in the domain of f.

The axis of symmetry

The idea of symmetry, as it applies to quadratic functions, is that for every y-value there are two different x-values that are paired with it.

For example, in Example 4.23, $(0,-5)$ corresponds to $(4,-5)$. They are said to be symmetrical points. All points, except the vertex, will have a corresponding point.

When answering the question about the axis of symmetry, the answer must be given as an '$x =$' equation. For example, $x = 2$.

A quadratic function can be classified as an 'even' function. An even function is a set of ordered pairs that are symmetric with respect to a vertical line. This line is called the **axis of symmetry**.

In a parabola:

- the axis of symmetry must pass through the vertex
- the equation of the axis of symmetry is: $x = \dfrac{-b}{2a}$, where $f(x) = ax^2 + bx + c, a \neq 0$
- the equation can also be found by observing the x-part of the vertex.

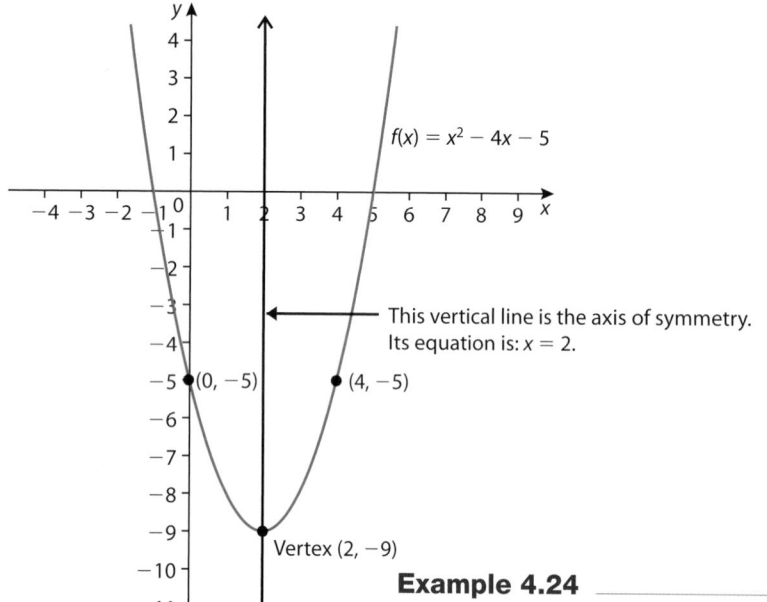

$f(x) = x^2 - 4x - 5$

This vertical line is the axis of symmetry. Its equation is: $x = 2$.

$(0, -5)$ $(4, -5)$

Vertex $(2, -9)$

Example 4.24

Find the equation for the axis of symmetry for $f(x) = (x - 3)^2 + 1$.

Solution

$$f(x) = (x - 3)^2 + 1$$
$$= x^2 - 6x + 9 + 1$$
$$\therefore f(x) = x^2 - 6x + 10$$

Since $x = \dfrac{-b}{2a} = \dfrac{-(-6)}{2 \cdot 1} = 3$, the equation for the axis of symmetry is: $x = 3$.

Example 4.25

If $f(x) = 2x^2 + 3x - 2$ and $(-3, 7)$ lies on the graph of f, but is not the vertex, what other point **must** lie on the graph of f?

Solution

Since f is a quadratic and hence an even function, there must be an ordered pair whose y-value is 7.

Therefore, $7 = 2x^2 + 3x - 2$
$$2x^2 + 3x - 9 = 0$$
$$(2x - 3)(x + 3) = 0$$
$$\therefore \quad 2x - 3 = 0 \Rightarrow x = \frac{3}{2} = 1.5$$

Therefore, the other ordered pair that **must** exist is $(1.5, 7)$.

Graphing quadratic functions using translations

The primary quadratic function is: $f(x) = x^2$. The graph is shown on the right.

Another way to graph a quadratic function is to compare the given function to $f(x) = x^2$ and then make the necessary horizontal and/or vertical translations.

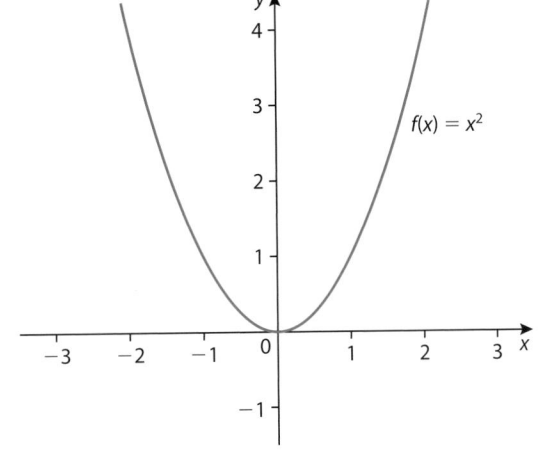

To translate a point is to change its coordinates by moving either horizontally or vertically or both.

- We will call a horizontal translation a 'horizontal slide'.
- We will call a vertical translation a 'vertical shift'.

Example 4.26

Draw the graph of $y = x^2 + 1$ by comparing it to $y = x^2$.

Solution

Since $y = x^2$, then you may correctly think of x^2 as the y-value.

Therefore $x^2 + 1$ can be thought of as 'one more than the y-value'.

Hence the y-value, for every ordered pair that lies on the graph of $y = x^2 + 1$, will be one more than the y-value for each ordered pair that lies on the graph of $y = x^2$. In other words, the entire graph of $y = x^2$ will be shifted vertically up one unit.

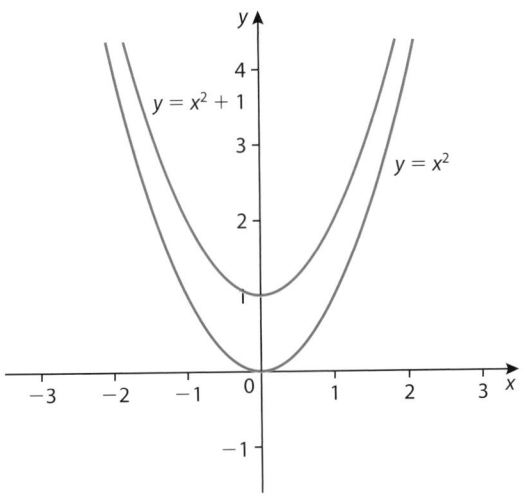

In Example 4.26, we say that $y = x^2 + 1$ has a vertical shift of 1 (when compared to $y = x^2$).

A horizontal slide occurs when the function looks like: $y = (x - 2)^2$ or $y = (x + 3)^2$.

- 'Subtracting 2' will slide the graph of $y = x^2$ two units to the **right.**
- 'Adding 3' will slide the graph of $y = x^2$ three units to the **left.**

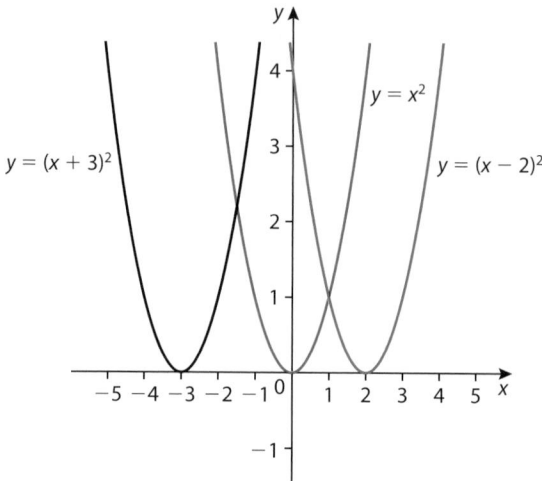

Example 4.27

For each of the following functions, give the horizontal slide and the vertical shift when compared to $y = x^2$.

a) $y = x^2 - 3$ b) $y = (x + 5)^2$ c) $y = (x - 1)^2 + 4$

Solution

a) Horizontal slide: None.   Vertical shift: 3 units down.
b) Horizontal slide: 5 units to the left.   Vertical shift: None.
c) Horizontal slide: 1 unit to the right.   Vertical shift: 4 units up.

Example 4.28

Draw the graph of $y = (x - 1)^2 + 4$ by comparing it to $y = x^2$.

Solution

Every ordered pair on the graph of $f(x) = x^2$ will be translated horizontally 1 unit to the right and vertically 4 units up. Therefore, choose at least three ordered pairs that lie on the graph of $y = x^2$ and translate them accordingly.

$(0, 0)$ will translate to $(0 + 1, 0 + 4) = (1, 4)$.
$(1, 1)$ will translate to $(1 + 1, 1 + 4) = (2, 5)$.
$(-1, 1)$ will translate to $(-1 + 1, 1 + 4) = (0, 5)$

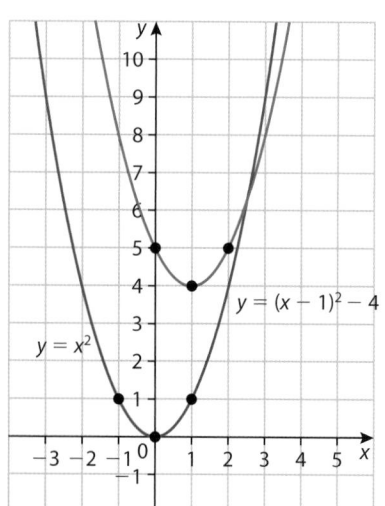

Graphing quadratic functions that are stretched

A 'stretch' occurs when the x^2 term is multiplied by a constant.

- $y = 3x^2$ is stretched by a factor of 3 when compared to $y = x^2$.
- $y = \frac{1}{2}x^2$ is stretched by a factor of $\frac{1}{2}$ when compared to $y = x^2$.
- $y = -x^2$ is stretched by a factor of -1 when compared to $y = x^2$.
- $y = -5x^2$ is stretched by a factor of -5 when compared to $y = x^2$.
- $y = -\frac{1}{4}x^2$ is stretched by a factor of $-\frac{1}{4}$ when compared to $y = x^2$.

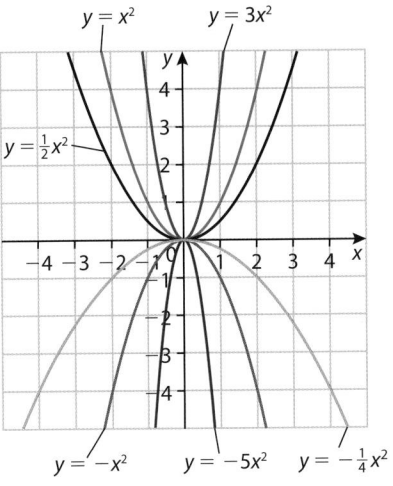

When the coefficient of x^2 is greater than 1 (or less than -1), the resulting graph will be narrower than the graph of $y = x^2$ (or $y = -x^2$).

When the coefficient of x^2 is between 0 and 1 (or 0 and -1), the resulting graph will be wider than the graph of $y = x^2$ (or $y = -x^2$).

Solving quadratic equations using accurately drawn graphs

There are several methods that can be used to solve quadratic equations. Three of these methods were studied in Chapter 2. They were:

- using the quadratic formula
- factorization
- using a GDC.

Another method that can be used to approximate the solutions to a quadratic equation involves using an accurately drawn graph that is associated with the equation.

Example 4.29

By drawing an accurate graph, approximate the solutions for $x^2 + x - 3 = 1$ correct to the nearest tenth.

Solution

Think of two functions, one being $y = x^2 + x - 3$ and the other being $y = 1$.

Step 1: Carefully draw the graph for both functions on the same 2-mm graph paper.

Step 2: Draw a vertical line from each point of intersection to the x-axis.

Step 3: Read the approximate answer for x where the vertical lines touch the x-axis.

- Therefore, $x = -2.6$ or $x = 1.6$ to the nearest tenth.

• **Examiner's hint:** Drawing lines on your graph is an excellent way to show your work and to score points on Paper 2.

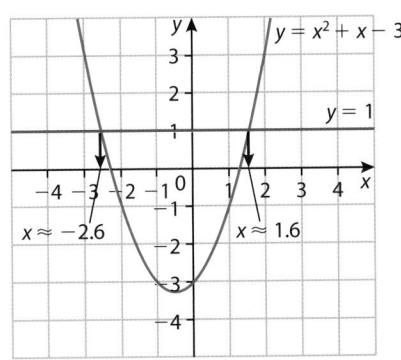

A GDC can be used to check the above solution.

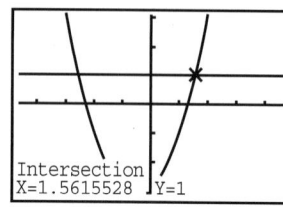

Intersection
X=1.5615528 Y=1

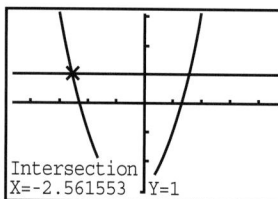

Intersection
X=-2.561553 Y=1

Example 4.30

Solve $x^2 + x - 3 = 1$ by using a GDC.

Solution

Use the following keystrokes:
- Y=
- $Y_1 = x^2 + x - 3$
- $Y_2 = 1$
- ZOOM 6 (or ZOOM 4)
- CALC (2ND TRACE)
- 5 : intersect
- $\triangleright, \triangleright, \triangleright$ (until you get close to the intersection point)
- ENTER, ENTER, ENTER
- $\therefore x = 1.56$ to 3 significant figures.

Repeat the above sequence of keystrokes and use the arrow keypad to get close to the other intersection point.
- $\therefore x = -2.56$ to 3 significant figures.

Exercise 4.6

1. Sketch the general shape of $y = ax^2 + bx + c$ where:
   a) $a > 0$ b) $a < 0$.

2. At least how many points are needed to draw a fairly accurate graph of a quadratic function?

3. What shape will a parabola have if it has a maximum value?

4. Why can't a quadratic function have this shape?

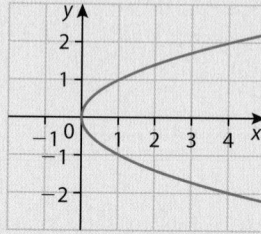

5. Sketch a parabola that does not have any x-intercepts.

6. Sketch a quadratic function that turns downward and has only one x-intercept.

7. Sketch a quadratic function that has two zeros.

8. Sketch a quadratic function that turns up, has only one zero and passes through $(0, 2)$. Draw the axis of symmetry.

9. Sketch a quadratic function that turns down, has two zeros at -3 and 5, and a y-intercept at 6.

10. Sketch a quadratic function that turns up, has two zeros at 1 and 5, and a y-intercept at 4.

11. Sketch a quadratic function that has all of the following characteristics:
    - has two zeros at -2 and 3
    - has the y-intercept at -6
    - shows the axis of symmetry as a solid line
    - has a vertex at approximately $(0.5, -6)$.

12. Sketch a quadratic function that has all of the following three characteristics:
    - has two x-intercepts at 0 and 5

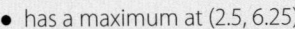

- has a maximum at (2.5, 6.25)
- shows the axis of symmetry as a solid line.

Find each of the following for Questions 13 and 14:

(i) the x-intercept
(ii) the y-intercept
(iii) the axis of symmetry
(iv) the vertex
(v) one extra ordered pair on each side of the axis of symmetry.

13. a) $f(x) = x^2 + 6x - 7$
b) $g(x) = 2x^2 - 2x + 15$
c) $y = 2x^2 - 15x + 18$
d) $y = -3x^2 - 2x + 8$

14. a) $y = (x - 1)^2 - 3$
b) $y = -2(x - 2)^2 + 2$
c) $f(x) = \frac{1}{2}(x + 3)^2 - 4$
d) $g(x) = \frac{-1}{3}(2x + 4)^2 + 1$

15. Using 2-mm graph paper and using the scale 2 units = 1 cm on both axes, draw the graph of each function in Question 13.

16. Using 2-mm graph paper and using the scale 1 unit = 1 cm on both axes, draw the graph of each function in Question 14.

17. If $g(x) = x^2 - 6x + 8$ and (1, 3) lies on the graph of g, but is not the vertex, what other ordered pair **must** lie on the graph?

18. If $f(x) = x^2 - 4x - 5$ and $(-2, 7)$ lies on the graph of f, but is not the vertex, what other ordered pair **must** lie on the graph?

19. For each of the following problems, first draw the graph of $y = x^2$ and then, on the same coordinate plane, draw the graph of the given function under the three conditions:

(i) Use the scale of 1 unit = 1 cm.
(ii) Draw the given quadratic function by using translations and/or stretches.
(iii) Show at least three points on the translated graph.

a) $y = x^2 + 1$
b) $y = x^2 - 3$
c) $y = (x - 1)^2$
d) $y = (x + 2)^2$
e) $y = (x - 3)^2 - 1$
f) $y = (x + 1)^2 + 3$
g) $y = -x^2$
h) $y = -2x^2$
i) $y = -\frac{1}{2}(x - 4)^2$
j) $y = -(x + 3)^2 - 1$

20. Draw an accurate graph of each function. Use 2-mm graph paper and the scale 1 unit = 1 cm on both axes. Locate all relevant information and plot at least five significant ordered pairs.

a) $y = \frac{1}{2}x^2 - x - \frac{5}{2}$
b) $y = -x^2 - 3x + 4$
c) $y = x^2 + 2x - 5$
d) $y = x^2 + 3x + 7$
e) $y = -x^2 + 4x - 7$
f) $y = x^2 - 4x + 4$

21. Using the method described in Example 4.29, solve each of the following using the graphs from Question 20.

a) $\frac{1}{2}x^2 - x - \frac{5}{2} = 5$

b) $-x^2 - 3x + 4 = 6$
c) $x^2 + 2x - 5 = -2$
d) $x^2 + 3x + 7 = 6$
e) $-x^2 + 4x - 7 = -3.5$
f) $x^2 - 4x + 4 = 8.4$

22. Using the method described in Example 4.30, solve each equation in Question 21 with your GDC.

23. The diagram shows a graph of the form $y = ax^2 + 12x + c$
The axis of symmetry is given as $x = -3$.
a) Find the value of a.

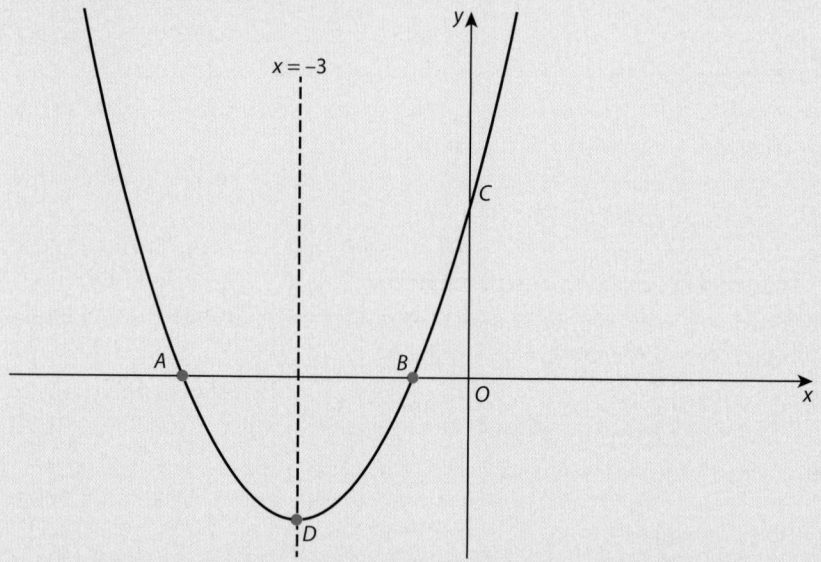

Given that the coordinates of the point D are $(-3, -8)$:
b) Determine the value of c, and write down the equation of the curve
c) Write the coordinates of the points A and B.

24.   A quadratic function of the form $y = ax^2 + bx + c$ is given below.

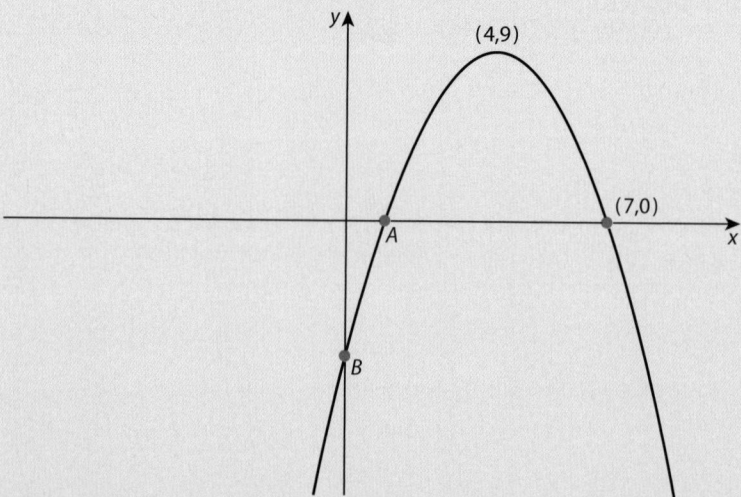

a) Find the coordinates of the point A.
b) Find the coordinates of the point B.
c) Write down the equation of the parabola.

4.7 Quadratic models

6.5 Quadratic models. Quadratic functions and their graphs (parabolas):
$f(x) = ax^2 + bx + c; a \neq 0$
Properties of a parabola: symmetry; vertex; intercepts on the x-axis and
y-axis. Equation of the axis of symmetry, $x = \frac{-b}{2a}$.

Some real-world phenomena that are associated with parabolas are: cables
that hold up suspension bridges, kicking a football, throwing a baseball,
dropping a rock from the top of a building, firing a pellet from a pellet gun,
making a reflective surface for a flashlight, and making a satellite disk to
receive a signal from space. As well, the shapes of spans of some bridges such
as the Sydney Harbour Bridge have a parabolic shape.

Example 4.31

A suspension bridge used by pedestrians and cyclists is shown below. The
bridge has a cable connecting the two pylons (uprights) together. The
shape of the cable is that of a parabola.

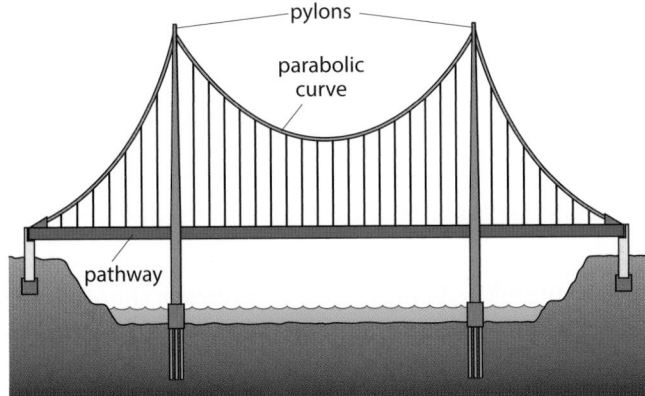

An equation for the cable is given by:

$f(x) = 0.01x^2 + 50$

with the minimum point on the curve having the coordinates $(0, 50)$. This
is at the lowest point of the curve above the footpath on the bridge.

Using your graphing display calculator.

a)   Sketch the graph of the curve.

b)   **Hence** use your graph to determine:

   i)   The minimum height of the cable above the path.

   ii) The height of the pylons above the path, given that the distance
        between the pylons is 200 metres.

Solution

a) Enter $Y1 = 0.01x^2 + 50$ into your calculator.

Set the window as shown.

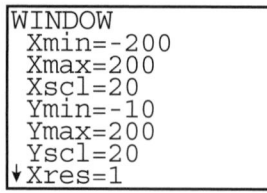

Then press GRAPH.

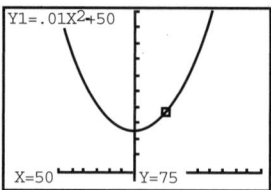

b) i) Use the following keystrokes to find the minimum height

CALC (2ND TRACE) 3:Minimum,

Choose a point to the left of the minimum, ENTER

Choose a point to the right of the minimum ENTER

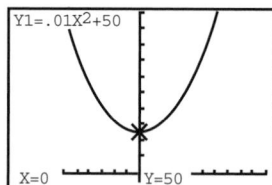

The minimum height of the cable above the path is 50 metres.

ii) To determine the height of the pylons, we can either use the graph or the table. As the graph is symmetric about the origin the distance of one pylon from the minimum is 100 metres.

Choose TABLE (2ND GRAPH) and scroll down the table to 100.

X	Y1	
60	86	
70	99	
80	114	
90	131	
100	150	
110	171	
120	194	
X=100		

CALC (2ND TRACE) 1:VALUE and Type in 100 for X = . The result will be given as Y = 150.

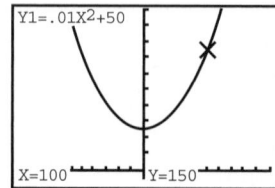

This shows that the height of the pylons above the path is 150 metres.

Example 4.32

A well-known formula from physics for describing the height H, in metres, of an object thrown upward with an initial velocity V, in metres per second $(m\,s^{-1})$, for T number of seconds, and from a starting height B, in metres, is given by:

$$H = -4.9T^2 + VT + B$$

a) Find the height that a ball will reach if it is thrown vertically upward, if the starting height is 2 m, the initial velocity is $30\,m\,s^{-1}$ and the ball stays in the air 1 second.

b) A toy rocket is shot vertically into the air, from ground level, with an initial velocity of $29.4\,m\,s^{-1}$. How many seconds will it take to reach its maximum height?

Solution

a) In the function $H = -4.9T^2 + VT + B$, let $V = 30$ and $B = 2$.

Therefore, $H = -4.9T^2 + 30T + 2$.

Hence, find H when $T = 3$: $H = -4.9(1)^2 + 30(1) + 2 = 27.1$ m.

b) In the function $H = -4.9T^2 + VT + B$, let $V = 29.4$ and $B = 0$ since the rocket left at ground level.

$$H = -4.9T^2 + 29.4T + 0 = -4.9T^2 + 29.4T$$

$0 = -4.9T^2 + 29.4T$ (This is true since H will be 0 again when the toy rocket hits the ground.)

$$0 = T(-4.9T + 29.4)$$

$$0 = T \text{ or } 0 = -4.9T + 29.4$$

Hence, $4.9T = 29.4$

Therefore, $T = \dfrac{29.4}{4.9} = 6$ seconds. (This is the total time for the rocket to go up and come back down.)

Therefore, the time to reach the maximum height is $\dfrac{6}{2} = 3$ seconds.

> (i) This is *not* the shortest solution. You could find the vertex and give the abscissa as the answer, or use any of the other methods described in the section.

Exercise 4.7

1. The owner of a bus company has been studying the number of passengers travelling on the bus each day. He has found a model for the relationship between the number of passengers and the time of day to be:

   $f(t) = 3.5t^2 - 100t + 800$ where t is the time of day and $0 \leqslant t \leqslant 24$.

   Graph the function and answer the following questions.

   a) At what time of the day does the bus company carry the least number of passengers?

   b) What is the least number of passengers at that time.

   c) Given that the buses only run between 6 am and 8 pm, when does the bus company carry the most passengers?

   d) If a bus can carry 60 passengers, sitting and standing, how many buses will be required at 8 am, and at 5 pm?

2. A scientist has been studying the growth of a population of rare marsupials in Australia. The scientist found that a possible model for the number (N) of marsupials over time in months is given by the formula

$N(t) = 50 + 4t + 0.2t^2$, where t is the time in months.

a) How many marsupials were there when the scientist began counting?

b) After 12 months what was the population of marsupials?

c) What increase had there been in 2 years?

d) It has been suggested that once the population reaches 500 marsupials, there will not be enough food for them to survive. After how many months will there be 500 marsupials?

3. A rectangular dog pen was built against the back wall of a house as shown below. The total amount of fencing used was 200 feet.

a) Write a linear equation describing the length, L, in terms of W.

b) Write a quadratic equation describing the area, A, in terms of W.

c) Find the width of the pen which will maximize the amount of area.

d) Find the maximum amount of area that can be fenced in.

4. Chris shoots a ball vertically upward with an initial velocity of 40 m s^{-1} from a 10 m building.

Give all answers correct to 1 decimal place. (Hint: Use your GDC.)

a) Sketch a diagram that describes the information given.

b) Find the time it will take to reach the maximum height.

c) Find the maximum height the ball will reach.

d) Find the time it will take the ball to reach the ground.

5. The gigantic pottery company makes vases in all sizes. The cost in dollars, C, to make vases is given by $C(x) = 2000 + 0.1x^2$, where x is the number of vases sold. The selling price for **one** vase, S dollars, is $S(x) = 200x - 0.25x^2$.

a) Write down the profit (P(x)) equation, that is $P(x) = S(x) - C(x)$.

b) Graph the profit function for $0 \leqslant x \leqslant 600$.

c) Write down the number of vases to be sold for maximum profit, and the profit.

d) Write down the point where the cost of manufacture will be greater than income from sales of vases.

6. It has been found that the formula for the stopping distance, C(v) of a car on a dry road in normal conditions is given by

$C(v) = 0.000208v^2 + 0.259v - 0.0476$

where v is the speed in kilometres per hour (km/h) and stopping distance C(v) is in metres.

a) Draw a graph of this function.

b) Write down the stopping distance when the car is travelling at 60 km/hr.

c) What is the maximum speed a driver can be travelling at if they see a traffic light change to red 30 metres in front of them?

It is apparent, however, that the stopping distance of a car is also related to the reaction time of the driver. It has been found that the reaction distance for the driver is given by

$R(v) = 0.00610v^2 - 0.0161v + 0.238$ where v is speed in km/hr.

d) Graph $R(v)$ on the same graph as $C(v)$, and describe the differences between the two curves.

e) Write down the reaction distance when the car is travelling at 60 km/hr.

The total stopping distance $T(v)$ of a car is a combination of the driver's reaction distance $R(v)$ and the car's stopping distance $C(v)$.

f) Write the new equation for $T(v)$.

g) Graph $T(v)$.

h) Write down the total stopping distance when the car is travelling at 60 km/hr.

4.8 Exponential functions

6.4 | Exponential models.
Exponential functions and their graphs: $f(x) = ka^x + c$; $a \in \mathbb{Q}^+$, $a \neq 1$, $k \neq 0$. $f(x) = ka^{-x} + c$; $a \in \mathbb{Q}^+$, $a \neq 1$, $k \neq 0$.
Concept and equation of a horizontal asymptote.

You might have heard the expression: 'The population is increasing exponentially'. In layman's terms, it means that the population is getting larger by larger amounts. The objective in this section is to explain the concept in mathematical terms.

Exponential expressions

An exponential expression is of the form a^b, where $a^b \in \mathbb{R}$, $a \neq -1, 0, 1$, and $b \in \mathbb{Q}$.

The following are examples of exponential expressions:

2^3

$\left(\dfrac{1}{2}\right)^4$

$(\sqrt{3})^{-2}$

$5^{\left(\frac{2}{3}\right)}$

The following examples are **not** exponential expressions:

1^5

$(-4)^{\frac{1}{2}}$, since $(-4)^{\frac{1}{2}} \notin \mathbb{R}$

$0^{2.3}$

An important algebraic law of exponents is: $a^{\frac{m}{n}} = \sqrt[n]{a^m}$, where $\sqrt[n]{a^m} \in \mathbb{R}$.

For example: $5^{\frac{2}{3}} = \sqrt[3]{5^2}$ and $3^{\frac{-1}{2}} = \sqrt[2]{3^{-1}}$.

 Even though $(-8)^{\frac{1}{3}} \in \mathbb{R}$, since $(-8)^{\frac{1}{3}} = \sqrt[3]{(-8)^1} = -2$, $(-8)^{\frac{1}{2}} \notin \mathbb{R}$, since $(-8)^{\frac{1}{2}} = \sqrt[2]{(-8)^1} \notin \mathbb{R}$.

$\sqrt[2]{-4096} = \sqrt[2]{-1 \times 64 \times 64}$
$= 64\sqrt[2]{-1} = 64i. \ 64i \in$
complex number system, but
not the real number system.

To find out more about
real numbers, visit www.
pearsonhotlinks.co.uk, enter
the ISBN for this book and click
on weblink 4.6.

Example 4.33

State whether each exponential expression is a real number. If it is, simplify
it.

a) $4^{\left(\frac{-3}{2}\right)}$

b) $(-16)^{\frac{3}{2}}$

Solution

a) $4^{\left(\frac{-3}{2}\right)}$ is a real number. $\therefore \ 4^{\left(\frac{-3}{2}\right)} = \sqrt[2]{4^{-3}} = \sqrt[2]{\frac{1}{4^3}} = \sqrt[2]{\frac{1}{64}} = \frac{1}{8}$.

b) $(-16)^{\frac{3}{2}}$ is not a real number since $\sqrt[2]{(-16)^3} = \sqrt[2]{-4096}$ and there is no
real number that can be multiplied by itself in order to get -4096.

Exponential functions

An exponential function is a set of ordered pairs in which the independent
variable is in the **exponent** position. $f(x) = 2^x$ is an example of such a
function.

There are two general shapes for the graphs of exponential functions.

One shape is when the y-value increases as the x-value increases from
left to right, as in Figure 4.4.

The other shape is when the
y-value decreases as the x-value
increases from left to right, as in
Figure 4.5.

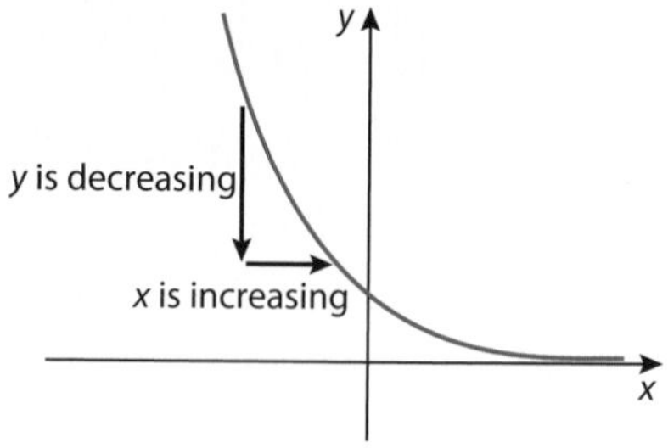

Figure 4.4 Increasing
exponentially.

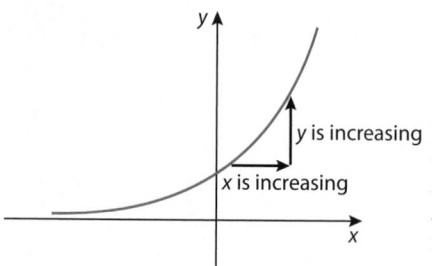

The graph above is said to
be increasing or to be an
increasing function.

The graph on the right is said
to be decreasing or to be a
decreasing function.

Figure 4.5 Decreasing exponentially.

In both graphs above, the x-axis acts as a boundary line for the curve.
The boundary line is a line that the curve approaches but never touches or
crosses.

The name of the boundary line is 'asymptote'. (The 'p' is silent in the
pronunciation.) Another way to describe the curve is to say that it is
asymptotic to the x-axis.

There are many ways to find the equation of an asymptote:
- graph the function and observe whether the graph approaches a line
- put your TI calculator into 'Ask' mode (see Example 4.8) and observe
  the y-values for larger and larger x-values in either the positive
  direction or in the negative direction
- observe patterns for specific functions that have asymptotes
- use techniques described in Chapter 5
- use calculus techniques (these are beyond the scope of this course).

Example 4.34

Find the equation of the horizontal asymptote for $y = 2^x + 3$.

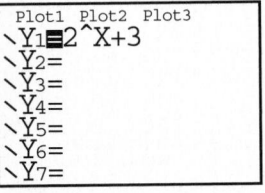

Solution

With your GDC: Y =

        $Y_1 = 2^x + 3$

        TBLSET

        Indpnt: Ask

        TABLE

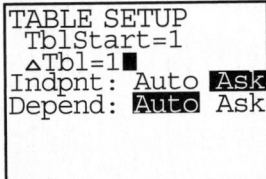

Now enter the following x-values one at a time: $-1, -2, -4, -6, -8, -10,$ -12

As you can see in the diagram below, the Y_1-values seem to be approaching 3.

This is enough information to write the equation of the horizontal asymptote as: $y = 3$.

The graph in Figure 4.6 supports the answer for the worked example.

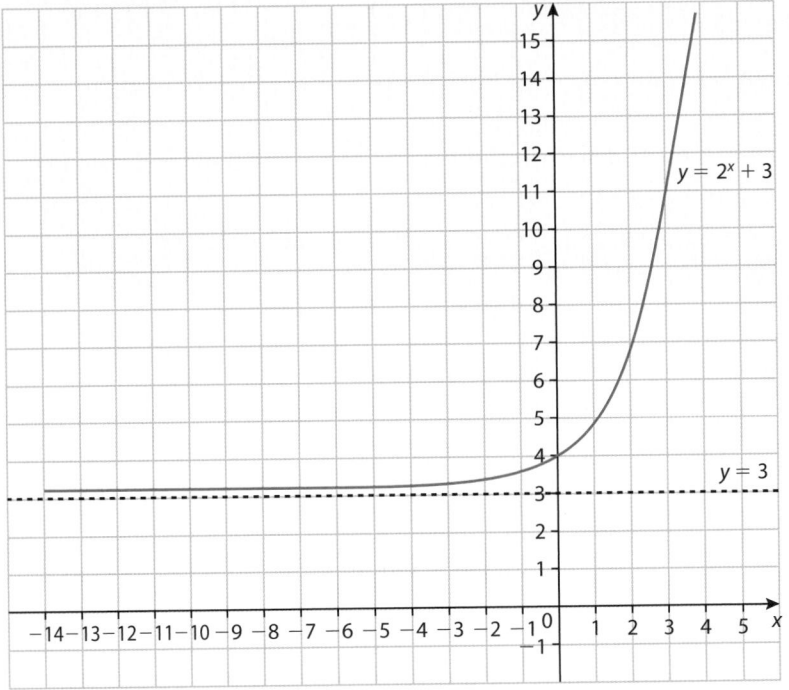

◀ **Figure 4.6** The graph for Example 4.34.

The symbol λ (lamda) is a Greek lower case letter.

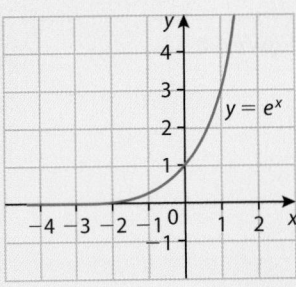
Arguably the most important exponential graph is $y = e^x$. The constant e is like π in that it is irrational. $e = 2.71828$ to 5 decimal places.

Graphing exponential functions

$f(x) = a^x$ is a general form of an exponential function, where $a > 0$, $a \neq 1$.

- $f(x) = 2^x$ is one example. See graph below.
- A function in this form is called the primary function, against which other functions are compared when using translations.

$f(x) = a^{\lambda x}$, where $\lambda \in \mathbb{Q}$, is one form of a function that can be compared to the primary function.

- $f(x) = 2^{3x}$ is one example.

• When λ is positive, then the exponential function is called a 'growth function' and is similar in nature to the primary graph.

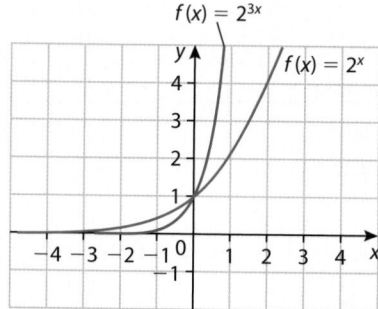

For example $f(x) = 2^{3x}$ is similar to the primary graph.

• Both graphs pass through $(0, 1)$.
• Both graphs are increasing.
• Both graphs have $y = 0$ as the horizontal asymptote.

The graph of $f(x) = 2^{3x}$ increases at a faster rate than $f(x) = 2^x$.

When λ is negative, then the exponential function is called a 'decay function'.

• One of its primary functions is $f(x) = 2^{-x}$. Its graph is shown in Figure 4.7.

• An example of a graph that is similar in nature to $f(x) = 2^{-x}$ is $f(x) = 2^{-1.5x}$.

Figure 4.7

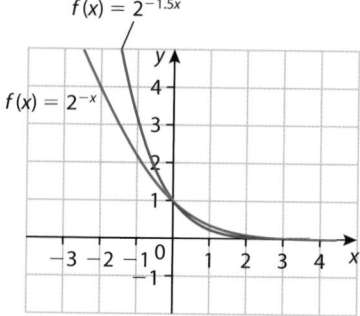

In Figure 4.8, a function of the form $f(x) = ka^{\lambda x}$, where $k, a, \lambda \in \mathbb{Q}$, represents a graph that is similar to the primary graph, $f(x) = a^x$, in that k 'stretches' the graph much like the coefficient of the term for quadratic functions (see Section 4.6).

Figure 4.8

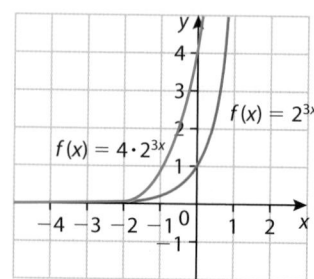

A function of the form $f(x) = ka^{\lambda x} + c$, where $k, a, \lambda, c, \in \mathbb{Q}$, represents a graph that is stretched and shifted vertically with respect to the primary graph, $f(x) = a^x$.

For example, $f(x) = 4 \cdot 3^{-x} - 1$ is a graph that has been stretched by a factor of 4 and has a vertical shift of -1 when compared to the primary graph $f(x) = 3^{-x}$. See Figure 4.9.

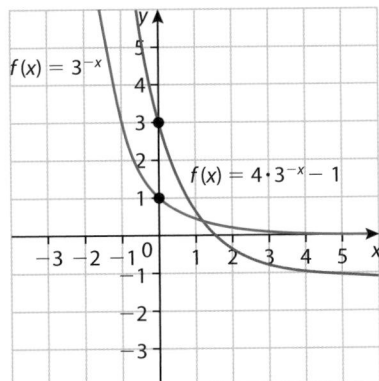

◀ Figure 4.9

From the ordered pair $(0, 1)$ on $f(x) = 3^{-x}$, the y-value, 1, was first multiplied by 4, putting a point at $(0, 4)$. 1 was then subtracted from 4 leaving an ordered pair at $(0, 3)$.

And so every ordered pair on $f(x) = 3^{-x}$ would be translated in the same manner.

The net result is the graph of the function $f(x) = 4 \cdot 3^{-x} - 1$.

Now that you know the basic shape of an exponential function, you can draw its graph using one of several methods:
- by point-plotting until the shape becomes evident
- using your GDC and a table of values
- by using knowledge of a primary graph and identifying stretches and shifts.

Example 4.35

Draw the graph of $f(x) = 2 \cdot 3^{(0.5x)} + 1$.

Solution
We know several bits of information:
- The horizontal asymptote is $y = 1$, since the vertical shift is 1 unit up.
- The graph is similar in shape to the primary graph $f(x) = 3^x$.
- It is increasing, since $\lambda = 0.5 > 0$.
- It is stretched by a factor of 2.

With the calculator in 'Ask' mode, the following ordered pairs have been produced:

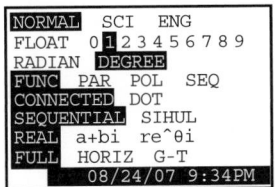

| NORMAL SCI ENG |
| FLOAT 0 **1** 2 3 4 5 6 7 8 9 |
| RADIAN **DEGREE** |
| FUNC PAR POL SEQ |
| CONNECTED DOT |
| SEQUENTIAL SIHUL |
| REAL a+bi re^θi |
| FULL HORIZ G-T |
| 08/24/07 9:34PM |

| TABLE SETUP |
| TblStart=1 |
| ΔTbl=1■ |
| Indpnt: Auto **Ask** |
| Depend: **Auto** Ask |

X	Y₁	
-4.0	1.2	
-2.0	1.7	
0.0	3.0	
1.0	4.5	
2.0	7.0	
3.0	11.4	

X=

The completed graph is shown below.

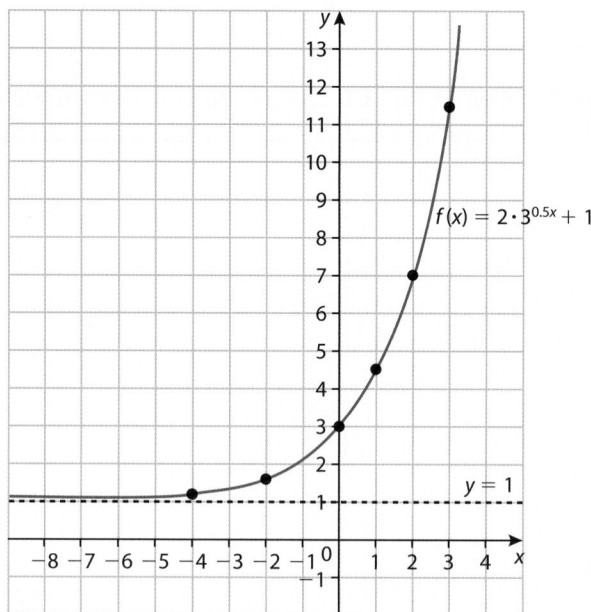

Example 4.36

By drawing an accurate graph, approximate the solution for $2^x = 6$ to the nearest tenth.

Solution

Think of two functions: $y = 2^x$ and $y = 6$.

Now, carefully graph both functions on the same set of axes.

Step 1: Draw a vertical line from the point of intersection to the x-axis.

Step 2: Read the answer where the vertical line touches the axis. Therefore, $x = 2.6$ to the nearest tenth.

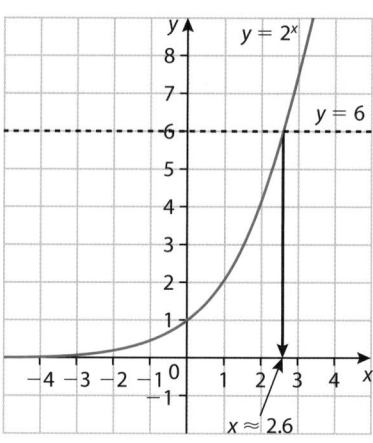

• **Examiner's hint:** Even though you don't need to know how to use logarithms, they can make finding the solutions to exponential equations more efficient.

The 'Guess and Check' method and logarithms

The Guess and Check method is an excellent way to solve a variety of problems. Almost all great science and mathematical discoveries have come by way of taking a guess and then verifying that guess.

Example 4.37

By using the Guess and Check method, solve $2^x = 7$, correct to 2 decimal places.

Step 1: Draw a 'T' chart.

Guess | Check

Step 2: Take a guess and write it in the 'Guess' column.
Step 3: Check your guess by substituting that value for x in $2^x = 7$ and write the answer in the 'Check' column.
Step 4: If the answer is too small, write 'S' by the number; if it is too large, write 'L' by it.
Step 5: Choose the next 'Guess' number between the last guess that was too small ('S') and the next guess that was too large ('L') or vice versa.
Step 6: Continue this process until the guess is correct to 2 decimal places.

Solution

Guess	Check	
0	$2^0 = 1$	S
1	$2^1 = 2$	S
2	$2^2 = 4$	**S**
3	$2^3 = 8$	**L**
2.5	$2^{2.5} \approx 5.6569$	S
2.6	$2^{2.6} \approx 6.0629$	S
2.7	$2^{2.7} \approx 6.4980$	S
2.8	6.9644	**S**
2.9	7.4643	**L**
2.85	7.2100	L
2.83	7.1107	L
2.81	7.0128	**L**
2.80	6.9644	**S**
2.807	6.9983	**S**
2.808	7.0031	**L**

(Note: If 0 was the answer, then 2^0 would be 7, but it isn't; therefore 0 is not the answer!)

(Note: 2.5 is halfway between 2 and 3. A calculator was used to approximate $2^{2.5}$.)
(Note: '$\approx$' is read as 'approximately equal to'.)

(Note: The better 'guesser' you are the fewer times you will have to guess.)

Instead of always retyping, for example, 2,^,2.7, use the keystrokes 2ND ENTER. This accesses the ENTRY key which returns the last equation entered. That equation can now be edited.

Since the too small 'S' number is 2.807 and the next too large 'L' number is 2.808, we can deduce that the solution to the equation $2^x = 7$ must be 2.81 correct to 2 decimal places.

For a detailed explanation of why the logarithm method works, ask your teacher or visit www.pearsonhotlinks.co.uk/ enter the ISBN for this book and click on weblink 4.7 or 4.8.

Although the above 'Guess and Check' solution produces the correct answer, it is somewhat laborious.

The logarithm method described below is much faster and you can learn it quickly.

Example 4.38

Solve the equation $2^x = 7$ using logarithms.

Solution

Step 1: $2^x = 7$ Write the equation.
Step 2: $\log 2^x = \log 7$ 'log' both sides. ('log' is short for logarithm.)
Step 3: $x \cdot \log 2 = \log 7$ 'bring x down in front'. (x is now a factor with log 7.)

On your GDC press: LOG, 2,),
ENTER.
You should see 0.301 029 9957.
This is the number, or at least
the approximate number,
that you must raise 10 to in
order to get 2. In other words,
$10^{0.301\,029\,9957} = 2$.

Make sure to enclose the 7 in
parentheses before dividing
by log 2.

Step 4: $x = \dfrac{\log 7}{\log 2}$ Divide both sides by log 2. (Log 2 is a real number.)

$$\left(\text{Note: } \dfrac{\log 7}{\log 2} \text{ is the exact answer.}\right)$$

Step 5: $x = \dfrac{\log 7}{\log 2} \approx 2.807 = 2.81$ correct to 2 decimal places.

Example 4.39

Solve the equation $2 \cdot 3^x - 1 = 29$ using logarithms. Find the exact answer
and the approximate answer correct to 2 decimal places.

Solution

$$2 \cdot 3^x - 1 = 29$$
$$2 \cdot 3^x = 30$$
$$3^x = 15$$
$$\log 3^x = \log 15$$
$$x \cdot \log 3^x = \log 15$$
$$\therefore x = \dfrac{\log 15}{\log 3} \text{ is the exact answer.}$$

Hence, $x = 2.46$ correct to 2 decimal places.

Exercise 4.8

1. Which of the following is not an exponential expression?

a) $3^{\frac{1}{2}}$ b) 1^6 c) $(-2)^2$ d) $(-25)^{\frac{1}{2}}$ e) $(-8)^{\frac{1}{8}}$

f) $\pi^{\frac{1}{2}}$ g) $(-36)^{\frac{1}{4}}$ h) 0^2 i) $\left(\frac{3}{4}\right)^{-2}$ j) $(\sqrt{3})^{\pi}$

k) $5^{\sqrt{2}}$ l) $3^{1.5}$ m) $(-1)^2$ n) $(0.2)^{\frac{5}{2}}$ o) $(\sqrt{2})^3$

2. Rewrite each exponential expression as an expression involving radicals.

a) $3^{\frac{1}{2}}$ b) $5^{\frac{-1}{3}}$ c) $7^{\frac{2}{3}}$

d) $(-5)^{\frac{2}{5}}$ e) $\left(\frac{2}{3}\right)^{-\frac{1}{2}}$ f) $\left(\frac{3}{4}\right)^{-\frac{2}{3}}$

3. Rewrite each radical expression as an exponential expression.

a) $\sqrt{3}$ b) $\sqrt[3]{5^2}$ c) $\sqrt[3]{-29}$

d) $\sqrt{\frac{3}{4}}$ e) $\sqrt[4]{\frac{1}{2}}$ f) $\sqrt[5]{(-13)^3}$

4. Simplify each of the following. If it is not a real number, say so.

a) $16^{\frac{3}{4}}$ b) 2^{-1} c) $(25)^{\frac{3}{2}}$

d) $\left(\frac{3}{4}\right)^{-2}$ e) $(-4)^{\frac{1}{2}}$ f) $(-27)^{\frac{2}{3}}$

5. Sketch a single exponential function that has the following three characteristics:
a) passes through (0, 2)
b) has $y = 0$ as a line of asymptote
c) is increasing.

6. Sketch a single exponential function that has the following three characteristics:

a) passes through (0, 3)

b) has $y = 1$ as a line of asymptote

c) is decreasing.

In Questions 7 and 8, state whether the graph is increasing or decreasing and give a reason why.

7.

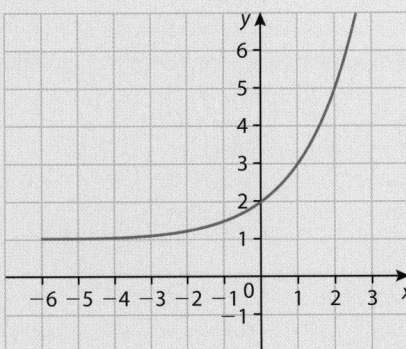

8.

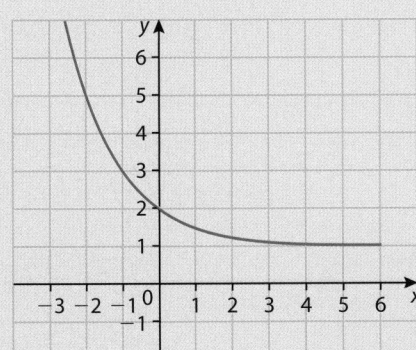

In Questions 9–12, find the equation of the horizontal asymptote.

9. $y = 2^x + 3$

10. $y = 3^{-x} - 1$

11. $f(x) = \left(\frac{1}{2}\right)^x + 5$

12. $g(x) = 2 \cdot \left(\frac{1}{3}\right)^{-2x} - 2.5$

In Questions 13 and 14, state the intervals (x-value intervals) for which the graph is increasing and for which it is decreasing.

13.

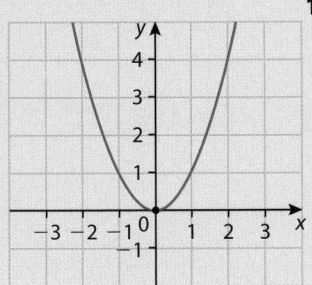

14.

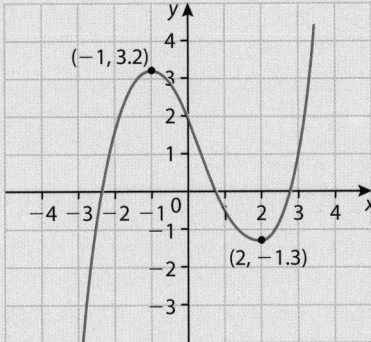

15. For each of the following problems, first draw the primary graph (e.g. $y = 2^x$) and then, on the same coordinate plane, draw the graph of the given function under these conditions:

(i) Use the scale of 1 unit = 1 cm.

(ii) Draw the given exponential function by using translations or stretches.

   (iii) Show the line of asymptote as a dashed line.

   (iv) Plot the y-intercept and at least three other points on the translated graph.

a) $y = 2^x + 1$ b) $y = 2 \cdot 2^x - 1$

c) $f(x) = 2^{-x} - 3$ d) $f(x) = \left(\frac{1}{2}\right)^x + 3$ $\left(\text{Hint: } \frac{1}{2} = 2^{-1}\right)$

e) $g(x) = 2 \cdot 3^x$ (Hint: The primary graph is $y = 3^x$)   f) $y = 3^x - 2$

16. Draw an accurate graph of each exponential function showing all relevant information. Use the scale 1 unit = 1 cm for a-d.

a) $y = 5^x + 1$ b) $y = 2^{1.5x} + 3$

c) $f(x) = -2.7^x - 2$ d) $f(x) = -1.4^{2x} + 1$

e) $g(x) = 100(1.08)^x$ (Hint: Change the WINDOW on your GDC.
                              e.g. Ymax = 500. Use a scale of 20 units = 1 cm on the
                              y-axis and 2 units = 1 cm on the x-axis.)

f) (g) $x = 5000\left(1 + \frac{0.09}{12}\right)^{(12x)}$ (Hint: Change the WINDOW on your GDC.
                                     e.g. Ymax = 15000. Use a scale of 1 cm = 1000
                                     units on the y-axis and let 1 cm = 1 unit on
                                     the x-axis.)

17. Draw an accurate graph of $y_1 = 10^x$ and $y_2 = 10^{-x}$ on the same set of axes. Use the scale of 1 unit = 1 cm.

18. Draw an accurate graph of $f(x) = e^x$ and $g(x) = e^{-x}$ on the same set of axes. Use the scale of 1 unit = 1 cm.

19. Draw an accurate graph of $y = 2^{-x}$ under the conditions:

   (i)   1 unit = 2 cm

   (ii) Domain = $\{x|-3 \leqslant x \leqslant 1, x \in \mathbb{R}\}$

   (iii) TblStart = 0, ΔTbl = .1, Indpnt:Auto.

20. Draw an accurate graph of $y = 2^{-x}$ under the conditions:

   (i)   1 unit = 2 cm

   (ii) Domain $\{x|0 \leqslant x \leqslant 3, x \in \mathbb{R}\}$

   (iii) TblStart = 0, ΔTbl = .1, Indpnt:Auto

For Questions 21–24, solve each equation by graphing the two associated functions and drawing lines on your graph to the x-axis. Follow Example 4.29.

21. $3^{-x} - 1 = 5$ **22.** $2(1.07)^x = 4$

23. $2^x = 11$ **24.** $\left(\frac{1}{2}\right)^{0.5x} = 8$

For Questions 25–28, solve each equation using your GDC. Follow Example 4.30.

25. $3^{-x} - 1 = 5$ **26.** $2(1.07)^x = 4$

27. $2^x = 11$ **28.** $\left(\frac{1}{2}\right)^{0.5x} = 8$

For Questions 29–32, solve each equation, correct to 2 decimal places, using the 'Guess and Check' method. Follow Example 4.37.

29. $3^x = 17$ **30.** $1.05^x = 2$

31. $10^x = 313$ **32.** $1.12^x = 3$

For Questions 33–42, solve each equation, both exactly and correct to 3 significant figures, using the logarithm method. Follow Example 4.38.

33. $5^x = 37$ **34.** $17^x = 107$

35. $1.06^x = 2$ **36.** $3^x + 1 = 7$

37. $4^{-x} = 91$ **38.** $2^x + 7 = 27$

39. $3 \cdot 2^x - 2 = 19$ **40.** $4 \cdot 3^{-x} + 1 = 20$

41. $1000\left(1 + \frac{0.08}{12}\right)^{12x} = 5000$ **42.** $45000\left(1 - \frac{15}{100}\right)^x = 10000$

4.9 Exponential models

6.4   Exponential models.
Exponential functions and their graphs: $f(x) = ka^x + c$; $a \in \mathbb{Q}^+$, $a \neq 1$, $k \neq 0$. $f(x) = ka^{-x} + c$; $a \in \mathbb{Q}^+$, $a \neq 1$, $k \neq 0$.
Concept and equation of a horizontal asymptote.

There is a formula that is used to find the amount of interest earned when an amount of money is deposited into an account in which the interest is compounded yearly. It assumes that the deposit occurs a single time and no money was withdrawn. It is given as:

$$I = C\left(1 + \frac{r}{100}\right)^n - C$$

where, I = the amount of interest earned, C = the amount of money deposited, r = the annual rate of interest (e.g. 7.25%) and n = the number of years that the investment was left in the account.

Example 4.40

a) Write a function, in terms of n, for the amount of interest earned after leaving \$1000 in an account that pays 7.25% per year compounded yearly for n years.

b) Find the amount of interest earned after the \$1000 was left in the account for 30 years.

Solution

a) In this case, $I = C\left(1 + \frac{r}{100}\right)^n - C$ can be written as

$$f(n) = 1000\left(1 + \frac{7.25}{100}\right)^n - 1000$$

$$f(n) = 1000(1.0725)^n - 1000, \text{ where } I = f(n).$$

b) $f(30) = 1000(1.0725)^{30} - 1000 = \$7164.30$ to the nearest cent.

The general form of the above equation is $y = A(1 + r)^x$, where A is the initial amount, y is the final amount and x is the number of years.

- $f(x) = 1000(1 + 0.0725)^x$ is an example of a function when r is positive. This is called a **growth function**.

Example 4.41

Alison buys a new sports car for \$28 000. Suppose that the value of the car depreciates 15% each year.

a) Write a function, in terms of x, describing the depreciation.

b) Find the value of the car after 5 years.

Solution

a) $f(x) = 28\,000(1 - 0.15)^x \Rightarrow f(x) = 28\,000(0.85)^x$.

b) $f(x) = 28\,000(0.85)^5 = \$12\,423.75$ to the nearest cent.

- $f(x) = 28\,000(1 - 0.15)^x$ is an example of a function when r is negative. This is called a **decay function**.

Example 4.42

Two students were discussing how quickly their cups of coffee cooled.

A friend then suggested that an equation to represent the data was of the form,

$$T(t) = k \times 0.90^t$$

where T is temperature in degrees Celsius, t is the time, and k is a constant.

Given that the temperature at time zero was 85°C

a) Calculate the value of k.
b) Using the formula determine the temperature after 5 minutes of cooling.
c) When would the temperature be less than 40°C?

Another group of students measured the rate at which a cup of tea cooled and collected the following data, where t is the time in minutes and T is the temperature in degrees Celsius.

Time (t)	0	2	4	6	8	10	12	14	16	18	20
Temperature °C	90	81.2	73.3	66.2	59.7	53.9	48.6	43.9	39.6	35.7	32.3

Using this information draw the graph of T against t for $0 \leqslant t \leqslant 20$. Use a scale of 1 cm to represent 2 minutes on the horizontal axis and 1 cm to represent 10 degrees on the vertical axis.

d) Using your graph determine the temperature of the cup of tea after 15 minutes of cooling.

Solution

a) When time $t = 0$, and $T = 85°$
$$T(0) = k \times 0.90^0 = 85$$
$k = 85$, as $0.90^0 = 1$

b) With $k = 85$ and $t = 5$ then
$$T(5) = 85 \times 0.90^5 = 50.19°C$$

c) Graph the functions $T(t) = 85 \times 0.90^t$ and $T = 40$. Then determine the point of intersection, which is 7.15 minutes.

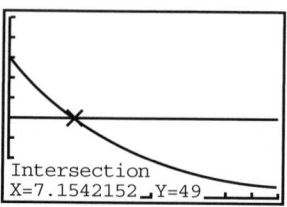

Intersection
X=7.1542152 Y=49

d) Graphing the data and drawing a smooth curve provides the resulting graph.

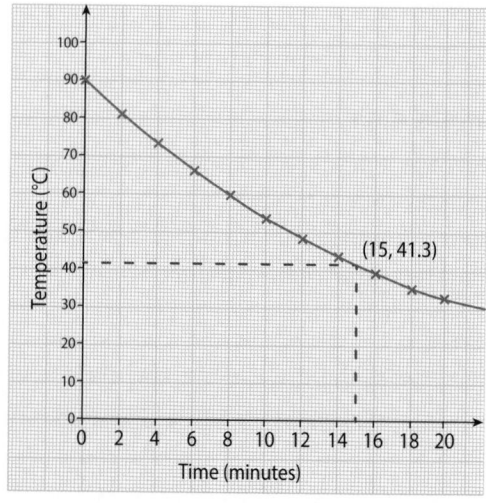

1. The population of a small town in 2000 was 200 and the increase in population can be described by the function:

   $p(t) = 200 \times 1.07^t$

   where $p(t)$ represents the number of persons living in the village after t years, and t is the year number after 2000 (e.g. for 2003, $t = 3$).

   a) Write down a table of values in the domain $0 \leqslant t \leqslant 5$.

   b) Draw a graph to show the growth of the population for the first 15 years. Use 1 cm to represent 1 year on the horizontal axis and 1 cm to represent 1 person on the vertical axis. (Hint: start the vertical axis at 200.)

   c) **Using your graph,** find approximately:

      (i)   when the population is 250;

      (ii) the population in 2010.

   d) In what year will the population exceed 1000?

2. Billy deposited 2000 euros in an account that pays 8% per year compounded yearly for n years. (Assume that no more money was added nor none was withdrawn.)

   a) Write an exponential function that finds the **total amount** of euros that would be in the account after n years.

   b) How many euros would be in the account at the end of 10 years? 20 years? 30 years?

3. Anna purchases a new car. The value of the car depreciates at the rate of 13% per year, and can be described by the formula $A(t) = 22\,000(0.87)^t$, where t is the time in years.

   a) Write down the initial purchase price of the car.

   b) Determine the value of the car after 6 years.

   Shana purchases a new car at the same time as Anna, and the purchase price was $25\,000. The depreciation rate for Shana's car is 15%.

   c) Write down the formula for the depreciation of Shana's car.

   d) Sketch a graph of both Anna's and Shana's depreciation equations, where $0 \leqslant t \leqslant 10$. Label the y-intercepts for each graph.

   e) Determine the time when both cars have the same value.

   f) Determine whose car's value will first drop below $5000 and when this will occur.

More Mathematical Models

Overview

By the end of this chapter, you will be able to:

- graph polynomial functions of degree greater than 2
- graph rectangular hyperbolas
- graph unfamiliar functions
- solve equations involving simple and unfamiliar functions
- approximate the local maximum, minimum points, and inflexion points using a GDC
- approximate the intervals for increasing, decreasing, and concavity using a GDC

5.1 Higher order polynomial functions

The general forms for the first four polynomial functions are given below:

- 1st degree: $f(x) = ax + b$
- 2nd degree: $f(x) = ax^2 + bx + c$
- 3rd degree: $f(x) = ax^3 + bx^2 + cx + d$
- 4th degree: $f(x) = ax^4 + bx^3 + cx^2 + dx + e$

The constants a, b, c, d, e, etc. are elements of the real ($\mathbb{R}$) numbers, and the pattern you see continues forever, theoretically. Many chapters in many mathematics books have been dedicated to understanding the theory and concepts surrounding the study of polynomial functions.

This section will condense that information to a necessary minimum while emphasizing the information you need to perform well in your IB exams. In the previous chapter, we have already discussed the first and second degree functions. They were known as linear and quadratic functions respectively.

Third degree polynomial functions

The primary function

A third degree polynomial function is known as a cubic function.

Its primary function is: $y = x^3$.

The graph of $y = x^3$ is shown in Figure 5.1.

The function has many characteristics:

- There are no lines of asymptote.
  - As x becomes very large in the positive direction, the y-value also becomes larger in the positive direction.
  - As x becomes very large in the negative direction, the y-value also becomes larger in the negative direction.
- It is common language to say that the curve 'starts in the third quadrant and ends in the first quadrant'.
- The domain is the set of real numbers.
- The range is the set of real numbers.
- The curve is not 'flat' near zero, it continues to curve up to zero from the left and then it curves away to the right of zero.
- The curve is increasing on the intervals: $(-\infty, 0]$ and $[0, \infty)$.
- The curve is never decreasing.
- The x-intercept is 0.
- The y-intercept is 0.

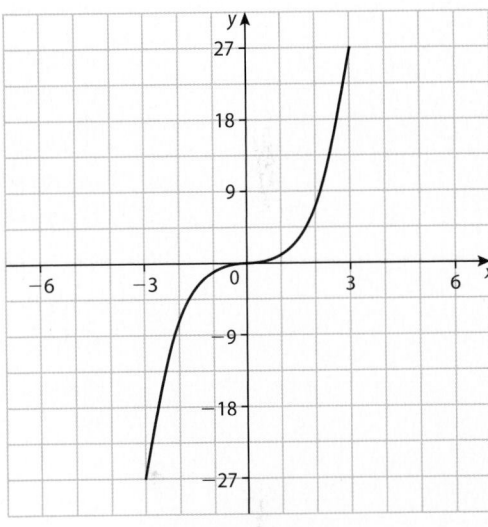

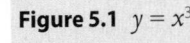
Figure 5.1 $y = x^3$

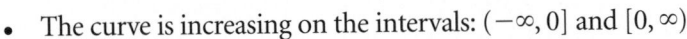

Example 5.1

Using translations, draw the graph of $y = (x - 3)^3 + 9$.

Solution

The horizontal slide is 3 units to the right and the vertical shift is 9 units up.

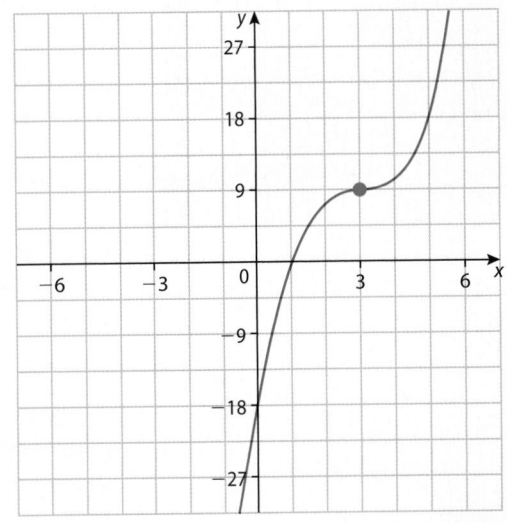

The general form of a third degree polynomial

A third degree polynomial of the form $f(x) = ax^3 + bx^2 + cx + d$, where not all of b, c, and d are 0, will often yield a graph that has 'humps' and 'valleys'.

- At most there can be only be one hump and one valley.
- There will never be a hump without a valley.

The general shape with a hump and valley looks like the graph in Figure 5.2.

A third degree polynomial function will have at most three zeros. For the graph in Figure 5.2 they are: $x = -3, -1$, and 2.

A third degree polynomial function will always have at least one zero, since if the graph starts in the third quadrant and ends in the second quadrant, it must pass through the x-axis.

Depending on the values of a, b, c, d, the valley might be above or below the x-axis and likewise for the hump.

If the graph does have a hump and a valley, then depending on the value of a, two sections will be increasing and one decreasing, or two sections will be decreasing and one increasing.

The graph in Figure 5.2 is:
- increasing on the closed intervals: $(-\infty, -2.12]$ and $[0.786, \infty)$
- decreasing on the closed interval: $[-2.12, 0.786]$.

A third degree polynomial function, whose domain is the set of real numbers, has no minimum or maximum values.
- If the function has a hump and valley (as in Figure 5.2), we say that a 'local' maximum and a 'local' minimum exist.

Figure 5.2 A third degree polynomial function

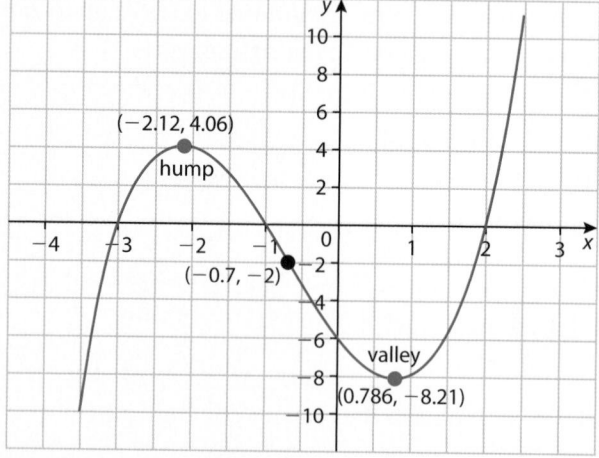

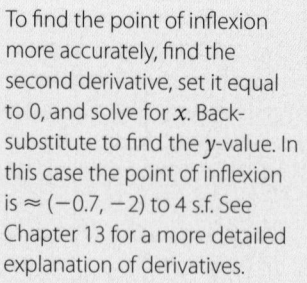

To find the point of inflexion more accurately, find the second derivative, set it equal to 0, and solve for x. Back-substitute to find the y-value. In this case the point of inflexion is $\approx (-0.7, -2)$ to 4 s.f. See Chapter 13 for a more detailed explanation of derivatives.

- In Figure 5.2, the local maximum value of the function is the y-value of the ordered pair on the very top of the hump, i.e. $y = 4.06$ to 3 s.f.
- The local minimum occurs at the very bottom of the valley at the ordered pair $(0.786, -8.21)$ and is therefore $y = -8.21$ to 3 s.f.

Points of inflexion are more difficult to find. To find them exactly, a calculus technique is needed. That technique is beyond the scope of this course; however we can still approximate them.

A point of inflexion is the point where the curve stops being 'cupped down' and starts becoming 'cupped up' or vice versa.

Think of 'cupped down' as a boomerang held in this position:

Imagine that if rain were to fall on this surface it would simply roll off the cup.

The technical name for this shape is 'concave down'.

Think of 'cupped up' as a boomerang held in this position:

Imagine that if rain were to fall on this surface, it would collect in the cup.

The technical name for this shape is 'concave up'.

It appears that the point of inflexion is about $(-0.7, -2)$.

Example 5.2

Using only your GDC, draw the graph of $f(x) = -x^3 + 4x^2 + 4x - 16$ and find:
a) the local maximum value of the function
b) the point where the local minimum value of the function occurs
c) the interval(s) where the function is decreasing
d) the interval(s) where the function is increasing
e) the zero(s) of the function
f) the approximate point of inflexion
g) the approximate interval where the function is concave up
h) the approximate interval where the function is concave down.

Solution
$Y_1 = -x^3 + 4x^2 + 4x - 16$
ZOOM 6
WINDOW
Xmin $= -5$, Xmax $= 7$, Ymin $= -20$, Ymax $= 10$
GRAPH

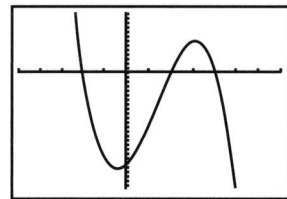

a) CALC, 4:maximum, 2, ENTER, 4, ENTER, 3, ENTER. $\therefore y = 5.05$ to 3 s.f.
b) CALC, 3:mimimum, -1, ENTER, 0, ENTER, $-.1$, ENTER. $\therefore$
    $(-0.430, -16.9)$ 3 s.f.
c) $(-\infty, -0.430)$ and $(3.10, \infty)$ to 3 s.f.
d) $(-0.430, 3.10)$ to 3 s.f.
e) CALC, 2:zero, -2.1, ENTER, -1.9, ENTER, -2, ENTER. $\therefore x = -2, 2, 4$.
    (Repeat for other zeros.)
f) TRACE (about halfway between the local maximum and local minimum):
    $(1.5 \pm 0.5, -6.5 \pm 0.5)$.
g) Using $(1.5, -6.5)$ as the point of inflexion, the interval for concave up is
    $(-\infty, 1.5)$.
h) Using $(1.5, -6.5)$ as the point of inflexion, the interval for concave down
    is $(1.5, \infty)$.

 The actual point of inflexion is $(1.333, -5.926)$ to 4 s.f.

Below are examples of other cubic functions:

$f(x) = 2x^3 + 2x^2 - 4x + 4$

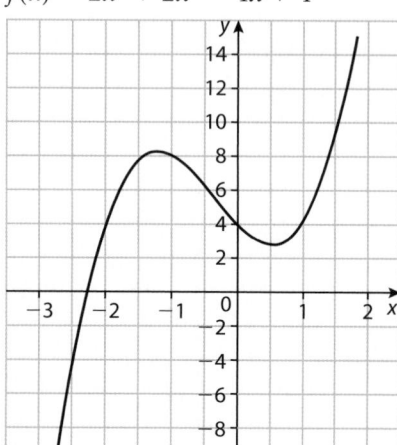

$f(x) = -x^3 + 9x^2 - 24x + 20$

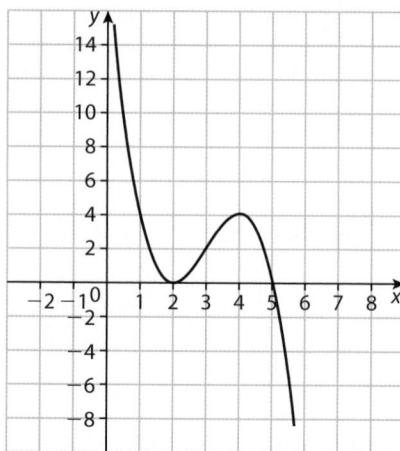

$f(x) = x^3 - x^2 - 5x - 3$

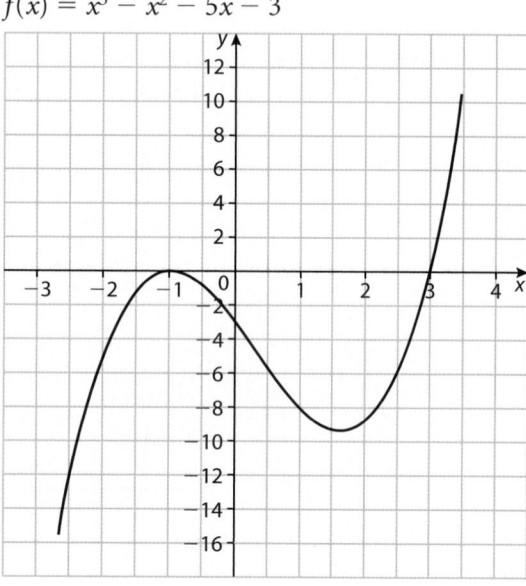

$f(x) = -x^3 - 2x^2 + 5x + 6$

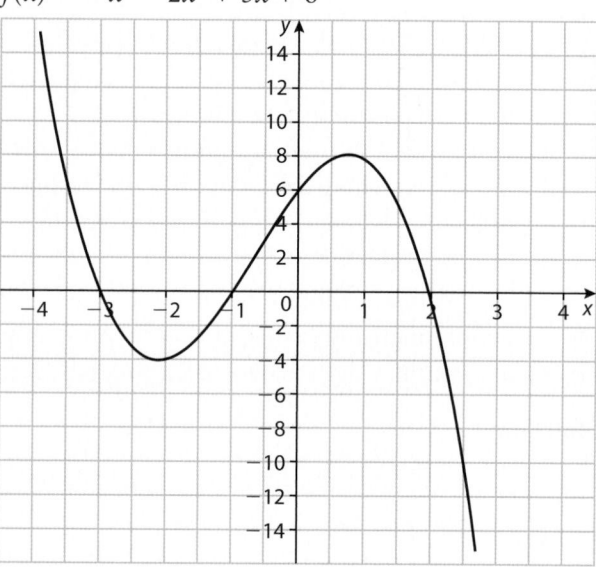

Fourth degree polynomial functions and beyond

Fourth degree, fifth degree, and nth degree polynomial functions behave similarly to third degree functions. The concepts of how to find zeros, maxima, minima, increasing intervals, decreasing intervals, points of inflexion, and concavity intervals all stay the same. The major difference concerns an understanding of the general shape of each higher degree polynomial function.

Facts about fourth degree polynomial functions:
- They have a general form of: $y = ax^4 + bx^3 + cx^2 + dx + e$.
  (For example: $y = x^4 - 2x^3 - 13x^2 + 14x + 24$.)
- The primary function is: $y = x^4$.

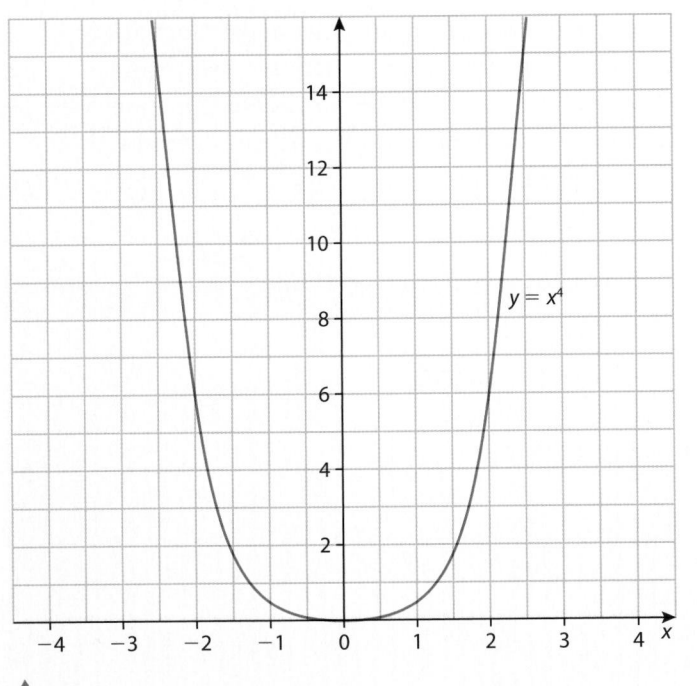

Figure 5.3

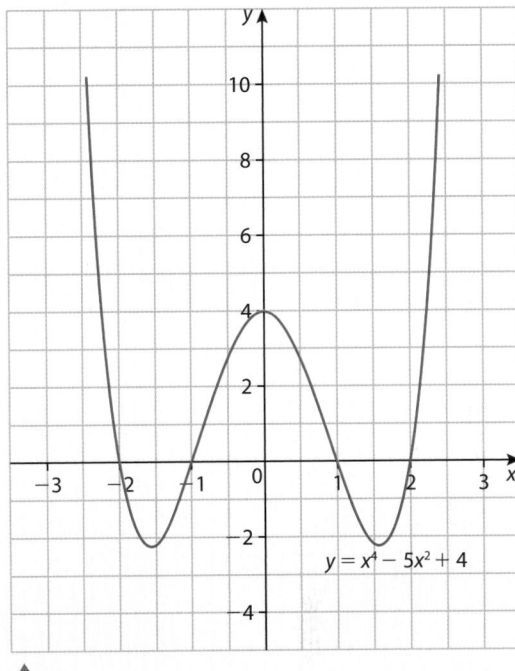

Figure 5.4

Figure 5.5

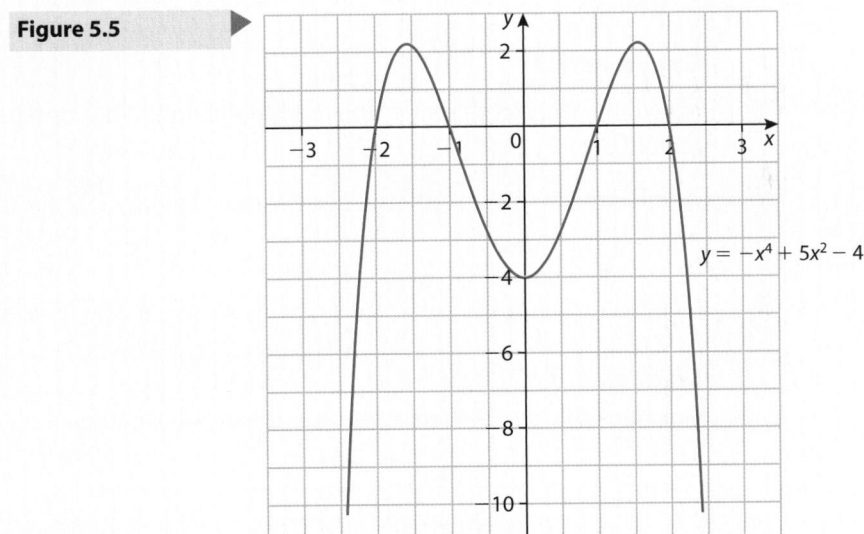

A fourth degree polynomial function:

- starts high in the second quadrant and ends high in the first quadrant, when *a* is positive. (See Figure 5.4.)
- starts low in the third quadrant and ends low in the fourth quadrant, when *a* is negative. (See Figure 5.5.)
- has at most four zeros and could have no zeros at all.
- has one hump and two valleys if *a* is positive, assuming it is not the primary function.
- has two humps and one valley if *a* is negative, assuming it is not the primary function.

Beyond fourth degree polynomials there are polynomial functions of degree 5, 6, 7, 8, etc. All such polynomial functions have very similar characteristics to the third and fourth degree polynomial functions we studied above. Listed below are characteristics which are common to all polynomial functions studied at this level.

Table 5.1

Degree	General function	Primary function	Max. no. of zeros, $a \neq 0$	Min. no. of zeros, $a \neq 0$	Starts if $a > 0$ $b, c, \dots \neq 0$	Ends if $a > 0$ $b, c, \dots \neq 0$	Max. no. of humps $a > 0$	Max. no. of valleys $a > 0$
1	$y = ax + b$	$y = x$	1	1	3rd Q	1st Q	none	none
2	$y = ax^2 + bx + c$	$y = x^2$	2	0	2nd Q	1st Q	0	1
3	$y = ax^3 + bx^2 + cx + d$	$y = x^3$	3	1	3rd Q	1st Q	1	1
4	$y = ax^4 + bx^3 + cx^2 + dx + e$	$y = x^4$	4	0	2nd Q	1st Q	1	2
5	$y = ax^5 + bx^4 + cx^3 + dx^2 + ex + f$	$y = x^5$	5	1	3rd Q	1st Q	2	2
6	$y = ax^6 + bx^5 + cx^4 + dx^3 + ex^2 + fx + g$	$y = x^6$	6	0	2nd Q	1st Q	2	3
7	$y = ax^7 + bx^6 + cx^5 + dx^4 + ex^3 + fx^2 + gx + h$	$y = x^7$	7	1	3rd Q	1st Q	3	3
8	$y = ax^8 + bx^7 + cx^6 + dx^5 + ex^4 + fx^3 + gx^2 + hx + i$	$y = x^8$	8	0	2nd Q	1st Q	3	4

If $a < 0$, the information in the 'start', and 'end' as well as in the 'hump' and 'valley', columns will be slightly different. 3rd Q will change to 2nd Q and 1st Q will change to 4th Q in the start/ end columns. The numbers in the hump/valley columns will interchange.

● **Examiner's hint:** In IB testing, it will be unusual for polynomial functions of degree 5 and higher to be tested.

Example 5.3

Sketch the general shape of what each polynomial function below could look like.

Assume that $b, c, \dots \neq 0$. Also, assume that the function has the maximum number of zeros.

a) $y = x^5 + bx^4 + cx^3 + dx^2 + ex + f$

b) $y = -x^4 + bx^3 + cx^2 + dx + e$

Solution

Answers will vary, but examples are shown below.

a)

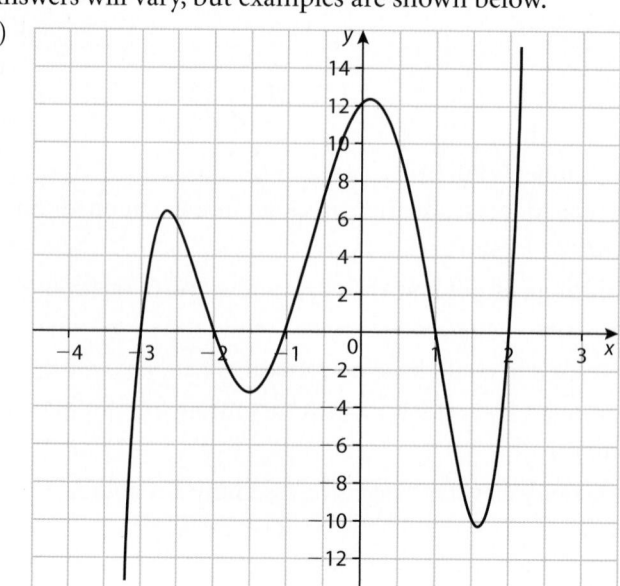

b)

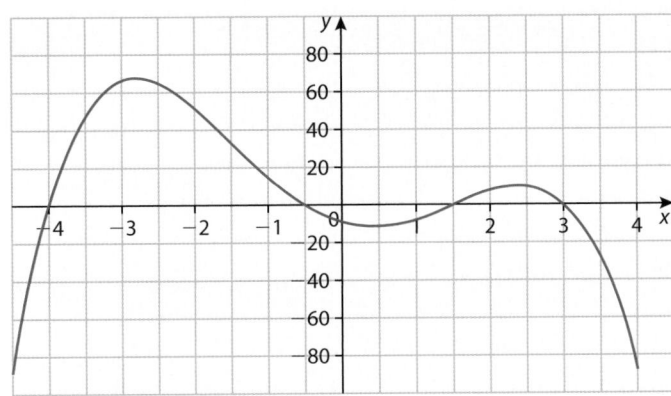

Example 5.4

Draw an accurate graph of $y = (x + 3)(x + 1)(x - 3)(x - 3)$.

Here we are referring to a local maximum as a 'hump' and a local minimum as a 'valley'.

Solution

List the information that can easily be deduced. Plot those ordered pairs. Use your GDC to find and plot any local maxima and minima that exist. Connect the points in a smooth curve.

- The function is a fourth degree polynomial:
  $y = x^4 - 2x^3 - 12x^2 + 18x + 27$.
- $a > 0$ $(a = 1)$.
- The graph starts by decreasing in the second quadrant and ends by increasing in the first quadrant.
- It has at most one hump and two valleys (i.e. one local maximum and two local minima).
- Let $y = 0$ to find the zeros of the function:
  $0 = (x + 3)(x + 1)(x - 3)(x - 3)$
  - $\therefore x = -3, -1, 3,$ or 3.
- Let $x = 0$ to find the y-intercept:
  $y = 0^4 - 2 \cdot 0^3 - 12 \cdot 0^2 + 18 \cdot 0 + 27 = 27$.
- The local minima are:
  $(-2.2, -25.0)$ to 1 d.p. and $(3,0)$.
- The local maximum is:
  $(0.7, 33.3)$ to 1 d.p.
- Three other ordered pairs are:
  $(-3.3, 27.4)$, $(3.5, 7.3)$ to 1 d.p. and $(4, 35)$.

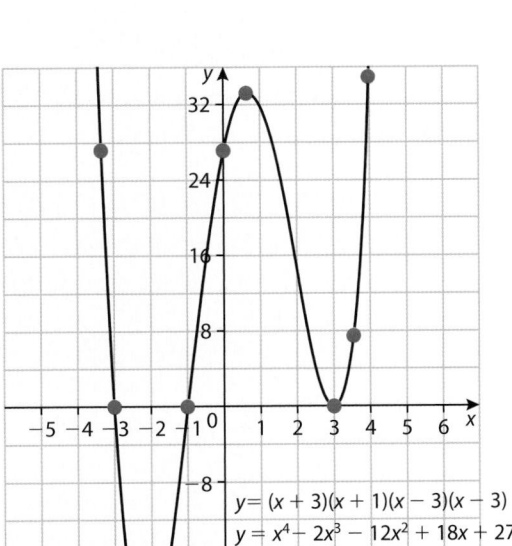

Since 3 occurs twice as a zero, it is said to be a double-zero, a double-root, or a root of multiplicity 2.

1. True or False?
   a) A 3rd degree polynomial in expanded form could have five terms.
   b) A cubic function will always be increasing.
   c) The function $y = x^3$ has a vertical asymptote.
   d) The function $y = x^3$ is perfectly flat near $x = 0$.
   e) The domain of any 3rd degree function is the set of real numbers.
   f) The function $y = x^3$ is always increasing.
   g) A cubic function will always have a y-intercept.
   h) A cubic function will always have an x-intercept.
   i) A cubic function will always have a point of inflexion somewhere.
   j) In Figure 5.5 the part of the graph that is in the second and third quadrants is a parabola that turns downwards.

2. True or False?
   a) A cubic function will have at most three real zeros.
   b) If a 3rd degree function has a local maximum, it must also have a local minimum.
   c) The graph of a cubic function eventually 'goes' straight up.
   d) All cubic functions will have both types of concavity: concave up and concave down.
   e) Knowledge of concavity will be tested on the IB exams.
   f) The function $y = -x^3$ starts in the second quadrant and ends in the fourth quadrant.
   g) The function $y = (x - 1)^3$ has only one zero at $x = 1$.
   h) The function $y = x^3 + 1$ has only one x-intercept at $x = 1$.
   i) The function $f(x) = (2x - 1)(x + 1)(x - 2)$ has three rational zeros.
   j) $y = \dfrac{2}{x^3}$ is an example of an advanced 3rd degree polynomial function.

3. True or False?
   a) A 4th degree polynomial function has at least one term and at most four terms.
   b) The primary 4th degree polynomial function can be called a parabola.
   c) A 4th degree polynomial function must have at least one real zero.
   d) A 4th degree polynomial function could always be concave up.
   e) A 4th degree polynomial function could start in the third quadrant and end in the first quadrant.
   f) The zeros of a 4th degree polynomial function could be: $x = -4, -1$, and 2.
   g) If a 4th degree polynomial function has one hump, it must have two valleys.
   h) A 4th degree polynomial function could have 0, 1, or 2 points of inflexion.
   i) The domain and range of all 4th degree polynomial functions is the set of real numbers.
   j) Some 4th degree polynomial functions have a horizontal asymptote.

4. True or False?
   a) A 5th degree polynomial function could have a maximum of five zeros.
   b) A 5th degree polynomial function must have at least one real zero.
   c) A 6th degree polynomial function must have at least one real zero.
   d) A 6th degree polynomial function could start in the third quadrant and end in the fourth quadrant.

e) A 7th degree polynomial function could have a maximum of three humps and three valleys.

f) A 7th degree polynomial function in expanded form could have only three terms.

g) An 8th degree polynomial function has a maximum of three humps and four valleys.

h) $y = 3x^{-8} + x^{-4} + 2x^{-2} + 1$ is an example of an 8th degree polynomial function.

i) A 10th degree polynomial function could have as many as eight points of inflexion.

j) All polynomial functions must have a y-intercept.

For Questions 5−8 use your GDC, graph the polynomial function and make sure that it is completely displayed in the viewing window. Answer each question using approximations to 1 decimal place when necessary.

a) Find the zero(s).

b) Find the y-intercept.

c) Find the interval(s) where the function is increasing.

d) Find the interval(s) where the function is decreasing.

e) Find the local maxima and the ordered pair(s) where they occur.

f) Find the local minima and the ordered pair(s) where they occur.

g) Approximate the point(s) of inflexion.

5. $y = x^3 - x^2 - 2x$

6. $y = -x^3 - 1.8x^2 + 4.35x + 6.5$

7. $y = x^4 + 3x^3 - 4x^2 - 12x$

8. $y = -0.1x^4 - 0.7x^3 + 1.3x^2 + 6.5x - 11$

For Questions 9–12, draw the graph of each polynomial function by comparing each to the primary function and then using the basic translations of horizontal slides and vertical shifts. Use the IB 2-mm graph paper and appropriate scaling.

9. $y = (x - 1)^3 + 2$ **10.** $y = (x + 2)^3 - 3$

11. $y = (x - 3)^4$ **12.** $y = (x + 4)^5 + 2$

For Questions 13–16, draw the graph of each function on IB 2-mm graph paper. Use your GDC to help you find the: zeros, y-intercept (if needed), local maxima and minima. Plot enough ordered pairs to be able to sketch a smooth curve.

13. $y = (x - 1)(x - 3)(x - 4)$

14. $y = -x^3 - 9x^2 - 23x - 15$

15. $y = (x - 1)(x - 4)^2$

16. $y = (-x - 2)(x + 2)(x + 4)(x + 5)$

For Questions 17–20, sketch a graph of each of the following polynomial functions. Make sure that each sketch includes the correct maximum number of humps and valleys. (Answers will vary somewhat.) $b, c, d, e, \ldots \neq 0$.

17. $y = -x^5 + bx^4 + cx^3 + dx^2 + ex + f$

18. $y = x^6 + bx^5 + cx^4 + dx^3 + ex^2 + fx + g$

19. $y = x^7 + bx^6 + cx^5 + dx^4 + ex^3 + fx^2 + gx + h$

20. $y = -x^8 + bx^7 + cx^6 + dx^5 + ex^4 + fx^3 + gx^2 + hx + i$

5.2 Rectangular hyperbolic functions

For information and visuals for the conic sections, visit www.pearsonhotlinks.co.uk, enter the ISBN for this book and select weblinks 5.1 to 5.3.

6.5 Models using functions of the form $f(x) = ax^m + bx^n + ...$; $m, n \in \mathbb{Z}$.
Functions of this type and their graphs. The y-axis as a vertical asymptote.

6.7 Use of a GDC to solve equations involving combinations of the functions.

A hyperbola is one of the four conic sections. The other three conic sections are: circle, ellipse, and parabola. The word 'conic' refers to the geometric figure − cone. The word 'section' refers to a view of the cone when a piece is sliced off.

The graph of a typical hyperbola, together with its equation, is shown right.

Figure 5.6 A hyperbolic function

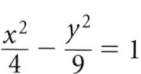

$$\frac{x^2}{4} - \frac{y^2}{9} = 1$$

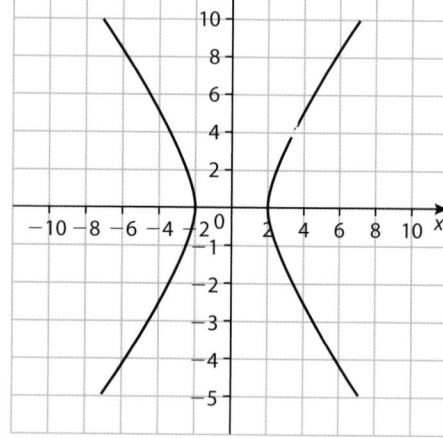

A hyperbola is one curve with two branches.

This hyperbola (Figure 5.6) represents a relation but not a function.

If the hyperbola in Figure 5.6 was squeezed and rotated so that the right branch was entirely contained in the first quadrant and the left branch was contained entirely in the third quadrant, then it would look like the graph in Figure 5.7.

Figure 5.7

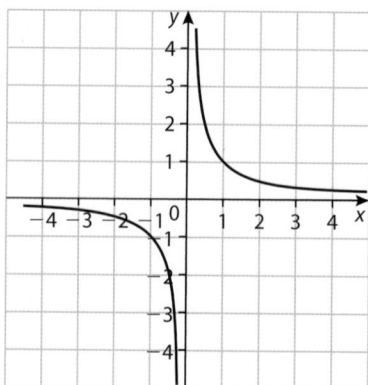

When the x- and y-axes are asymptotes, this set of ordered pairs does represent a function.

The general form for the function is $xy = c$, where $c \in \mathbb{R}$.

Functions of this type are called **rectangular hyperbolic** functions.

The word 'rectangular' is used since the branches are contained in the first and third quadrants (or the second and fourth quadrants), of the 'rectangular' coordinate plane.

For example, the equation of the graph in Figure 5.7 is: $xy = 1$.

$xy = 1$ is considered the primary equation.

Dividing both sides by x yields the primary functional form, $f(x) = \frac{1}{x}$.

The basic properties of a **primary** rectangular hyperbolic function are listed below:

- There is no maximum value.
- There is no minimum value.
- Both branches are decreasing.
- The x-axis, $y = 0$, is the horizontal asymptote.
- The y-axis, $x = 0$, is the vertical asymptote.
- The domain is the set of real numbers, except 0; i.e. $D(f) = \{x \mid x \in \mathbb{R}, x \neq 0\}$.
- The range is the set of real numbers, except 0; i.e. $R(f) = \{y \mid y \in \mathbb{R}, y \neq 0\}$.
- The branches do *not* connect together.

When using the GDC, the graph of $y = \frac{1}{x - 1}$ is shown in Figure 5.8 after pressing ZOOM 6.

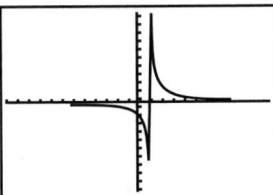

Figure 5.8

It appears that the two branches are connected with a straight, vertical line. In fact, this is not correct. The vertical line, $x = 1$, is a line of asymptote. Pressing ZOOM 4, Figure 5.9, gives a more accurate picture, except that the vertical asymptote is not shown.

The general form of a rectangular hyperbolic function is: $y = \frac{a}{x - b} + c$.

- a represents the 'stretch' with respect to the primary graph.
  - If $a > 0$, the curve is decreasing and has a shape similar to the primary graph. (See Figure 5.7)
  - If $a < 0$, the curve is increasing and has a shape similar to Figure 5.10.

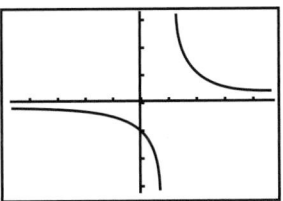

Figure 5.9

- b represents the 'horizontal slide' with respect to the primary graph.
- c represents the 'vertical shift' with respect to the primary graph.

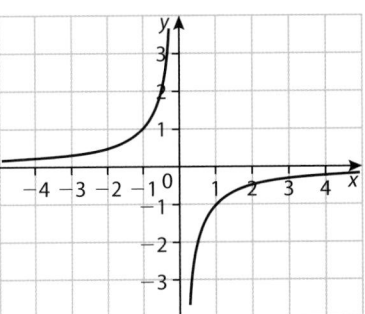

Figure 5.10

143

Example 5.5

Given the function $y = \dfrac{2}{x - 3} + 4$, identify the:

a) stretch, a
b) horizontal slide, b
c) vertical shift, c
d) vertical asymptote
e) horizontal asymptote
f) domain
g) range.

Solution

a) The stretch: $a = 2$.
b) The horizontal slide: $b = 3$.
c) The vertical shift: $c = 4$.
d) The vertical asymptote: $x = 3$.
e) The horizontal asymptote: $y = 4$.
f) $D = \{x \mid x \in \mathbb{R}, x \neq 3\}$.
g) $R = \{y \mid y \in \mathbb{R}, y \neq 4\}$.

Although the hyperbola is perhaps the least known of the conic sections, it has applications that are very well known, for example, the shadow a lampshade casts on a wall. For other applications, visit www.pearsonhotlinks. co.uk, enter the ISBN for this book and select weblink 5.4.

Graphing the rectangular hyperbola

The graphing methods (point-plotting, GDC, translations) previously discussed can be applied in the graphing of rectangular hyperbolic functions. At this point, perhaps the easiest and quickest is a combination of those methods.

Example 5.6

Draw the graph of $y = \dfrac{2}{x - 3} + 4$. Use 2-mm graph paper and the scale of 1 unit = 1 cm on both the x- and y-axes.

Solution

Step 1: Using information from the solution in Example 5.5, enter the function into your GDC and use ASK mode, then ZOOM 4 and/or ZOOM 6 to check your working.

Step 2: Draw the lines of asymptotes as dashed lines.
- Vertical asymptote: $x = 3$.
- The horizontal asymptote: $y = 4$.

Step 3: Recognize the function as a rectangular hyperbolic function that is decreasing and similar to the primary graph.

Step 4: Plot some ordered pairs:
- some close to the vertical asymptote on both sides
- a few to the right of the vertical asymptote
- a few to the left of the vertical asymptote
- the y-intercept.

x	y
3.01	204
3.1	24
3.2	14
5	5
10	4.2
2.99	−196
2.9	−16
2.8	−6
1	3
0	≈3.3
−5	3.75
−10	≈3.8

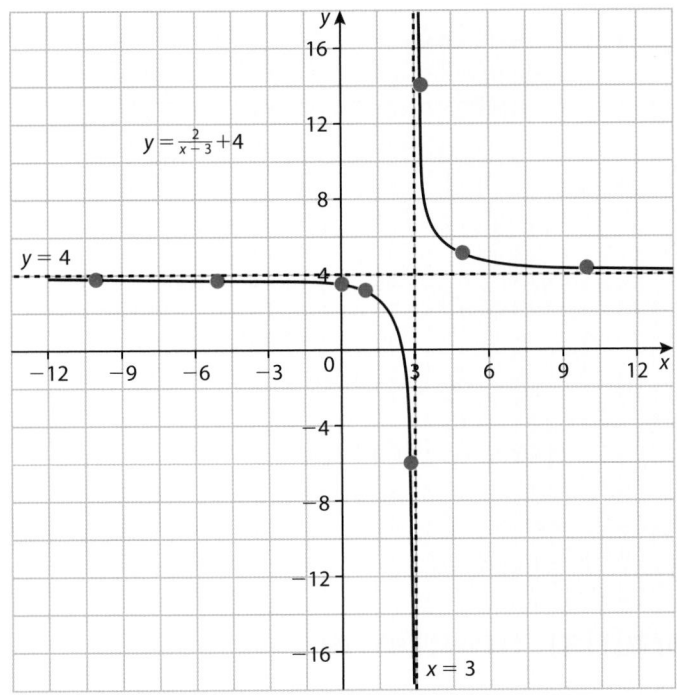

$y = \frac{2}{x+3} + 4$

$y = 4$

$x = 3$

Attempting to plot the points $(3.01, 204)$, $(3.1, 24)$, $(2.99, -196)$, and $(2.9, -16)$ would be impractical due to the extreme y-values. As you become better at choosing x-values, you will be able to look at the pattern of y-values and decrease the number of points you need to plot in order to draw an accurate graph.

Exercise 5.2

1. What is an asymptote?

2. What is a horizontal asymptote?

3. What is a vertical asymptote?

4. What value do asymptotes have when drawing the graph of a relation?

5. When trying to determine the equation of the horizontal asymptote, choose x-values that are very _____.

6. When trying to determine the vertical asymptote, set the expression in the _____ equal to _____.

7. When choosing points to plot in order to draw the graph of a function that has a vertical asymptote, choose x-values that are very _____ to the asymptote.

8. True or False? A function will sometimes intersect a vertical asymptote.

9. True or False? A function will never intersect a horizontal asymptote.

10. True or False? If a function has a vertical asymptote then the domain must be restricted.

For each function in Questions 11–14, identify the following when the function is compared to the primary function, $y = \frac{1}{x}$

    a) the stretch
    b) the horizontal slide
    c) the vertical shift
    d) the vertical asymptote
    e) the horizontal asymptote
    f) the domain
    g) the range.

11. $y = \frac{1}{x} + 2$

12. $f(x) = \frac{3}{x-1} - 2$

13. $(g)\ x = \frac{-2}{3+x}$

14. $y = \frac{-1}{2+x} - 3$

Draw the graph for each of the functions in Questions 15–18. Use 2-mm graph paper, an appropriate scale, such as 1 unit = 1 cm, on both axes, and your GDC. Make sure to clearly show the ordered pairs that you plot.

15. $y = \frac{1}{x} + 2$ **16.** $f(x) = \frac{3}{x-1} - 2$

17. $g(x) = \frac{-2}{3+x}$ **18.** $y = \frac{-1}{2+x} - 3$

Draw the graph for each of the functions in Questions 19–22. Use 2-mm graph paper, an appropriate scale, such as 1 unit = 1 cm, on both axes, and your GDC. Make sure to clearly show the ordered pairs that you plot.

19. $y = \frac{x}{x-1}$ **20.** $y = \frac{x-3}{x-2}$

21. $y = \frac{4x}{x+2}$ **22.** $y = \frac{3x+2}{x+1}$

5.3 Unfamiliar functions

> 6.6 Drawing accurate graphs. Creating a sketch from information given. Transferring a graph from GDC to paper.
> Reading, interpreting, and making predictions using graphs. Include all the functions above as well as additions and subtractions.
> 6.7 Use of a GDC to solve equations involving combinations of functions.

There are many categories of functions, and many other functions that do not necessarily fit into any category at all. Some examples are listed below:

(I) Examples of unfamiliar functions that can be classified into categories.

- Logarithmic: $f(x) = \log x$
- Rational: $f(x) = \frac{x^2 - x - 6}{x + 1}$
- Absolute value: $f(x) = |x|$
- Greatest integer: $f(x) = [\![x]\!]$
- Piecewise: $f(x) = \begin{cases} x^2, x \geqslant 2 \\ 2x + 1, x < 2 \end{cases}$

(II) Examples of functions that cannot be classified into categories.

- $f(x) = x \cdot e^x$
- $f(x) = \frac{e^x}{x}$
- The glog function

To learn more about the glog function, visit www.pearsonhotlinks.co.uk, enter the ISBN for this book and select weblink 5.5 or 5.6.

Example 5.7

a) Create a function that can be added to the list of examples of categorized functions.
b) Create a function that can be added to the list of examples of non-categorized functions.

Solution

Answers will vary, but examples are given below.

a) Semicircular: $f(x) = \sqrt{4 - x^2}$

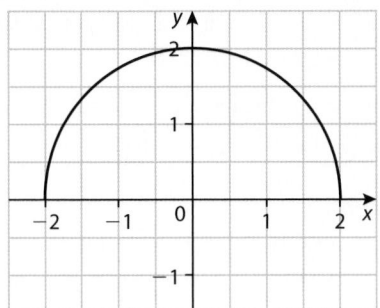

b) $f(x) = e^x(x^2+1)$

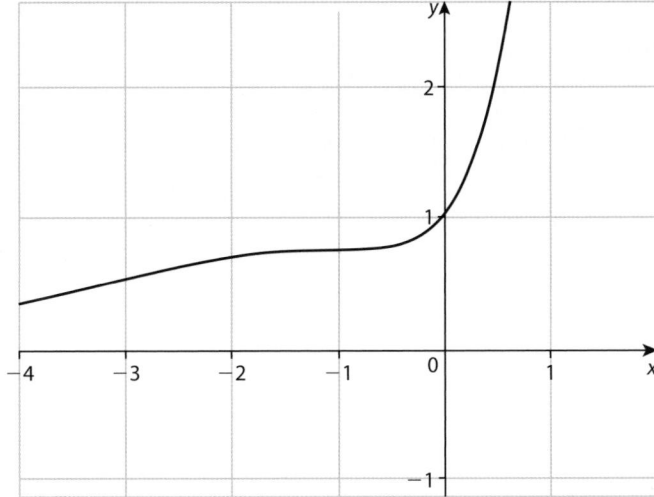

Graphing unfamiliar functions

Drawing the graph of *familiar* functions requires fewer ordered pairs since the shape of the graph is already known reasonably well. Drawing the graph of an *unfamiliar* function can be much more tedious, since the final shape is not necessarily known at all. Therefore, there are two methods that can be used to draw the graph of an unfamiliar function:

1. point-plotting
2. using your GDC.

Although the two methods can be used independently, in Example 5.8 they are used together. Knowledge and understanding of both methods are required to be successful on the IB exams.

 Points of inflexion can be found exactly by using a calculus method involving the second derivative. Find the second derivative, set it equal to zero, and solve for each x-value. Back-substitute to find the y-value(s).

Knowledge of increasing, decreasing, inflexion points, and concavity can be useful in graphing unfamiliar functions. These concepts, as they apply to graphing functions in this chapter, are beyond the scope of this course.

Example 5.8

Draw the graph of $f(x) = x^2 e^x$ by:

a) point-plotting

    (i) Use TBLSET and TABLE to produce ordered pairs

    (ii) Use 2-mm graph paper and the scale 1 cm = 1 unit on both axes

b) using your GDC and WINDOW to produce a large, detailed graph in the display.

X	Y₁	
-6	.08924	
-5.8	.10185	
-5.6	.11597	
-5.4	.1317	
-5.2	.14917	
-5	.16845	
-4.8	.18961	

X=-6

X	Y₁	
-4.6	.2127	
-4.4	.23769	
-4.2	.26452	
-4	.29305	
-3.8	.32303	
-3.6	.35412	
-3.4	.3858	

X=-3.4

X	Y₁	
-3.2	.4174	
-3	.44808	
-2.8	.47675	
-2.6	.50209	
-2.4	.52254	
-2.2	.53629	
-2	.54134	

X=-2

X	Y₁	
-1.8	.53557	
-1.6	.51686	
-1.4	.48333	
-1.2	.43372	
-1	.36788	
-.8	.28757	
-.6	.19757	

X=-.6

X	Y₁	
-.4	.10725	
-.2	.03275	
0	0	
.2	.04886	
.4	.23869	
.6	.65596	
.8	1.4243	

X=.8

X	Y₁	
1	2.7183	
1.2	4.781	
1.4	7.9482	
1.6	12.68	
1.8	19.601	
2	29.556	
2.2	43.601	

X=2.2

Solution

a)   1.   TBLSET (2ND WINDOW), TblStart −6, ΔTbl = 0.2, Indpnt: Auto, Depend: Auto

     2.   TABLE (2ND GRAPH)

     3.   Plot each ordered pair.

     4.   Connect the points with a smooth curve.

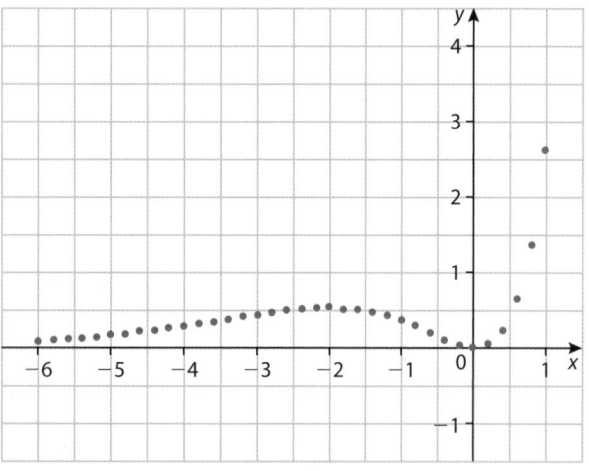

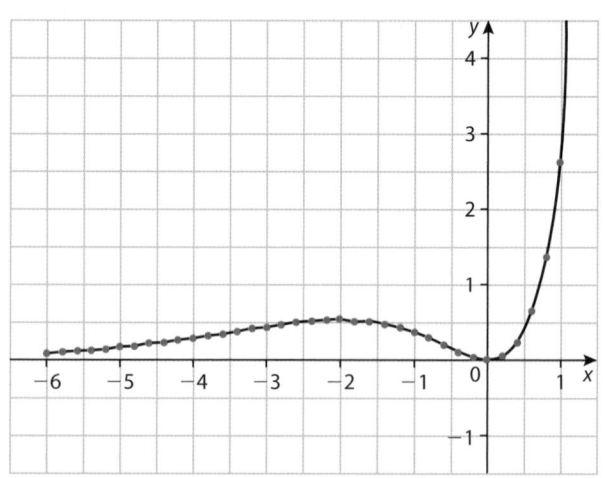

b) $Y_1 = x^2 e^x$, ZOOM 6, WINDOW, Xmin = −6, Xmax = 2, Ymin = −0.5, Ymax = 2, GRAPH

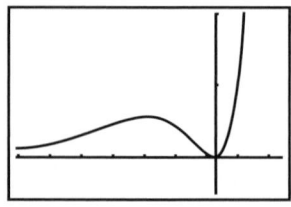

1. Say whether each of the following is an example of a function that be categorized or a function that cannot be categorized, according to the examples in I and II (page 143). If it can be categorized, state which category.
   a) $f(x) = 2 \log x$
   b) $y = \dfrac{2e^{x+1}}{3x}$
   c) $h(x) = 3|x - 2|$
   d) $f(x) = \dfrac{2x^3 - x}{x + 1}$
   e) $h(x) = \begin{cases} 2x^2 + 1, x \geqslant 1 \\ x - 2, x < 1 \end{cases}$
   f) $f(t) = 0.5gt^2$, where g is a constant

2. Create a function that can be added to the list of examples of categorized functions and draw its graph.

3. Create a function that can be added to the list of examples of non-categorized functions and draw its graph.

4. Draw the graph of each of the following functions within the given domain. Use an appropriate scale, such as 1 unit = 1 cm, on both the x- and y-axes.
   a) $f(x) = \log x, 0 < x \leqslant 10$
   b) $f(x) = \dfrac{x^2 - x - 6}{x + 1}, -10 \leqslant x \leqslant 10$
   c) $f(x) = 2|x - 3| + 1, -2 \leqslant x \leqslant 8$.

5. Draw the graph of the following function within the given domain. Use an appropriate scale, such as 1 unit = 1 cm, on both the x- and y-axes.
   a) $f(x) = xe^x, -4 \leqslant x \leqslant 1.5$ and let 2 cm = 1 unit on both axes.

The Greatest Integer Function is defined as: y is the greatest (largest) integer that is less than or equal to x, where x is any real number. The symbol for the GIF is $[\![\ ]\!]$ and the function is written as $f(x) = [\![x]\!]$. For example, $f(2.1) = [\![2.1]\!] = 2$, since 2 is the greatest integer that is less than **or** equal to 2.1.

6. If f is the Greatest Integer Function, find each functional value.
   a) $[\![2.001]\!]$
   b) $[\![2.01]\!]$
   c) $[\![2.5]\!]$
   d) $[\![2.9]\!]$
   e) $[\![2.99]\!]$
   f) $[\![2.999]\!]$
   g) $[\![3]\!]$

7. Let $f(x) = [\![x]\!]$, for $-3 \leqslant x \leqslant 3$.
   a) What is the domain of f?
   b) What is the range of f?
   c) Draw the graph of f for the required domain.

 To access the absolute value function on the TI calculator, press the following keys: MATH,⟩ (NUM), ENTER (abs).

• **Examiner's hint:** Questions 6 and 7 should be done as the GIF concept has been tested on the IB exams.

 To access the Greatest Integer Function on the TI calculator, press the following keys: MATH, ⟩ (NUM), V,V,V,V, (5: int().

On some TI calculators when graphing $f(x) = [\![x]\!]$ with your TI calculator by pressing ZOOM 4, the graph looks like:

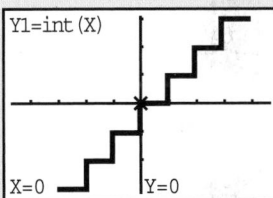

The screen seems to suggest that the horizontal segments are connected with vertical line segments. This is **not** correct. Put your calculator in DOT mode by pressing MODE, V,V,V,V, ⟩, ENTER. You will see the screen below.

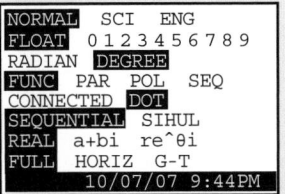

Now press ZOOM 4 and you will see the screen below.

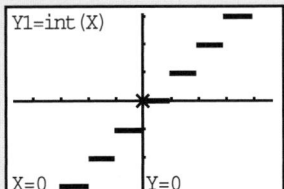

Even with this screen the graph is a little misleading. It appears that, for example, (2,1) exists, where in fact it does not.

Solving equations involving combinations of unfamiliar functions

6.7	Use of a GDC to solve equations involving combinations of functions.

Many equations are solvable using algebraic, logarithmic or trigonometric techniques. Below are a few examples.

- $2x + 1 = 5$
- $3x^2 = 27$
- $x^2 + 7x = -12$
- $3^x = 41$
- $x^{\frac{1}{2}} + 5 = 9$

There are also many equations that cannot be solved using algebraic or logarithmic techniques. Below are a few examples.

- $2^x = 2x$
- $3x = \dfrac{2}{x}$
- $x^5 + 3x^4 + 2x^3 - x^2 - x + 7 = 0$

There are specific cases when an insightful substitution will allow a solution to an equation that at first glance appears not to be solvable. These techniques are beyond the scope of this course; however, an example is given below.

Example 5.9

Find a substitution that will lead to an easy algebraic solution for $x^{\frac{2}{3}} - x^{\frac{1}{3}} = 6$.

Solution

$$\text{Let } z = x^{\frac{1}{3}} \Rightarrow z^2 - z - 6 = 0$$
$$(z - 3)(z + 2) = 0$$
$$\text{hence, } z = 3 \text{ or } z = -2.$$
$$\text{Therefore, } x^{\frac{1}{3}} = 3 \text{ or } x^{\frac{1}{3}} = -2$$
$$x = 27 \text{ or } x = -8.$$

Further methods for solving equations

Listed below are a few equations involving a combination of unfamiliar functions.

- $2^x = \dfrac{1}{x}$
- $x^2 = 2^x$
- $x^2 e^x = \dfrac{2x^2}{x - 1}$

Two methods that can be used to solve equations such as these are:

- point-plotting
- using your GDC.

Both methods involve:

- thinking of the left side of the equation as one function
- thinking of the right side of the equation as another function
- graphing both functions
- finding the point(s) of intersection by:
  - drawing lines on the graph paper, or
  - using the intersection function on your GDC.

Example 5.10

Solve the equation $x^2 = 2^x$ using your GDC.

Solution

Press the following keys:
- $Y_1 = x^2$
- $Y_2 = 2^x$
- ZOOM 6
- WINDOW
  - Xmin $= -2$, Xmax $= 5$
  - Ymin $= -5$, Ymax $= 18$
- GRAPH

Looking carefully, you can see that there are three points of intersection.
- CALC
- 5: intersect
- $\langle, \langle$, to the leftmost point of intersection
- ENTER, ENTER, ENTER
  - $\therefore x = -0.767$ to 3 s.f.

Repeat the above keystrokes two more times, each time locating the next point of intersection.
  - $\therefore x = 2$
  - $\therefore x = 4$

An interesting method to zoom in on a section of a graph is to use the 'Zbox' function. Using the above equation, press the following keys after the WINDOW has been set.
- ZOOM
- 1: ZBox

Now find the new cursor. It looks like a blinking '+' sign. Position the new cursor above and to the left of the section of the graph you want to box in (in this case the middle point of intersection of the two functions).
- ENTER (Notice that the cursor changed to a blinking square.)
- $\rangle, \rangle$, etc. until a line is drawn above and beyond the section of graph you want to zoom in on.
- $\vee, \vee$, etc. until the section of the graph you need is inside the 'box'.
- ENTER

You may now use the TRACE, CALC, 5: intersect.

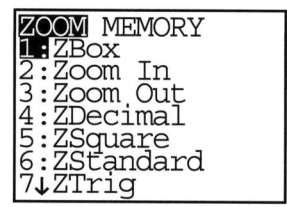

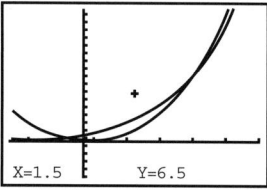
X=1.5   Y=6.5

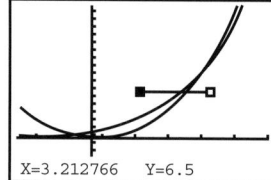
X=3.212766   Y=6.5

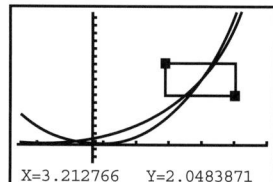
X=3.212766   Y=2.0483871

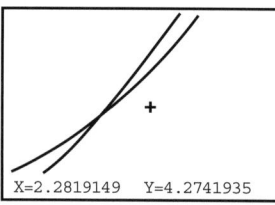
X=2.2819149   Y=4.2741935

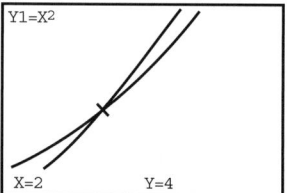
X=2   Y=4

Exercise 5.4

1. Solve each equation using well-known algebraic techniques.

a) $2x + 1 = 5$ b) $3x^2 = 27$

c) $x^2 + 7x = -12$ d) $3^x = 41$

e) $x^{\frac{1}{2}} + 5 = 9$

2. Explain why there is no simple algebraic solution to each equation below. What makes each equation in this problem set different to those in Question 1 above?

a) $2^x = 2x$ b) $3^x = \dfrac{2}{x}$

3. Find a substitution that will lead to an easy algebraic solution for the following equation.

$$x^{\frac{2}{3}} - 7x^{\frac{1}{3}} = -12$$

For Questions 4–7, solve each equation by using your GDC in conjunction with the method in Example 5.10.

4. $3x = \dfrac{2}{x}$ **5.** $2^{-x} = 3x + 5$

6. $[\![x]\!] = 2x$ **7.** $x \cdot e^x = \dfrac{e^x}{x}$

8. $g(x) = -0.5x^2 + 4$

$h(x) = \dfrac{1}{x - 1}, \ x \in \mathbb{R}, x \neq 1$

a) Draw, on the same axes, the graph of g and the graph of h for $-4 \leqslant x \leqslant 4$. Label your graphs clearly.

b) Draw the vertical asymptote of the graph of h. Write down the equation of the vertical asymptote.

c) Write down the solutions for $h(x) = g(x)$.

9. An archer is located at ground level and shoots an arrow at a target in the shape of a ring located 30 metres away at a height of 23 metres.

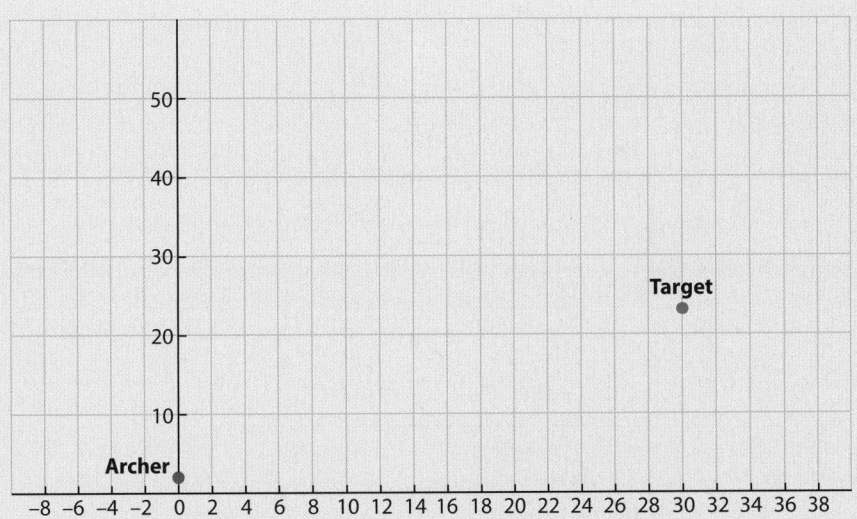

Assuming that the arrow is fired by the archer at a height of 2 metres

a) write down the coordinates for the archer
b) write down the coordinates of the target.

The arrow follows a path that can be modelled by a part of the graph

$$y = -0.18x^2 + 6.1x + 2, y \geqslant 0$$

x is the horizontal distance of the ball from the origin

y is the height above the ground of the arrow. Both x and y are measured in metres.

c) Draw an accurate graph showing the path of the arrow from the point where it is shot to the point where it hits the ground again. Use 1 cm to represent 4 m on the horizontal axis and 1 cm to represent 4 m on the vertical scale.
d) Use your graph to calculate the horizontal distance the arrow has travelled from the archer when the arrow is at its maximum height.
e) Find the maximum vertical height reached by the arrow.
f) How far from the archer will the arrow land?

10. A rollercoaster has the shape as shown in the following diagram.

Two students who are visiting the theme park suggest that the curve looks similar to one they were working on in their class. They return to class and start working on an equation for the rollercoaster track.

The students determined that the rollercoaster's path could be modelled by a part of the graph

$$f(x) = -0.0005x^3 + 0.07x^2 - 2x + 80$$
$$0 \leqslant x \leqslant 100$$

x is the horizontal distance of the car from the origin.

y is the height above the ground of the roller coaster track. Both x and y are measured in metres.

a) Draw an accurate graph showing the rollercoaster track. Use 1 cm to represent 10 m on the horizontal axis and 1 cm to represent 10 m on the vertical scale.

b) What is the starting height of the rollercoaster?

c) Calculate the lowest point the rollercoaster reaches.

d) Find the maximum height of the rollercoaster track.

e) Determine the horizontal distances where the rollercoaster is at the same height as its starting height.

11. The height (cm.) of a particular grass along the banks of a river has been found to be modelled by $h(t) = 1.5^{0.25t} + 10$, where t is the time in weeks and height is in centimetres.

a) Copy and complete the table

t (weeks)	0	1	2	3	4	5	6	7	8
$h(t)$									

b) On graph paper draw axes for t and $h(t)$, placing t on the horizontal axis and $h(t)$ on the vertical axis.

c) **Using your graph** write down

   i) The initial height of the grass
   ii) The height of grass at the end of 10 weeks.

d) A laser beam has been installed to measure the height of the grass. The laser beam travels in a straight line and measures the height of the grass. The height of the laser beam is modelled by the function $g(t) = 0.5t + 9$.

e) Determine at which times the beam will measure the height of the grass.
   Hint: it may be appropriate to sketch the laser beam function on the graph.

6 More Geometry

Assessment statements

5.2 Use of sine, cosine, and tangent ratios to find the sides and angles of right-angled triangles. Angles of elevation and depression.

5.3 Use of the sine rule: $\dfrac{a}{\sin A} = \dfrac{b}{\sin B} = \dfrac{c}{\sin C}$

Use of the cosine rule: $a^2 = b^2 + c^2 - 2bc \cos A$; $\cos A = \dfrac{b^2 + c^2 - a^2}{2bc}$

Use of area of triangle: $\frac{1}{2}ab \sin C$

Construction of labelled diagrams from verbal statements.

5.4 Geometry of three-dimensional solids: cuboid; right prism; right pyramid; right cone; cylinder; sphere; hemisphere and combination of these solids.

The distance between two points; e.g. between two vertices or vertices with midpoint or midpoints with midpoints. The size of an angle between two lines or between a line and plane.

5.5 Volume and surface areas of the three-dimensional solids defined in 5.4.

Overview

By the end of this chapter you will be able to:
- use the sine, cosine, and tangent ratios
- use the sine and cosine rules
- find side lengths and angles of triangles
- find areas of triangles and other two-dimensional shapes
- apply principles of trigonometry to problems involving three-dimensional shapes.

The Eiffel Tower is constructed from a lattice that contains thousands of triangles.

6.1 Advanced right-angled trigonometry

 What other combinations of the ratio of the three sides are there? Why do you think we choose these three examples to teach?

5.2 Use of sine, cosine, and tangent ratios to find the sides and angles of right-angled triangles. Angles of elevation and depression.

Many years ago, mathematicians discovered a fact about any right-angled triangle. The ratio of any two sides is the same for any given angle, regardless of the size of the triangle, as long as the shape of the triangle remains constant.

Let $\triangle ABC$ be a right triangle as shown below.

 If you choose to use trigonometry for your project for this course, and you need to create diagrams for your work, visit www.pearsonhotlinks. co.uk, enter the ISBN for this book and click on weblink 6.1. This links to an excellent open source geometry program.

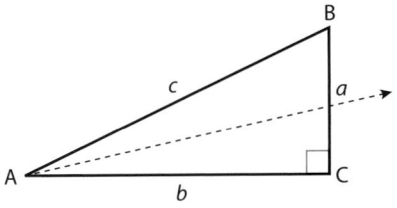

◀ **Figure 6.1**

To find a side opposite an angle, put your pencil on a vertex and draw a line through the middle of the triangle until it meets a side. That side is said to be opposite the angle. For example, [BC] is opposite angle A. (See diagram.)

The hypotenuse is the side opposite the right angle. In this triangle [AB] is the hypotenuse.

The adjacent side of an angle is the side that is neither the opposite side nor the hypotenuse.

The sine, cosine, and tangent of an angle

'sine' is abbreviated as 'sin'
'cosine' is abbreviated as 'cos'
'tangent' is abbreviated as 'tan'

Briefly speaking, each one of these words (sin, cos, and tan) is a concept that is used to describe the relationship between the size of an angle and the lengths of the sides of the triangle.

For example, the correct wording for the use of the sine would be: 'The sine of thirty degrees is one-half.'
The symbolic trigonometric form of those words would be: $\sin 30° = \frac{1}{2}$.

The definitions follow:

- The sine of an angle equals the length of the opposite side divided by the length of the hypotenuse.

$$\text{Sin} = \frac{\text{Opposite}}{\text{Hypotenuse}}$$

- The cosine of an angle equals the length of the adjacent side divided by the length of the hypotenuse.

$$\text{Cos} = \frac{\text{Adjacent}}{\text{Hypotenuse}}$$

- The tangent of an angle equals the length of the opposite side divided by the length of the adjacent side.

$$\text{Tan} = \frac{\text{Opposite}}{\text{Adjacent}}$$

Considering angle A from above, the symbolic forms would be:

- $\sin A = \frac{a}{c}$

- $\cos A = \frac{b}{c}$

- $\tan A = \frac{a}{b}$

These definitions will only work with right-angled triangles.

To help you remember the definitions, there is a mnemonic: SOH – CAH– TOA.

● **Examiner's hint:** In order to make sure to have your calculator in 'Degree Mode', type the following keystrokes: MODE, ∨, ∨, >, ENTER, 2ND, QUIT

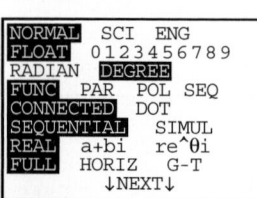

- **Sin Opposite Hypotenuse**
- **Cos Adjacent Hypotenuse**
- **Tan Opposite Adjacent**

Example 6.1

Given the right triangle below, find each of the following:

a) $\sin A$

b) $\cos A$

c) $\tan A$

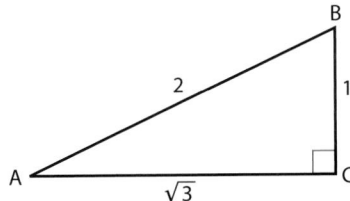

● **Examiner's hint:** Although there is a radical in the denominator for part c) and it is customary to consider it unsimplified, the IBO examiners will accept this as the correct answer.

Solution

a) $\sin A = \dfrac{1}{2}$

b) $\cos A = \dfrac{\sqrt{3}}{2}$

c) $\tan A = \dfrac{1}{\sqrt{3}}$

 The simplified version follows:

$$\frac{1}{\sqrt{3}} = \frac{1}{\sqrt{3}} \cdot \frac{\sqrt{3}}{\sqrt{3}} = \frac{\sqrt{3}}{\sqrt{9}} = \frac{\sqrt{3}}{3}$$

Example 6.2

Solve for x correct to 3 significant figures: $\sin 39° = \dfrac{x}{7}$

Solution

$$\sin 39° = \frac{x}{7}$$

$$7 \cdot \sin 39° = x \quad \textbf{Hint: } 7 \cdot \sin 39° \text{ is often written as simply } 7 \sin 39°$$

$$\therefore \qquad x = 7(0.62932) = 4.40524 = 4.41 \text{ to 3 s.f.}$$

Make sure to use more significant figures than required with intermediate answers and to only round off as required with the final answer.

Exercise 6.1

1. Given the right-angled triangle, find the following:

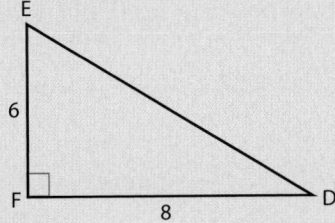

a) $\sin D$

b) $\cos D$

c) $\tan D$

d) $\sin E$

e) $\cos E$

f) $\tan E$

2. Let $\triangle ABC$ be a right triangle with vertex C at the right angle. Also let $\hat{A} = 40°$, $\hat{B} = 50°$ and a, b, c be the lengths of the sides opposite angles A, B, C respectively. Set up six equations representing the six trigonometric equations we have discussed. $\left(\text{Hint: } \sin 40° = \dfrac{a}{c}\right)$

3. Find each of the following correct to 4 decimal places.

a) $\sin 37°$ b) $\cos 85°$ c) $\tan 7°$

d) $\sin 42.7°$ e) $\cos 60°$ f) $\tan 45°$

g) $\sin 30°$ h) $\cos 53°$ i) $\tan 90°$

4. Solve each equation expressing your answer correct to 3 significant figures.

a) $\sin 15° = \dfrac{x}{23}$ b) $\cos 73° = \dfrac{21}{a}$ c) $17 = \dfrac{b}{\tan 52°}$

d) $\dfrac{\sin 58°}{33} = \dfrac{\sin 43°}{c}$ e) $17 = \frac{1}{2}(5b \sin 41°)$ f) $2\tan 78° = 3(5x + 1)$

6.2 Further right-angled trigonometry

> **5.2** Use of sine, cosine, and tangent ratios to find the sides and angles of right-angled triangles. Angles of elevation and depression.

We can use the trigonometric ratios to solve a variety of problems. Occasionally we will have to use this fact about triangles two or more times to solve a problem, because there may be several unknown sides and angles.

● **Examiner's hint:** The notation for an angle can vary. An alternative for the notation $\angle ABC$ often used in IB exams is $A\hat{B}C$.

Example 6.3

Given $d = 10\,\text{cm}$, $c = 6\,\text{cm}$, and $\angle CAB = 30°$, find the length of e.

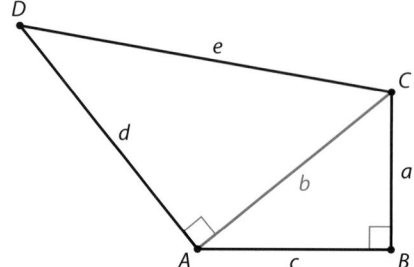

Solution

First we find b since it is common to both triangles, and use this with the length of d to find e.

$$\cos 30° = \frac{6}{b}$$

$$b = \frac{6}{\cos 30°}$$

$$b = 4\sqrt{3}\ \text{cm}$$

To find e, we can use the Pythagorean theorem, hence:

$$(10)^2 + (4\sqrt{3})^2 = e^2$$

$$10^2 + 48 = e^2$$

$$e = \sqrt{148}\ \text{cm}$$

We can also use right-angled trigonometry to find an unknown angle, given two sides of the triangle.

Example 6.4

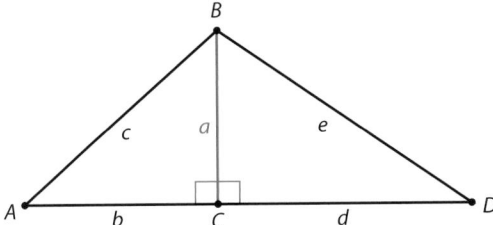

If $c = 4.5$ m, $\angle BAC = 35°$, and $\angle CBD = 60°$, find the length of e.

Solution

Since a is common to both triangles, we find it first, so that we can then find e.

$$\sin 35° = \frac{a}{4.5 \text{ m}}$$

$$a \approx 2.5811 \text{ m}$$

Now we use this to find e.

$$\cos 60° \approx \frac{2.5811}{e}$$

$$e \approx 5.16 \text{ m}$$

Often angles are given as either an angle of elevation (or inclination), or an angle of depression (or declination). An angle of elevation is above the horizontal, and an angle of depression is below the horizontal, as shown in the following diagrams.

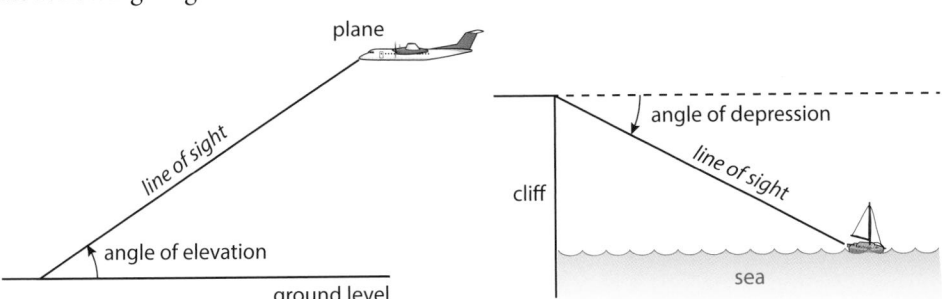

Example 6.5

Bjorn is on a boat in the bay which is located 200 metres from the shoreline. On the banks of the shoreline sits an old windmill. Bjorn looks from a height of 1.8 metres at the top of the windmill and notes that it has an angle of elevation of 10 degrees.

a) Draw a diagram to represent the angle of elevation to the top of the windmill building.

b) Determine the height of the top of the windmill building from the waterline.

Solution

a) Diagrammatically

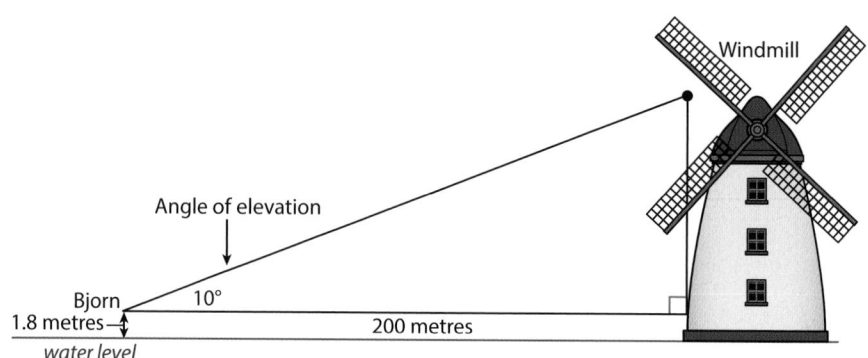

Windmill

Angle of elevation

↓

Bjorn 10°
1.8 metres →
water level

200 metres

b) The height of the top of the triangle is calculated by

$$\text{tangent } 10° = \frac{\text{height of triangle}}{200}$$

height of triangle = 200 tan 10° = 35.3 (3 s.f.) metres

As he was looking from a height of 1.8 metres, the height of the windmill above the waterline is 35.3 + 1.8 = 37.1 metres.

Exercise 6.2

1. Use the diagram below for the problems that follow.

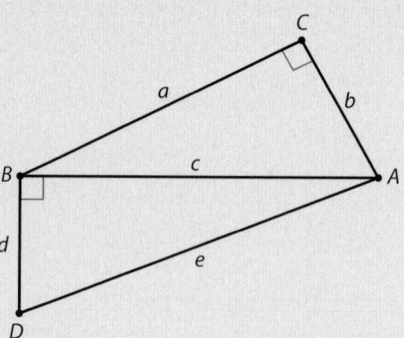

a) If $b = 10$, $\angle CBA = 20°$, and $\angle BAD = 20°$, find the length of e.
b) If $a = 12$, $\angle CAB = 65°$, and $d = 6$, find the length of e.
c) If $e = 14$, $\angle BAD = 15°$, and $\angle CBA = 25°$, find the length of a.
d) If $\angle BDA = 55°$, $d = 5.5$, and $\angle BAC = 62°$, find the length of b.

2. Use the diagram below for the problems that follow.

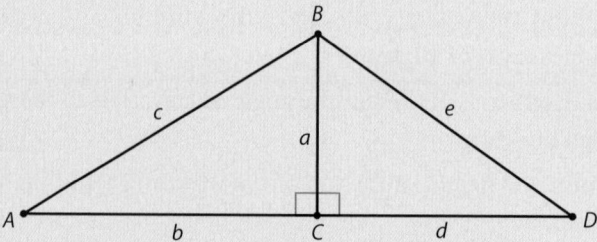

a) If $\angle BDA = 62°$, $d = 5.5$, and $\angle BAC = 55°$, find the length of b.
b) If $c = 12$, $\angle BAC = 30°$, and $\angle CBD = 40°$, find the length of e.

c) If $b = 120$, $c = 155$, and $\angle CBD = 55°$, find the length of d.

d) If $\angle BDC = 35°$, $d = 27\,300$, and $b = 23\,500$, find the length of c.

3. Use the diagram below for the problems that follow.

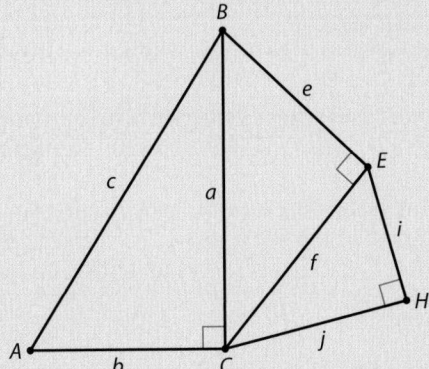

a) If $c = 13$, $b = 5$, $\angle BCE = 35°$, and $\angle ECH = 30°$, find the length of j.

b) If $b = 8$, $\angle BAC = 65°$, $e = 8$, and $\angle CEH = 65°$, find the length of i.

c) If $\angle ABC = 30°$, $c = 10$, $\angle CBE = 60°$, and $j = 5$, find the size of $\angle ECH$.

d) If $\angle ECH = 28°$, $i = 220$, $e = 300$, and $\angle BAC = 55°$, find the length of c.

4. Use the diagram below for the problems that follow.

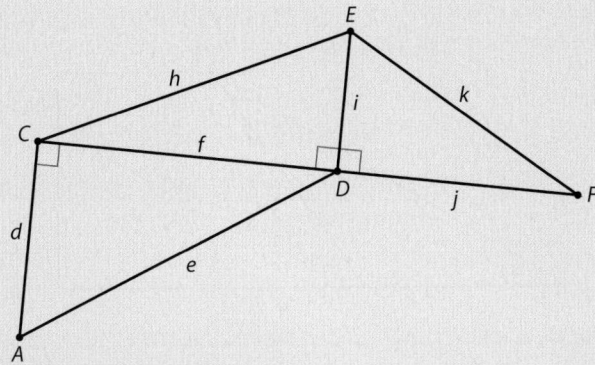

a) If $e = 1500$, $d = 900$, $E\hat{C}D = 20°$, and $E\hat{F}D = 30°$, find the length of k.

b) If $C\hat{A}D = 50°$, $e = 133$, $C\hat{E}D = 70°$, and $D\hat{E}F = 65°$, find the length of j.

c) If $d = 13.2$, $e = 17.6$, $E\hat{C}D = 10°$, and $j = 11.3$, find the size of $E\hat{F}D$.

d) If $C\hat{D}A = 39°$, $d = 192$, $h = 300$, and $k = 265$, find the size of $D\hat{E}F$.

5. Use the diagram below for the problems that follow.

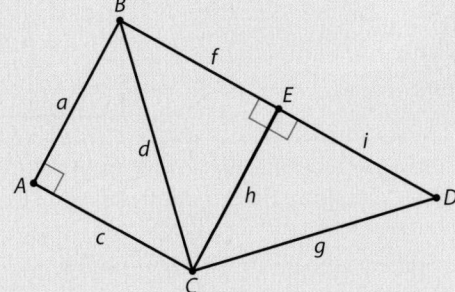

a) If $B\hat{C}A = 45°$, $c = 84$, $B\hat{C}E = 45°$, and $E\hat{D}C = 45°$, find the length of g.

b) If $a = c = f = h$, $E\hat{C}D = 40°$, and $g = 64$, find d.

6. Malachi looks up at the top of a tower and notices that the top of the tower is 100 m away and has an angle of elevation of 30° with the horizontal. Abdullah looks at the top of the same tower from the opposite side to Malachi, and notices that the base of the tower is 80 m away.
   a) How tall is the tower?
   b) How far is Abdullah from Malachi?
   c) From Abdullah's position, what is the angle up to the top of the tower?

7. Christie rides her bicycle 100 m at 45° east of north, then turns to 15° east of north and rides another 120 m. How far has she travelled from her starting position?

8. In the diagram below, Julie is standing at point C, and Manjeet is standing at point D. From Julie's perspective, the lake appears to subtend an angle of 60°, and from Manjeet's perspective, the lake appears to subtend an angle of 50°. The lake is actually 100 m wide.
   a) Find the length of [AD].
   b) Find the length of [BC].
   c) Find the distance between Julie and Manjeet
      (i) as the crow flies
      (ii) walking.

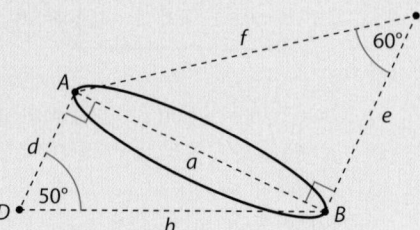

9. From Wu's position, the angle of elevation to the top of a building is 53°. From Tom's position the angle of elevation to the top of the building is 30°. If the building is 42 m tall, how far apart are Wu and Tom?

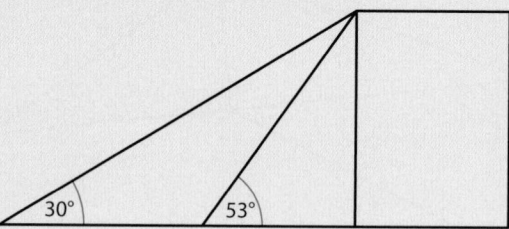

10. From the top of an office building, Irene notices that the angle of elevation to the top of a building 80 feet away is 45°, and the angle of depression to the bottom of the nearby building is 70°.
   a) Find the difference in height of the two buildings.
   b) Find the height of the building Irene is in.
   c) Find the height of the nearby office building.

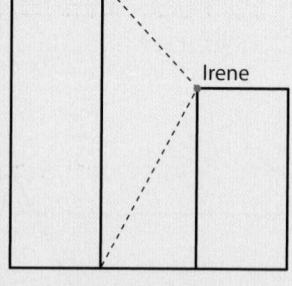

11. A rescue worker runs up a ladder that is placed at an angle of 65° with the ground and reaches a third storey window of a building, 32 feet above the ground. She enters the building and runs along three ramps to the roof, each of which is at an incline of 15°, and travels the entire width of the 80 feet wide building.
   a) How long is the ladder?
   b) How long is each of the ramps?

At the roof she finds a student
and jumps with him down a
slide to the ground 120 feet
below, at an angle of 40° with
the building.

c) How long is the slide?
d) How far does the rescue worker
   travel in total?

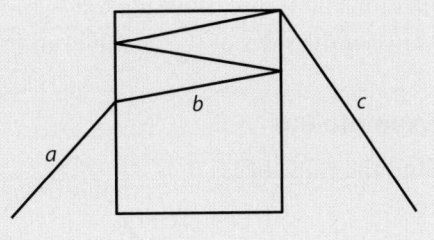

12. Three triangular plots of land are joined
    together as shown opposite. Five friends,
    Ali, Beth, Charles, Danika and Edward, live
    on the corners of the lots as shown by
    letters A, B, C, D, and E.
    a) If the distance between Ali and Beth is
       1.2 km, find the distance between:
       (i) Beth and Charles
       (ii) Ali and Charles.
    b) If the distance between Charles and
       Edward is 1.4 km, find the distance
       between:
       (i) Charles and Danika
       (ii) Edward and Danika.
    c) Find the distance between Ali and Edward.

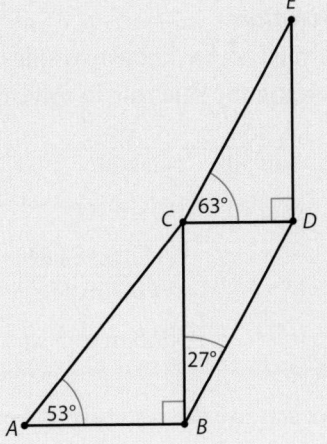

The sine rule

| 5.3 | Use of sine rule: $\dfrac{a}{\sin A} = \dfrac{b}{\sin B} = \dfrac{c}{\sin C}$ |

Not all triangles are right-angled triangles, so mathematicians have
developed other strategies that work with any triangle (including right-
angled triangles).

One of these strategies is known as the sine rule: $\dfrac{a}{\sin A} = \dfrac{b}{\sin B} = \dfrac{c}{\sin C}$

 When humans have had to
solve a problem, different
cultures have often found
different solutions. Why do you
think this happens?

◀ **Figure 6.2** The relationship
between the sides of a triangle and
its angles.

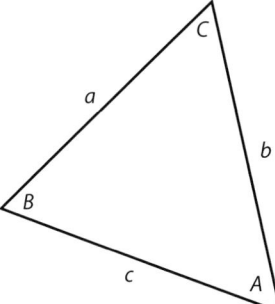

This equation represents three possible relationships between the sides
of the triangle, a, b, and c, and the angles of the triangle, A, B, and C.
Depending on which sides of the triangle are known, we choose which

relationship to use. If we look at Figure 6.2, we can see the relationship between the sides of the triangle, and the angles of the triangle.

Example 6.6

Find the value of x.

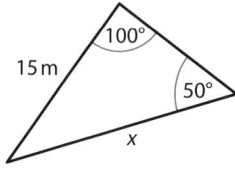

Solution

Given that we know two sides and their corresponding opposing angles we can use the sine rule to solve for x. Hence:

$$\frac{15}{\sin 50°} = \frac{x}{\sin 100°}$$

$$x \sin 50° = 15 \sin 100°$$

$$x = \frac{(15)(0.984\,81)}{0.766\,04}$$

$$x = 19.3 \text{ m to 3 s.f.}$$

● **Examiner's hint:** One of the ways that you earn points in your exam is by showing the formula you used with the variables correctly substituted.

We can use the sine rule to find any one of the following unknowns, given the other three are known: a, b, A, or B.

Example 6.7

Find the value of x.

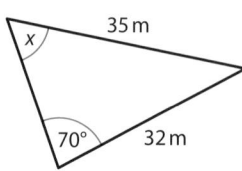

Solution

We know the values of two sides and one angle, so to find the angle opposite one of our known sides, we can use the sine rule. Hence:

$$\frac{32}{\sin x} = \frac{35}{\sin 70°}$$

$$35 \sin x = 32 \sin 70°$$

$$\sin x = \frac{32 \sin 70°}{35}$$

$$\sin x = \frac{32(0.93969)}{35}$$

$$x = \sin^{-1}(0.859\,15)$$

$$x = 59.2° \text{ to 3 s.f.}$$

To use the sine rule, you must know any three values of two of the sides of a triangle and their opposing angles.

Use the diagram below for the Questions 1 to 4.

1. a) If $AB = 10$, $\angle BAC = 40°$, $\angle BCA = 45°$, find AC.
   b) If $AC = 20$, $\angle ABC = 70°$, $\angle BAC = 60°$, find BC.

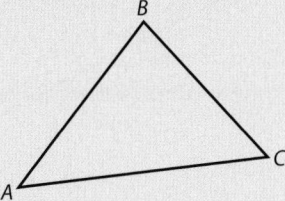

2. a) If $AB = 10$, $BC = 8$, $\angle BAC = 40°$, find $\angle BCA$.
   b) If $AC = 20$, $BC = 15$, $\angle ABC = 70°$, find $\angle BAC$.

3. a) If $AB = 10$, $BC = 8$, $\angle BAC = 40°$, find $\angle ABC$.
   b) If $AC = 20$, $BC = 15$, $\angle ABC = 70°$, find $\angle BCA$.

4. a) If $AB = 110$, $BC = 82$, $\angle BAC = 47°$, find AC.
   b) If $AC = 604$, $BC = 340$, $\angle ABC = 67°$, find AB.

5. Use the diagram below to answer the questions.

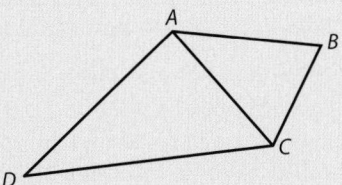

   a) Given $AD = 35\,cm$, $DC = 30\,cm$, $\angle DCA = 60°$, find $\angle DAC$.
   b) Use your answer from part a) to find AC.
   c) $\angle BAC = 32°$, $\angle ABC = 75°$. Use your answer from part b) to find BC.

6. The following diagram shows a triangle ABC. $AB = 8\,m$, $AC = 14\,m$, $BC = 18\,m$, and $\angle BAC = 110°$.

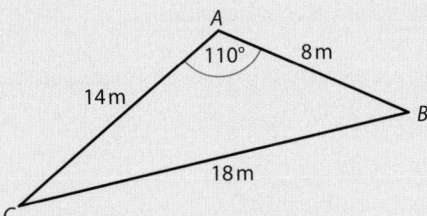

Calculate
   a) The size of angle ACB to 2 significant figures.
   b) The length of time (to the nearest second) taken to walk around triangle ABC at a rate of $1.5\,ms^{-1}$.

7. In the diagram, $AB = BC = 3\,cm$, $DC = 4.5\,cm$, angle $DAC = 88°$, and angle $ACD = 25°$.
   a) Calculate the length of $[AC]$.
   b) Calculate the perimeter of quadrilateral $ABCD$.

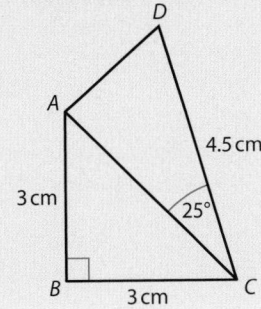

8. Salomé takes 3 minutes and 45 seconds running at a constant rate to run around a triangular park. The length of [AB] is 340 m and the length of [AC] is 450 m.

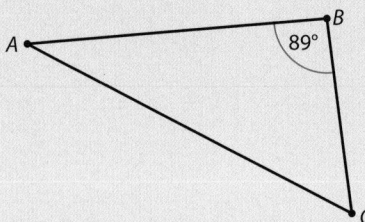

    a) If ∠ABC = 89°, find the size of ∠ACB.

    b) Find the length of [BC].

    c) Calculate Salomé's speed in ms⁻¹.

9. Andy lives 5 km from Billy, and Billy lives 3 km from David. The angle between the path from Andy's house to Billy's house and David's house is 32°.

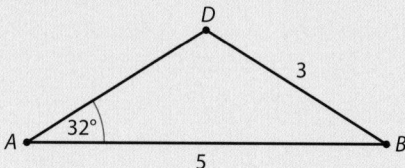

    a) Find the angle between the paths from Billy's house (as in the diagram).

    b) Find the distance between Andy's and David's houses.

    c) Find how many minutes it would take Andy to run from David's house to Billy's house and back to his house, following the paths shown above, at a speed of 6 ms⁻¹.

10. A surveyor finds the following measurements, recording his results in a diagram. Unfortunately, an emergency arises and he is called away before he can finish the measurements.

Calculate the rest of the missing angles and sides for the surveyor.

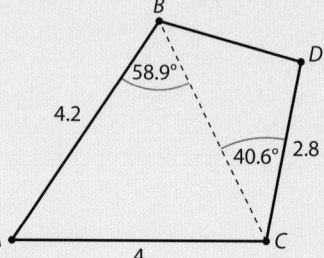

11. Given the diagram on the right:

    a) (i) Calculate the length of [BC].

        (ii) Calculate the length of [AC].

        (iii) Calculate the perimeter of triangle ABC.

    b) (i) Write down the information given in the diagram rounded to the nearest whole number.

        (ii) Calculate the length of [BC] using your answer to part b) (i).

        (iii) Calculate the length of [AC] using your answer to part b) (i).

        (iv) Calculate the perimeter of triangle ABC using your answers to b) (i) and b) (ii).

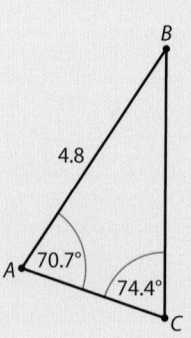

    c) Calculate the percent error in your answers from a) (iii) and b) (iv).

12. The relationship between three stars in our galaxy is shown below (all of the

stars lie in the same two-dimensional plane).
Star A is 6.2×10^4 light years from star B, and
8.9×10^4 light years from star C.

a) Calculate $\angle ACB$.

b) Calculate the distance between star
   B and star C. Write your answer in
   the form $a \times 10^k$ where $1 \leqslant a < 10$
   and $k \in \mathbb{Z}$.
   A light year is defined to be the distance light travels in one year. The speed
   of light is 3×10^8 ms^{-1}.

c) Using your answer to part b), calculate the distance between star B and star
   C in kilometres.

6.4 The cosine rule

| 5.3 | Use of cosine rule: $a^2 = b^2 + c^2 - 2bc \cos A$; $\cos A = \dfrac{b^2 + c^2 - a^2}{2bc}$ |

 The cosine rule is similar to
Pythagoras' theorem. Why do
you think this is? What does
the formula simplify to when
A is 90°?

Another rule mathematicians have developed to work with any triangle
(including right-angled triangles) is the cosine rule. It allows us to solve
for the unknown sides or angles in cases where the sine rule will not work.
If we know two sides of a triangle, and the included angle, we can use the
cosine rule to calculate the third side of the triangle. Also, if we know all
three sides of a triangle, we can use the cosine rule to find any angle of the
triangle.

Example 6.8

Find the value of x.

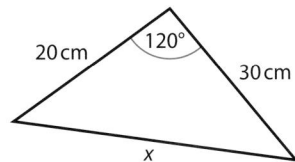

 When solving for an angle, you
can use another version of the
cosine rule:

$$\cos C = \frac{a^2 + b^2 + c^2}{2ab}$$

Solution

Given that we know the values of two adjacent sides and the angle between
them, we can use the cosine rule. Hence:

$x^2 = (20)^2 + (30)^2 - 2(20)(30) \cos 120° \text{ cm}^2$

$x^2 = 400 + 900 - (1200) \cos 120° \text{ cm}^2$

$x^2 = 1300 - (1200)(-0.5) \text{ cm}^2$

$x^2 = 1900 \text{ cm}^2$

$x = 43.6 \text{ cm to 3 s.f.}$

Example 6.9

Find the value of x.

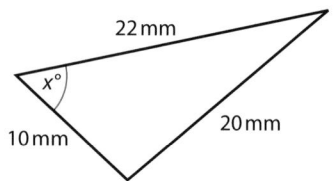

Solution

Since we know the values of all three sides of the triangle and we are looking for the value of one of the angles, we are able to use the cosine rule. Hence:

$$\cos x = \frac{(22)^2 + (10)^2 - (20)^2}{2(22)(10)}$$

$$\cos x = \frac{(484) + (100) - (400)}{440}$$

$$\cos x = \frac{184}{440}$$

$$\cos x = 0.4\dot{1}\dot{8}$$

$$x = \cos^{-1}(0.4\dot{1}\dot{8})$$

$$x = 65.3° \text{ to 3 s.f.}$$

We can use the cosine rule to solve for any angle of the triangle, when we know either two sides of a triangle and the angle between them, or all three sides of a triangle.

Exercise 6.4

1. Use the diagram below for all parts.

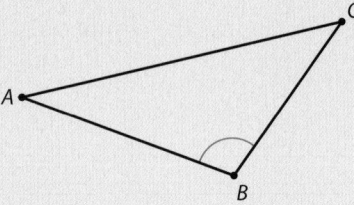

  a) If $\angle ABC = 110°$, $AB = 12$ cm, $BC = 12$ cm, find AC.
  b) If $\angle ABC = 95°$, $AB = 4.5$ m, $BC = 6.2$ m, find AC.

2. Use the diagram on the right for all parts.

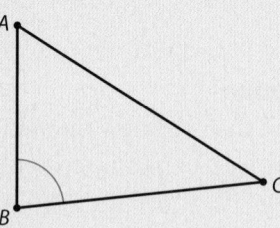

  a) If $\angle ABC = 85°$, $AB = 1200$ m, $BC = 1500$ m, find AC.
  b) If $\angle ABC = 80°$, $AB = 55$ feet, $BC = 36$ feet, find AC.

3. Use the diagram on the right for all parts.

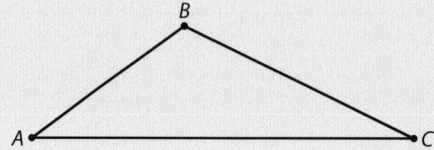

a) If $AB = 13$, $BC = 15$, and $AC = 21$, find $\angle BAC$.
b) If $AB = 0.48$, $BC = 0.65$, and $AC = 0.82$, find $\angle ABC$.

4. Use the diagram on the right for all parts.

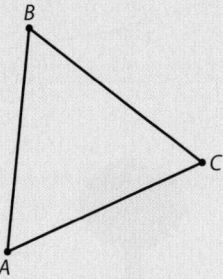

a) $\angle ABC = 65°$, $\angle BCA = 55°$, $AC = 22$ mm, $AB = 18$ mm. Find BC.
b) $\angle ABC = 70°$, $\angle BCA = 45°$, $AC = 250$ m, $AB = 164$ m. Find BC.

5. Use the diagram on the right for all parts.

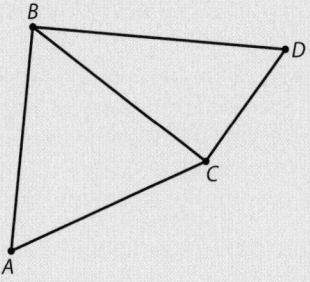

a) $AB = 13.4$ cm, $AC = 14.1$ cm, $\angle BAC = 62°$, $CD = 8.9$ cm, $\angle BCD = 92°$. Find BD.
b) $AB = 25$ m, $AC = 24$ m, $\angle BAC = 70°$, $CD = 20$ m, $BD = 27$ m. Find angle $\angle BDC$.

6. While riding her motorcycle, Natalia travels 10 km, then turns 86° to her right and travels another 9 km.

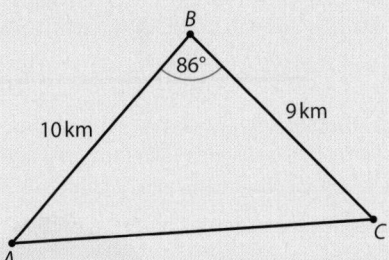

a) How far from her starting position is she?

b) If she travels at 30 km/h and takes no time to make turns, how long will it take her to travel around triangle ABC? Give your answer to the nearest minute.

7. Joseph is designing a house, and needs to calculate how much his roof will cost. He starts by drawing a picture of his roof, shown right.

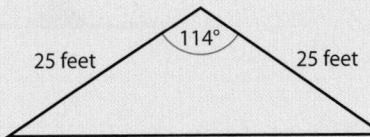

a) Find the angle between the base of the roof and the top of the roof.
b) Find the height of the roof.
c) Find the length of the base of the roof.
d) If the cost of the wood for the roof is $3.55 per foot, calculate the cost to build the frame of the roof shown.

8. A forestry worker notices that a baby tree is growing underneath a power line.

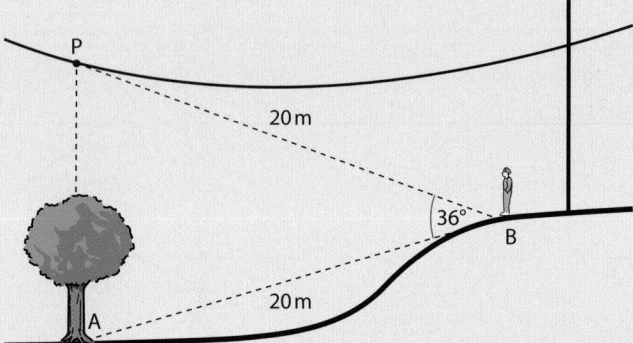

The distance from the forestry worker to the base of the tree is 20 m and the distance from the forestry worker to the point on the power line above the tree is also 20 m.

The angle between the two lines of sight to the base of the tree and the power line above the tree is 36°, as shown in the diagram.

a) Find the height, to the nearest cm, the tree will be when it first touches the power line.

b) If the tree is currently 3 m tall and grows at a rate of 20 cm a year, calculate how long, to the nearest year, it will take for the tree to first touch the power line.

9. Carli is flying a kite with 4.9 m of string let out. Jamal can see the kite 6.2 m away.

The angle between the string and Jamal's line of sight is 30°.

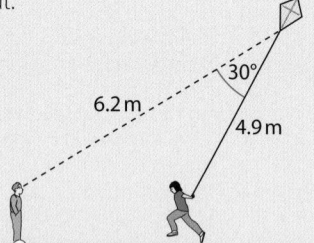

a) Find the distance between Carli and Jamal to the nearest tenth of a metre.

b) To the nearest second, how long will it take for Jamal to walk over to Carli at a rate of 0.5 ms⁻¹?

10. Isaac notices an apple fall from a tree. As the apple falls, he notices that it passes through an angle of 55°. At the beginning of the apple's fall it is located 6 m from Isaac's eye. When the apple hits the ground, it is 4 m from Isaac's eye.

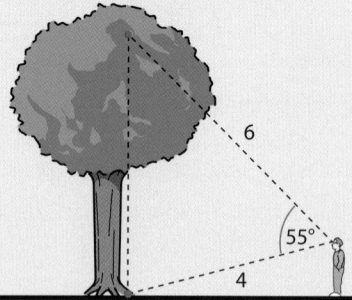

a) Find the distance the apple falls.

b) If the apple takes 0.71 seconds to fall, what is its average speed while falling?

11. Abiel is playing pool and decides to take a long bank shot from one corner of the table to a corner on the same side of the table.

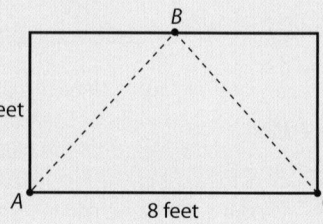

a) Find the length of AB if point B is in the middle of the side of the pool table.

b) Write down the length of BC.

c) Find ∠ABC.

d) Hence, find the angle the ball needs to make with the side of the pool table in order for Abiel to make his shot.

12. The figure shows two adjacent triangular fields ABC and ACD where $AD = 30$ m, $CD = 80$ m, and $BC = 50$ m; $\angle ADC = 60°$ and $\angle BAC = 30°$.
    a) Using triangle ACD, calculate the length of AC.
    b) Calculate the size of $\angle ABC$.

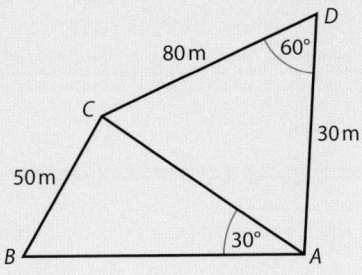

6.5 The area of a triangle

5.3   Use of area of triangle: $\frac{1}{2} ab \sin C$

What is the relationship between our traditional area formula and this alternative formula?

For a right-angled triangle, we can use the formula $A = \frac{1}{2}b \times h$ to find the area of a triangle. For any triangle (including right-angled triangles) we use the formula, $A = \frac{1}{2}ab \sin C$, referring to Figure 6.2 for the definitions of C, a, and b. Therefore if we have a triangle where we know two sides and the included angle, then we are able to find the area of the triangle.

Example 6.10

Find the area of triangle ABC.

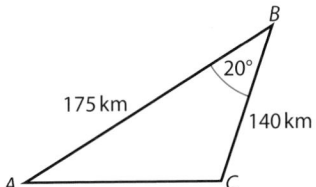

● **Examiner's hint:** Make sure the units of your answer for area are squared, and cubed for volume. This is a very common mistake students make on their exams.

Solution
Since triangle ABC is not a right-angled triangle, we need to use our alternative area formula for a triangle. We know the lengths of two sides of the triangle, and the angle between them. Hence:

$A = \frac{1}{2}(175)(140) \sin 20 \text{ km}^2$

$A \approx \frac{1}{2}(24\,500)(0.342\,02) \text{ km}^2$

$A \approx 4190 \text{ km}^2$

To use the formula $A = \frac{1}{2}ab \sin C$ to find the area of a triangle, we need to know two sides and the angle between them.

Example 6.11

Find the area of an equilateral triangle with a side length of 28 m.

Solution
In an equilateral triangle, all angles are equal to 60°. Therefore:

$A = \frac{1}{2}(28)(28) \sin 60° \text{ m}^2$

$A \approx 339 \text{ m}^2$

Exercise 6.5

1. Use the diagram below for all parts.
   a) $AB = 12$ m, $AC = 12$ m, $\angle BAC = 60°$.
      Find the area of $\triangle ABC$.
   b) $BC = 810$ km, $AC = 775$ km,
      $\angle BCA = 56°$. Find the area of $\triangle ABC$.

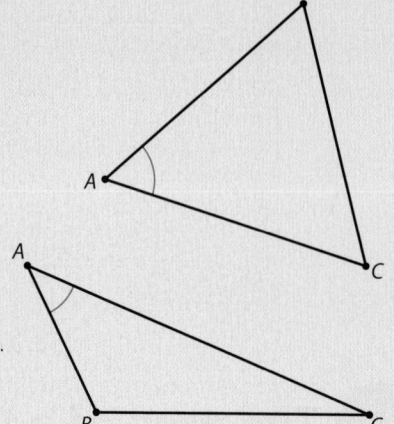

2. Use the diagram right for all parts.
   a) $AB = 37$ cm, $AC = 75$ cm,
      $\angle BAC = 29°$. Find the area of $\triangle ABC$.
   b) $AB = 3.2$ m, $BC = 4.1$ m, $\angle ABC = 125°$.
      Find the area of $\triangle ABC$.

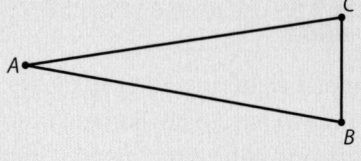

3. Use the diagram right for all parts.
   a) If $\angle BAC = 20°$, $\angle ABC = 80°$, $AC = 2.4$
      feet, $BC = 0.8$ feet, find the area of $\triangle ABC$.
   b) If $\angle BAC = 15°$, $\angle ABC = 100°$,
      $AC = 19$ inches, $BC = 6$ inches, find
      the area of $\triangle ABC$.

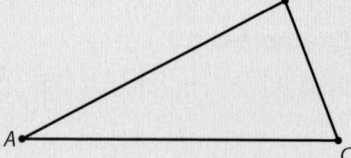

4. Use the diagram right for all parts.
   a) If $AB = 7.8$ m, $AC = 8$ m, $\angle ABC = 86°$.
      Find the area of $\triangle ABC$.
   b) If $AB = 45$ dm, $BC = 22$ dm, $\angle BAC = 27°$.
      Find the area of $\triangle ABC$.

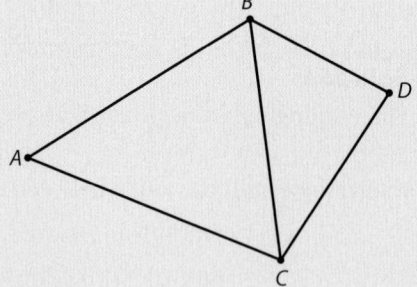

5. Use the diagram right for all parts.
   a) $AB = 2.3$ miles, $AC = 2.4$ miles,
      $\angle BAC = 50°$, $CD = 1.6$ miles, and
      $\angle BCD = 40°$. Find the area of
      quadrilateral $ABDC$.
   b) $AB = 5$ cm, $AC = 4.6$ cm,
      $\angle BAC = 60°$, $CD = 3$ cm, and
      $\angle BCD = 43°$. Find the area of
      quadrilateral $ABDC$.

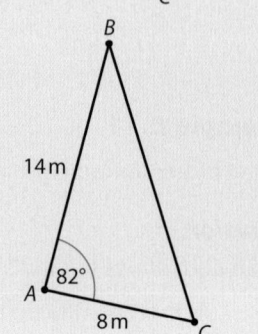

6. A triangular cloth with the dimensions
   right is used to make clothes for some children.
   a) Find the area of the cloth.
   b) If each 1.2 m² can be used to create one
      outfit, how many complete outfits can be
      created with the cloth?

7. The triangular sail of a boat has horizontal length 3.2 m and vertical length 10.4 m. The angle between these two sides is 89°.

a) Find the area of the sail to the nearest tenth of a metre.
b) If each 3 m² of sail contributes 1 km/h of speed in a strong wind, how fast is the boat in this picture moving (to the nearest kilometre per hour)?

8. A sign company has a contract to create 100 000 signs. Each of the signs has the shape shown in the diagram below.

Each sign is approximately an equilateral triangle with a side length of 40 cm.
a) Calculate the area of the equilateral triangle shown in the diagram to the nearest cm².
b) If the actual area of the signs is 680 cm², find the percent error in your estimate from part a) to the nearest tenth of a percent.

9. Quadrilateral *ABCD* is defined right.
a) Find *BD*.
b) Find the area of △*ABD*.
c) Find the size of ∠*BCD*.
d) Find the area of △*BCD*.

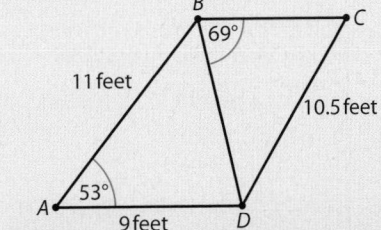

10. In this photo right we see one face of The Great Pyramid in Egypt.

This one slanted face has a diagonal height of 588 feet, and a base of 450 feet. The angle between the slant height and the base is 69°.

a) Find the area of this face of the Great Pyramid.

Each block that originally covered the face of the pyramid was a square about 8.1 feet across.

b) Calculate how many blocks would be required to cover the face of the pyramid again.

11. The quadrilateral *ABCD* is defined by the points *A*(−2,−2), *B*(−3,2), *C*(2,3), and *D*(1,−1).

a) Plot *ABCD* on 1 cm square graph paper.

b) Find the length of:

(i) *AB*   (ii) *AD*   (iii) *BD*   (iv) *D*

c) Find the size of:

(i) ∠*BAD*   (ii) ∠*BDC*

d) Hence find the area of quadrilateral *ABCD*.

12. A regular pentagon *ABCDE* is shown below.

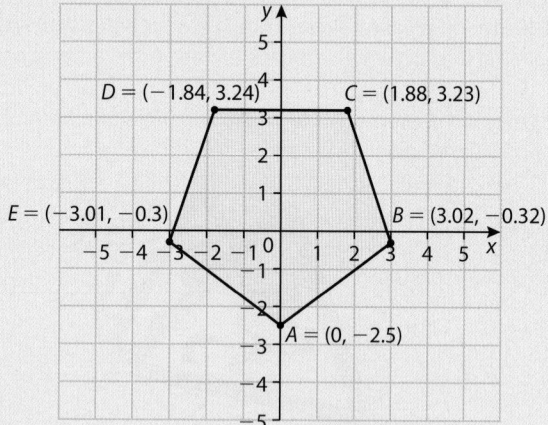

● **Examiner's hint:** When doing a question with multiple parts, make sure to record each answer with more accuracy than is required, in order to prevent rounding errors later in your work.

a) (i) Find *AD*.   (ii) Find *AC*.

(iii) Find *AB*.

b) Round your answers the nearest whole number.

(i) Find the size of ∠*DAE*.   (ii) Find the size of ∠*DAC*.

(iii) Find the size of ∠*CAB*.

c) Use your answers from part b).

(i) Find the area of ∠*DAE*.   (ii) Find the area of ∠*DAC*.

(iii) Find the area of ∠*CAB*.

d) Find the area of pentagon *ABCDE*.

6.6 The geometry of three-dimensional solids

5.4	Geometry of three-dimensional solids: cuboid; right prism; right pyramid; right cone; cylinder; sphere; hemisphere; and combinations of these solids.
5.5	Volume and surface areas of the three-dimensional solids.

Figure 6.3 Some three-dimensional solids.

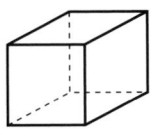 Cuboid

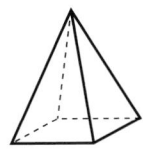 Right prism

Square pyramid

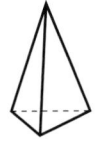 Triangular pyramid

 Cylinder

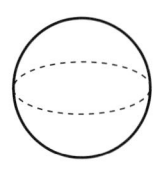 Right cone

Sphere

 Hemisphere

The square pyramid is seen around the world in the architecture of many different cultures. The Mayan temples of South America and the Great Pyramids of Egypt are two very famous examples.

 Plato of ancient Greece thought that all things in the universe could be formed from the five platonic solids. Was he right or wrong?

There are seven basic three-dimensional shapes that you need to know for this course: cuboid, right prism, right pyramid, cylinder, sphere, hemisphere, and a right cone. Refer to Figure 6.3 for a picture of each.

Each of these solids has a special formula for volume and surface area, as in the table below. Remember that the volume of the solid is a measure of how much 3-dimensional shape it fills, and the surface area is a measure of the total external area of the shape.

Name of shape	Volume formula	Surface area formula
Cuboid	$V = l \times w \times h$	$A = 2lw + 2lh + 2wh$
Right prism	$V = $ area of base $\times$ vertical height	$A = $ total area of all faces
Right pyramid	$V = \frac{1}{3}$(area of base $\times$ vertical height)	$A = $ area of each face + area of base
Cylinder	$V = \pi r^2 h$	$A = 2\pi r^2 + 2\pi rh$
Sphere	$V = \frac{4}{3}\pi r^3$	$A = 4\pi r^2$
Hemisphere	$V = \frac{2}{3}\pi r^2$	$A = 2\pi r^2 + \pi r^2$
Right cone	$V = \frac{1}{3}\pi r^2 h$	$A = \pi rl + \pi r^2$, where l is the slant height and r is the radius

Table 6.1 Volume and surface area formulae.

The surface area of a 3-dimensional solid is defined to be the total area of all of the external surfaces of the solid. So when in doubt, we can calculate each area separately, and total the individual areas.

Example 6.12

If the radius of a right cone is 12 cm and the height is 16 cm, find the volume and surface area of the cone to 3 s.f.

Solution

For the volume:

$$V = \tfrac{1}{3}\pi(12)^2(16) \text{ cm}^3$$
$$V \approx 2410 \text{ cm}^3$$

For the surface area, we need to first find the slant height l, which is the diagonal length of the cone. We can find this using Pythagoras' theorem. Hence:

$$l^2 = (12)^2 + (16)^2 \text{ cm}^2$$
$$l^2 = 400 \text{ cm}^2$$
$$l = 20 \text{ cm}$$

Now we can use l in our area formula, and add on the area of the base. Hence:

$$A = \pi(12)(20) + \pi(12)^2 \text{ cm}^2$$
$$A = \pi(240) + \pi(144) \text{ cm}^2$$
$$A \approx 1210 \text{ cm}^2$$

Example 6.13

Given the square pyramid right, with a height of 20 m and a base length of 10 m, find the volume and surface area of the pyramid.

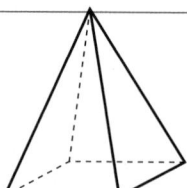

Solution

For the volume:

$$V = \tfrac{1}{3}(\text{area of the base}) \times (\text{height})$$
$$V = \tfrac{1}{3}(10\,\text{m})(10\,\text{m})(20\,\text{m})$$
$$V \approx 667 \text{ m}^3$$

For the surface area, we have 4 triangles and 1 square. We need more information to find the area of one of the triangles, such as the height of the triangle, but we already have the length of the base.

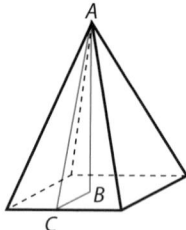

From the diagram above, we can see that $AB = 20$ m and $BC = 5$ m, and we can use this information to calculate AC using the Pythagorean theorem. Hence:

$$AC^2 = (20)^2 + (5)^2 \text{ m}^2$$
$$AC^2 = 400 + 25 \text{ m}^2$$
$$AC^2 = 425 \text{ m}^2$$
$$AC = \sqrt{425} \text{ m}$$
$$AC \approx 20.615\,53 \text{ m}$$

Now that we have the height of the triangles, we can calculate the area of one of the triangles, multiply it by four and then add on the area of the square base. Hence:

$$A \approx \tfrac{1}{2}(10)(20.615\,53) \times 4 + (10)(10) \text{ m}^2$$
$$A \approx 412.311 + 100 \text{ m}^2$$
$$A = 512 \text{ m}^2 \text{ to 3 s.f.}$$

Exercise 6.6

 To solve volume and surface area problems, we need to first find all of the dimensions of the object, which often requires trigonometry or Pythagoras' theorem.

1. For each cuboid, find the volume and surface area correct to 3 significant figures.
   a) $AB = 10\,\text{m}, BD = 12\,\text{m}, DG = 15\,\text{m}$
   b) $BF = 1.75\,\text{cm}, FG = 1.6\,\text{cm}, DC = 1.2\,\text{cm}$
   c) $AC = 290\,\text{mm}, CD = 240\,\text{mm}, DG = 260\,\text{mm}$

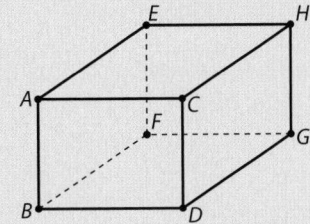

2. For each cylinder, find the volume and surface area correct to 2 significant figures.
   a) $r = 3.2$ feet, $h = 4.5$ feet
   b) $r = 410\,\text{dm}, h = 620\,\text{dm}$
   c) $r = 2$ inches, $h = 4$ inches

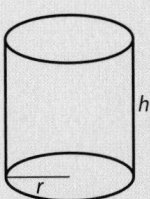

3. For each sphere, find the volume and surface area correct to 2 decimal places.
   a) $r = 9.1\,\text{m}$
   b) $r = 12$ inches
   c) $r = 2\,\text{cm}$

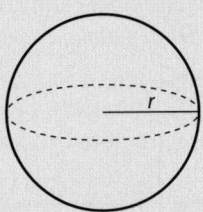

4. Find the volume and surface area of the following correct to 3 significant figures.
   a) $r = 3.4\,\text{cm}, h = 5.1\,\text{cm}$

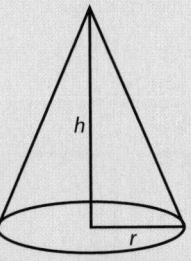

   b) $h = 23\,\text{mm}, s = 18\,\text{mm}$

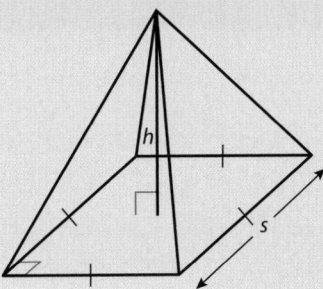

5. Find the volume and surface area of the prisms to the nearest tenth.
  a) $h = 2.4\,cm$, $b = 3.7\,cm$, $l = 3.7\,cm$
  b) $h = 7.7\,m$, $b = 8.4\,m$, $l = 7.0\,m$

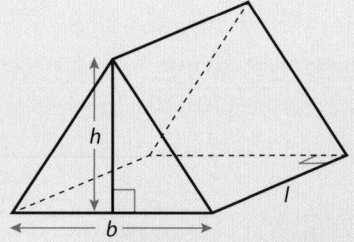

6. The Earth is approximately a sphere with a diameter of 12 756 km at its equator.
  a) Find the volume of the Earth to 3 significant figures.
  b) Write your answer to part a). in the form $a \times 10^k$ where $1 \leqslant a < 10$ and $k \in \mathbb{Z}$.
  c) Find the surface area of the Earth to 2 significant figures.

7. A glass company decides they want to create a new line of prisms for scientific research.

The prisms have an equilateral triangle as their base with side length of 2 cm.
  a) Find the area of the base of the prisms.
The height of each prism is 4 cm.
  b) Find the volume of each prism.
    The company decides to create 25 000 prisms, with a cost of $0.35 per cm³.
  c) Find the total cost of the prisms.

8. A traffic cone has a height of 30 cm from its base to its top. However, the top of the traffic cone is cut off as shown and an additional 8 cm of the cone is missing. The radius of the cone at its base is 6 cm and the radius at its top is 2.8 cm.
  a) Find the volume of the missing portion of the cone.
  b) Find the volume of the traffic cone.
  c) Find the surface area of the outside of the traffic cone.

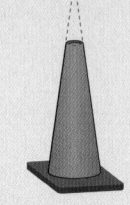

9. A grain silo is a giant cylindrical building, with a cone for a roof, used to store grain.

The grain silos shown have a diameter of 20 feet and a height of 50 feet to the base of their roof.
  a) Find the volume of the cylindrical portion of a grain silo.
    The total height of the silos is 58 feet.
  b) Find the total volume of a grain silo.
    One cubic foot of grain can feed a family of three for a week.
  c) How many people could be fed for a month by the grain in these six silos?

10. Find the volume of this prism.

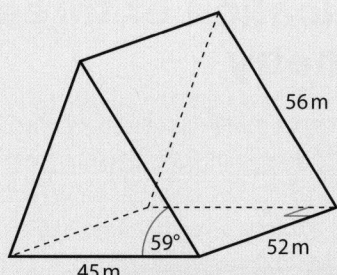

56 m

59°

52 m

45 m

11. A typical golf ball has a diameter of 3.4 cm.

a) Find the volume of a golf ball (assuming they are perfect spheres). The hole shown in the picture has a diameter 4 times that of the golf ball and is 3.8 times as deep.

b) Calculate the volume of the hole.

c) Find the surface area of a golf ball.

d) Estimate how many golf balls will fit in the hole.

e) It is estimated that at a particular golf course, one out of every 20 000 shots at this hole results in a 'hole-in-one'. At this rate, approximately how many people will have to play this hole in order for it to be filled only by 'hole-in-one's?

12. In the diagram, *ABEF*, *ABCD*, and *CDFE* are all rectangles.

$AD = 12$ cm, $DC = 20$ cm, and $DF = 5$ cm.

M is the midpoint of [*EF*] and *N* is the midpoint of [*CD*].

a) Calculate (i) the length of [*AF*];

(ii) the length of [*AM*].

b) Calculate the angle between [*AM*] and the face *ABCD*.

c) Calculate the volume of *ADFBCE*..

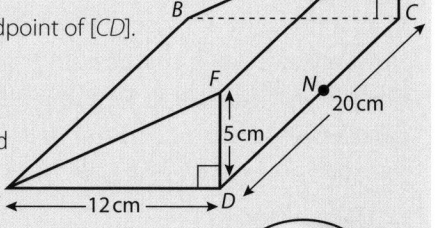

E

B

M

C

F

N

20 cm

5 cm

A

12 cm

D

13. A hemisphere sits on top of an 8 cm tall cylinder, as shown and the diameter of the cylinder is 12 cm.

a) Determine the volume of the combined shape.

b) Determine the surface area of the shape.

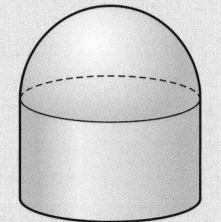

14. A barn is built in the shape of a rectangular prism, as shown and the roof is in the shape of a half cylinder.

The dimensions of the base of the barn are 9 metres by 20 metres and the heights of the walls are 5 metres.

a) Write down the height of the roof section.

b) Hence write down the total height of the barn.

c) The farmer intends to paint the entire barn in the same colour but needs to know how much paint is required. Determine the total surface area that will require painting.

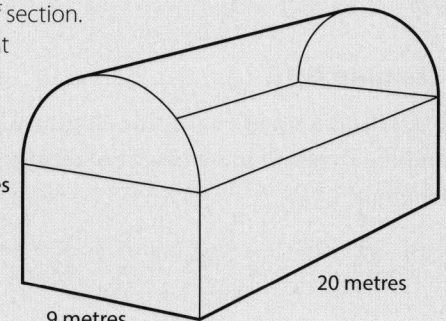

5 metres

9 metres

20 metres

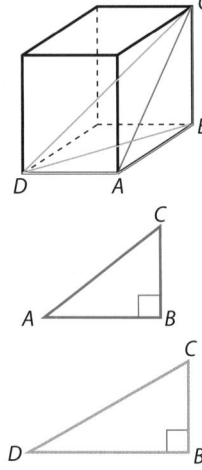

Figure 6.4 Every three-dimensional shape has two-dimensional shapes within it.

6.7 Application of three-dimensional geometry

5.4 Geometry of three-dimensional solids: cuboid; right prism; right pyramid; right cone; cylinder; sphere; hemisphere; and combinations of these solids. The distance between two points; e.g. between two vertices, or vertices with midpoints, or midpoints with midpoints. The size of an angle between two lines or between a line and a plane.

We know from Chapter 1 how to find the length of a line using Pythagoras' theorem. We can also use Pythagoras' theorem to solve problems involving three-dimensional shapes. If we look at Figure 6.4, we can see that triangle ABC is a right-angled triangle, which means we can find the length of AC using Pythagoras. Similarly, we can see that triangle DBC is a right-angled triangle.

Example 6.14

Find the length of $[DC]$ from Figure 6.4, if $AD = 10$ feet, $AB = 12$ feet, and $BC = 14$ feet.

Solution

Triangle DBC is a right-angled triangle, so we could use Pythagoras' theorem to solve for side $[DC]$, but we only know the length of side $[BC]$. However, side DB is shared with triangle DBA, and we know the lengths of side AD and side AB, so we can use this information to solve for the length DB. Hence:

$$DB^2 = (10)^2 + (12)^2 \text{ feet}^2$$
$$DB^2 = 100 + 144 \text{ feet}^2$$
$$DB^2 = 244 \text{ feet}^2$$
$$DB \approx 15.6205 \text{ feet}$$

Next:

$$DC^2 \approx (15.6205)^2 + (14)^2 \text{ feet}^2$$
$$DC^2 = 244 + 196 \text{ feet}^2$$
$$DC^2 = 440 \text{ feet}^2$$
$$DC \approx 21.0 \text{ feet}$$

Example 6.15

If $DEBC$ is a square with side length of 20 cm and the pyramid has a height of 30 cm, find the length of $[AB]$.

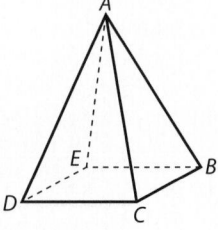

Solution

Our first step is to draw two right-angled triangles onto our diagram.

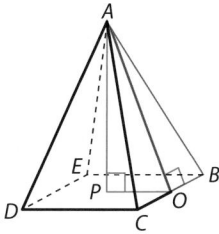

From our drawing, we can see triangle ABO contains side AB, and shares side AO with triangle AOP. So first we calculate the length of AO, and use this to calculate the length of AB. Hence:

$$AO^2 = AP^2 + OP^2$$
$$AO^2 = (10\,\text{cm})^2 + (30\,\text{cm})^2$$
$$AO^2 = 1000\,\text{cm}^2$$
$$AO \approx 31.622\,78\,\text{cm}$$

Now we use this to find the length of AB. Note that we have already calculated the value of AO^2, so we use this in our further calculations for more accuracy.

$$AB^2 = AO^2 + OB^2$$
$$AB^2 = (1000\,\text{cm}^2) + (10\,\text{cm})^2$$
$$AB^2 = 1100\,\text{cm}^2$$
$$AB \approx 33.2\,\text{cm}$$

We can also use the right-angled triangles we have just identified in the three-dimensional shapes, to find the angles between lines or planes in the shape, using trigonometry discussed earlier in this text.

 What are some relationships between two-dimensional and three-dimensional shapes?

Example 6.16

The shape on the right is a cube with all edges having length 25 km.

Find the angle between plane $ABFE$ and plane $HGFE$.

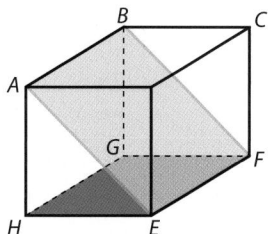

Solution

From the diagram, we can see that we are looking for $\angle AEH$. This is an acute angle of triangle AHE. Since we know the length of $[AH]$ and $[HE]$, we can use right-angled trigonometry to find the angle. Hence:

$$\tan x = \frac{25}{25}$$
$$\tan x = 1$$
$$x = \tan^{-1} 1$$
$$x = 45°$$

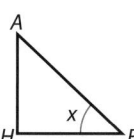

Example 6.17

ABC is an equilateral triangle with side length 15 cm, and the height of pyramid *ABCD* is 28 cm. Find the area of triangle *DBC*.

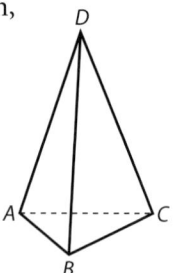

Solution

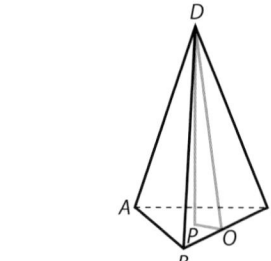

We know the length of the base of triangle *DBC*. We just need to calculate its height, and then we can find its area. From the diagram above, we can see we need to first find the length of $[PO]$, so that we can next calculate the length of $[OD]$. If we look more closely at triangle *ABC*, we can see the following:

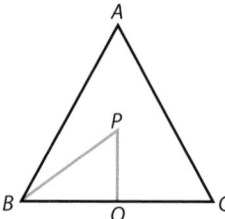

Triangle *BPO* is also a right-angled triangle, with $\angle PBO = \frac{1}{2} \angle ABC = 30°$. Since we know an angle and a side of this triangle, we can solve for *PO* using right-angled trigonometry. Let $PO = x$.

$$\tan 30° = \frac{\text{opp}}{\text{adj}} = \frac{x}{BO} = \frac{x}{7.5}$$

$$x = 7.5 \tan 30°$$

$$x \approx 4.3301 \text{ cm}$$

Now we use this information and find *OD*. Hence:

$$OD^2 = x^2 + (28)^2$$
$$OD^2 = 18.75 + 784$$
$$OD^2 = 802.75$$
$$OD \approx 28.33 \text{ cm}$$

The area formula we need to use is $A = \frac{1}{2}b \times h$. Hence:

$$A \approx \frac{1}{2} \times 15 \times 28.33 \text{ cm}^2$$
$$A \approx 212.5 \text{ cm}^2$$

1. Use the diagram below for parts a) and b) to find the length of *AG* to the nearest tenth.

   a) *AC* = 12 cm, *CD* = 12 cm, *DG* = 20 cm

   b) *AC* = 4.5 m, *DG* = 6.2 m, *FG* = 4.1 m

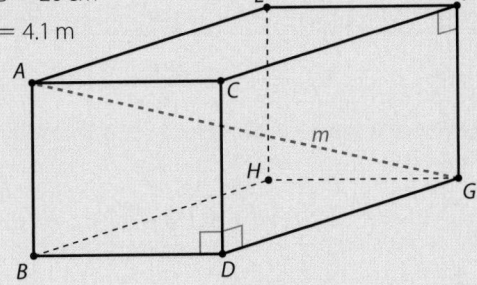

2. Use the diagram below to find the angle between plane *AEGD* and plane *BDGH*.

   a) *BD* = 8 feet, *AB* = 6 feet

   b) *AD* = 13 cm, *AB* = 5 cm

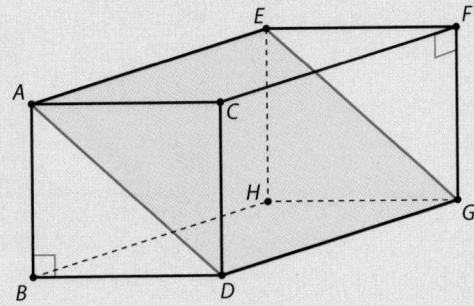

3. Use the diagram below to find the angle between [*FA*] and the base *ABCD*.

   a) *FE* = 10 inches, *AE* = 8 inches

   b) *AD* = *DC* = *CB* = *BA* = *FE* = 60 mm

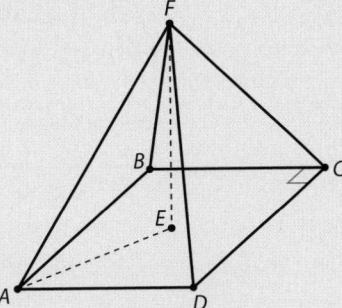

4. Use the diagram of the square pyramid below for parts a) and b).

   a) If *FC* = 24 mm and *BC* = *CD* = 20 mm, find the angle between plane *FBC* and plane *FCD*.

   b) If *AD* = 9.2 cm and *FD* = *FC* = 11.3 cm, find the angle between plane *ABCD* and plane *FCD*.

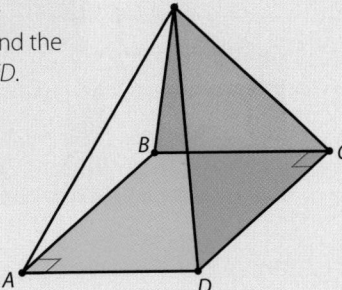

5. Use the diagram of the rectangular prism on the right for parts a) and b).

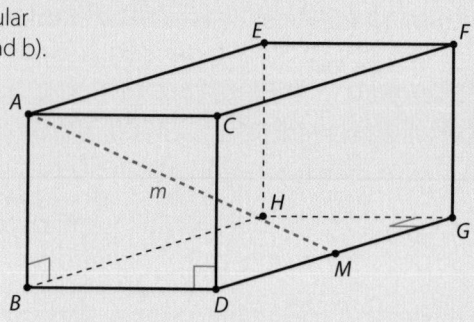

a) $AB = 6\,\text{m}$, $BD = 8\,\text{m}$, $DG = 10\,\text{m}$. Find $\angle AMD$.
b) $AB = 12\,\text{m}$, $BD = 18\,\text{m}$, $DG = 10\,\text{m}$. Find $\angle AMC$.

6. An ancient Mayan temple is shown in the picture below.

This temple is approximately a square pyramid with a base 200 feet by 200 feet, and height of 180 feet.

Find the angle the staircase makes with the ground.

7. During a storm, a 20 m tree is blown over by the wind but is prevented from falling completely by another tree. The top of the fallen tree is suspended 5 m off the ground.

Find the angle the tree makes with the ground.

8. The Brooklyn Bridge uses large cables to help support its columns, which are nearly straight lines.

The length of the large cable from the top of the tower to the road beneath is 250 m and the height of the column is 100 m. Find the angle the cable makes with the road.

9. Find the angle between plane *ABE* and the base of the triangular prism right.

AC = 4 inches, *CF* = 8 inches, and *BC* = 6 inches.

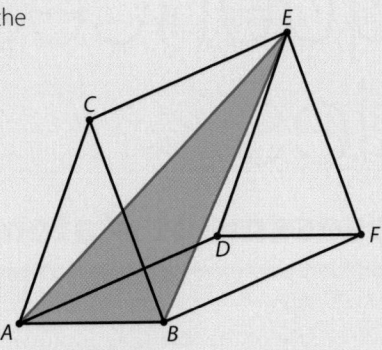

10. In the diagram of the house below, the total height of the house is 25 feet, and the width of the house is 20 feet.
The slope of the roof is 14 feet long.
a) Find the angle the roof makes with the ground.
b) Find the length of a ladder whose top meets the base of the roof, and which is at the same angle with the ground as the roof.

11. Plane *ACGH* and plane *BCFH* intersect on *CH*.

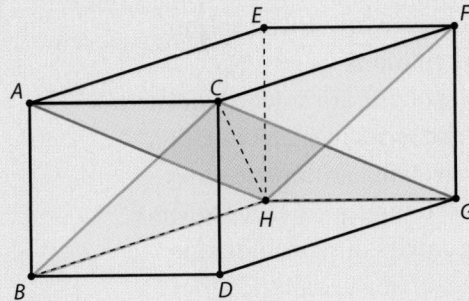

AB = 6, *BD* = 8, *DG* = 8
a) Find *AG*.
b) Find *BF*.
c) Copy the diagram, draw [*AG*] and [*BF*] and mark their intersection as *M*.
d) Find ∠*BMA* and hence the angle between *ACGH* and *BCFH*.

12. In the diagram, *PQRS* is the square base of a solid right pyramid with vertex *V*. The sides of the square are 8 cm, and the height [*VG*] is 12 cm. *M* is the midpoint of *QR*.
a) (i) Write down the length of [*GM*].
(ii) Calculate the length of [*VM*].
b) Find
(i) the total surface area of the pyramid
(ii) the angle between the face *VQR* and the base of the pyramid
(iii) the volume of the pyramid.

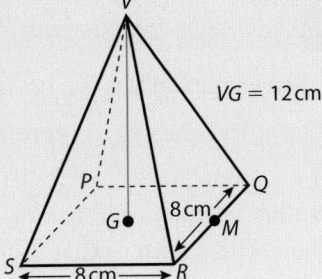

Sequences and Series

Assessment statements

1.7	Arithmetic sequences and series, and their applications. Use of the formulae for the nth term and the sum of the first n terms of the sequence.
1.8	Geometric sequences and series. Use of the formulae for the nth term and the sum of the first n terms of the sequence.
1.9	Financial applications of geometric sequences and series: compound interest, annual depreciation.

Overview

By the end of this chapter you will be able to:
- write recursive definitions
- solve problems involving arithmetic sequences
- write an arithmetic series in sigma notation
- solve problems involving arithmetic series
- solve problems involving geometric sequence
- write a geometric series in sigma notation
- solve problems involving geometric series
- solve problems involving compound interest
- solve problems involving appreciation and depreciation

There are many well known sequences. One of those is the Fibonacci sequence:
1, 1, 2, 3, 5, 8, 13, 21, 34, ...

Leonardo Fibonacci, also known as Leonardo of Pisa (c.1170 – c.1250) was perhaps the most talented mathematician of the 13th century.

To learn more about the Italian mathematician, Leonardo Fibonacci, visit www. pearsonhotlinks.co.uk, enter the ISBN for this book and click on weblink 7.1

7.1 Arithmetic sequences

1.7	Arithmetic sequences and series, and their applications. Use of the formulae for the nth term and the sum of the first n terms of the sequence.

A sequence of numbers is a list of numbers that are ordered in a pattern.

For example, 2, 4, 6, 8, ... represents a sequence with a pattern that is easy to see.

As another example, 3, 3.1, 3.14, 3.141, ... represents a sequence of numbers with a pattern that is not so easy to see.

In general, a sequence can be written as

$u_1, u_2, u_3, \ldots, u_n, u_{n+1}, \ldots$, where
- u_1 represents the first term, u_2 represents the second term, and so on
- u_1 is read as 'u subscript one', 'u sub one', or simply 'u one'
- u_n represents the nth term.

For example, in the first sequence above, $u_1 = 2$, $u_2 = 4$, $u_3 = 6$.

When considering the sequence 2, 4, 6, 8, ... it common to say that the 'next' term is 10. It follows, when considering the sequence 2, 4, 6, 8, 10, ..., that 8 can be thought of as the 'previous' term with respect to 10.

Sequences can often be worded as: the next term is equal to the previous term plus a constant number. In symbols:

$$\underbrace{u_{n+1}}_{\text{The next term}} = \underbrace{u_n}_{\text{The previous term}} + \underbrace{d}_{\text{The constant number}}$$

This is called a recursive definition.

$\therefore u_{n+1} - u_n = d$

The constant number d is referred to as 'the common difference'.

An arithmetic sequence is defined as a sequence whose next term is always the previous term plus a real number. The formal definition for an arithmetic sequence is given below.

> An arithmetic sequence is a sequence such that $u_1 = a$ and $u_{n+1} = u_n + d$ where $n \in \mathbb{Z}^+$ and $d \in \mathbb{R}$.

Example 7.1

Let $u_1 = 1$ and $d = 3$. Find the first 5 terms of the sequence and then write the sequence.

Solution

$$u_1 = 1$$
$$u_2 = 1 + 3 = 4$$
$$u_3 = 4 + 3 = 7$$
$$u_4 = 7 + 3 = 10$$
$$u_1 = 10 + 3 = 13$$

$\therefore$ The sequence is: 1, 4, 7, 10, 13

Example 7.2

Write a recursive definition for the sequence: 1, 5, 9, 13, 17, ...

Solution

By inspection, each successive term is 4 more than the previous term. Therefore, let $u_1 = 1$ and $u_{n+1} = u_n + 4$.

Now suppose we wanted to find the 100th term of the arithmetic sequence, 1, 5, 9, 13, 17, ...

We could find the term by:

Method I: Listing the first 100 terms by hand (pencil and paper)

Method II: Using the GDC and the ENTER key

Method III: Using a formula

Method I makes use of the recursive definition by adding 4 to each previous term until the number of terms reaches 100. This list, although possible, is not very practical.

• **Examiner's hint:** This formula $(u_n = u_1 + (n-1)d)$ can be found in the Mathematical Studies SL Formula Booklet (MSSLFB).

The formula for the nth term of an arithmetic sequence can be arrived at by using a method called inductive reasoning. It can be proved by a method called 'Proof by Mathematical Induction'.

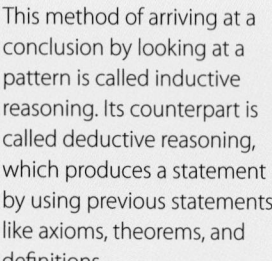

This method of arriving at a conclusion by looking at a pattern is called inductive reasoning. Its counterpart is called deductive reasoning, which produces a statement by using previous statements like axioms, theorems, and definitions.

How do we make use of inductive reasoning in daily lives? Is this a valid form of reasoning?

What is the difference between mathematical inductive reasoning and proof by mathematical induction? Why do we need proofs?

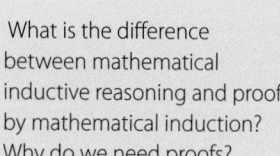

Choosing an appropriate formula, listing the information term-by-term, substituting back into the formula, and then solving, is a problem solving technique that is very useful. We advise using this technique whenever possible throughout this chapter.

Method II makes use of the fact that when the same operation ($\div$, $\times$, $-$, $+$) is repeated at least twice, the operation is 'remembered' by the TI calculator and pressing the ENTER key activates the operation again.

1, ENTER, $+$, 4, ENTER, $+$, 4, ENTER, ENTER, ENTER, and so on 95 more times. (You should see the results below.)

1   5   9   13   17   ...

Even though this method is faster, it does require that you count accurately and so may not be reliable when many terms are involved.

Method III involves the use of the nth term formula for an **arithmetic sequence**.

Consider the arithmetic sequence:

$$u_1, \underbrace{u_1 + d}_{\text{2nd}}, \underbrace{(u_1 + d) + d}_{\text{3rd}}, \underbrace{((u_1 + d) + d) + d}_{\text{4th} \ldots}, \ldots$$

The number of common differences between each term and u_1 is always one less than the term number.

We can simplify the sequence as: $u_1, \underbrace{u_1 + 1d}_{\text{2nd}}, \underbrace{u_1 + 2d}_{\text{3rd}}, \underbrace{u_1 + 3d}_{\text{4th}}, \ldots, \underbrace{u_1 + (n-1)d}_{n\text{th}},$

Using this idea, we can write the formula for the nth term of an arithmetic sequence:

$$u_n = u_1 + (n-1)d, \text{ where}$$

- u_n is the nth term of the sequence
- u_1 is the first term of the sequence
- n is the number of terms in the sequence
- d is the common difference between consecutive terms of the sequence.

Example 7.3

Using the formula, $u_n = u_1 + (n-1)d$, find the 100th term of the sequence 1, 5, 9, 13, 17, ...

Solution

$$u_n = u_1 + (n-1)d$$
$$u_1 = 1$$
$$n = 100$$
$$d = 5 - 1 = 9 - 5 = 13 - 9 = 17 - 13 = 4$$
$$u_{100} = ?$$
$$\therefore \quad u_{100} = 1 + (100 - 1) \cdot 4 = 397$$

Example 7.4

Given the sequence, $u_{50} = 6$, $u_{51} = 11$, $u_{52} = 16$, find u_1.

Solution

Draw the diagram. $\underbrace{\quad}_{u_1}, \underbrace{\quad}_{u_2}, \underbrace{\quad}_{u_3}, \ldots, \underbrace{6}_{u_{50}}, \underbrace{11}_{u_{51}}, \underbrace{16}_{u_{52}}$

Write an appropriate formula. $u_n = u_1 + (n-1)d$

List the information: $u_1 = ?$
$$d = 11 - 6 = 5$$
$$n = 52$$
$$u_{52} = 16$$

Substitute and solve: $16 = u_1 + (52 - 1) \cdot 5 = u_1 + 255$
$$\therefore \quad u_1 = 16 - 255 = -239$$

Example 7.5

For the arithmetic sequence, $u_{10} = 25$, $u_{15} = 70$, find the first term.

Solution

Draw a diagram: __, __, __, __, __, __, __, __, __, 25, __, __, __, __, 70

Write an appropriate formula: $u_n = u_1 + (n - 1)d$

List the information: $u_1 = 25$ (Do this in order to find d.)

$d = ?$

$n = 6$ (From the 10th term to the 15th term)

$u_6 = 70$

Substitute and solve: $70 = 25 + (6 - 1) \cdot d$

$45 = 5d$

$\therefore \quad d = 9$

Since $d = 9$ is the common difference for the **entire** sequence you can use the procedure again.

$u_1 = ?$

$d = 9$

$n = 15$

$u_{15} = 70$

Hence, $70 = u_1 + (15 - 1) \cdot 9$

$70 = u_1 + 126$

$\therefore \quad u_1 = -56$

Since the common difference is the same for every pair of consecutive terms, a popular technique for some problems is to think of a middle term in the sequence as the first term.

• **Examiner's hint:** You could let $n = 10$ as long as $u_{10} = 25$.

Example 7.6

Susie decides to deposit an initial amount of $1200 into a savings account on January 1.

The account pays 5% simple interest on the initial amount of money. Assuming that she does not deposit or withdraw any money, how much will be in the account at the end of 10 years?

Solution

The interest that is earned each year is $1200(0.05) = 60$. (See section 7.5 for more information.)

Method I: Draw the diagram:

Begin Year 1	End Year 1	EY2	EY3	...	EY10
1200	1200+60=	1260+60=	1320+60		
	1260	1320	1380	...	?

Redraw the diagram: 1260, 1320, 1380, ..., ?

Write an appropriate formula: $u_n = u_1 + (n - 1)d$

List the information. $u_1 = 1260$

$n = 10$

$d = 60$

$u_{10} = ?$

Substitute and solve: $u_{10} = 1260 + (10 - 1) \cdot 60 = 1260 + 9 \cdot 60$

$$\therefore \quad u_{10} = \$1800$$

Method II: Use the information given in the problem.
Method I uses the concept of arithmetic sequences and their properties. A faster and more efficient way to solve the problem is to use only the information given in the problem: 1200, 10, and 0.05

Rearranging the information:
$$u_{10} = 1260 + (10 - 1) \cdot 60$$
$$u_{10} = 1260 + 10 \cdot 60 - 60$$
$$u_{10} = 1200 + 10 \cdot 60$$
$$u_{10} = 1200 + 10\,(1200 \cdot 0.05)$$
$$u_{10} = 1200 + 1200 \cdot 0.05 \cdot 10$$

For ease of recalling, substitute appropriate formulae variables for the numbers:

Hence, $A = C + \dfrac{Crn}{100}$, where A is the final amount, C is the capital invested, r is the rate of simple interest (e.g. if the interest rate is 5%, then $r = 5$), and n is the time in number of years.

$$\therefore \quad A = 1200 + \frac{1200(5)(10)}{100} = 1800$$

Exercise 7.1

1. State which of the following are arithmetic sequences.
   a) 10, 20, 30, 40, …
   b) 3, 3.1, 3.14, 3.141, …
   c) 100, 50, 25, 12.5, …
   d) 1, 1, 1, 1, …
   e) 5, −2, −9, −16, …
   f) $3\frac{3}{4}, 3\frac{7}{8}, 4, 4\frac{1}{8}, 4\frac{1}{4}, …$

2. Write the first four terms of each sequence. If it is arithmetic, say so.
   a) $u_1 = 2$ and $u_{n+1} = u_n + 8$
   b) $u_1 = 5$ and $u_{n+1} = 2u_n$
   c) $u_1 = -3$ and $u_{n+1} = 3 - u_n$
   d) $u_1 = 1$ and $u_{n+1} = (-1)u_n$
   e) $u_1 = -3$ and $d = 5$
   f) $u_1 = 4$ and $u_{n+1} = \sqrt{u_n}$

3. Write each sequence in recursive definition form.
   a) 2, 4, 6, 8, …
   b) 15, 11, 7, 3, …
   c) 2, 4, 8, 16, …
   d) $1, \frac{1}{2}, \frac{1}{4}, \frac{1}{8}, …$
   e) 1, 8, 27, 64, …
   f) 5, 7, 9, 11, …

4. For each of the following arithmetic sequences, find the 8th term and the common difference.
   a) 5, 13, 21, 29, …
   b) 50, 35, 20, 5, …
   c) $u_1 = 11$ and $u_{n+1} = u_n - 12$
   d) $u_1 = 5$ and $u_{n+1} = u_n + 7$
   e) $u_3 = 9$ and $u_{n+1} = u_n + 10$
   f) $u_1 = -4$ and $u_{n+1} = u_n - 4$

5. For each of the following find:
   a) The 10th term of: 3, 11, 19, 27, …
   b) u_{12} for the sequence: 16, 13, 10, 7, …
   c) The 100th term of the sequence where, $u_1 = 7$ and $u_{n+1} = u_n + 19$
   d) u_{150} for the sequence: 2, 8, 14, 20, …
   e) The 313th term for: 500, 473, 446, 419, …
   f) u_{75} for the sequence: $1, \frac{3}{2}, 2, \frac{5}{2}, …$

6. For the nth term formula for an arithmetic sequence, $u_n = u_1 + (n - 1)d$, three of the four unknowns are given; find the fourth.
   a) $u_1 = 4, n = 50, d = 3$

b) $u_{17} = 1001, n = 17, d = 5$

c) $u_n = 2121, u_1 = 252, d = 3$

d) $u_{15} = \frac{252}{3}, u_1 = \frac{1}{3}, n = 15$

7. Each of the following represents an arithmetic sequence. Find u_1.

a) ___, ___, ___, ___, 17, ___, ___, 35

b) ___, ___, ___, ___, ___, ___, −27, ___, ___, ___, ___, ___, ___, ___, −123

c) $u_{37} = 145, u_{85} = 673$

d) $u_{52} = 70, u_{125} = -149$

8. Each of the following represents an arithmetic sequence. Find the missing terms.

a) 3, ___, ___, 24

b) ___, ___, 5 ___, ___, ___, 117

c) $\frac{1}{4}$, ___, ___, 4

d) $\frac{2}{3}$, ___, $\frac{3}{16}$

9. Find x if each of the following sequences is arithmetic.

a) $8 - x, x, x + 8$

b) $x, 2x + 2, x - 5$

10. Insert the required number of terms between the given values so that the resulting sequence is arithmetic.

a) $u_1 = 2$ and $u_6 = 17$; 4 terms

b) $u_4 = -29$ and $u_{10} = -53$; 5 terms

11. Write the first 15 terms for the sequences given by the recursive definition:
$$u_1 = 1, u_2 = 1 \text{ and } u_{n+2} = u_{n+1} + u_n.$$

12. a) Using the sequence generated in Question 11, create 14 terms of a new sequence by finding, either exactly or to 7 s.f., the ratio of successive terms:
$t_n = \frac{u_{n+1}}{u_n}$.

For example, $t_1 = \frac{1}{1} = 1, t_2 = \frac{2}{1} = 2, t_3 = \frac{3}{2} = 1.5$, etc.

b) What pattern do successive terms seem to be following?

c) What number do the terms seem to be approaching?

13. Plot the ordered pairs for the t function: $t(n) = t_n$. (See Question 12.) Label the horizontal axis as n, the term numbers.
Label the vertical axis as t_n the value of the nth term of the t function and round t_n to the nearest tenth. For example: (1,1), (2,2), (3,1.5), (4,1.7), etc.
Let 2 cm = 1 unit on both axes.
(Note: Do not connect the ordered pairs.)

14. Find the exact value for the Golden Ratio and compare that value to the number that Questions 12 and 13 seem to be presenting.

15. Jason deposits 2000 euros into an account on January 1 that pays 7% simple interest on the initial deposit. Assuming that he does not deposit or withdraw any money, how much will be in the account at the end of a) 5 years? b) 10 years? c) 30 years?

16. An artist decides to make a small sculpture of marbles in the shape of a triangle. She plans to have 45 marbles on the first (bottom row) of the triangle and one less marble in each successive row. How many marbles will be in the 24th row?

17. Brad agrees with his son's request for an increase in his allowance of $0.75 per week for 24 weeks. If the son's allowance is $3.00 per week now, what will be his allowance 24 weeks later?

18. A long-term care insurance policy pays $80 per day and will increase by 5% simple interest per year to a maximum $140 per day. How many years will it take to reach the maximum?

7.2 **Arithmetic series**

The harmonic series is written as:
$1 + \frac{1}{2} + \frac{1}{3} + \frac{1}{4} + \dots$.
The Gregory-Leibniz series is written as:
$\frac{\pi}{4} = \frac{1}{1} - \frac{1}{3} + \frac{1}{5} - \frac{1}{7} + \frac{1}{9} - \frac{1}{11} + \dots$

1.8 Arithmetic sequences and series, and their applications.
Use of the formulae for the nth term and the sum of the first n terms of the sequence.

A series is a sequence whose terms are being added. There are many well-known series:

- Arithmetic series
- Harmonic series
- Taylor series
- Maclaurin series
- Gregory-Leibniz series

For more information on harmonic series, visit www.pearsonhotlinks.co.uk, enter the ISBN for this book and click on weblink 7.2.
For more information on how sin 5.27(for example) is calculated using a Taylor series, click on weblink 7.3

Example 7.7

If the arithmetic sequence is 1, 4, 7, 10, 13, 16, then write the associated arithmetic series.

Solution
$1 + 4 + 7 + 10 + 13 + 16$

There are several methods that can be used to find the sum of an arithmetic series.

When in elementary school, a young German boy, Carl Friedrich Gauss, was able to find the sum of the first 100 counting numbers within a few seconds. It is reported he was 7 years old at time. He found the sum of 50 pairs of 101 to get 5050.

Example 7.8

Find the sum of the arithmetic series, $1 + 4 + 7 + 10 + 13 + 16$

Method I: Simply add the terms with (or without) your calculator.
$$1 + 4 + 7 + 10 + 13 + 16 = 51$$

Method II: Add pairs of terms from the outside in and multiply by the number of pairs you see.

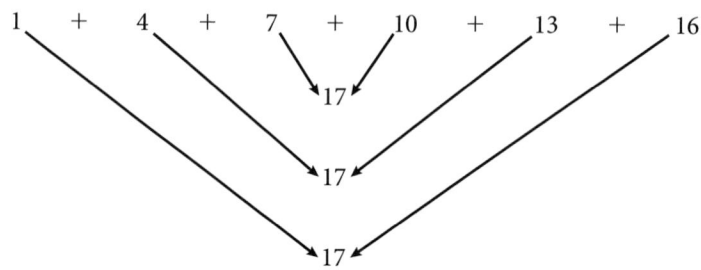

Therefore, $1 + 4 + 7 + 10 + 13 + 16 = 3 \times 17 = 51$

Method III: Use your TI calculator. Press the following keys:

STAT, ENTER, $\wedge$, CLEAR, $\vee$, 1, ENTER, 4, ENTER, 7, ENTER, 10, ENTER, 13, ENTER, 16, ENTER, 2ND ,MODE (QUIT), 2ND ,STAT (LIST), $\rangle$, $\rangle$ (MATH), $\vee$, $\vee$, $\vee$, $\vee$ (5:SUM()), ENTER 2ND , 1 (L$_1$),), ENTER

 To learn more about the famous German mathematician, Carl Friedrich Gauss, visit www.pearsonhotlinks.co.uk, enter the ISBN for this book and click on weblink 7.4.

The TI screen looks like:

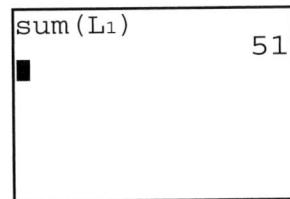

The calculator returns the value 51.

Method IV: Use your TI calculator and the nth term formula for an arithmetic sequence.

1. Use $u_n = u_1 + (n - 1)d$
2. Substitute values for u_1 and d leaving u_n and n as unknowns.
3. Hence, $u_n = 1 + (n - 1) \cdot 3 = 3n - 2$.
4. Now press the following keys:
   - 2ND STAT (LIST)
   - $\rangle$, $\rangle$ (MATH)
   - $\vee$, $\vee$, $\vee$, $\vee$, (5: sum())
   - ENTER
   - 2ND STAT (LIST)
   - $\rangle$ (OPS)
   - $\vee$, $\vee$, $\vee$, $\vee$, (5: seq())
   - ENTER
   - 3X$-$2,X,1,6,1,)
   - Paste: ENTER,ENTER

   (Note: The numbers 1 and 6 tell the calculator to start with the 1st term and end with the 6th term.)

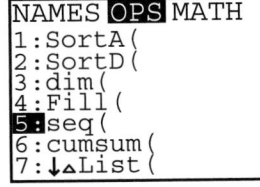
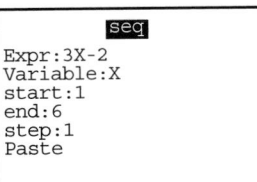

5. The calculator returns the value 51.

Method V: Use of the formula for the sum of an arithmetic series.

$$S_n = \frac{n}{2}(u_1 + u_n)$$

List the information: $n = 6$

$$u_1 = 1$$
$$u_6 = 16$$
$$S_6 = ?$$

Substitute and solve: $S_6 = \frac{6}{2}(1 + 16) = 3 \cdot 17 = 51$

 The two sum formulae can be found in the MSSLFB.

• **Examiner's hint:** Per the Mathematical Studies SL GUIDE, 'Students may use a GDC for calculations, but they will be expected to identify clearly the first term and common difference.'

The other formula to find the sum of an arithmetic series is:
$$S_n = \frac{n}{2}(2u_1 + (n-1)d)$$
This formula is used when d is known and u_n is not known.

Example 7.9

Find the sum of the first 1000 terms of the arithmetic series,
$3 + 7 + 11 + 15 + \ldots$

Solution

Choose an appropriate formula. Since d is known and u_{1000} is not known, choose:
$$S_n = \frac{n}{2}(2u_1 + (n-1)d)$$
List the information: $u_1 = 3$
$$n = 1000$$
$$d = 7 - 3 = 4$$
$$u_{1000} = ?$$
$$S_{1000} = ?$$

Substitute and solve: $S_{1000} = \dfrac{1000}{2}(2 \cdot 3 + (1000 - 1) \cdot 4) = 500\,(6 + 999 \cdot 4)$

$$\therefore \ S_{1000} = 2\,001\,000$$

There are many applications for which it is important to find the sum of a series.

Example 7.10

A teacher earned \$35 000 at the beginning of the school year. She estimates that she will receive \$1750 step increase each year for the next 20 years. How much will she have earned in total after she has worked 20 years?

Solution

Draw the diagram.

Year 1	Y2	Y3	Y4	...	Y20
35 000	36 750	38 500	40 250	...	?

Write an appropriate formula. $S_n = \frac{n}{2}(2u_1 + (n-1)d)$

List the information. $u_1 = 35\,000$
$$n = 20$$
$$d = 1750$$
$$u_{20} = ?$$
$$S_{20} = ?$$

Substitute and solve: $S_{20} = \dfrac{20}{2}(2 \cdot 35\,000 + (20 - 1) \cdot 1750)$

$$= 10\,(70\,000 + 19 \cdot 1750)$$

$$\therefore \ S_{20} = \$1\,032\,500$$

Sigma notation

The symbol that is often used to express the concept of summation is the upper case Greek letter, sigma Σ.

The Σ notation is used in the following form: $\displaystyle\sum_{i=1}^{n} u_i$

The notation is read as 'the summation of all the u_i's from $i = 1$ to $i = n$, where:

- n is the number of terms as long as $i = 1$
- u_i represents the terms that are being added
- i is the variable that is used to increment to the next term.

Example 7.11

Expand $\displaystyle\sum_{i=1}^{5} (2i + 1)$

Solution

$$\sum_{i=1}^{5}(2i + 1) = (2 \cdot 1 + 1) + (2 \cdot 2 + 1) + (2 \cdot 3 + 1) + (2 \cdot 4 + 1) + (2 \cdot 5 + 1)$$
$$= \quad 3 \quad + \quad 5 \quad + \quad 7 \quad + \quad 9 \quad + \quad 11$$

Example 7.12

Find the sum of the arithmetic series, $\displaystyle\sum_{i=1}^{20}(3i - 7)$.

Solution

Draw the diagram: $\displaystyle\sum_{i=1}^{20}(3i - 7) = (3 \cdot 1 - 7) + \ldots + (3 \cdot 20 - 7)$
$$= -4 + \ldots + 53.$$

Write a formula: $\quad S_{20} = \dfrac{n}{2}(u_1 + u_n)$

List the information: $\quad u_1 = -4$
$$n = 20$$
$$u_{20} = 53$$
$$S_{20} = ?$$

Substitute and solve: $S_{20} = \dfrac{20}{2}(-4 + 53) = 10 \cdot 49 = 490$

Writing an arithmetic series using sigma notation involves using the nth term formula and leaving n and u_n as unknowns.

Example 7.13

Write the following arithmetic series using sigma notation.
$$3 + 10 + 17 + 24 + 31 + 38$$

Fraternities and sororities use upper case Greek letters as their name. For example, $X\Phi$ is the symbol designation for the Chi Phi fraternity.

For a list of the upper and lower case symbols for the Greek alphabet, visit www.pearsonhotlinks.co.uk, enter the ISBN for this book and click on weblink 7.5.

The harmonic series can be written

as: $\displaystyle\sum_{k=1}^{\infty} \dfrac{1}{k}$

This summation is read as 'the summation of $2i + 1$ from $i = 1$ to $i = 5$'.

- **Examiner's hint:** In Topic 2, in the MSSLFB, there is one formula that deals with sigma notation. You are expected to be able to read and apply this formula. See Chapters 11 and 12 for more information.

Solution

Write a formula: $\quad u_n = u_1 + (n-1)d$

List the information: $u_1 = 3$

$$d = 10 - 2 = 7$$
$$n = ? \ (n = 6, \text{ but leave it as an unknown})$$
$$u_n = ? \ (u_n = 38, \text{ but leave it as an unknown})$$

Substitute and solve: $u_n = 3 + (n-1) \cdot 7 = 3 + 7n - 7$
$$\therefore \ u_n = 7n - 4$$

Now let $n = i$, and write, $u_i = 7i - 4$.

Hence, $3 + 10 + 17 + 24 + 31 + 38 = \displaystyle\sum_{i=1}^{6} (7i - 4)$.

Exercise 7.2

1. Find the sum of each arithmetic series given that:

a) The series is: $1 + 2 + 3 + \dots + 998 + 999 + 1000$

b) The series is: $5 + 17 + 29 + \dots + 581 + 593 + 605$

c) There are 200 terms in the series: $-25 + -17 + -9 + \dots$

d) There are 125 terms in the series: $100 + 84 + 68 + \dots$

e) $u_1 = 37, u_{98} = 1104, n = 98$

f) $u_1 = -23, u_{50} = 222, n = 100$

2. Write each sigma notation in expanded form.

a) $\displaystyle\sum_{i=1}^{4} (i + 7)$
b) $\displaystyle\sum_{n=1}^{5} (2n - 1)$
c) $\displaystyle\sum_{k=1}^{6} (2 - 3k)$
d) $\displaystyle\sum_{m=1}^{7} (5m)$

3. Write each arithmetic series in sigma notation.

a) $2 + 15 + 28 + 41 + 54 + 67$
b) $-2 - 21 - 40 - 59 - 78$

c) $1 + 3 + 5 + \dots + 95 + 97 + 99$
d) $17 + 22 + 27 + \dots + 387$

4. Find the first term, last term and the common difference for each arithmetic series.

a) $\displaystyle\sum_{i=1}^{100} (3i - 1)$
b) $\displaystyle\sum_{n=1}^{80} (5 - 7n)$
c) $\displaystyle\sum_{k=1}^{50} (9k)$

d) $\displaystyle\sum_{m=1}^{120} (-12m)$
e) $\displaystyle\sum_{n=1}^{n} n$
f) $\displaystyle\sum_{n=1}^{n} (2n - 1)$

5. Find the sum of each arithmetic series in 4a) – f).

6. Find the sum of the arithmetic series given that:

a) $u_1 = 22, n = 18, d = 3$
b) $u_1 = -100, n = 95, d = -20$

c) $u_1 = 31, n = 20, d = 11$
d) $u_1 = \frac{5}{2}, n = 34, d = \frac{1}{3}$

7. An artist decides to make a small sculpture of marbles in the shape of a triangle. She plans to have 45 marbles on the first (bottom row) of the triangle and one less marble in each successive row until the triangle has been completed. How many marbles will be in the **entire sculpture**?

8. Ron agrees with his son, Brett's, request for an increase in his allowance of $0.75 per week for 24 weeks. If Brett's allowance is $3.00 per week now, what will be the **total amount** of allowance he has accumulated 24 weeks later?

9. The sum of the first six terms of an arithmetic sequence is 42 and the sum of the next eight terms is 168.
   a) Given that *a* is the first term in the sequence and that *d* is the common difference write an equation for
      i)   the sum of the first **six** terms of the sequence
      ii) the sum of the first **fourteen** terms of the sequence.
   b) i)   Write down the value of *a.*
      ii) Write down the common difference, *d.*
      iii) Describe the set of numbers in the sequence.

10. Javid wants to save for a new musical instrument and his parents have suggested a novel way of saving money. They have suggested that he uses a chess board. On day 1, he places $0.50 coin on the first square, on day 2, he places 2 $0.50 coins on the second square. This continues until all 64 squares have coins on them.
   a) Explain why this saving plan forms an arithmetic sequence.
   b) Write down the values of u_1 and *d*.
   c) How many coins will there be on the chessboard after the sixteenth (16th) day?
   d) Javid needs to save at least $1000. Will there be enough squares to save $1000? If so, how many days will it take?

7.3 Geometric sequences

1.8   Geometric sequences and series.
      Use of the formulae for the *n*th term and the sum of the first *n* terms of the sequence.

A geometric sequence has similar characteristics to an arithmetic sequence. They are:
- It is a sequence of numbers that has a pattern.
- It can be defined as a recursive definition.
- An explicit *n*th term formula is used to find a specific term.

There are, however, some significant differences. They are:
- Multiplying, instead of adding, produces the next term.
- The number that is used to multiply by to get the next term is called the 'common ratio'. The variable used for the ratio is '*r*'.

For example, 2, 4, 8, 16, 32, 64, 128, is a geometric sequence with a common ratio,

$$r = \frac{4}{2} = \frac{8}{4} = \frac{16}{8} = \frac{32}{16} = \frac{64}{32} = \frac{128}{64} = 2.$$

The definition of a geometric sequence is:

> A geometric sequence is a sequence such that $u_1 = a$ and $u_{n+1} = u_n \cdot r$, where $n \in \mathbb{Z}^+$ and $r \in \mathbb{R}$, $r \neq 0$.

● **Examiner's hint:** Per the Mathematical Studies SL GUIDE, 'Students may use a GDC for calculations, but they will be expected to identify clearly the first term and common ratio.'

Notice that $\frac{6}{2} = \frac{18}{3} = \frac{54}{18} = 3$.

This is the reason that **'common ratio'** is used.

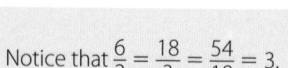

Example 7.14

Write the first four terms of the geometric sequence where $u_1 = 2$ and $r = 3$.

Solution

$u_1 = 2$
$u_2 = u_1 \cdot 3 = 2 \cdot 3 = 6$
$u_3 = u_2 \cdot 3 = 6 \cdot 3 = 18$
$u_4 = u_3 \cdot 3 = 18 \cdot 3 = 54$

$\therefore$ The first terms of the sequence are 2, 6, 18, 54.

Example 7.15

Given the geometric sequence, $1, \frac{1}{2}, \frac{1}{4}, \frac{1}{8}, \ldots$, write a recursive definition.

Solution

By inspection, each successive term is $\frac{1}{2}$ times the previous term. Therefore, let

$$u_1 = 1 \text{ and } u_{n+1} = u_n \cdot \frac{1}{2}$$

The nth term formula for a geometric sequence can be developed by way of an inductive process in much the same way that the nth term formula for an arithmetic sequence was developed.

● **Examiner's hint:** This formula, $u_n = u_1 \cdot r^{n-1}$, can be found in the Mathematical Studies SL Formula Booklet (MSSLFB).

The nth term formula for a geometric sequence is:

$$u_n = u_1 \cdot r^{n-1}, \text{ where } n \in \mathbb{Z}^+ \text{ and } r \in \mathbb{R}, r \neq 0$$

Example 7.16

Find the 10th term of the geometric sequence, 2, 4, 8, 16, …

Solution

Method I.   Continue the sequence by observing the pattern until you reach the 10th term.

$$2, 4, 8, 16, 32, 64, 128, 256, 512, 1024$$

Method II.   Use your calculator.

Press: 2, $\times$ 2, ENTER, $\times$, 2, ENTER, ENTER, E, E, E, E, E, E,

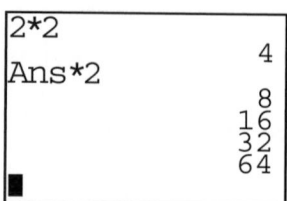
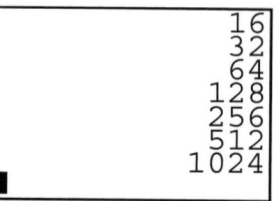

The calculator returns 1024

Method III. Use the nth term formula.

List the information: $u_1 = 2$

$$r = \frac{4}{2} = 2$$

$$n = 10$$

$$u_{10} = ?$$

Write an appropriate formula: $u_n = u_1 \cdot r^{n-1}$

Substitute and solve: $u_{10} = 2 \cdot 2^{10-1} = 2 \cdot 2^9$

$$\therefore \ u_{10} = 1024$$

 $2 \cdot 2^9$ is **not** equal to 4^9. Order of operations dictates that exponent operations must be performed first and then the operation of multiplication.

Example 7.17

The number of students studying a one-year online mathematics course is increasing each year by 5%. In 2012 there were 15 000 students who were studying the course.

a) How many students studied the course in 2013?

b) Assuming the rate of increase does not change how many students will be studying the course in 2020?

c) In which year will the number of students studying the course exceed 30 000?

Solution

a) We can write the problem as a geometric sequence since there is a percentage increase each year.

$u_1 = 15\,000$ i.e. the number of students studying the course in 2012.

The common ratio $r = 1.05$ i.e. $\left(1 + \dfrac{5}{100}\right)$

The number of students in 2013 is

$u_2 = u_2 \cdot r^{2-1} = 15\,000 \cdot 1.05^{2-1} = 15\,750$

 The common ratio is **not** 0.05.

b) 2020 is year 9, therefore we need to calculate u_9.

$u_9 = 15\,000 \cdot 1.05^{9-1} = 15\,000 \cdot 1.05^8 = 22\,162$ to the nearest integer.

c) Need to solve $30\,000 = 15\,000 \cdot 1.05^n$

Simplifying $\dfrac{30\,000}{15\,000} = 2 = 1.05^n$

 This formula is used since multiplication is used to get the next term.

Using the GDC we can solve the equation $1.05^n = 2$.

Enter in $Y1 = 1.05^x$ and $Y2 = 2$ and set window as shown:

```
WINDOW
 Xmin=0
 Xmax=20
 Xscl=1
 Ymin=0
 Ymax=3
 Yscl=1
↓Xres=1
```

Press GRAPH, 2nd, TRACE (GRAPH), 5, ENTER, ENTER, ENTER. The resulting screen will appear.

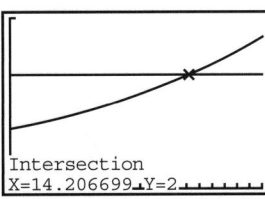

Intersection
X=14.206699 Y=2

x indicates that it will take 14.2 years to reach 30 000 students. Thus it will be 2026 before there are 30 000 students enrolled on the course.

Exercise 7.3

1. State which of the following are geometric sequences.
   a) 10, 20, 30, 40, ...
   b) 1.5, 3, 6, 12, ...
   c) 1, 1, 1, 1, ...
   d) 2, −4, 8, −16, 32, ...
   e) 1, 4, 9, 16, 25, ...
   f) 120, 128.4, 137.388, ...

2. Write the first four terms of each sequence. If geometric, say so.
   a) $u_1 = 3$ and $u_{n+1} = 2u_n$
   b) $u_1 = 100$ and $u_{n+1} = \frac{1}{2}u_n$
   c) $u_n = n^2$
   d) $u_1 = 4$ and $u_{n+1} = -3u_n$
   e) $u_1 = 2, u_2 = 4$ and $u_{n+2} = u_{n+1} \cdot u_n$
   f) $u_n = 1$ and $u_{n+1} = \frac{1}{3}u_n$

3. Write each sequence in recursive definition form.
   a) 2, 6, 18, 54, ...
   b) $1, \frac{1}{10}, \frac{1}{100}, \frac{1}{1000}, \ldots$
   c) 64, −32, 16, −8, ...
   d) 100, 108, 116.64, 125.9712, ...

4. For each of the following geometric sequences, find the 6th term and the common ratio.
   a) −4, 8, −16, ...
   b) 0.2, 0.1, 0.05, ...
   c) $u_1 = \frac{1}{2}$ and $u_{n+1} = \frac{3}{2}u_n$
   d) $u_1 = \frac{1}{3}$ and $u_{n+1} = -2u_n$

5. For each of the following, find:
   a) 10th term of: $\frac{1}{8}, \frac{1}{4}, \frac{1}{2}, \ldots$
   b) u_{10} for the sequence: 1, 2, 4, ...
   c) The 30th term of the sequence where $u_1 = 100$ and $u_{n+1} = 1.07u_n$
   d) u_{12} for the sequence: 5, 25, 125, ...

6. For the nth term formula for a geometric sequence, $u_n = u_1 r^{n-1}$, three of the four unknowns are given. Find the fourth.
   a) $u_1 = 13, r = 2, n = 4$
   b) $u_{14} = 4\,782\,969, r = 3, n = 14$
   c) $u_n = 40\,353\,607, r = 7, u_1 = 1$
   d) $u_1 = \frac{3}{4}, u_6 = \frac{729}{4096}, n = 6$

7. Solve for n exactly or correct to 3 significant figures.
   a) $2^n = 8192$
   b) $\frac{243}{1024} = \left(\frac{3}{4}\right)^n$
   c) $3^n = 30$
   d) $100 = 5 \cdot 2^n$

8. Solve for n exactly or correct to 3 significant figures.
   a) $2000 = 1000(1.06)^n$
   b) $50 = 17(1.055)^n$
   c) $900 = 300(1 + r)^{15}$
   d) $850\,000 = 225\,000\left(1 + \frac{r}{12}\right)^{12 \cdot 30}$
   e) $3^{n-1} = 17$
   f) $3^{n+1} = 37$
   g) $r^4 = 16$
   h) $r^5 = 112$
   i) $r^7 = 317$
   j) $r^{3.5} = 200$

9. Each of the following represents a geometric sequence. Find u_1.
   a) ___, ___, ___, 135, ___, ___, 3645
   b) ___, ___, $\frac{4}{9}$, ___, ___, ___, ___, $\frac{128}{2187}$
   c) ___, ___, ___, ___, 1024, ___, 16 384
   d) ___, ___, ___, $\frac{1}{16}$, ___, $\frac{1}{64}$

10. Each of the following represents a geometric sequence. Find the missing terms.
   a) 7, ___, ___, 3584
   b) ___, ___, ___, 0.4, ___, ___, ___, ___, −12.8

c) ____, ____, 6400, ____, ____, ____, ____, ____, 1677.7216

d) ____, 39, ____, ____, ____, 3159

11. Find x, if each of the following sequences is geometric.

a) $x, x + 2, x + 8$

b) $\sqrt{x}, \sqrt{x + 4}, \sqrt{x + 16}$

c) If $x, 5, y$ is a geometric sequence, then find three pairs of possible values for x and y.

12. Insert the required number of terms between the given values, so that the resulting sequence is geometric.

a) $3,$ ____, ____, 24

b) $2,$ ____, ____, 12

c) $1,$ ____, ____, ____, 16

d) $1,$ ____, ____, ____, 10

13. A father decides on a savings plan for his daughter. He will deposit $0.01 (1 US penny) on the first day of the month, $0.02 (2 pennies) on the second day, $0.04 (4 pennies) on the third day, and so on, doubling the amount deposited on each successive day of the month. He agrees, with his wife, to do this for an entire month of 30 days. How much will the father have to deposit into his daughter's savings account on the:

a) 10th day?

b) 15th day?

c) 20th day?

d) 21st day?

e) 30th day?

14. The population of Bangor is growing each year. At the end of 1996, the population was 40 000. At the end of 1998, the population was 44 100. Assuming that these annual figures follow a geometric progression, calculate

a) the population of Bangor at the end of 1997

b) the population of Bangor at the end of 1992.

7.4 Geometric series

| 1.8 | Geometric sequences and series, and their applications. Use of the formulae for the **n**th term and the sum of the first **n** terms of the sequence. |

ⓘ Geometric series and fractal geometry are connected and have many applications in the realms of physics, engineering, and economics.

A geometric series is a geometric sequence whose terms are added.

The general form for a geometric series can be expressed using summation notation.

$$u_1 + u_1 r + u_1 r^2 + u_1 r^3 + \ldots + u_1 r^{n-1} = \sum_{i=1}^{n} u_1 r^{i-1}.$$

For more about geometric series and fractal geometry, visit www.pearsonhotlinks. co.uk, enter the ISBN for this book and click on weblinks 7.6 and 7.7.

Example 7.18

Express the geometric series, $\displaystyle\sum_{i=1}^{5} 2(3^{i-1})$, as a sum of terms.

Solution

$$
\begin{aligned}
\sum_{i=1}^{5} 2(3^{i-1}) &= 2\cdot3^{1-1} + 2\cdot3^{2-1} + 2\cdot3^{3-1} + 2\cdot3^{4-1} + 2\cdot3^{5-1} \\
&= 2\cdot3^{0} \quad\; + 2\cdot3^{1} \quad\; + 2\cdot3^{2} \quad\; + 2\cdot3^{3} \quad\; + 2\cdot3^{4} \\
&= 2 + 6 + 18 + 54 + 162
\end{aligned}
$$

Example 7.19

Express the geometric series, $7 + 14 + 28 + 56 + 112 + 224$, using summation notation.

Solution

Write a formula: $\qquad u_n = u_1 r^{n-1}$

List the information: $u_1 = 7$

$$r = \frac{14}{7} = 2$$

$n = 6$ (use 6 on the Σ symbol, but leave it unknown in the argument.)

$$u_n = ?$$

Substitute and solve: $u_n = 7(2^{n-1})$

Apply the sigma notation: $\displaystyle\sum_{n=1}^{6} 7(2^{n-1})$

Therefore, $7 + 14 + 28 + 56 + 112 + 224 = \displaystyle\sum_{n=1}^{6} 7(2^{n-1})$.

• **Examiner's hint:** Since the sum of many geometric series will become very large after only a few terms, many questions limit the number of terms. Therefore the fastest (and often the most accurate) method to find the sum is to use your calculator and add the terms.

Example 7.20

Find the sum of the geometric series, $\displaystyle\sum_{i=1}^{4} 2(4^{i})$.

Solution

Method I. Draw the diagram: $\displaystyle\sum_{i=1}^{4} 2(4^{i}) = 2\cdot4^{1} + 2\cdot4^{2} + 2\cdot4^{3} + 2\cdot4^{4}$

$$= 8 + 32 + 128 + 512$$

Simply use your calculator and add the terms. Hence, the sum $= 680$

Method II. Finding the sum using a method without the use of a formula.

Method II may serve as a hint for the development of the formula to find the sum of a geometric series.

Let S be the sum, $S_4 = 8 + 32 + \quad 128 + \quad 512$

$$-4 \cdot S_4 = \quad -4 \cdot 8 + \ -4 \cdot 32 \ + \ -4 \cdot 128 \ + \ -4 \cdot 512$$
$$S_4 + \ -4 \cdot S_4 = 8 + 0 \quad + \quad 0 \quad + \quad 0 \quad + \ -4 \cdot 512$$

$$S_4(1 + -4) = 8 + -8 \cdot 256 = 8 - 8 \cdot 4^4$$

$$\therefore \ S_4 = \frac{(8 - 8 \cdot 4^4)}{(1 - 4)} = \frac{8(1 - 4^4)}{(1 - 4)} = 680$$

• **Examiner's hint:** Also found in the Mathematical Studies SL Formula Booklet is an equivalent form of this formula:

$$S_n = \frac{u_1(r^n - 1)}{r - 1}, r \neq 1.$$

One of the formulae used to find the sum of a geometric series is:

$$S_n = \frac{u_1(1 - r^n)}{1 - r}, r \neq 1$$

Example 7.21

Find the sum of the first 10 terms of the geometric series: $4 + 2 + 1 + \ldots$

Solution

Method I. Use the geometric series summation formula.

Write a formula: $\quad S_n = \dfrac{u_1(1 - r^n)}{1 - r}$

List the information: $u_1 = 4$

$$r = \tfrac{1}{2}$$

$$n = 10$$

$$S_{10} = ?$$

Substitute and solve: $S_{10} = \dfrac{4\left(1 - \left(\frac{1}{2}\right)^{10}\right)}{1 - \dfrac{1}{2}} = \dfrac{4\left(1 - \dfrac{1}{1024}\right)}{\dfrac{1}{2}} = 8\left(\dfrac{1023}{1024}\right)$

Hence, $S_{10} = \dfrac{1023}{128}$

Method II. Use your TI calculator and the 'sum(seq(' functions.

- Press the following keys
  - 2ND STAT (LIST)
  - ⟩, ⟩ (MATH)
  - V,V,V,V, (5: SUM())
  - ENTER
  - 2ND STAT (LIST)
  - ⟩ (OPS)
  - V,V,V,V, (5: seq())
  - ENTER
- Clearly $u_n = 4\left(\dfrac{1}{2}\right)^{x-1} = 4(0.5)^{x-1}$
- Continue by entering the following
  - 4,*,.5^(X–1),X,1,10,1))
- Paste: ENTER, ENTER

```
NAMES OPS  MATH
2↑max(
3:mean(
4:median(
5:sum(
6:prod(
7:stdDev(
8:variance(
```

```
NAMES OPS  MATH
1:SortA(
2:SortD(
3:dim(
4:Fill(
5:seq(
6:cumSum(
7↓▲List(
```

```
              seq
Expr:4*.5^(X-1)
Variable:X
start:1
end:10
step:1
Paste
```

- MATH
- ENTER (1: ▷Frac)
- ENTER
- Calculator returns $\dfrac{1023}{128}$

There is a story of an eccentric maths teacher who claimed that once in his classroom his students would never be able to leave. When asked why, he responded: 'Before you can go through the doorway you must first go halfway. Once you get halfway, you must once again go halfway to the doorway, and so on ad infinitum. Thus, once you are in my classroom you will never be able leave!' Discuss the logic behind the story and the paradoxical nature of actually being able to exit the classroom. This is based on Zeno's paradox.

• **Examiner's hint:** It would be advised to practise using the two methods described in Section 7.4 to find each sum: using 'seq(sum(' in preparation for Paper 1 and using the formula in preparation for Paper 2.

Exercise 7.4

1. Find the sum of each geometric series, exactly or to 2 decimal places, given that:
   a) The series is: $1 + 2 + 4 + 8 + \ldots + 32\,768$
   b) The series is: $81 + 27 + 9 + \ldots + \dfrac{1}{243}$
   c) There are 12 terms in the series: $-2 + 10 - 50 + \ldots$
   d) There are 8 terms in the series: $1 + \dfrac{1}{2} + \dfrac{1}{4} + \ldots$
   e) $u_1 = 4, u_6 = 67\,228, n = 6$
   f) $u_1 = 1, u_7 = 15\,625, n = 10, r < 0$

2. Write each sigma notation in expanded form.
   a) $\displaystyle\sum_{i=1}^{5} 2^i$
   b) $\displaystyle\sum_{n=1}^{6} 3^{n-1}$
   c) $\displaystyle\sum_{k=1}^{4} (2 \cdot 5^{k+1})$
   d) $\displaystyle\sum_{m=1}^{7} \left(\dfrac{1}{2}\right)^{m-1}$
   e) $\displaystyle\sum_{k=1}^{n} u_k$
   f) $\displaystyle\sum_{k=1}^{n} (u_1 r^{k-1})$

3. Write each geometric series in sigma notation.
   a) $9 + 18 + 36 + 72 + 144 + 288 + 576$
   b) $-3 - 12 - 48 - 192 - 768 - 3072$
   c) $13 - 78 + 468 - \ldots - 3\,639\,168$
   d) $\dfrac{8}{3} + 2 + \dfrac{3}{2} + \ldots + \dfrac{243}{512}$

4. Find the first term, last term, and the common ratio for each geometric series.
   a) $\displaystyle\sum_{i=1}^{4} 3^i$
   b) $\displaystyle\sum_{n=1}^{5} 0.2^{n-1}$

c) $\displaystyle\sum_{k=1}^{9} (2 \cdot 2^{k+1})$

d) $\displaystyle\sum_{m=1}^{6} (-3 \cdot 2^{-m})$

e) $\displaystyle\sum_{k=1}^{n} (u_1 r^{k-1})$

5. Find the sum of each geometric series in 4a) – e).

In a series, the sum is designated by S. The sum of the first term is S_1, the sum of the first two terms is S_2, the sum of the first three terms is S_3, and so on. Each of S_1, S_2, S_3, is called a partial sum. For example if $S = 1 + 2 + 4 + 8$, then $S_1 = 1$, $S_2 = 1 + 2$, $S_3 = 1 + 2 + 4$, and $S_4 = 1 + 2 + 4 + 8$.

6. For the (finite) geometric series, $1 + 3 + 9 + 27 + 81 + 243$,
   a) find S_1 through S_6
   b) what do you notice about the successive partial sums?
   c) do the partial sums seem to be getting larger?
   d) if the partial sums are getting larger, is this by larger or smaller amounts?

7. For the (finite) geometric series, $1 + \frac{1}{2} + \frac{1}{4} + \frac{1}{8} + \frac{1}{16} + \frac{1}{32} + \frac{1}{64} + \frac{1}{128}$,
   a) find S_1 through S_8 (Hint: to speed up the process, use the x^{-1} key.)
   b) what do you notice about the successive partial sums?
   c) do the partial sums seem to be getting larger?
   d) if the partial sums are getting larger, is this by larger or smaller amounts?
   e) what number do the partial sums seem to be approaching?

8. For the (infinite) geometric series, $1 + \frac{1}{3} + \frac{1}{9} + \frac{1}{27} + \frac{1}{81} + \ldots$,
   a) find S_1 through S_8
   b) what do you notice about the successive partial sums?
   c) do the partial sums seem to be getting larger?
   d) if the partial sums are getting larger, is this by larger or smaller amounts?
   e) what number do the partial sums seem to be approaching?
   f) do you think that the nth partial sum will ever reach that number (in e)) above?
   g) explain your answer to f).

9. $0.33\dot{3}$ can be written as a geometric series: $0.33\dot{3} = 0.3 + 0.03 + 0.003 + \ldots$
   a) What is the common ratio?
   b) Find S_4.
   c) Find the percentage error between your answer in b) and the exact value $\frac{3}{9}$.

10. For $n = 0.77\dot{7}$
   a) Write n as a geometric series.
   b) Find S_5.
   c) Find the percentage error between your answer in b) and the exact value $\frac{7}{9}$.

11. Make a conjecture based on what you have observed in Questions 9 and 10.

12. Find the exact value in fraction form for:
   a) $0.55\dot{5}$
   b) $0.99\dot{9}$
   c) $0.25\dot{2}5\dot{5}$
   d) $0.36\dot{3}6\dot{6}$

13. Suppose you are about to enter your mathematics classroom from a point in the hallway that is 4 metres from the doorway.

a) Paradoxically, why won't you be able to enter the classroom?
(Hint: see page 204.)

b) Mathematically, why will you be able to enter the classroom?

14. A father decides on a savings plan for his daughter. He will deposit $0.01 (1 US penny) on the first day of the month, $0.02 (2 pennies) on the second day, $0.04 (4 pennies) on the third day, and so on doubling the amount deposited on each successive day of the month. He agrees, with his wife, to do this for an entire month of 30 days. What is the total amount the father deposited into his daughter's savings account at the end of the month?

15. Measure a stack 500 sheets (a ream) of copier paper to the nearest cm. Calculate the thickness of one sheet of paper.
Now suppose that you placed one sheet of paper on the floor on Day 1, 2 sheets of paper on top of that one on Day 2, 4 sheets of paper on top of those on Day 3, doubling the number of sheets each successive day.

a) Estimate (take your best guess) as to the height of the stack of paper after 50 days.

b) Calculate the height of the stack after 50 days.

c) Convert the answer in b) to km.

d) Convert the height to terms we can better comprehend.

16. Alex is planning to walk a quarter-mile track in support of fighting cancer. She plans to walk in a geometric progression. On the first day, she will walk 4 miles (16 laps) and then will increase the mileage each day by 20%.

a) (i) She decides to walk for 21 days. How many miles will she walk on the 21st day?

  (ii) Mow many *miles* in total will she walk?

  (iii) How many *laps* in total will she walk?

b) Does Alex have a well-thought-out plan? Explain your answer.

c) How could Alex improve her plan?

17. A ball is dropped vertically. It reaches a height of 2 m on the first bounce and 1.9 m on the second bounce. Assume that the heights form a geometric sequence.

a) Determine the common ratio.

b) Calculate the height of the twelfth (12th) bounce.

c) When will the ball have a height less than 10 cm?

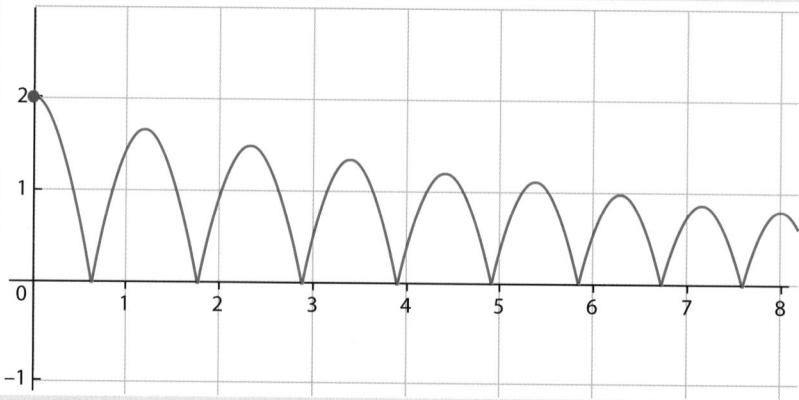

7.5 Compound interest

1.9 Financial applications of geometric sequences and series: compound interest, annual depreciation.

When money is borrowed from a bank, they charge a fee for this service through increasing the amount of money owed. This increase is known as the interest. When we put money in a bank in a savings account, we earn interest on that money as an encouragement from the banks to lend them our money.

 Is simple interest realistic in the real world?

When the interest earned each year is always dependent on the original amount of money, we call it **simple interest**. For example, if we put $1000 in a bank account and are given 5% simple interest by the bank, we would earn 5% of $1000 each year we left the money in the bank.

The table below shows the result of 4 years of 5% simple interest earned on $1000.

Year	Interest per year	Total money
1	$50	$1050
2	$50	$1100
3	$50	$1150
4	$50	$1200

◀ **Table 7.1** Simple interest example

However, it is often not only the original amount of money that earns interest. Usually, interest is earned on the principal plus the interest already earned. When this happens, we say that the money is earning **compound interest**.

 What is the link between compound interest and population growth?

Table 7.2 shows the effect of 4 years of 5% compound interest on $1000.

Year	Interest per year	Total money
1	$50	$1050
2	$52.50	$1102.50
3	$55.13	$1157.63
4	$57.88	$1215.51

◀ **Table 7.2** Compound interest example.

● **Examiner's hint:** This formula $FV = PV \times \left(1 + \dfrac{r}{100k}\right)^{kn}$ can be found in the Mathematical Studies SL Formula Booklet (MSSLFB).

Notice that in Table 7.2, the amount of interest earned each year increases. The total amount of money after 4 years is more than if the money had been earning simple interest.

The formula for the total amount of compound interest earned is

$FV = PV \times \left(1 + \dfrac{r}{100k}\right)^{kn}$, where FV = future value, PV = present value, n = number of years, k = number of compounding periods per year, $r\%$ = nominal annual rate of interest. A compounding period is the length of time between times the money earns interest. This can be yearly, half-yearly, quarterly, or monthly.

Example 7.22

Find the total compound interest earned on $4500 over 6 years if the money earns 7% interest annually.

Solution

We use the compound interest equation, and in our example $PV = \$4500$, $r = 7$, $k = 1$, $n = 6$.

$$FV = PV \times \left(1 + \frac{r}{100k}\right)^{kn}$$
$$FV = \$4500 \times \left(1 + \frac{7}{100 \times 1}\right)^{1 \times 6}$$
$$FV = \$6753.29$$

To calculate the interest earned we need to subtract the Present value (PV) from the Future value (FV).

Interest earned = $6753.29 − $4500 = $2253.29

When the compounding period is not a year, then the value for k will not be one, as shown in the following example.

The compound interest equation is
$$FV = PV \times \left(1 + \frac{r}{100k}\right)^{kn}$$
This equation is used when the interest earned is included in the principal for the following interest earning periods.

Example 7.23

Samantha puts €15 000 in a bank account earning 6% annual interest compounded monthly. How much money in total will she have after 20 years?

Solution

We use the compound interest equation, and in our example $PV = €15\,000$, $r = 6$, $k = 12$ (as the interest will be added monthly), $n = 20$.

$$FV = PV \times \left(1 + \frac{r}{100k}\right)^{kn}$$
$$FV = €15\,000 \times \left(1 + \frac{6}{100 \times 12}\right)^{12 \times 20}$$
$$FV = €49\,653.07$$

So the total amount of money in the account after 20 years is €49 653.07.

1. Calculate the compound interest on $10 000 if the interest is compounded annually at a 5% annual interest rate for:
   a) 5 years
   b) 2 years
   c) 10 years
   d) 3.5 years.

2. Find the future value of £30 000 earning compound interest for 10 years compounded annually at an interest rate of:
   a) 3%
   b) 10%
   c) 1.2%
   d) 20%.

3. Find the future value of ¥5000 that earns 6% annual interest for 15 years compounded:
   a) annually
   b) quarterly
   c) monthly.

4. Given a future value of $12 000 and an annual interest rate of 12%, calculate the principal when compounded monthly over:
   a) 2 years
   b) 5 years
   c) 8 months
   d) 3.5 years.

5. Which earns more money, £5000 earning 4% simple interest for 10 years, or £5000 earning 3.5% interest compounded yearly for 9 years?

6. Vasili invests $25 000 in an account earning 5.75% simple interest for 3 years. At what annual interest rate would the same amount of money have to be earning monthly compound interest over 3 years?

7. On January 1, 2009, Manuel deposited £1500 into an account paying 6% interest compounded annually. If he did not make any further deposits or withdrawals, how much would be in the account at the end of 30 years?

8. On January 1, 2009 Tara deposited £1500 into an account that offered 6% compounded monthly. If she did not make any further deposits or withdrawals, how much would be in the account at the end of 30 years?

9. Kurt wants to invest 2000 euros in a savings account for his new grandson.
   a) Calculate the value of Kurt's investment based on a simple interest rate of 4% per annum, after 18 years.
   Inge tells Kurt about a better account which offers interest rate of 3.6% per annum, compounded monthly.
   b) Giving your answer to the nearest euro, calculate the value of Kurt's investment after 18 years, if he follows Inge's advice.

10. John invests X USD in a bank. The bank's stated rate of interest is 6% per annum, compounded monthly.
    a) Write down, in terms of X, an expression for the value of John's investment after 1 year.
    b) What rate of interest, when compounded annually (instead of monthly) will give the same value of John's investment as in part a)? Give your answer correct to three significant figures.

11. The table belows shows the deposits, in Australian dollars (AUD), made by Vicki in an investment account on the *first* day of each month for the first four months in 1999. The interest rate is 0.75% per month compounded monthly. The interest is added to the account at the end of each month.

Month	Deposit (AUD)
January	600
February	1300
March	230
April	710

a) Show that the amount of money in Vicki's account at the end of February is AUD 1918.78.
b) Calculate the amount of Australian dollars in Vicki's account at the end of April.
   Vicki makes no withdrawals or deposits after 1st April 1999.
c) How much money is in Vicki's account at the end of December 1999?
   From 1st January 2000 the bank applies a new interest rate of 3.5% per annum compounded annually.
d) In how many full years after December 1999 will Vicki's investment first exceed AUD 3300?

7.6   **Financial maths on the calculator**

What effect has technology had on the financial aspects of our lives?

1.9	Financial applications of geometric sequences and series: compound interest, annual depreciation

On some calculators, there is a very useful program called 'TVM Solver' which is a financial maths program with some useful applications. One of the more useful ones is the 'compound interest solver'.

Step 1: First navigate, using the arrow keys, to the TVM solver.
Press ENTER.

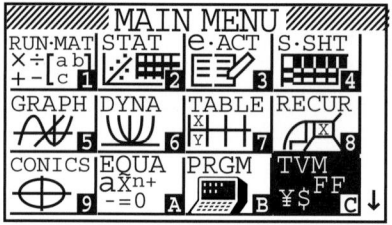

Step 2: On the next menu, press F2 for the Compound Interest application.

```
Financial(1/2)
F1:Simple Interest
F2:Compound Interest
F3:Cash Flow
F4:Amortization
F5:Conversion
F6:Next Page
SMPL CMPD CASH AMT CNUT   D
```

Step 3: On the next screen, we can enter the information from the problem. Suppose we have an annual interest rate of 8% on an investment of $5000 over a period of 5 years, compounded yearly.

n = number of compounding periods

I% = annual interest rate

PV = principal

PMT = monthly payment

FV = future value (or the result after the interest has been added)

```
Compound Interest:End
n    =0
I%   =0
PV   =0
PMT  =0
PV   =0
P/Y  =12               ↓
 n   I%  PV  PMT  FV  AMT
```

P/Y = payments per year

C/Y = compounding periods per year

Step 4: For this problem, we enter n = 5, I% = 8, PV = −5000, PMT = 0, FV = 0, P/Y = 1, and C/Y = 1. We enter PV as a negative because while our money is invested, we do not have access to it. We then press F5 to solve for FV.

```
Compound Interest:End
n    =5
I%   =8
PV   =-5000
PMT  =0
PV   =0
P/Y  =1                ↓
  n  | I% | PV |PMT| FV | AMT
```

Step 5: Now we can see that that the future value of our investment will be $7346.64. If we want to solve another problem we press F1, otherwise we can press EXIT.

```
Compound Interest
FV =7346.640384




 REPT         AMT      GRPN
```

We can use this program to solve for any of the variables. We just need to make sure that all of the other variables in the TVM solver are filled in. This program is one of the more efficient ways to solve compound interest equations. This is definitely the easiest way to find the payment required per month to pay back a loan or to save a certain amount of money.

Example 7.24

Find the monthly payment required to save $4000 when earning 4% annual interest compounded monthly within 5 years.

Solution

Using the TVM Solver, enter n = 60, I% = 4, PV = 0, PMT = 0, FV = 4000, P/Y = 12, C/Y = 12, then solve for PMT. This gives a result of −60.332 754, which indicates the answer is $60.34. We round our answer up to make sure we save at least $4000 in the time required.

Example 7.25

Hans intends to invest $5000 in a bank account for 5 years. He has two banks to choose from.

i) Bank A pays 7% interest compounded yearly.

ii) Bank B pays 5% interest compounding quarterly.

Determine which bank will provide the greatest interest payment for Hans at the end of 5 years.

● **Examiner's hint:** Always substitute the values from the question into the compound interest equation, even if you use a financial maths application on your calculator.

• **Examiner's hint:** All answers involving money should be given correct to two decimal places, otherwise points may be deducted.

Solution

i) We will use the TVM solver on the TI. The steps are:

APPS
Finance ...
TVM Solver ...
N = 5 (Number of years)
I% = 7 (Interest rate)
PV = −5000 (Present Value. Invested or given away, hence the negative sign.)
PMT = 0 (Payment amount. This does not apply to this problem.)
FV = 0 (Future Value)
P/Y = 1 (Payments per Year)
C/Y = 1 (Compounding periods per Year)
PMT: END

Now put the cursor on FV = 0 and press,

ALPHA, ENTER (SOLVE).

The calculator returns $7012.76

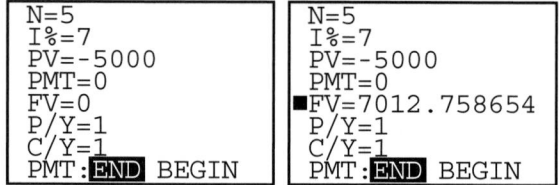

ii) Return to the TVM solver and the values are

N = 20 (5 years compounding quarterly)
I% = 5
PV = −5000
PMT = 0
FV = 0
P/Y = 4 (as interest is earned quarterly)
C/Y = 4 (Interest compounding quarterly)
PMT: END

Now put the cursor on FV = 0 and press,

ALPHA, ENTER (SOLVE).

The calculator returns $6410.19

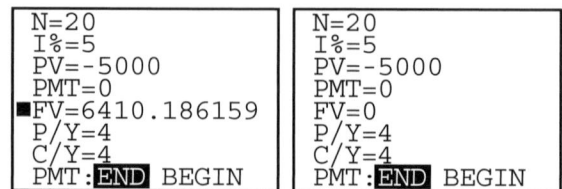

Hans should choose Bank A.

Bank A will pay $602.57 more than Bank B.

1. Find the missing values in the table below.

	a.	b.	c.	d.
N	8	36	30	12
I%	6		5	9
PV	−9000	−26 000		6500
PMT	0	0	0	
FV		32 000	95 000	0
P/Y	1	12	6	4
C/Y	1	12	6	4

We can use a TVM Solver to solve compound interest problems. On the Casio calculator, this program is included. On the TI calculator, the TVM Solver is part of the finance application.

2. Given an annual interest rate of 3.95% , a principal of $9000, and a time period of 4 years, find the interest earned if the money is compounded:

a) yearly b) quarterly c) monthly.

3. Given a principal of £1300, compounded monthly over 3.5 years, find the future value if the annual interest is:

a) 4.5% b) 6% c) 8% d) 10%.

4. Given an annual interest rate of 8% compounded monthly over 6 years, find how much interest would be earned by:

a) $500 b) $800 c) $1500 d) $3000.

5. Calculate the minimum number of months it will take for a principal of $40 000 to double in value when it earns interest compounded quarterly at an annual rate of:

a) 2% b) 5% c) 10% d) 20%.

6. In 7 years, Ricardo has earned €1300 interest on an initial investment of €3500. He is receiving interest compounded monthly.

a) What is the annual interest rate of his investment?

b) How much longer will it take Ricardo to at least double his investment? Write your answer in the form a years, b months.

7. Every month, since her 12th birthday, Hana has deposited $200 in a bank account earning 6.2% annual interest compounded monthly.

a) If Hana has just had her 32nd birthday, how much money is in the account?

b) When Hana turns 40, she doubles the amount of money invested. How much money, to the nearest thousand, will she have when she retires at age 65?

8. On January 1, 2009 Craig bought 5 acres of land in North Carolina for $75 000. At the end of 2018 the value of the land was $128 110.83. What was the annual rate of interest?

9. Allessandro has three options for choosing a bank in which to invest his $3000.

Bank A offers 2.75% interest compounded monthly.

Bank B offers 3% interest compounded quarterly.

Bank C offers 3.25% interest compounded annually.

In which bank should Allessandro invest his money?

10. In one bank Martin deposits $2000 and in another bank he deposits $1500. The first bank offers 3.75% annual interest compounded monthly. The second bank compounds the interest quarterly. After 6 years, both accounts have the same amount of money.

What annual interest rate does the second bank offer?

11. Miranti deposits $1000 into an investment account that pays 5 % interest per annum.

a) What will be the value of the investment after 5 years if the interest is reinvested?

b) How many years would it take Miranti's investment of $1000 to double in value?

Brenda deposits $1000 into a different type of investment account which doubles in value in 10 years when compounded monthly.

c) To the nearest tenth of a percent, what is the annual interest rate for the second investment account?

12. At the ABC Bank there are four different types of accounts.

Account A: Earn 4% annual interest compounded monthly.

Account B: Earn 4.75% annual interest compounded annually.

Account C: Earn 4.5% annual interest compounded quarterly.

a) Which of these accounts is the best deal?

b) If Hank chooses the best account, how much interest will he earn on a deposit of $23 500 in 5 years?

Hank chooses account A, and at the end of each month, deposits an additional $400 on an initial deposit of $23 500.

c) How much money does he have in the account after 10 years?

13. a) The total population of human beings on our planet doubles every 30 years and there are currently 6.7×10^9 people on Earth.

    (i) How many people will live on Earth in 30 years?

    (ii) If this trend continues, how many people will live on Earth in 150 years?

    (iii) How many more people will live on Earth in 5 years?

b) Ms V. Rich has an account with 6.7×10^9 deposited in it. Her account balance due to interest will double every 30 years.

    (i) Find her account balance in 30 years.

    (ii) Find her annual interest rate if her account earns interest compounded yearly.

c)   (i) Compare your answer to a) (i) to your answer to b) (i). What statement can you make about these two answers?

    (ii) By what annual percent is our planet's human population increasing each year?

7.7 Inflation and depreciation

> **1.9** Financial applications of geometric sequences and series: compound interest, annual depreciation.

Inflation is a general increase in the price of all commodities. This happens because of a variety of different reasons. For example, a business owner needs to give her employees a raise because the employees have been with her for a long time and have excellent work experience. She then needs to raise the costs of her goods to offset the cost of the higher employee salaries. Anyone else who relies on her goods needs to do the same thing in response, hence inflation occurs.

 Does inflation happen because of human nature?

When we calculate the effect of inflation, we treat the original price in the past as the principal and the inflation rate as an interest rate. We then use the compound interest formula to calculate the effect of the inflation rate on the price. Typically the inflation rate is given as an annual percent rate.

 The nominal interest rate of an investment is the annual interest rate offered which includes an offset to inflation.

Example 7.26

If the annual inflation rate over the past 30 years has been constant at 2.5%, find (to the nearest thousand dollars) the original cost, in the year 1978, of a house that cost $400 000 in 2008.

Solution

We use the compound interest formula with $FV = \$400\,000$, $r = 2.5$, $k = 1$, $n = 30$

$$FV = PV \times \left(1 + \frac{r}{100k}\right)^{kn}$$

$$400\,000 = PV \times \left(1 + \frac{2.5}{100 \times 1}\right)^{1 \times 30}$$

$$400\,000 \approx PV(2.097\,57)$$

$$PV \approx \$191\,000 \text{ (to the nearest thousand dollars)}$$

 After World War I, inflation was so bad in Germany that a loaf of bread could cost as much as a wheelbarrow full of money.

Depreciation occurs over time as the value of a commodity or object decreases. For example, the value of a new car typically decreases by 10% of the car's current value each year. This happens because, as the car is used, it has a greater chance of breaking down or stopping working completely. We calculate depreciation using the compound interest equation *but with a negative interest rate*.

Example 7.27

The value of clothing decreases by 15% each year. Find the value of a £60 pair of blue jeans in 5 years.

Solution

We use the compound interest equation with a *negative* annual interest rate of 15% with $PV = £60$ $r = -15$, $k = 1$, $n = 5$.

$$FV = PV \times \left(1 + \frac{r}{100k}\right)^{kn}$$

$$FV = 60 \times \left(1 + \frac{-15}{100 \times 1}\right)^{1 \times 5} = 60 \times \left(1 - \frac{15}{100}\right)^{5}$$

$$FV = £26.62$$

Appreciation is the opposite of depreciation. An object gains value over time for a variety of reasons. For example, it might become rarer over time. Often this is the case with rare coins and stamps and other collectable items. We can use the compound interest equation again, using the original price as the principal and the appreciation rate as the interest rate.

Example 7.28

A rare coin has increased in value 800% since it was created 70 years ago. Find the annual rate of appreciation to the nearest tenth of a percent.

Solution

We use the TVM solver, since we are solving for the interest rate. In our calculator we use $n = 70$, $PV = 1$, $PMT = 0$, $FV = 9$ (since it has increased 800% and we need to write it as a decimal), $P/Y = 1$ and $C/Y = 1$. Once we have entered this information, we solve for $I\%$, which is 3.2% when rounded to the nearest tenth of a percent.

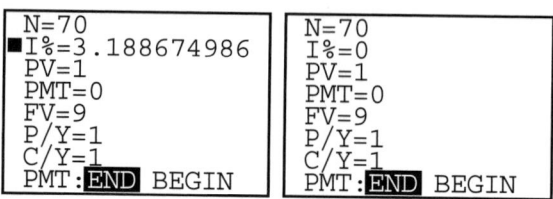

```
N=70                      N=70
■I%=3.188674986            I%=0
PV=1                      PV=1
PMT=0                     PMT=0
FV=9                      FV=9
P/Y=1                     P/Y=1
C/Y=1                     C/Y=1
PMT:END BEGIN             PMT:END BEGIN
```

Exercise 7.7

1. An item originally cost $10 000. Calculate the price of the item after 2% annual inflation over:

   a) 2 years   b) 5 years   c) 10 years   d) 20 years.

2. Calculate the price of a $800 washing machine after 10 years of inflation at an annual rate of:

   a) 2%   b) 3%   c) 5%   d) 10%.

3. At an annual 2.5% depreciation rate over 7 years, calculate the final price of the following items:

   a) $400 watch   b) £580 laptop   c) $15 000 car   d) € 9.80 book.

4. The current cost of 2 litres of milk is $2.40. Assume the annual rate of inflation has remained constant at 2.5% and hence calculate the price of the milk:
   a) 5 years ago   b) 10 years ago   c) 25 years ago   d) 50 years ago.

5. Over a 10 year period, find the annual rate of appreciation if the value of a rare stamp has increased by:
   a) 20%   b) 35%   c) 75%   d) 100%.

6. A baseball card collector buys a rare Mickey Mantle baseball card for $400. Every year the value of his card increases by 10%.
   a) Find the value of his card after 10 years.
   b) The rate of inflation over the 10 years is 2%. Find the increase in the value of the card adjusted for inflation.

7. The rate of inflation between 1990 and the year 2000 remained constant at 2.5%.
   If the value of a television in the year 2000 was $400, find the value of the same television, adjusted for inflation, in 1990.

8. Sharlene buys a used 2003 Chevrolet Corvette for US$24 000. It is known that the value of Corvettes decreases at the rate of 10% per year. What will be the value of the car at the end of 5 years?

9. Lateasia buys a house in South Africa for 200 000 SAR. The value of houses in South Africa is increasing at the rate of 5% per year. What will be the value of the house in 10 years?

10. A computer depreciates at the same rate each year for 3 years. If the value of the computer has halved during those 3 years, find the annual rate of depreciation.

11. William bought an apartment in 1985 for $117 000 which appreciated at an annual rate of 5.4%.
   a) Find the value of his apartment in 2008.
   b) If William had instead invested his $117 000 in a bank account earning 7% annual interest compounded monthly, but paid $350 per month in rent, would he have earned more or less money, and by how much?

12. a) If the cost of goods has increased by a factor of 10 during the past 25 years, what has been the rate of inflation during this time? Assume the rate of inflation has remained constant and round your answer to the nearest tenth of a percent.
   b) At this same rate of inflation, calculate to the nearest whole number by how many times will the cost of goods increase during the next 15 years.

13. The cost of a new car increases by 2% each year due to inflation.
   a) Find the price of a $17 950 car in 6 years.
       Sarah bought a car for $20 000 in January 2001, and in January 2007 she sold it.
   b) If the car depreciated by 10% each year, find the value of the car when she sold it.
   c) If she bought a new car that would have cost $20 000 in 2001 to replace her old car, find the difference in the new car's price and your answer to part b).
       In 2001, Sarah started saving money each month in an account earning 5% annual interest compounded monthly.
   d) How much money per month did she save?

14. Jake invests € 30 000 in an account earning compound interest at an annual rate of 4.5% compounded monthly.
   a) (i) What will the value of Jake's investment be in 10 years?
       (ii) Calculate the investment's *effective* annual rate to the nearest thousandth.
   b) How many *whole* months will it take for Jake's investment to triple?
       After 10 years Jake takes €20 000 of his money and puts it in a different account where after 10 years, it earns €10 000 interest compounded annually.

c) What is the new account's annual interest rate?

During a stay in the US, Jake converts € 50 000 to $ US at a rate of $1 US = 0.667 euro.

d) How many $ US does he receive?

In 2007, Jake decides to use all of this money to buy a car in the US. Each year he is in the US, his car depreciates in value 5 %.

e) What will the value of his car be in the year 2017?

The annual rate of inflation in the US during Jake's stay is 2 %.

f) (i) What will the cost of a car be in 2017 that has the same original value as his car in the year 2007?

(ii) If Jake sells his car to help pay for an equivalent new car in 2017, how much more additional money will he need to make up the difference in value?

Sets

Assessment statement

3.5 Basic concepts of set theory: elements $x \in A$; subsets $A \subset B$; intersection $A \cap B$; union $A \cup B$; complement A'.
Venn diagrams and simple applications.

Overview

By the end of this chapter you will be able to:
- understand set theory
- represent sets as Venn diagrams
- use Venn diagrams in solving problems.

8.1 Introduction to set theory

3.5 Basic concepts of set theory: elements $x \in A$; subsets $A \subset B$; intersection $A \cap B$; union $A \cup B$; complement A'.
Venn diagrams and simple applications.

● **Examiner's hint:** Make sure that you are familiar with the real number sets listed – they are tested on nearly every exam.

A set is a group of elements that have all been classified in an identical way. For instance, a very simple set is the set of all even numbers {2, 4, 6, 8, ...}. We generally use curly brackets to denote a set, and ellipses (…) to show that a pattern continues. Other examples of number sets include $\mathbb{N}$, $\mathbb{Z}$, $\mathbb{Q}$, $\mathbb{Q}'$, and $\mathbb{R}$, which are grouped according to the properties of the real numbers.

The number of elements in the set A is written as $n(A)$. This number can be finite or infinite. If the number of elements in a set is 0, we call that set 'the empty set' which is represented by either the symbol $\varnothing$ or {}.

Example 8.1

A is the set of all odd square numbers less than 100.

a) Write down the elements of A.

b) Find $n(A)$.

Solution

a) An odd square number is an odd number whose square root is a whole number.
   Therefore, A = {1, 9, 25, 49, 81}.

b) $n(A) = 5$

To indicate that a specific element belongs to a set, we use notation such as: $x \in$ A to mean x is an element in A. To show that an element is *not* in a set we use notation like: $x \notin$ A.

When we create a more complicated set, we can use the notation $\{x : x \in \mathbb{Z}, 1 \leqslant x < 10\}$. This indicates that x belongs to the universal set $\mathbb{Z}$ and is greater than or equal to 1, and less than 10. Another way of writing this set is $\{1, 2, 3, 4, 5, 6, 7, 8, 9\}$. In general, this notation starts with the universal set for all of the elements, followed by any exceptions or restrictions.

Which set has more elements, the set of natural numbers or the set of integers?

For a good summary of set theory, visit www. pearsonhotlinks.co.uk, enter the ISBN for this book and click on weblink 8.1.

Example 8.2

Set A is the set of all positive integers less than 15.

a) Write set A in set notation.

b) List all of the elements in A.

c) Find $n(A)$.

Solution

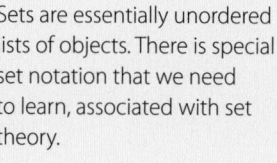

Sets are essentially unordered lists of objects. There is special set notation that we need to learn, associated with set theory.

a) Since A is the set of positive integers, it makes sense to use $\mathbb{Z}^+$ as the universal set. Then A $= \{x : x \in \mathbb{Z}^+, x < 15\}$. Since we use $\mathbb{Z}^+$ as our universal set, we already know that $x > 0$ and so we do not need to include it in our set definition.

b) A $= \{1, 2, 3, 4, 5, 6, 7, 8, 9, 10, 11, 12, 13, 14\}$

c) $n(A) = 14$

Exercise 8.1

1. Write down the elements of each set.
   a) All prime numbers less than 40.
   b) All multiples of 3 less than or equal to 36.
   c) The continents of the world.
   d) The vowels in the English alphabet.

2. Find the number of elements of each set.
   a) The positive square numbers less than or equal to 144.
   b) The factors of 60.
   c) The possible orders of the letters M, A, T and H.
   d) All composite numbers between 40 and 55.

3. Find the number of elements in each set.
   a) $\{x : x \in \mathbb{Z}^+, 10 < x \leqslant 100\}$
   b) $\{x : x \in \mathbb{Z}, -20 < x \leqslant 20, x \text{ is even}\}$
   c) $\{x : x \in \mathbb{Z}^+, x \leqslant 100, x \text{ is prime}\}$
   d) A is the set of all x such that x is a positive multiple of 3 less than 300.

4. Write down the elements in each set.
   a) $\{x : x \in \mathbb{N}, x < 8\}$
   b) $\{x : x \in \mathbb{Z}, -3 \leqslant x \leqslant 3\}$

c) $\{x : x \in \mathbb{Z}^+, 1 < x \leqslant 8\}$

d) $\{x : x \in \mathbb{N}, x \text{ is prime}, x < 32\}$

5. Write down each description of a set in set notation.

a) The set of all integers between -10 and 10.

b) The set of all fractions between -1 and 2.

c) The set of all numbers greater than or equal to 7.

d) The set of all positive irrational numbers less than π.

6. Write down the elements in each set.

a) The positive odd numbers less than or equal to 16 and greater than 3.

b) The factors of 100.

c) The different possible orders of the letters A, C, and E.

d) All composite numbers between 20 and 35.

7. $A = \{x : x \in \mathbb{Z}, 0 < x \leqslant 12\}$, $B = \{x : x \in \mathbb{Z}, 5 \leqslant x < 15\}$

Determine if the following statements are true or false.

a) Sets A and B have 8 elements in common.

b) $5 \in A$ and $5 \in B$.

c) $12 \notin A$ or B.

d) There are 12 unique elements in sets A and B.

8.2 Subsets and complements of sets

3.5   Basic concepts of set theory: elements $x \in A$; subsets $A \subset B$; intersection $A \cap B$; union $A \cup B$; complement A'.

Venn diagrams and simple applications.

A **subset** of set A is another set B that contains any number of elements from set A. Therefore, the number of elements in subset B is less than or equal to the number of elements in set A. By definition, both the empty set and the original set A itself are subsets of set A. If B is a subset of A, we write $A \subseteq B$.

Example 8.3

A is the set of all of the composite numbers less than or equal to 20.

$B = \{2, 4, 6, 8, 9, 10, 13\}$

C is the set of all even numbers less than 10.

a) Is B a subset of A?

b) Is C a subset of A?

c) Is C a subset of B?

Solution

We recall that a composite number is any number that has more than two divisors. Therefore, set $A = \{4, 6, 8, 9, 10, 12, 14, 15, 16, 18, 20\}$.

 Georg Cantor, the creator of set theory, was a German mathematician, but was awarded the Sylvester Medal by the Royal Society of London, 10 years before the start of the First World War.

 Which set has more elements, the set of natural numbers or the set of integers?

 There are more subsets of the set of all positive integers less than 100 than there are stars in the universe.

a) Since 13 $\notin$ A, B is not a subset of A.

b) Since 2 $\notin$ A, C is not a subset of A.

c) Since every element of C is contained within B, C is a subset of B.

A **proper subset** B of another set A is a set that has fewer elements than A. So, for example, the set {2, 4} is a proper subset of {2, 4, 6, 8}.

The **universal set U** is the set of which all other sets are subsets. For many problems the universal set is not explicitly given to us. We must use the context of the problem to find the definition of the universal set. When we are solving a problem involving the number sets discussed in section 8.1, the universal set is usually the real numbers.

All of the elements of a subset or a proper subset of a given set are within that given set. All of the elements of the complement of a set are *not* in the given set.

The **complement** of set A is the set of all elements which are in the universal set U, but *not* in set A. We can think of the complement of a set as being the set's opposite. We write the complement of set A as either A^C or A'.

Example 8.4

U = {1, 2, 3, 4, 5, 6, 7, 8, 9, 10, 11, 12}

A = {3, 6, 9, 12}

Find A'.

Solution

The elements in U, but not in A, are {1, 2, 4, 5, 7, 8, 10, 11} so therefore this is the complement of A.

Exercise 8.2

1. Which of the following are subsets of the set of composite numbers less than or equal to 18?

a) {2, 4, 5, 10}

b) {4, 6, 10, 12, 16, 18}

c) {18}

d) {}

2. Find the proper subsets of each of the following:

a) {1, 2, 3}

b) {A, B, C}

c) {10, 20}

d) {blue, red, yellow, green}

3. Find the complement of each of the following sets, given U is the set of positive integers less than or equal to 20.

a) A is the set of prime numbers less than 20.

b) B is the set of even numbers less than or equal to 16.

c) A = {1, 2, 3, 4, 5, 6, 7, 8}

d) C is the set of composite numbers less than 20.

4. Write down the complement of each set.

a) $\{x : x \in \mathbb{Z}, x > 0\}$ with U = $\mathbb{Z}$.

b) $\{x : x \in \mathbb{Z}, 10 < x < 20\}$ with U = $\mathbb{Z}$.

c) $\{x : x \in \mathbb{R}, 1 \leqslant x \leqslant 2\}$ with U = $\mathbb{R}$.

d) $\{x : x \in \mathbb{Q}, x < -4\}$ with U = $\mathbb{R}$.

5. Find the number of elements in the *complement* of each of the following sets, given U is the set of positive integers between 10 and 100.
   a) A is the set of even integers in U.
   b) B is the set of composite numbers in U.
   c) C is the set of integers between 30 and 60 inclusive.
   d) D is the set of perfect squares in U.

6. Find the number of possible subsets of the set $\{2, 3, 5, 7, 11\}$.

7. Find the number of subsets of the set of prime numbers less than or equal to 100.

8. $U = \{x : x \in \mathbb{N}, x < 16\}$
   $A = \{x : x \in U, x \text{ is even}\}$
   $B = \{x : x \in U, 3 < x < 10\}$
   $C = \{x : x \in U, x \leq 9\}$

   Determine if the following statements are true or false.
   a) $\{2, 4, 6\} \subseteq A$ b) $A \subset B$
   c) If $x \in B$ then $x \in C$.   d) $-1 \in C$

8.3 The union and intersection of sets

> **3.5** Basic concepts of set theory: elements $x \in A$; subsets $A \subset B$; intersection $A \cap B$; union $A \cup B$; complement A'.
> Venn diagrams and simple applications.

The **union** of two sets is the set that contains all of the original two sets' elements. For example, the union of $\{1, 2, 3, 4\}$ and $\{1, 3, 5, 7\}$ is $\{1, 2, 3, 4, 5, 7\}$. We write the union of the set A and the set B as $A \cup B$. We also notice that if set B is a subset of A, then $A \cup B = A$.

 The union of a set and its complement is the universal set.

Example 8.5

A is the set of all multiples of 4 that are less than or equal to 20.
B is the set of all of the factors of 24.
Find $A \cup B$.

Solution

$A = \{4, 8, 12, 16, 20\}$
$B = \{1, 2, 3, 4, 6, 8, 12, 24\}$
Therefore, $A \cup B = \{1, 2, 3, 4, 6, 8, 12, 16, 20, 24\}$

The **intersection** of two sets A and B is the largest set that is a subset of both sets. In other words, this intersection is the set of all of the elements that are in both sets A and B. If two sets have no elements in common, their subset is the empty set. The intersection of A and B is written $A \cap B$.

 The intersection of a set and its complement is the empty set.

Example 8.6

Set A is defined to be all of the odd numbers greater than 1 and less than 12. Set B is defined to be all of the prime numbers less than 12. Find the intersection of A and B.

Solution

Set A = {3, 5, 7, 9, 11} and set B = {2, 3, 5, 7, 11}. Therefore, their intersection is {3, 5, 7, 11}.

Exercise 8.3

1. Find the union of A and B if:
   a) A = {2, 3, 4} and B = {1, 2, 5, 6}.
   b) A = {4, 8, 12, 16} and B = {3, 4, 5, 6, 7}.
   c) A is the set of the factors of 36 and B is the square numbers less than 10.
   d) A = $\mathbb{Q}$ and B = $\mathbb{Q}'$.

2. Find the intersection of A and B if:
   a) A = {2, 3, 4} and B = {1, 2, 5, 6}.
   b) A = {4, 8, 12, 16} and B = {3, 4, 5, 6, 7}.
   c) A is the set of the factors of 36 and B is the square numbers less than 10.
   d) A = $\mathbb{Q}$ and B = $\mathbb{Q}'$.

3. U is the set of all integers greater than -2 and less than 11. Find A ∩ B' if:
   a) A = {2, 4, 6, 8, 10} and B = {-1, 0, 1, 2, 3, 4}.
   b) A = {$x : x \in \mathbb{Z}, 0 \leqslant x \leqslant 8$} and B = {1, 3, 5, 7, 9}.
   c) A = {-1, 0, 1, 2} and B = {$x : x \in \mathbb{N}, x < 11$}.
   d) A is all of the composite numbers less than 11 and B is all of the positive even numbers less than 11.

4. U is the set of positive integers less than or equal to 30. A is the set of natural numbers that are multiples of 5 in U. B is the subset of all of the even integers in U.
   a) Find $n(A \cup B)$.
   b) Find $n(A \cap B)$.
   c) Find $n(A \cup B')$.
   d) Find $n(B \cap A')$.

5. U = $\mathbb{N}$
   A = {$x : x \in \mathbb{Z}, x < 10$}
   B = {$x : x \in \mathbb{Q}, 0 < x < 10$}
   Determine if the following statements are true or false.
   a) A ⊆ U
   b) (A ∩ B) ⊂ U
   c) (A ∪ B) ⊂ U
   d) (A ∩ B') = ø

6. U = $\mathbb{Z}$
   A = {$x : x \in \mathbb{Z}^+, 2x + 5 \leqslant 19$}
   B = {$x : x \in \mathbb{Z}^-, 3x - 1 > -22$}
   Use the above sets to solve the questions below.

a) Write down the elements in A.

b) Write down the elements in B.

c) Write down the elements of A ∩ B.

d) Find $n(A \cup B)$.

7. U = {$x : x \in \mathbb{N}, 0 < x < 30, x$ is prime}

a) Give an example of A such that A ⊂ U and $n(A) = 6$.

b) Write down two sets A and B such that:

  (i) A ∩ B = ∅ and $n(A \cup B) = 8$

  (ii) $n(A \cup B') = n(U)$.

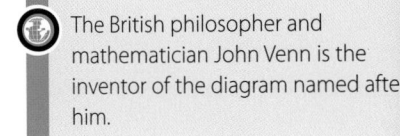
The British philosopher and mathematician John Venn is the inventor of the diagram named after him.

8.4 Introduction to Venn diagrams

> 3.5   Basic concepts of set theory: elements $x \in A$; subsets A ⊂ B; intersection A ∩ B; union A ∪ B; complement A′.
> Venn diagrams and simple applications.

A Venn diagram is a graphical way of representing sets. Using a Venn diagram you can very easily show the relationships between sets, including their intersections, unions, complements and more.

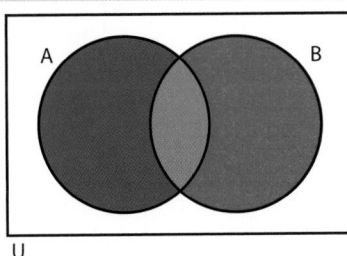

Figure 8.1 Set A and set B are subsets of the universal set U.

If we look at Figure 8.1, we can see that sets A and B overlap each other, shown in purple in the figure. This overlap corresponds to the intersection of A and B. We also notice the union of A and B, which is the portion that is coloured. The complement of A will be the portion of B that is shaded red, plus all of the unshaded part of the set U.

Example 8.7

U = {1, 2, 3, 4, 5, 6, 7, 8, 9, 10}

A is the set of odd numbers less than 10.

B is the set of square numbers less than 10.

Create a Venn diagram to represent the relationship between the sets A, B, and U.

Solution

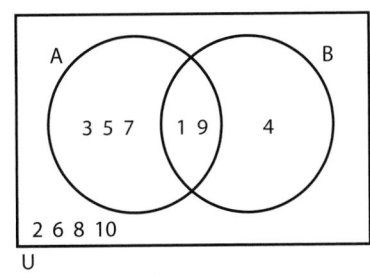

Figure 8.2 The total number of elements in U is equal to the sum of the elements in the Venn diagram, in this case 50.

When we use a Venn diagram, we can either show the elements and their relationships in the sets or the number of elements. What we show will depend on the question we have been asked.

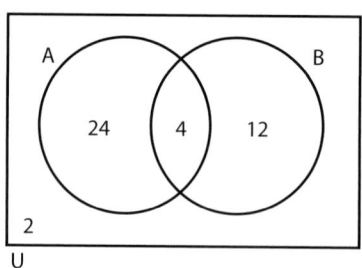

Example 8.8

Given the Venn diagram below:

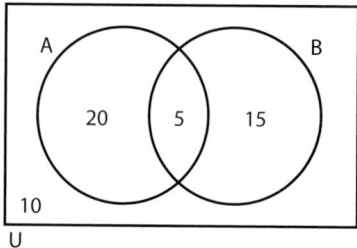

a) Find the number of elements in U.
b) Find the number of elements in A ∩ B.
c) Find the number of elements in A ∪ B.
d) Find the number of elements in A′.

Venn diagrams are visual representations of sets. In a Venn diagram, we must always include a rectangle to represent the universal set.

Solution

a) This is the sum of all of the elements in U, which is 42.
b) This is the number of elements in the overlap of A and B, which is 4.
c) This is the sum of all of the elements in A and B, which is 40.
d) This is the sum of all of the elements not in A, which is 18.

Venn diagrams do not always have to include overlapping circles representing intersecting sets. For an example of this, see Figure 2.1 from Chapter 2.

Exercise 8.4

1. Use the Venn diagram below for the following questions.

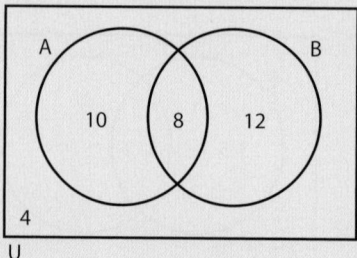

a) Find $n(A)$.   b) Find $n(B)$.   c) Find $n(A ∪ B)$.   d) Find $n(A ∩ B)$.

2. Use the Venn diagram below for the following questions.

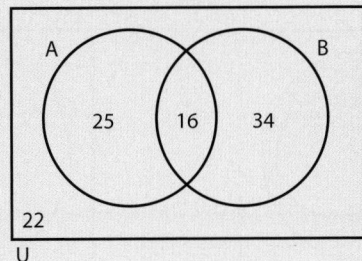

a) Find $n(A')$.   b) Find $n(B')$.   c) Find $n((A \cup B)')$.   d) Find $n((A \cap B)')$.

3. Given $n(A \cap B) = 20$, $n(U) = 45$, $n(A) = 25$, $n(B) = 35$, and $n(A \cup B) = 40$, construct a Venn diagram to represent this information.

4. Given $n(A \cap B) = 20$, $n(U) = 94$, $n(A') = 35$, $n(B') = 57$, and $n(A \cup B) = 76$, construct a Venn diagram to represent this information.

5. Given $A \subset B$, $n(U) = 61$, $n(A') = 31$, and $n(B') = 39$, construct a Venn diagram to represent this information.

6. Given $U = \{x : x \in \mathbb{Z}, 0 \leqslant x \leqslant 12\}$, $A = \{0, 2, 4, 6, 8, 10\}$, and $B = \{2, 6, 8, 9, 11, 12\}$:
a) Construct a Venn diagram to represent the relationships between these sets.
b) Find $n((A \cup B)')$.

7. Given U is the set of integers between −6 and 6, A is the set of even integers in U, and B is the set of prime numbers in U:
a) List the elements in U, A, and B.
b) Create a Venn diagram to represent these sets.
c) Find $n(A \cup B)$.

8. Draw and *shade* a Venn diagram to represent each of the following subsets of U.
a) $A \cap B$ b) $A \cup B$ c) A' d) $(A \cap B)'$

9. Write down an expression to describe the shaded region on each of the following Venn diagrams:

a)
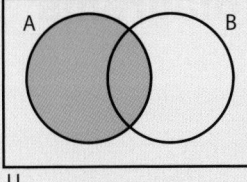

b)
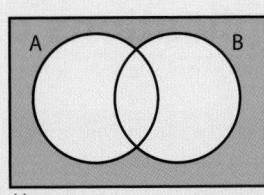

c)
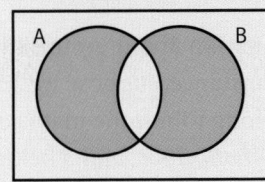

d)
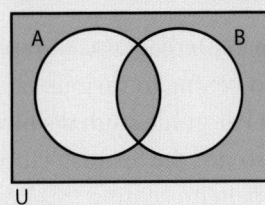

10. Write down an expression to describe the shaded region on each of the following Venn diagrams:

a)
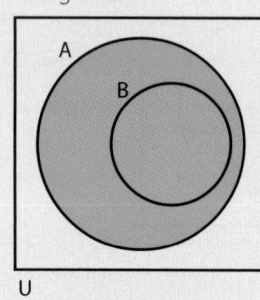

b)
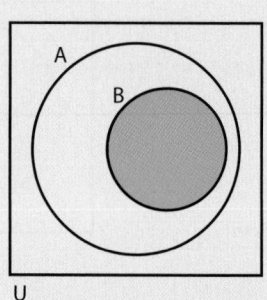

c)
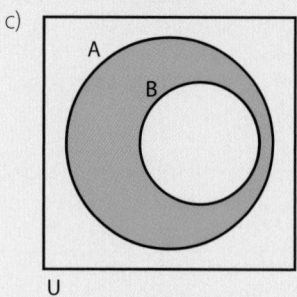

d)
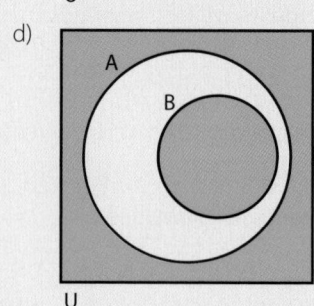

11. Given $U = \{x : x \in \mathbb{Z}, 1 \leqslant x \leqslant 25\}$, A is a prime number in U, and B is a factor of 24 in U:

a) (i) Write down the elements of A.
   (ii) Write down the elements of B.

b) Draw a Venn diagram to represent this information.

c) On your diagram, shade the portion that represents A ∩ B.

d) What percentage of U is A ∪ B?

(9.5) Applications of Venn diagrams

> **3.2**   Basic concepts of set theory: elements $x \in A$; subsets $A \subset B$; intersection $A \cap B$; union $A \cup B$; complement A'.
> Venn diagrams and simple applications.

> Venn diagrams and set theory are abstract concepts. What other abstract human inventions can be used to solve real problems?

We can use Venn diagrams to represent information about groups of people who belong to various categories. For instance, suppose we have fifty-four 11th-grade students of whom 25 belong to *just* the maths club, 20 belong to *just* the Spanish club, 6 belong to *both* clubs, and 3 belong to *neither* club. Representing this information on a Venn diagram we see:

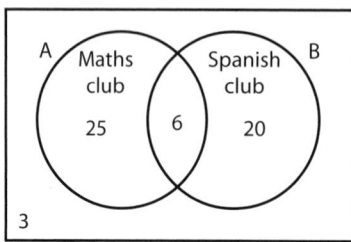

Students in 11th grade

From the diagram we can see that the total number of maths club members is 25 + 6 = 31 and the total number of Spanish club members is 20 + 6 = 26.

Example 8.9

There are two artist guilds in a city, the ABC Artist's Association and the Professional Association of Artists (PAA). 300 artists in the city are members of both guilds and 200 artists are members of neither. There are 1500 total members of the ABC guild and 2000 total members of the PAA.

a) How many artists are only members of the ABC Artist's Association?
b) How many artists are only members of the PAA?
c) How many artists does this city have in total?

Solution

If you know a total for a category in a Venn diagram, it is useful to indicate this total on the edge of the circle that represents the group.

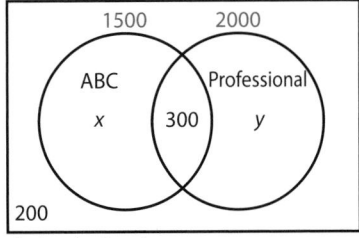
Artists in the city

a) $x + 300 = 1500$; therefore, $x = 1200$.
b) $y + 300 = 2000$; therefore, $y = 1700$.
c) This is straightforward once we know the values of x and y.

$1200 + 300 + 1700 + 200 = 3400$ artists in total.

We can also solve problems when we know the total number of elements in two groups, the number of elements in neither group and the number of elements in the universal set, but not the number of elements in both. We call the number of elements in both groups x, then create an equation and solve for x.

Example 8.10

There are 65 golf players at a charity tournament. 45 of these players will play 9 holes and 40 will play 18 holes. There are 5 people at the tournament who have decided not to play at all.

a) How many people will play both 9 holes and 18 holes of golf?
b) How many people will not play 9 holes of golf?

Solution

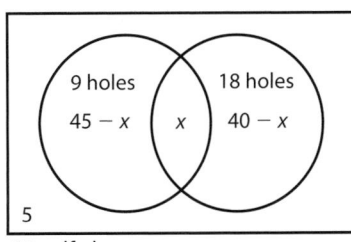
65 golf players

a) Since we know the *total* number of players who will play 9 holes and the *total* number of players who will play 18 holes, we can use this information to find the number of players who will play both. From the diagram we see that:

$$45 - x + x + 40 - x + 5 = 65$$
$$90 - x = 65$$
$$x = 25 \text{ players}$$

 Using Venn diagrams to solve set word problems relies on an understanding of the relationships between the union, intersection and complement of the sets involved.

b) This is the sum of all of the numbers outside of the 9 holes circle, which is $(40 - 25) + 5 = 20$ players.

Exercise 8.5

1. The Aquatic Society has 65 members. 40 of these members enjoy using individual submarines to explore the underwater world. 50 members enjoy scuba diving around sunken treasure ships. All members of the Aquatic Society enjoy doing at least one of these two activities.
   a) How many members of the Aquatic Society enjoy doing both activities?
   b) How many members of the Aquatic Society enjoy only scuba diving?

2. A survey was done of Lincoln Elementary School to find out which kind of fruit the students prefer to eat.
   a) Copy and fill in the missing values in the Venn diagram right, given that there are 175 students in total.
   b) How many students do not like apples?

   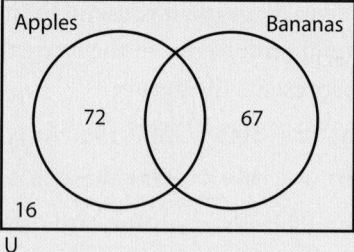

3. In a cooking class of 32 students, there are 24 people who have cooked a pie before, 28 who have cooked a cake before, and 2 who have cooked neither.
   a) Construct a Venn diagram to represent this information.
   b) How many students in this class have already cooked both a pie and a cake before?

4. In a company of 400 people, 280 use their work email addresses for their correspondence and 200 use their private email addresses. Another 40 company employees do not use email at all.
   a) Draw a Venn diagram to represent this information.
   b) How many employees use both their work email and their private email for their correspondence?
   c) How many employees are in violation of company policy to use only their company emails for their correspondence?

5. U is the set of objects that orbit the sun. A is the set of planets which orbit the sun and B is the set of giant gas planets orbiting our sun.
   a) Draw a Venn diagram to represent the relationship between U (the universal set), A and B.
   b) Write down a sentence which describes the specific relationship between set A and set B.
   c) (i) Shade the portion of your Venn diagram which corresponds to A ∩ B′.
      (ii) What does this shaded portion of your Venn diagram represent?

6. B and C are subsets of a universal set U such that
   $U = \{x : x \in \mathbb{Z}, 0 \leqslant x < 10\}$
   $B = \{\text{prime numbers} < 10\}$
   $C = \{x : x \in \mathbb{Z}, 1 < x \leqslant 6\}$
   a) List the elements of: (i) B   (ii) C ∩ B   (iii) B ∪ C′
   b) Draw a Venn diagram to represent this information.

7. There are 40 students in foreign language classes at an NYC public high school. Of these, 25 of the students are taking French and 30 of the students are taking Spanish.

a) How many students are taking both French and Spanish?

b) Create a Venn diagram to represent this information.

c) Shade the portion of your Venn diagram that represents the students who are not taking Spanish.

8. $U = \{x : x \in \mathbb{N}, x < 15\}$
$A = \{x : x \in U, \frac{2}{3}x + 5 < 14\}$
$B = \{x : x \in U, 2x - 3 \geqslant 4\}$
Use the above sets for the questions below.

a) (i) Write down the elements of A.   (ii) Write down the elements of B.

b) Draw a Venn diagram to represent this information.

c) Shade the portion of the Venn diagram that corresponds to (A ∪ B′).

9. In Vancouver, in any given year there are, on average, 166 days of the year with precipitation for at least part of a day. However, Vancouver is a coastal town and the weather changes frequently, so there are, on average, 289 days of the year with at least some sunshine. Assume that any day in Vancouver has either precipitation or sunshine.

a) Draw a Venn diagram to represent this information.

b) How many days of the year is there both precipitation and sunshine in Vancouver during a typical year?

c) What percentage of the year is Vancouver *without* rain?

10. In a recent survey of the latest 50 movies released from Hollywood, movie watchers claimed that:

- 35 of the movies were shorter than two hours
- 20 of the movies were interesting
- 8 of the movies were two hours or more long and were not interesting.

a) How many movies were both interesting and shorter than two hours?

b) Draw a Venn diagram to represent this information.

Each Hollywood movie grosses $100 million dollars if it is interesting and $30 million if is not interesting.

c) What was the total earned on these 50 movies?

The movies cost $50 million to make if they are shorter than two hours and $70 million otherwise.

d) How much money did these 50 movies cost to make in total?

11. After a poll taken in a major metropolitan area it is discovered that the city's recycling programme needs improvement. Only 28% of the city's residents recycle both paper and glass, and 30% of residents do not recycle at all! The numbers of people who recycle only glass or only paper are equal.

a) Draw a Venn diagram to represent this information.

b) How many of the city's 8 million residents recycle only glass?

The city creates an ad campaign and convinces 200 000 of the non-recycling residents to start recycling both glass and paper.

c) What percentage of the city is now recycling both glass and paper?

d) What is the total percentage increase, to the nearest tenth of a per cent, of any type of recycling in the city?

e) Given that the polls taken by the city have an error of ±1% for each category, find the maximum possible number of people who could be recycling in this city. Write your answer in the form $a \times 10^k$, where $1 \leqslant a < 10$ and $k \in \mathbb{Z}$.

8.6 Venn diagrams involving three sets

> **3.2** Basic concepts of set theory: elements $x \in A$; subsets $A \subset B$; intersection $A \cap B$; union $A \cup B$; complement A'.
> Venn diagrams and simple applications.

Figure 8.3 A visual representation of three sets ▶

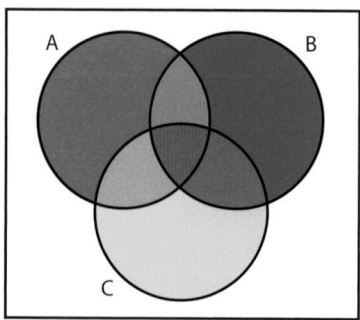

A Venn diagram need not always be composed of two circles carefully overlapping each other. We can also very easily use a Venn diagram when presented with a problem with three sets.

If we look at Figure 8.3, we can see that this Venn diagram represents sets A, B, and C. We also notice that the intersection of any two of the sets is shown by where their different colours overlap. The intersection of all three sets is shown where all three colours overlap.

Example 8.11

A web software company decides to do a survey of its 220 employees to find out which web programming languages its employees know. They discover:

- 20 of their employees know no programming languages
- 40 know PHP, JavaScript, and Flash
- 60 know PHP and JavaScript
- 50 know PHP and Flash
- 80 know JavaScript and Flash
- 120 know PHP
- 120 know JavaScript
- 110 know Flash.

a) Create a Venn diagram to represent this data.

b) How many of their employees know exactly two languages?

Solution

a) First we record as much information as we can from the problem, being careful to use the correct locations.

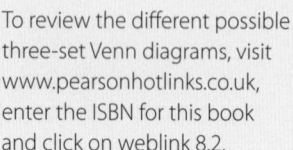

To review the different possible three-set Venn diagrams, visit www.pearsonhotlinks.co.uk, enter the ISBN for this book and click on weblink 8.2.

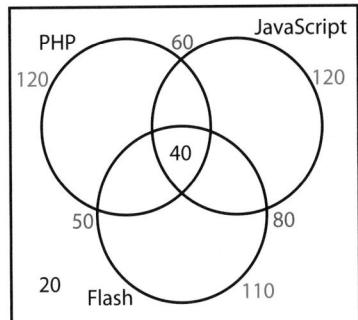

From this we see that the number of people who know *only* PHP and Flash is 10, the number that know *only* JavaScript and PHP is 20 and the number that know *only* Flash and JavaScript is 40. We get each of these numbers by subtracting the number of people who know all three programming languages from each of the number of people who know two programming languages.

Now we have:

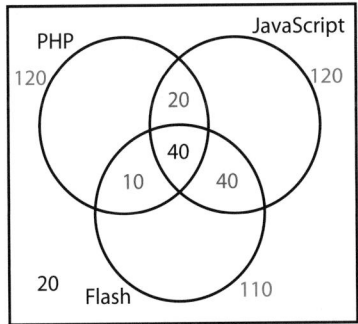

We now use a similar procedure to find out how many people know *only* PHP, JavaScript or Flash. For example, for PHP this is $120 - (10 + 20 + 40) = 50$ people.

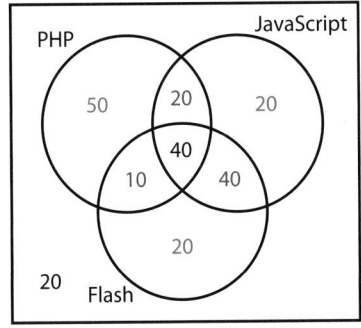

● **Examiner's hint:** Always check that the sum of all of the numbers inside your Venn diagram is the same as the number of elements in the universal set.

We then do one final check to make sure these numbers add up to 220.

$50 + 20 + 10 + 40 + 20 + 40 + 20 + 20 = 220$ so we know it is likely that we have done this problem correctly.

b) From the final diagram, we can see that this is $10 + 20 + 40 = 70$ people.

Venn diagrams including three sets involve more information than less complicated Venn diagrams, but we use the same procedure to solve them.

Exercise 8.6

1. Use the diagram below to answer the following questions.
   a) Find $n(A)$.
   b) Find $n(A \cup B)$.
   c) Find $n(A \cap B \cap C)$.
   d) Find $n(C')$.

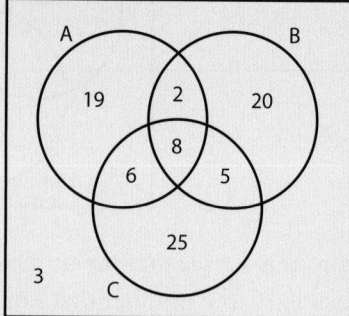

2. Use the diagram below to answer the following questions.
   a) Find $n(B)$.
   b) Find $n(A \cap B)$.
   c) Find $n(B')$.
   d) Find $n(A' \cap B')$.

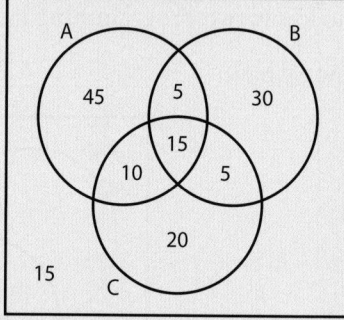

3. Copy and shade the indicated region of the Venn diagram below.
   a) $A \cup B$
   b) $A \cap C$
   c) $A \cup B \cap C$
   d) $A' \cap (B \cup C)$

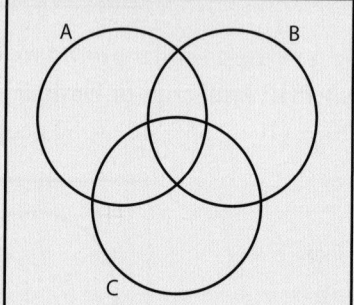

4. Copy and shade the indicated region of the Venn diagram below.
   a) $(A \cap B) \cup C$
   b) $(A \cup B)'$
   c) $(A \cap C)'$
   d) $(A \cap B) \cup (A \cap C) \cup (B \cap C)$

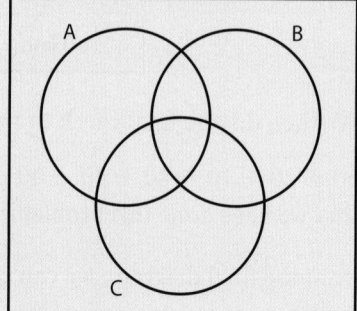

5. Use the diagram right to answer the following questions.

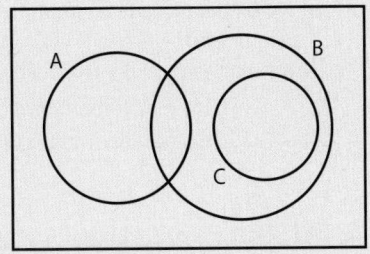

a) Find A ∩ C.

b) Shade the portion of the Venn diagram that represents (A ∪ B)′ ∩ C.

6. Write down an expression to describe the shaded area on the following Venn diagrams:

a)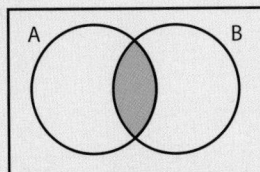

b)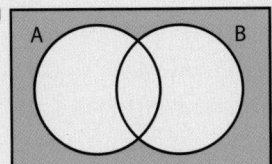

c)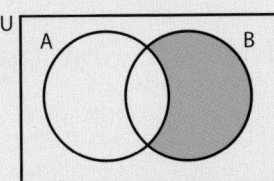

d)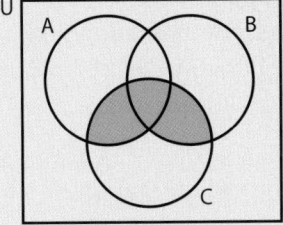

7. Use the diagram below to fill in the given information for the problem.

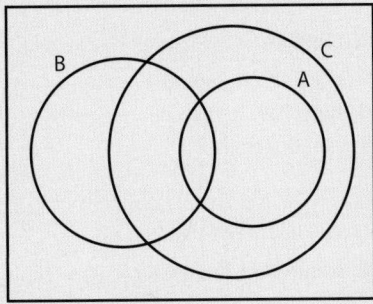

a) Shade the portion of the diagram that represents A ∩ B.

b) Shade the portion of the diagram that represents (A′ ∩ B) ∩ C.

c) Describe in words the relationship between set A and set C.

8. The diagram below shows the percentage of high school students at Richmond High that read, watch TV, or use the internet regularly.

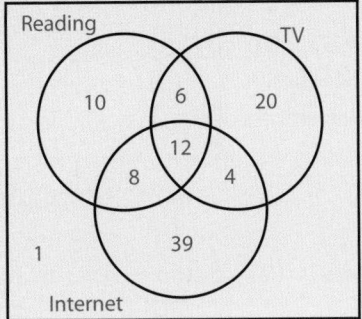

a) What percentage of students watch TV and read books regularly at Richmond?

There are 1200 students at Richmond High.

b) How many of them only watch TV and use the internet?

c) How many of them either do not watch TV or do not use the internet?

9. Of 165 first year computer science students:
- 95 know the Windows operating system
- 60 know the Mac operating system
- 50 know the Linux operating system
- 10 know all 3 operating systems
- 20 know Windows and Linux
- 15 know Windows and Mac
- 15 know Mac and Linux.

All of the students know at least one of the Windows, Mac, or Linux operating systems.

a) Create a Venn diagram to represent this information.

b) How many students only know one operating system?

c) How many students know exactly two operating systems?

10. There are three choices for science classes at an international High School, where all 225 students are required to take at least one science class:
- 150 take biology
- 100 take chemistry
- 80 take physics
- 15 take biology and physics
- 60 take biology and chemistry
- 40 take physics and chemistry
- 10 students take all three courses.

a) Draw a Venn diagram to represent this information.

b) How many students take all three sciences?

c) How many students do not take chemistry?

d) How many students do not take biology or chemistry?

11. Let U be the set of all positive integers from 1 to 21 inclusive.

A, B, and C are subsets of U such that:

A contains all the positive integers that are factors of 21,

B is the set of multiples of 7 contained in U,

C is the set of odd numbers contained in U.

a) Draw a Venn diagram to represent this information.

b) List all the members of set A.

c) Write down all the members of
  (i) $A \cup B$
  (ii) $C' \cap B$
  (iii) $(A \cap B)' \cap C$.

12. The No-Name shoe company makes some shoes with synthetic leather and some shoes with red shoelaces. Of their 60 different models, 40 of them use synthetic leather, 35 use red shoelaces, and 5 use neither.

   a) Draw a Venn diagram to represent this information.

   b) How many shoes does this company make with synthetic leather and red laces?

   After two years of research, the No-Name shoe company introduces gel packed soles in 20 of their existing shoe models. Five models of shoes have all three features, 12 shoes have both red laces and gel packed soles, and 10 shoes have both gel packed soles and use synthetic leather.

   c) Draw a Venn diagram to represent this new information.

   d) Shade the portion of your Venn diagram that corresponds to the shoes that have synthetic leather, but do not have either red laces or gel packed soles.

   e) To the nearest tenth, what percentage of the shoes has neither synthetic leather nor red laces?

9 Logic

Assessment statements

3.1 Basic concepts of symbolic logic: definition of a proposition; symbolic notation of propositions.

3.2 Compound statements: implication, ⇒; equivalence, ⇔; negation, ¬; conjunction, ∧; disjunction, ∨; exclusive disjunction, ⊻.

Translation between verbal statements and symbolic form.

3.3 Truth tables: concepts of logical contradiction and tautology.

3.4 Converse; inverse; contrapositive.

Logical equivalence. Testing the validity of simple arguments through the use of truth tables.

Overview

By the end of this chapter you will be able to:
- understand the basic logical connectives that all arguments share
- show how to write these statements using symbolic logic
- analyze the truth value of logical statements using truth tables
- define some properties of logical arguments
- explain how we can use these properties to further examine logical statements
- test the validity of a simple argument with or without truth tables.

9.1 Introduction to symbolic logic

3.1 Basic concepts of symbolic logic: definition of a proposition; symbolic notation of propositions.

In the fourth century BC, Aristotle formalized a system of logical analysis, in which a complicated argument was reduced into simpler statements, which were themselves used to determine the truth of the argument. Leibniz is credited as the philospher who first developed symbolic logic. The basis of this method is to take the sub-statements and represent them with a single capital letter.

For example, consider the two statements:

'It is raining.'
'It is sunny.'

We use P to represent the first statement and Q to represent the second statement.

If we want to represent the sentence 'If *it is raining* then *it is sunny*' we can use instead, 'If P then Q.'

The German mathematician Kurt Gödel proved that any system of logic sufficient to prove the existence of all of the natural numbers would have statements that are true, but that it would be unable to prove.

Example 9.1

P: I will do well on the tests tomorrow.

Q: I will study tonight.

Replace P and Q for the appropriate proposition below.

a) If I will study tonight then I will do well on the tests tomorrow.

b) I will study tonight and I will do well on the tests tomorrow.

c) I will not study tonight.

Solution

a) If Q then P.

b) Q and P.

c) Not Q.

The most basic types of logical statements are called **propositions**. These are statements that can either be true or false, and which themselves contain no logical connectives such as 'or', 'and', etc.

Some examples of statements that are propositions are:

'The cow is green.'

'I went to the movie.'

'She will travel to Thailand.'

'$2x + 3 \leqslant 13$'

These are all propositions because the truth value of the statement can be determined, although in some cases it will be difficult to do so. In the last statement, the truth value of the statement depends on the value of x. Some examples of statements that are not propositions are:

'I like watermelon.'

'That puppy is cute.'

Statements that are opinions are not considered propositions, for the purposes of this course, because they cannot have their truth value determined.

 A truth value indicates the extent to which a proposition is true. In mathematics, 'true' has a truth value of 1, and 'false' has a truth value of 0.

 True or false? 'This statement is false.'

 Propositions are statements that can be proved true or false.

Exercise 9.1

1. Identify which statements are opinions and which are propositions.

a) The taco is delicious.   b) The banana is ripe.

c) Monkeys are funny.   d) The sky is blue.

2. P: The chair is brown.
Q: The couch is comfortable.
Replace for P and Q in each of the following.

a) The chair is brown or the couch is comfortable.

b) The chair is not brown and the couch is not comfortable.

c) If and only if the chair is brown then the chair is comfortable.

3. P: The clock has stopped.
Q: I will be late.
Replace P and Q in the following with the propositions.
a) If P then Q.
b) P and Q.
c) P or Q.
d) If not P then Q.

4. P: $2x - 4 \geqslant 10$
Q: $10 - 3x \leqslant 4$
Set A is the set of values of x for which P is true.
Set B is the set of values of x for which Q is true.
a) Write down set A in set notation.
b) Write down set B in set notation.
c) Find $(A \cap B)$.

5. Write down the propositions in the statements below.
a) If the slope is wet then Jane will slip.
b) The car can drive fast and is red.
c) Stars are hot and far away.

6. Given P is $x = 3$ and Q is $2x^2 + 2 = 20$, replace for P and Q in the following statements.
a) If P then Q.
b) P and Q.
c) P or Q.
d) Not Q and not P.

7. P: The piano is in tune.
Q: The choir is in key.
Translate each of the following symbolic logic statements into words.
a) If P then Q.
b) If not P then not Q.
c) Either not P or not Q.
d) If and only if P then Q.

9.2 Implication and equivalence

3.2   Compound statements: implication, $\Rightarrow$; equivalence, $\Leftrightarrow$; negation, $\neg$; conjunction, $\wedge$; disjunction, $\vee$; exclusive disjunction, $\underline{\vee}$.
Translation between verbal statements and symbolic form.

When proposition P being true means that another proposition Q must be true as well, we say that P implies Q. We can then write 'If P *then* Q', creating a compound statement from the two propositions called an **implication**. We can write this in symbolic form as $P \Rightarrow Q$.

The first proposition of an implication statement is called the **antecedent** of the statement, and the second proposition is called the **consequent**.

An implication is false when the antecedent is true, but the consequent is false. This is because of the definition of the implication statement. An implication is true in all other cases.

● Examiner's hint: When translating symbolic statements of the form $P \Rightarrow Q$ into words, make sure to include the word *'If'* as omitting this will cause you to lose marks.

Example 9.2

Write down the cases when the statement:
'When I say I am going to do something, then I do it.'
is true.

● **Hint:** The false case is the only one that 'breaks the promise.'

Solution
The only time this statement will be false is when I say I am going to do something (the antecedent is true) and I do not do it (the consequent is false).

1. When I say I am going to do something and I do it.
2. When I don't say I am going to do it and I do it.
3. When I don't say I am going to do it and I don't do it.

Two propositions P and Q are called **equivalence statements** if each statement implies the other. We can say that '*If and only if* P *then* Q'. In symbolic form we write P ⇔ Q. Therefore, if P is true, Q is true. Similarly, if P is false, Q must also be false. If one proposition is true and the other is false, then P ⇔ Q is a false statement.

 Is it possible to prove every proposition true or false?

Example 9.3

Determine the truth value of 'If and only if I study for my test then I will do well', if I studied for my test but I did not do well.

Solution
Since one of the sub-propositions of 'If and only if I study for my test then I will do well' is true, and one is false, we know the overall statement must be false. Otherwise the statement is true.

 In the implication P ⇒ Q, P is the antecedent and Q is the consequent. P ⇔ Q is the symbolic representation of an equivalence statement.

Exercise 9.2

1. Determine the truth of P ⇒ Q, given:
   a) P is true, Q is true.
   b) P is true, Q is false.
   c) P is false, Q is true.
   d) P is false, Q is false.

2. Determine the truth of P ⇔ Q, given:
   a) P is true, Q is true.
   b) P is true, Q is false.
   c) P is false, Q is true.
   d) P is false, Q is false.

3. Determine the truth value of 'If it is light then I can see', given:
   a) It *is* light, but I *cannot* see.
   b) It *is* dark and I *can* see.
   c) It *is* dark and I *cannot* see.
   d) It *is* light and I *can* see.

4. Consider the following logical propositions:
   P: x is a factor of 6.
   Q: x is a factor of 24.
   Write each of the following in words.
   a) P ⇒ Q
   b) Q ⇒ P
   c) P ⇔ Q
   d) Not P ⇔ not Q

5. Consider the following logical propositions:
P: x is a prime number.
Q: x is a multiple of 3.
a) Determine the value of x which makes both of these statements true.
b) Write P $\Rightarrow$ Q in words.
c) If $x = 6$, determine the truth value of Q $\Rightarrow$ P.
d) If $x = 7$, determine the truth value of P $\Leftrightarrow$ Q.

6. Write each of the following in symbolic logic form.
a) If the wind is strong then the waves will be large.
b) If and only if the wind is strong then the waves will be large.
c) If the waves are large then the wind is strong.
d) If and only if the waves are large then the wind is strong.

7. Consider the following logical propositions:
P: The temperature is less than 0 degrees Celsius.
Q: The water is frozen.
Write each of the following in words.
a) P $\Rightarrow$ Q   b) Q $\Rightarrow$ P   c) P $\Leftrightarrow$ Q   d) Not P $\Leftrightarrow$ not Q

9.3 Negation and conjunction

> **3.2** Compound statements: implication, $\Rightarrow$; equivalence, $\Leftrightarrow$; negation, $\neg$; conjunction, $\wedge$; disjunction, $\vee$; exclusive disjunction, $\underline{\vee}$.
> Translation between verbal statements and symbolic form.

A proposition and its negation have opposite truth value. So if the proposition P is true, the negation of P, written $\neg$P, is false. Similarly, if proposition P is false, $\neg$P is true. If the proposition P is 'It is raining', the negation of P is written as 'It is *not* raining'. The negation of $\neg$P is P.

When we look at an algebraic expression such as $2x + 3 = 9$, the solution set of this proposition is $\{x : x \in \mathbb{R}, x = 3\}$. The negated proposition has the complement of this set as its solution set. This means that the negation of $2x + 3 = 9$ is $2x + 3 \neq 9$. Similarly, the negation of $2x \leqslant 10$ is $2x > 10$.

• **Examiner's hint:** When translating negation sentences, make sure not to change the tense of the sentence.

Example 9.4

Write down the negation of the following statements.
a) It is sunny.   b) $2(x + 3) > 6$
c) The sun is not a star.   d) $\neg$Q

Solution

For each of the following statements, we replace the proposition with its negation.
a) It is not sunny.   b) $2(x + 3) \leqslant 6$
c) The sun is a star.   d) Q

The **conjunction** of two propositions is true when both propositions are true, and false otherwise. We form the conjunction of two propositions by combining them together with the word 'and'. So if P is 'I will go to the movies' and Q is 'I will have dinner', the conjunction of P and Q is 'I will go to the movies and I will have dinner'. In symbolic form, we write P ∧ Q.

Example 9.5

P: The cat is black.
Q: The cat has a short tail.
a) Write down the conjunction of P and Q in symbolic form.
b) Write down the conjunction of P and Q in words.

Solution
a) P ∧ Q
b) The cat is black and the cat has a short tail.

We can combine together different propositions using logical connectives to produce compound statements. For example, using the statements 'It is raining' and 'It is cold' with 'I will wear a jacket', we can produce the statement '*If* it is raining *and* it is cold *then* I will wear a jacket'.

To determine the truth value of this compound statement, we first determine the truth value of each sub-statement, then use these truth values to determine the overall truth value.

Example 9.6

P: The apples are ripe.
Q: The harvest season is started.
R: The current month is July.
Translate (P ∧ Q) ⇒ ¬R into words.

Solution
We replace P, Q and R with their meanings, and translate ∧ and ⇒ into the conjunction and implication respectively.

If the apples are ripe and the harvest season has started, the current month is not July.

Exercise 9.3

1. Determine the truth value of each of the following statements, given that P is true and Q is false.
   a) ¬Q
   b) ¬P
   c) P ∧ Q
   d) P ∧ ¬Q

2. Determine the truth value of each of the following statements, given that P is false and Q is true.
   a) P ∧ Q
   b) ¬P ∧ Q
   c) ¬P ⇒ Q
   d) P ⇔ Q

The negation of proposition P is written ¬P and has opposite truth value of P. The conjunction of P and Q is written P ∧ Q. The conjunction is only true when both P and Q (which are called the conjuncts) are true.

3. Consider the following logical propositions:
P: x is an even number.
Q: x is a composite number.
Write each of the following in words.
a) $P \wedge Q$ b) $\neg P$ c) $\neg Q$ d) $\neg P \wedge Q$

4. P: It is sunny.
Q: It is warm.
Write each of the following logical statements in symbolic form.
a) It is sunny and it is warm.
b) It is not sunny.
c) It is not warm and it is not sunny.
d) If it is not warm and it is sunny then it is not warm.

5. P: The speed limit is 50 kilometres per hour.
Q: The car is travelling at 60 kilometres per hour.
R: The car is speeding.
Translate each of the following into words.
a) $P \Rightarrow \neg Q$ b) $P \wedge \neg Q$ c) $\neg P \Rightarrow R$ d) $(P \wedge Q) \Rightarrow R$

6. P: Birds lay eggs.
Q: Fish swim.
Write each of the following statements in words.
a) $\neg P \Rightarrow \neg Q$ b) $\neg P \wedge \neg Q$ c) $(P \wedge Q) \Rightarrow P$ d) $(P \wedge Q) \Rightarrow \neg Q$

7. P: $\frac{x}{3} - 4 \leqslant 1$
Q: $3x > 9$
a) (i) Write down the solution set of $\frac{x}{3} - 4 \leqslant 1$.
    (ii) Write down the solution set of $3x > 9$.
b) (i) Write down the solution set for $\neg P$.
    (ii) Write down the solution set for $\neg Q$.
c) Translate $P \wedge Q$ into words.
d) Find the solution set of $P \wedge Q$.

9.4 Disjunction and exclusive disjunction

> **3.2** Compound statements: implication, $\Rightarrow$; equivalence, $\Leftrightarrow$; negation, $\neg$; conjunction, $\wedge$; disjunction, $\vee$; exclusive disjunction, $\underline{\vee}$.
>
> Translation between verbal statements and symbolic form.

What words have a mathematical meaning which differs slightly from everyday use?

A **disjunction** is a statement created by forming two propositions together, such that the compound statement formed is true whenever either or both sub-statements are true, but false when both are false. If proposition P is 'The couch is brown' and proposition Q is 'The chair is red' then the disjunction of P and Q is 'The couch is brown *or* the chair is red'. We write the disjunction of P and Q as $P \vee Q$.

Example 9.7

P: It is snowing.

Q: It is cold.

a) Write $P \vee Q$ in words.

b) Determine the truth of $P \vee Q$ if P is true and Q is false.

Solution

a) It is snowing or it is cold.

b) $P \vee Q$ is a disjunction and is therefore true if either P or Q is true. Since P is true, we know $P \vee Q$ is true.

An **exclusive disjunction** is a disjunction where not both of the sub-propositions can be true at once. It is therefore true when either proposition is true, and false otherwise. If P is 'The sky is blue' and Q is 'The sky is overcast' then the exclusive disjunction is '*Either* the sky is blue *or* the sky is overcast'. Given propositions P and Q, their exclusive disjunction is written $P \veebar Q$.

Note that this is the definition that is most commonly incorrectly assumed of the word '*or*'. We generally imply the use of the word '*either*' when we make a statement including the word '*or*'.

> 🔒 The disjunction and the exclusive disjunction are represented with the symbols $\vee$ and $\veebar$ respectively. They have similar truth values, except the exclusive disjunction cannot have both P and Q true.

Example 9.8

P: The sky is blue.

Q: Grass is green.

a) Write $P \veebar Q$ in words.

b) Determine the truth value of $P \veebar Q$ in the following circumstances.

    (i) P is true, Q is true.

    (ii) P is true, Q is false.

    (iii) P is false, Q is true.

    (iv) P is false, Q is false.

Solution

To rewrite $P \veebar Q$ in words, we replace P and Q for 'The sky is blue.' and 'Grass is green.' into the structure of the exclusive disjunction, which is 'Either P or Q'.

a) Either the sky is blue or grass is green.

The exclusive disjunction is true when either disjunct is true, but not when both are false or both are true.

b)   (i) Both P and Q are true, so $P \veebar Q$ is false.

    (ii) Only P is true, so $P \veebar Q$ is true.

    (iii) Only Q is true, so $P \veebar Q$ is true.

    (iv) Both P and Q are false, so $P \veebar Q$ is false.

> ● **Examiner's hint:** The hardest part of translating the difference between the **exclusive or** and the **regular or** is remembering when to use the word 'either'. We can just remember that the symbol $\veebar$ includes the extra bar underneath and must require an extra word in its translation.

Exercise 9.4

1. Consider the following logical propositions:
   P: x is a multiple of 5.
   Q: x is a factor of 100.
   Write each of the following in words.
   a) P ∨ Q
   b) P ⊻ Q
   c) ¬P ∨ Q
   d) P ⊻ ¬Q

2. Determine the truth value of each of the following statements, given that P is false and Q is false.
   a) P ∨ Q
   b) P ⊻ Q
   c) ¬P ∨ Q
   d) ¬P ⊻ ¬Q

3. Determine the truth value of each of the following statements, given that P is true and Q is true.
   a) P ∨ ¬Q
   b) ¬P ∨ ¬Q
   c) (P ∨ Q) ⇒ Q
   d) (P ⊻ Q) ⇒ ¬Q

4. P: It is raining.
   Q: It is snowing.
   Write each statement below in words.
   a) P ∧ ¬Q
   b) P ⊻ Q
   c) ¬P ∨ ¬Q
   d) P ∧ Q

5. Determine whether the following statements are true or false.
   a) P ∧ Q, given that P is true and Q is false.
   b) P ⊻ Q, given that both P and Q are true.
   c) ¬P ∨ ¬Q, given that P is true and Q is true.
   d) (P ∧ Q) ⇒ P, given that P is false and Q is false.

6. P: A square is a type of rhombus.
   Q: A rhombus has all four sides equal.
   R: A parallelogram has opposite sides parallel.
   a) Translate the following into words.
       (i) P ∧ Q
       (ii) (P ∨ Q) ⇔ R
   b) Assuming P, Q and R are all true, find the truth value of:
       (i) Q ⇒ R
       (ii) Either a parallelogram has opposite sides parallel or a square is a type of rhombus.

7. P: The Sun rises in the east.
   Q: The Earth rotates on its axis.
   R: The Moon orbits the Earth.
   a) Write each of the following in symbolic form.
       (i) If the Earth rotates on its axis then the sun rises in the east.
       (ii) Either the Earth rotates on its axis or the Moon does not orbit the Earth.
   b) Write each of the following in words.
       (i) P ∧ ¬Q
       (ii) (P ∨ Q) ⇒ R
   c) Determine the truth value of P ∧ ¬Q if:
       (i) P is true and Q is true.
       (ii) P is true and Q is false.

9.5 Truth tables

Truth tables: concepts of logical contradiction and tautology.

A **truth table** is a way of organizing the possible combinations of truth values of two or more propositions. We can use these tables to determine under what circumstances compound statements are true or false.

p	q	$\neg p$	$p \wedge q$	$p \vee q$	$p \underline{\vee} q$	$p \Rightarrow q$	$p \Leftrightarrow q$
T	T	F	T	T	F	T	T
T	F	F	F	T	T	F	F
F	T	T	F	T	T	T	F
F	F	T	F	F	F	T	T

To use truth tables to determine the truth of a compound statement, we can work from left to right in a table. We then use the columns we have figured out to deduce the truth values of the further columns.

In what way are truth tables similar to the language of computers?

◀ **Table 9.1** A complete reference of the truth values of the six basic logical connectives This table can also be found in the Mathematical studies SL formula booklet (MSSLFB).

● **Examiner's hint:** Never leave any blanks in a truth table when asked to solve one of these problems on the IB exam; if you do not know the answer, guess! You have a 50% chance of being right.

Example 9.9

Create a truth table for $(P \vee Q) \Rightarrow P$.

Solution

We will need a column for P, Q, $(P \vee Q)$ and finally $(P \vee Q) \Rightarrow P$.

● **Hint:** Use Table 9.1 above.

P	Q	$(P \vee Q)$	$(P \vee Q) \Rightarrow P$
T	T	T	T
T	F	T	T
F	T	T	F
F	F	F	T

We can also use truth tables when we have three different sub-propositions, P, Q, and R. We construct the table with 9 rows: 1 for the row of propositions and 8 for the different possible combinations of the truth values of P, Q, and R. After that, we use the same procedure as we would with two sub-propositions.

Example 9.10

Create the truth table for $(P \wedge Q) \vee (P \wedge \neg R)$.

Solution

We need to break the statement into its component propositions as below.

P	Q	R	¬R	P ∧ Q	P ∧ ¬R	(P ∧ Q) ∨ (P ∧ ¬R)
T	T	T	F	T	F	T
T	T	F	T	T	T	T
T	F	T	F	F	F	F
T	F	F	T	F	T	T
F	T	T	F	F	F	F
F	T	F	T	F	F	F
F	F	T	F	F	F	F
F	F	F	T	F	F	F

We use truth tables to determine the circumstances under which a complex proposition is true.

Exercise 9.5

1. Complete the truth table.

P	Q	¬P	¬P ∧ Q
T	T		
T	F		
F	T		
F	F		

2. Complete the truth table.

P	Q	¬P	¬Q	¬P ⇒ ¬Q
T	T			
T	F			
F	T			
F	F			

3. Complete the truth table.

P	Q	¬P	¬Q	(Q ⇒ P)	(¬P ⇒ ¬Q)	(Q ⇒ P) ⇔ (¬P ⇒ ¬Q)
T	T					
T	F					
F	T					
F	F					

4. Complete the truth table.

P	Q	R	(P $\underline{\vee}$ Q)	(P $\underline{\vee}$ Q) $\wedge$ R
T	T	T		
T	T	F		
T	F	T		
T	F	F		
F	T	T		
F	T	F		
F	F	T		
F	F	F		

5. Show, using a truth table, that $\neg(P \wedge Q)$ has the same truth value as $\neg P \vee \neg Q$ in all circumstances.

6. Use a truth table to determine under what circumstances $(P \wedge Q) \Rightarrow R$ is false.

7. Let P, Q and R be the statements:
   P: x is a multiple of four.
   Q: x is a factor of 36.
   R: x is a square number.
   a) Write a sentence, in words, for the statement:
      $(P \vee R) \wedge \neg Q$
   b) Use a truth table to determine the truth values of $(P \vee R) \wedge \neg Q$.
   c) Write down one possible value of x for which $(P \vee R) \wedge \neg Q$ is true.

8. Complete the truth table for the compound proposition $(P \wedge \neg Q) \Rightarrow (P \vee Q)$.

P	Q	$\neg Q$	(P $\wedge \neg$Q)	(P $\vee$ Q)	(P $\wedge \neg$Q) $\Rightarrow$ (P $\vee$ Q)
T	T	F	F		
T	F	T	T		
F	T	F		T	
F	F		F	F	

9. $[(P \Leftrightarrow Q) \wedge P] \Leftrightarrow Q$
   Complete the truth table below for the compound statement above.

P	Q	P $\Leftrightarrow$ Q	(P $\Leftrightarrow$ Q) $\wedge$ P	[(P $\Leftrightarrow$ Q) $\wedge$ P] $\Leftrightarrow$ Q
T	T			
T	F			
F	T			
F	F			

10. Consider the following logic statements:

p: The train arrives on time.

q: I am late for school.

a) Write the expression $p \Rightarrow \neg q$ as a logic statement.

b) Write the following statement in logic symbols:

The train does not arrive on time and I am not late for school.

c) Complete the truth table.

p	q	$\neg p$	$\neg q$	$p \Rightarrow \neg q$	$\neg p \wedge \neg q$
T	T	F	F	F	F
T	F	F	T	T	
F	T	T	F		
F	F	T	T	T	T

d) What statement can be made about the truth values of $p \Rightarrow \neg q$ and $\neg p \wedge \neg q$?

9.6 Inverse, converse, contrapositive

3.4	Converse; inverse; contrapositive.
	Logical equivalence. Testing the validity of simple arguments through the use of truth tables.

With every implication, there are three other statements that are associated with that implication. If the implication is P $\Rightarrow$ Q then:

- the **inverse** of P $\Rightarrow$ Q is ¬P $\Rightarrow$ ¬Q
- the **converse** of P $\Rightarrow$ Q is Q $\Rightarrow$ P
- the **contrapositive** of P $\Rightarrow$ Q is ¬Q $\Rightarrow$ ¬P.

Table 9.2 Comparison of the truth values of the implication, inverse, converse, and contrapositive

P	Q	P $\Rightarrow$ Q	¬P $\Rightarrow$ ¬Q	Q $\Rightarrow$ P	¬Q $\Rightarrow$ ¬P
T	T	T	T	T	T
T	F	F	T	T	F
F	T	T	F	F	T
F	F	T	T	T	T

Example 9.11

Write the inverse, converse and contrapositive of P $\Rightarrow$ ¬Q.

Solution

Inverse: ¬P $\Rightarrow$ Q

Converse: ¬Q $\Rightarrow$ P

Contrapositive: Q $\Rightarrow$ ¬P

Often we have to find the inverse, converse and contrapositive of an implication written in words. If P is 'It is raining' and Q is 'I will bring an umbrella' then P $\Rightarrow$ Q is 'If it is raining then I will bring an umbrella'.

Therefore,

> the inverse of $P \Rightarrow Q$ is 'If it is not raining then I will not bring an umbrella',
> the converse of $P \Rightarrow Q$ is 'If I will bring an umbrella then it is raining', and
> the contrapositive of $P \Rightarrow Q$ is 'If I will not bring an umbrella then it is not raining'.

Example 9.12

Find the inverse, converse and contrapositive of 'If I studied for the test, then I did well on the test.'

Solution

Inverse: If I did not study for the test then I did not do well on the test.

Converse: If I did well on the test then I studied for the test.

Contrapositive: If I did not do well on the test then I did not study for the test.

● **Examiner's hint:** Remembering the meanings of inverse, converse and contrapositive is much easier if you know the definitions of the three terms. The word 'inverse' means to *turn upside down* and the word 'converse' means to *turn around*, which match their logical definitions. The word 'contrapositive' can be broken into two parts, 'contra' and 'positive'. **Contrapositive** means *in comparison with the affirmative*, which is another way of indicating it has the same truth value as the implication.

Given the implication $P \Rightarrow Q$, its inverse is $\neg P \Rightarrow \neg Q$, its converse is $Q \Rightarrow P$ and its contrapositive is $\neg Q \Rightarrow \neg P$.

Exercise 9.6

1. Write the inverse, converse, and contrapositive of the statement:
'If the basketball is flat then the basketball will not bounce.'

2. Write the inverse, converse, and contrapositive of the statement:
'If there is a drought then the crops will fail.'

3. Write the inverse, converse, and contrapositive of the following statements:
a) $\neg P \Rightarrow \neg Q$ b) $\neg P \Rightarrow Q$ c) $(P \wedge Q) \Rightarrow Q$ d) $P \Rightarrow (P \vee Q)$

4. Two propositions P and Q are defined as follows:
P: The number ends in zero.
Q: The number is divisible by 5.
a) Write in words:
(i) $P \Rightarrow Q$
(ii) the converse of $P \Rightarrow Q$.
b) Write in symbolic form:
(i) the inverse of $P \Rightarrow Q$
(ii) the contrapositive of $P \Rightarrow Q$.

5. Consider the statement: 'If a figure is a square then it is a rhombus.'
a) For this statement, write in words:
(i) its converse
(ii) its inverse
(iii) its contrapositive.
b) Only one of the statements in part a) is true. Which one is it?

6. Consider the statement: 'If Canada is in North America then British Columbia is in North America.'
a) For this statement, write in words:
(i) its converse
(ii) its inverse
(iii) its contrapositive.
b) Two of these statements have the same truth value. Which one does not?

7. Consider the statements
   P: The grass is green.
   Q: The grass is watered regularly.
   a) Write down, in words, the meaning of the statement: ¬Q ⇒ P.
   b) Complete the truth table.

P	Q	¬Q	¬Q ⇒ P
T	T		
T	F		
F	T		
F	F		

   c) Write down, in symbols, the contrapositive of ¬Q ⇒ P.

8. Consider the statements:
   P: Dolphins are porpoises.
   Q: Porpoises are mammals.
   a) Write down in words, the meaning of the statement: ¬P ⇒ ¬Q.
   b) Copy and complete the truth table.

P	Q	¬P	¬Q	¬P ⇒ ¬Q
T	T			
T	F			
F	T			
F	F			

   c) Write down, in symbols, the inverse of ¬P ⇒ ¬Q.

9. Let P and Q be the statements:
   P: Sarah eats lots of carrots.
   Q: Sarah can see well in the dark.
   Write the following statements in words.
   a) P ⇒ Q
   b) ¬P ∧ Q
   c) Write the following statement in symbolic form.
      If Sarah cannot see well in the dark then she does not eat lots of carrots.
   d) Is the statement in part c) the *inverse*, the *converse* or the *contrapositive* of the
      statement in part a)?

10. Let P and Q be the statements:
   P: Peter eats his vegetables.
   Q: Peter is healthy.
   a) Write down the implication of P and Q in words.
   b) Using symbolic logic, write:
      (i) the inverse of ¬P ⇒ Q
      (ii) the contrapositive of P ⇒ ¬Q.
   c) Use a truth table to show that P ⇒ Q and ¬P ∨ Q have the same truth value.

9.7 Logical equivalence, tautologies, and contradictions

3.4 Converse; inverse; contrapositive.
Logical equivalence. Testing the validity of simple arguments through the use of truth tables.

Two statements are said to be **logically equivalent** if they have the same truth value under all circumstances. For instance, an implication and its contrapositive are logically equivalent, and the inverse and converse of an implication are also logically equivalent.

Example 9.13

Prove that $\neg(Q \wedge P)$ is logically equivalent to $\neg Q \vee \neg P$.

Solution

We will prove by comparing the truth values for both statements.

P	Q	¬P	¬Q	Q ∧ P	¬(Q ∧ P)	¬Q ∨ ¬P
T	T	F	F	T	F	F
T	F	F	T	F	T	T
F	T	T	F	F	T	T
F	F	T	T	F	T	T

First we find the $\neg P$ and $\neg Q$ columns, which are easy because they have the opposite truth values of P and Q. The truth value of $Q \wedge P$ we determine by comparing when P and Q are both true. $\neg(Q \wedge P)$ simply has the opposite truth value of $Q \wedge P$. To find the truth value of $\neg Q \vee \neg P$, we remember that it is true when either $\neg Q$ or $\neg P$ is true and when both $\neg Q$ and $\neg P$ are true. Finally we notice that the statement $\neg(Q \wedge P)$ is true under the same circumstances as $\neg Q \vee \neg P$; hence, they are logically equivalent.

A statement is called a **tautology** if it is always true, under any circumstances. For example, the statement $P \Rightarrow P$ is a tautology, since it is not possible for the antecedent to be true and the consequent false, since they have the same truth value. If a statement is a tautology, it is logically valid. Thus a **tautology** is a **valid argument**.

 True or false? 'Logical arguments that are true because they are tautologies are fallacious because they use circular reasoning.'

Example 9.14

Prove that $(Q \wedge \neg Q) \Rightarrow P$ is a tautology.

Solution

We will prove this by confirming that it is true under any circumstance, using a truth table.

Investigate 'proof by contradiction' where the assumption of truth is made about a proposition, which leads to a contradiction; hence the opposite of the original proposition must be true.

P	Q	¬Q	Q ∧ ¬Q	(Q ∧ ¬Q) ⇒ P
T	T	F	F	T
T	F	T	F	T
F	T	F	F	T
F	F	T	F	T

First, the truth value of ¬Q will be the opposite of Q. Q ∧ ¬Q is always false since Q and ¬Q are never both true. (Q ∧ ¬Q) ⇒ P is always true, since the antecedent is never true and the consequent false. Therefore, (Q ∧ ¬Q) ⇒ P is a tautology.

Example 9.15

Consider the following proposition.

h: Helen finishes her extended essay
q: Helen plays football

a) Write in symbolic form the following proposition.

If Helen finishes her extended essay then she plays football.

b) Write in symbolic form the inverse of the following proposition

If Helen finishes her extended essay then she plays football.

c) Is the argument in part b) valid? Give your reasons.

Solution

a) If Helen finishes her extended essay then she can play football.

In symbols: $h \Rightarrow q$

b) The inverse of 'If Helen finishes her extended essay then she can play football' is 'If Helen does not finish her extended essay then she will not play football'. In symbols it is $\neg h \Rightarrow \neg q$.

c) It is not a valid argument because Helen may still play football though she has not finished her extended essay.

A statement is a **contradiction** if it is always false under any circumstances. For example, the statement ¬P ∧ P cannot ever be true, since it is not possible for both ¬P and P to be true at the same time.

Example 9.16

Prove that (¬Q ∧ ¬P) ∧ (Q ∨ P) is a contradiction.

Solution

We will show that $(\neg Q \wedge \neg P) \wedge (Q \vee P)$ is always false, using a truth table.

P	Q	¬P	¬Q	¬Q ∧ ¬P	Q ∨ P	(¬Q ∧ ¬P) ∧ (Q ∨ P)
T	T	F	F	F	T	F
T	F	F	T	F	T	F
F	T	T	F	F	T	F
F	F	T	T	T	F	F

i. A tautology is true under all circumstances; last column of the truth table contains all T's.
ii. A contradiction is false under every circumstance; last column of the truth table contains all F's.
iii. A valid argument is true under all circumstances, and is a tautology.
iv. An invalid argument is not true under all circumstances, i.e. there is at least one F in the last column of a truth table.

To find the truth value of ¬P and ¬Q, we reverse the truth values of P and Q as before. ¬Q ∧ ¬P is only true when both ¬P and ¬Q are true. Q ∨ P is true when either P or Q is true. We notice that ¬Q ∧ ¬P and Q ∨ P are never both true, so their conjunction, $(\neg Q \wedge \neg P) \wedge (Q \vee P)$, is never true and is a contradiction.

Exercise 9.7

1. Use a truth table to prove that P ∧ ¬P is a contradiction.

2. Use a truth table to prove that P ⇒ P is a tautology.

3. Use a truth table to prove that (P ∧ Q) ∧ (¬P ∧ ¬Q) is a contradiction.

4. Use a truth table to prove that (P ∨ Q) ∨ (¬P ∨ ¬Q) is a tautology.

5. P: The book is short.
   Q: The book is easy to read.
   a) Write (P ∨ ¬P) ⇒ (Q ∧ ¬Q) in words.
   b) Fill in the truth table below.

P	Q	¬P	¬Q	(P ∨ ¬P)	(Q ∧ ¬Q)	(P ∨ ¬P) ⇒ (Q ∧ ¬Q)
T	T	F	F			
T	F	F	T			
F	T	T	F			
F	F	T	T			

   c) Write a word that describes (P ∨ ¬P) ⇒ (Q ∧ ¬Q).

6. P: $3x + 4 = 13$
   Q: $4x + 2 = 14$
   a) (i) Solve $3x + 4 = 13$ for x.
      (ii) Solve $4x + 2 = 14$ for x.
   b) Write down the value of x which makes both P and Q true.
   c) Is it possible for P to be true and Q false?

255

P	Q	¬Q	P ∨ ¬Q
T	T		
T			
F			
F			

d) Fill in the missing values in the truth table above.

e) Write a word that describes the statement P ∨ ¬Q.

7. a) The following truth table contains two entries which are incorrect, one in column three and one in column four. Copy the table and circle the two incorrect entries.

   b) Fill in the two missing values in column five.

   c) Which **one** of the following words could you use to describe the statement represented by the values in the last column (number 6)?

   (i) converse   (ii) tautology   (iii) inverse

   (iv) contradiction   (v) contrapositive

1	2	3	4	5	6
P	Q	P ∧ Q	¬P	P ∨ Q	(P ∧ Q) ∧ (¬P ∧ ¬Q)
T	T	T	F	T	F
T	F	F	F		F
F	T	F	T	T	F
F	F	T	F		F

8. Use a truth table to show that the inverse and converse of P ⇒ Q are logically equivalent.

9. Use a truth table to show that P ⇒ Q and its contrapositive are logically equivalent.

10. Consider each of the following statements:
    P: Alex is from Uruguay.
    Q: Alex is a scientist.
    R: Alex plays the flute.

    a) Write each of the following arguments in symbols.
       (i) If Alex is not a scientist then he is not from Uruguay.
       (ii) If Alex is a scientist then he is either from Uruguay or plays the flute.

    b) Write the following argument in words.
       ¬R ⇒ ¬(Q ∨ P)

    c) Construct a truth table for the argument in part b). Test whether or not the argument is logically valid.

11. Consider the following statements:
    P: Good mathematics students go to good universities.
    Q: Good music students are good mathematics students.
    R: Students who go to good universities get good jobs.

    a) From these statements write 2 **valid** conclusions.

    b) Write each of the following in words.
       (i) ¬Q   (ii) P ∧ R

    c) Is your answer to b) (ii) a valid argument? Give a reason for your answer.

12. Let the propositions P, Q and R be defined as:

P: Matthew arrives home before six o'clock.

Q: Matthew cooks dinner.

R: Jill washes the dishes.

a) (i) Express the following in logical form.
If Matthew arrives home before six o'clock then he will cook dinner.

(ii) Write the following logic statement in words.
$\neg Q \Rightarrow \neg R$

b) (i) Copy and complete the truth table below.

P	Q	R	P⇒Q	Q⇒R	¬R	(P⇒Q)∧(Q⇒R)∧¬R	¬P	[(P⇒Q)∧(Q⇒R)∧¬R]⇒¬P
T	T	T						T
T	T	F						T
T	F	T						T
T	F	F						T
F	T	T						T
F	T	F						T
F	F	T						T
F	F	F						T

(ii) Explain the significance of the truth table above.

13. Let *p* stand for the proposition, 'The metro is running on time'. Let *q* stand for the proposition, 'Ali will arrive at school on time'.

a) Write down the following propositions in symbolic form.
(i) If the metro is on time then Ali will arrive at school on time.
(ii) If Ali does not arrive at school on time then the metro was not on time.

b) Write down the converse statement of
If the metro is on time then Ali will arrive at school on time.

c) Is the converse statement found in b) a valid argument?

14. Complete the table to determine whether the argument is valid.

P	Q	P⇒Q	(P⇒Q)∧P	[(P⇒Q)∧P]⇒Q
T	T	T		
T	F	F		
F	T	T		
F	F	T		

Probability

Assessment statements

> 3.6 Sample space; event A; complementary event, A'. Probability of an event. Probability of a complementary event. Expected value.
> 3.7 Probability of combined events, mutually exclusive events, independent events. Use of tree diagrams, Venn diagrams, sample space diagrams, and tables of outcomes. Probability using 'with replacement' and 'without replacement'. Conditional probability.

Overview

By the end of this chapter you will be able to:
- draw and use Venn diagrams, tree diagrams, sample space diagrams, and tables of outcomes to solve problems
- solve problems involving classical probability
- find the probability of a complementary event
- use the laws of probability to solve problems involving
  - mutually exclusive events
  - combined events
  - independent events
  - dependent events
- understand and use the concept of conditional probability to solve problems
- understand the concept of the probability of 'at least one' to solve problems
- understand the relationship between independent and mutually exclusive events.

10.1 Diagrams for experiments

> 3.6 Sample space; event A; complementary event, A'. Probability of an event. Probability of a complementary event. Expected value.
> 3.7 Probability of combined events, mutually exclusive events, independent events. Use of tree diagrams, Venn diagrams, sample space diagrams, and tables of outcomes. Probability using 'with replacement' and 'without replacement'. Conditional probability.

There is consensus among historians that the current mathematical theory of probability is attributed to correspondence between two French mathematicians, Blaise Pascal and Pierre de Fermat, concerning a gambling dispute posed by the Chevalier de Méré.

In 1494 Fra Luca Paccidi wrote the first printed work on probability, but the modern-day theory of probability began in the seventeenth century with a problem posed about gambling. Today we are surrounded with results from the solutions to probability problems. Those results affect areas such as insurance rates, environmental regulations, genetics, the kinetic theory of gases, and quantum mechanics in physics.

A diagram, or sample space, for an experiment is a listing or picture of all of the outcomes for that experiment. Diagrams can take several forms: a list, a table, a tree diagram, or a Venn diagram.

To view the two questions that the Chevalier de Méré posed to Blaise Pascal, visit www. pearsonhotlinks.co.uk, enter the ISBN for this book and click on weblink 10.1.

Example 10.1

List all of the possible outcomes for tossing a coin once.

Solution

Head, tail, or HT

Example 10.2

Let an event be tossing a single coin two times. Represent the sample space
a) in list form   b) in table form   c) as a tree diagram.

Solution

a) HH, HT, TH, TT

b)

Toss 1	Toss 2
Head	Head
Head	Tail
Tail	Head
Tail	Tail

or

Toss 1	Toss 2
H	H
H	T
T	H
T	T

c)

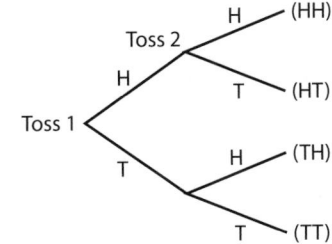

Example 10.3

Draw the sample space diagram for the combination of throws of an 8-sided die and spins of a 6-sided spinner.

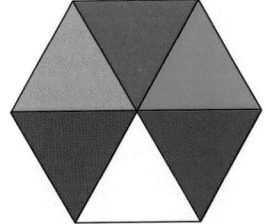

Solution

Die / Spinner	1	2	3	4	5	6	7	8
Orange	O, 1	O, 2	O, 3	O, 4	O, 5	O, 6	O, 7	O, 8
Blue	B, 1	B, 2	B, 3	B, 4	B, 5	B, 6	B, 7	B, 8
Red	R, 1	R, 2	R, 3	R, 4	R, 5	R, 6	R, 7	R, 8
Green	G, 1	G, 2	G, 3	G, 4	G, 5	G, 6	G, 7	G, 8
Purple	P, 1	P, 2	P, 3	P, 4	P, 5	P, 6	P, 7	P, 8
White	W, 1	W, 2	W, 3	W, 4	W, 5	W, 6	W, 7	W, 8

Example 10.4

Let the universal set $U = \{0, 1, 2, 3, 4, \ldots, 20\}$ and
set $A = \{0, 1, 2, 3, 4, 17, 18, 19, 20\}$.
Let the experiment be the selection of numbers that are not in A (i.e. A′).
Draw a Venn diagram to represent the sample space.

Solution

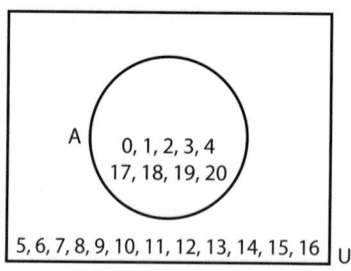

Exercise 10.1

1. Draw a sample space, in list form, for the event of selecting an integer from the set A:
$$A = \{x \mid -3 \leqslant x \leqslant 10, x \in \mathbb{Z}\}$$

2. An urn contains 5 purple and 3 white marbles. Two marbles are drawn. Draw a sample space for the event, using a tree diagram.

3. An urn contains 5 purple, 3 white and 4 red marbles. Two marbles are drawn. Draw a sample space for the event, using a tree diagram.

4. Draw a sample space for the number of different outfits you can wear to school if you have only two pairs of trousers and three shirts. Use a tree diagram to represent the outcomes.

5. Let $U = \{x \mid -5 \leqslant x \leqslant 5, x \in \mathbb{Z}\}$, $A = \{-4, -3, -2, -1, 0\}$, and $B = \{-1, 0, 1, 2, 3\}$. Draw a Venn diagram to represent the sample space.

6. Let $U = \{n \mid 0 \leqslant n \leqslant 20, n \in \mathbb{N}\}$, $A = \{\text{first 8 prime numbers}\}$, and $B = \{\text{first 8 positive even integers}\}$. Draw a Venn diagram to represent the sample space.

7. Draw a sample space using six rows and six columns for the event of rolling two dice. Use two colours, one for each die. (Note: This sample space will be used in later exercises.)

8. Draw a sample space for the event of drawing a single card from a standard 52-card deck. (Hint: Use 4 rows with 13 columns.) (Note: This sample space will be used in later exercises.)

9. Let the event be the rolling of three fair six-sided dice.
   a) How many outcomes will be in the sample space?
   b) How many of the outcomes will be triples?
   c) How many of the outcomes will have a sum of at least 16?

10. Let the event be tossing a coin and then rolling a fair six-sided die. Draw a sample space for the outcomes
    a) using a tree diagram
    b) by listing the outcomes.

11. Let the event be a couple having a child. Draw a tree diagram for the outcomes of a couple having 4 children.

12. Draw a sample space for the event of rolling a six-sided die and drawing a face card that is a diamond.

13. Let the event be answering a three-question True-False test. Draw a tree diagram showing all of the possible outcomes for the answers.

14. The town of Alright has a population of 200 adults: 60% are women and 10% of them are more than 6 feet tall. Of the men, 30% are more than 6 feet tall. Draw and label a tree diagram for the town's population.

15. Draw a sample space by using a four-column chart to represent all of the outcomes of the event of tossing 4 coins at the same time.

● **Examiner's hint:** Although practice problems will be devised to test understanding of the concepts of probability, no questions involving playing cards will be set in either Paper 1 or Paper 2 on the Mathematical Studies IB examinations.

10.2 Classical probability

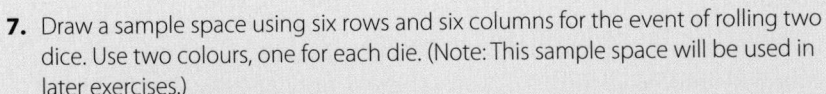
3.6 Sample space; event A; complementary event, A'. Probability of an event. Probability of a complementary event. Expected value.

The probability of an event occurring is the chance, usually in terms of a fraction or a percentage, that the event will happen. For example, you might say that there is a 30% chance that it is going to rain tomorrow or that there is a $\frac{2}{3}$ chance of winning the next game.

Basic probability theory, sometimes called classical probability, is defined as:

$$P(\text{an outcome for an event occuring}) = \frac{\text{the number of outcomes for the event that you desire to occur}}{\text{the total number of outcomes for the event that could occur}}$$

$$P(A) = \frac{n(A)}{n(U)}$$

● **Examiner's hint:** The formula for probability of an event is given on page 4 of the Mathematical Studies SL Formula Booklet.

For example, suppose that the experiment is tossing a coin and the event is the chance that the coin will land heads up.

P(coin will land heads up) $= \frac{1}{2}$, since there is only one way for a coin to land heads up and there are two ways for a coin to land: heads up or tails up.

It is very important to understand that just because the probability of a coin landing heads up is one-half, that *does not* mean that the coin will land heads up once every two tosses.

P(coin landing heads up) $= \frac{1}{2}$ means that in the long run (1000 tosses, 10 000 tosses or 1 000 000 tosses), the proportion of tosses that land heads up will be very close to (and in some trials, may be equal to) $\frac{1}{2}$. For example, a trial of ten tosses may yield H,H,T,T,H,H,H,T,H,H.

It is also important to understand several other concepts:

- Tossing one coin ten times or ten coins one time will mathematically yield the same result.
- Even if the first 100 tosses land heads up, the probability that the next toss will land heads up will still be $\frac{1}{2}$.
- The maximum value of any probability event is 1. This type of probability is considered a certainty.
- The minimum value of any probability event is 0. This type of probability is considered an impossibility.
- All other values for probability events will be between 0 and 1.
- The sum of all of the probabilities of all of the outcomes in the sample space of a event will always be 1.

Example 10.5

Find the probability that a 7 will show when rolling a fair six-sided die whose faces are numbered 1 through 6.

Solution

P(a 7 will show) $= 0$, since there is no 7 on any one of the six faces of the die.

Example 10.6

Let the experiment be tossing a coin three times. Represent the sample space in table form, hence find the probability of having all three tosses landing tails up.

Toss 1	Toss 2	Toss 3
H	H	H
H	H	T
H	T	H
H	T	T
T	H	H
T	H	T
T	T	H
T	**T**	**T**

Each of the outcomes, HHH, HHT, …, has a probability of $\frac{1}{8}$ of occurring. There are 8 different outcomes and therefore the sum of the probabilities is $8 \cdot \frac{1}{8} = 1$.

Solution

For the sample space use the table shown here:

∴ by way of inspection, P(all three tosses land tails up) $= \frac{1}{8}$.

Example 10.7

Find the probability that an even number shows when a fair six-sided die is rolled.

Solution

Since the sample space is 1, 2, 3, 4, 5, 6 and there are three even numbers, then

$$P(\text{an even number shows}) = \frac{3}{6} = \frac{1}{2}.$$

● **Examiner's hint:** An answer to a probability question can be left as a fraction, a decimal or as a percentage.

The probability of the **complement** of an event is defined as all of the outcomes in which the event under consideration **does not** occur:

$$P(A') = 1 - P(A).$$

For example, if there is a 35% chance it will rain tomorrow, there must be a 65% chance that it will *not* rain tomorrow.

● **Examiner's hint:** The formula for probability of the complement of an event is given on page 4 of the Mathematical Studies SL Formula Booklet.

Example 10.8

Find the probability that when rolling two dice they will not show doubles.

Solution

P(the roll does **not** show doubles) = 1 − P(the roll does show doubles)
$$= 1 - \frac{6}{36} = \frac{30}{36} = \frac{5}{6}.$$

1. Enter 'expected value' into a search engine to find out about the concept.
2. Visit www.pearsonhotlinks.co.uk, enter the ISBN for this book and click on weblink 10.2.
3. Read 'The Two Envelopes Problem', but not the 'Follow Up'.
4. Discuss the philosophy behind the decision as to whether you should take the other envelope.
5. Discuss the philosophy behind what your intuition is telling you.
6. Now read the 'Now, suppose …', 'Follow Up', and 'Explanation' boxes and discuss the paradox.

Exercise 10.2

1. If a coin is tossed 1000 times and each time it lands heads up, find the probability that it will land heads up on the 1001 toss.

2. Find the probability that an odd number will show when a fair six-sided die is rolled.

3. If an urn contains 6 red and 7 blue marbles, find the probability of:
   a) drawing a blue marble
   b) not drawing a blue marble.

4. A coin is tossed and a fair six-sided die is rolled. Find the probability that the die lands heads up and an even number shows on the roll of the die.

5. When three fair six-sided dice are rolled, find the probability
   a) that triple sixes show
   b) that triple sixes do not show.

6. When tossing three coins, find the probability that
   a) all three coins land tails up
   b) at least one coin lands heads up.

7. Two fair six-sided dice are rolled. Find the probability that:
   a) the sum is 7
   b) the sum is at least 10
   c) the sum is 13

Recall the keystrokes for rewriting a fraction as a decimal and vice versa. For example, to write $\frac{5}{7}$ press ALPHA, Y = , 1, 5, >, 7, ENTER.
To convert to decimal
Press MATH, 2, ENTER

```
5
7
                          5
                          7

Ans▶Dec
        .7142857143
```

● **Examiner's hint:** Use correct notation when showing your working during your IB examination; e.g. $P(\text{head}) = \frac{1}{2}$.

d) the sum is between 2 and 12 inclusive

e) a 5 shows.

8. A single card is drawn from a standard 52-card deck. Find the probability that:

a) a red card is drawn

b) a green card is drawn

c) a club is selected

d) a face card is selected

e) a card numbered 14 is selected

f) a 7 is drawn.

9. The probability that it will rain tomorrow is 60%. What is the probability that it will not rain tomorrow?

10. Find the probability that when a natural number is selected from the first 20 natural numbers it will be prime .

11. A six-sided die is rolled and then a diamond face card is drawn. Find the probability that the die shows 5 and the draw is a king. (Hint: Review the sample space you drew for Question 12, Exercise 10.1.)

12. Four coins are tossed at the same time. Find the probability that:

a) all four coins show heads

b) two of the coins show tails.   (Hint: Review the sample space you drew for Question 15, Exercise 10.1.)

13. By reviewing the sample space you drew for Question 15, Exercise 10.1, find the probability that when tossing four coins at the same time, at least one of them shows a tail.

14. When a single card is selected from a standard 52-card deck, find the probability that: (Hint: See sample space from Question 8, Exercise 10.1)

a) the card is red and a 9

b) the card is a 3 or a queen

c) the card is a club or a 10 (Hint: Don't double count!)

d) the card is black or a 2.

15. A coin is tossed and a six-sided die is rolled. Find the probability that the outcome is:

a) head-four (H4)   b) T7   c) H2 or T1.

16. Two seven-sided dice, with numbers 1, 2, 3, 4, 5, 6, 7 on the faces, are rolled.

a) What is the probability that the number on the uppermost face of the first die rolled is smaller than the number on the uppermost face of the second die?

b) What is the probability that the sum of the two numbers on the faces is even?

Two events are said to be mutually exclusive if it is not possible for the events to occur at the same time. For example, let event B be that a person is bearded and let event C be the event that a person is clean-shaven. Events B and C are said to be mutually exclusive, since a person cannot be both bearded and clean-shaven at the same time.

The Venn diagram showing this information is shown right:

The set language that would be used to describe the events would be: $B \cap C = \emptyset$.

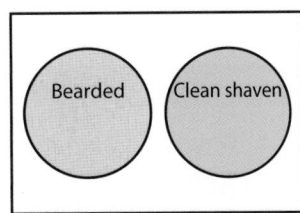

We would also say the events B and C are **disjoint**.

Example 10.9

Let A be the event that a person is a female and B be the event that a person is 1.6 metres tall.
a) Are the two events mutually exclusive?
b) Justify your answer by drawing a Venn diagram.
c) Write your answer in set language.

Solution
a) The two events are not mutually exclusive, since there are people who are both female and 1.6 metres tall.
b)

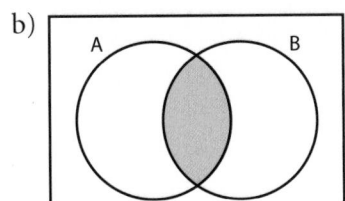

c) $A \cap B \neq \emptyset$

In Chapter 8 you learned the relationship $n(A \cup B) = n(A) + n(B)$, where the sets A and B had no elements in common.

We know that if A and B are considered as events instead of sets, then the events A and B are said to be mutually exclusive.

We can now write that result in the language of probability:

$$P(A \cup B) = P(A) + P(B), \text{ where } P(A \cap B) = 0.$$

It is important to identify key words when solving probability problems. The word that is used for union ($\cup$) is 'or'. When the word 'or' is used in a

● **Examiner's hint:** The formula for mutually exclusive events is found on page 4 of the Mathematical Studies SL Formula Booklet.

stated problem, or when you can naturally rephrase the problem so as to use the word 'or', then the operation of addition should be used.

$$P(A \cup B) = P(A) + P(B)$$

Example 10.10

Find the probability of having a 5 or an even number show when rolling a six-sided die.

Solution

Since 5 is not an even number, we know that the two events are mutually exclusive. Therefore, $P(A \cup B) = P(A) + P(B)$.
$P(5 \text{ or even}) = \frac{1}{6} + \frac{3}{6} = \frac{4}{6} = \frac{2}{3}$.

Exercise 10.3

1. Write an equation, in **set theory notation**, describing two events that are mutually exclusive.

2. Can two political parties be considered mutually exclusive?

3. If two events are mutually exclusive, write a **probability formula** that describes them.

4. Draw a Venn diagram that shows that two events are mutually exclusive.

5. State which of the following events are mutually exclusive. If the event is not mutually exclusive, give an example of why not.
   a) A die is thrown and a coin is tossed.
   b) A red 7 is drawn from a standard 52-card deck.
   c) A student is a football player and the student is a basketball player.
   d) A person is a Lutheran and the person is a Catholic.
   e) A quadrilateral is a square and the quadrilateral is a rectangle.
   f) A triangle is scalene and the triangle is equilateral.

6. Find the probability of each event. A card is drawn from a 52-card deck.
   a) The card is a 7 or a king.   b) The card is a club or a diamond.
   c) The card is an ace or a face card.   d) The card is black or red.

7. An urn contains 9 red, 4 blue, and 5 black marbles. A marble is selected at random. Find each probability.
   a) The marble is red or blue.   b) The marble is not red.

8. In an IB school in St Petersburg, there are 145 freshmen, 135 sophomores, 132 juniors, and 115 seniors. A student is selected at random. Find each probability.
   a) The student is a freshman or a sophomore.
   b) The student is not a senior.

9. A coin is tossed 3 times. Find each probability.
   a) All of the toses land tails up.   b) All of the tosses land tails up or heads up.

10. The probability that I will go golfing is 35%. The probability that I will go to a movie is 25%. Find the probability that I will go golfing or go to a movie.

11. Two fair six-sided dice are rolled. Find each probability.
    a) A sum of 5 or a sum of 11 shows.   b) A 3 shows or the sum is 10.

12. A couple plan to have 3 children. Assume that the gender of each child is equally

likely to happen. Find each probability.
a) All three children are boys.
b) One of the children is a boy and two of the children are girls.

13. Data were collected for which gender reads which type of book. A book was selected at random. Find each probability.
a) Romance or action books were read.
b) Action books were not read.
c) The book was read by a female.

	Action	Romance	Historical
Male	10	17	8
Female	3	4	8

10.4 Combined events

> **3.7** Probability of combined events, mutually exclusive events, independent events. Use of tree diagrams, Venn diagrams, sample space diagrams, and tables of outcomes. Probability using 'with replacement' and 'without replacement'. Conditional probability.

There are events that are **not** mutually exclusive. For example:

1) Rolling a six-sided die and noting whether the number that shows is a prime number **or** an even number.
   - 2 is a number that is prime and is also a number that is even.

2) Drawing a single card from a standard 52-card deck and noting whether the card is red **or** a seven.
   - There are two red sevens: the 7 of diamonds and the 7 of hearts.

Notice that the word 'or' is present in both of the above examples. This means that addition will be the correct operation to use. However, these types of events are called **combined events** and the operation of subtraction must also be used.

Consider a Venn diagram for the first example, the rolling of a die. The events are either a prime number shows or an even number shows.

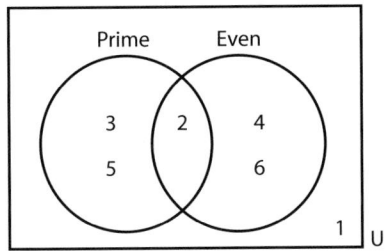

Again, from Chapter 8, $n(A \cup B) = n(A) + n(B)$

$$n(\text{prime or even}) = n(\text{prime}) + n(\text{even}) = 3 + 3 = 6.$$

However, by inspection, you can see that there are a total of only five elements within the sets prime and even.

The answer to the quandary is that the number 2 is both prime **and** even. It has been added twice to the total: once for the number of primes and once for the number of evens. In order to get the correct total, one of the 2's must be subtracted, since 2 is the **intersection** of both sets.

The correct formula to find the number of elements for two sets that are not disjoint is,

$$n(A \cup B) = n(A) + n(B) - n(A \cap B).$$

• **Examiner's hint:** The formula for combined events is found on page 4 of the Mathematical Studies SL Formula Booklet.

The formula used to find the probability for **combined events**, those for which the events are *not* mutually exclusive is,

$$P(A \cup B) = P(A) + P(B) - P(A \cap B), \text{ where } P(A \cap B) \neq 0.$$

When the wording of a probability problem involves the word 'or', it is important to ask if the events are mutually exclusive. A good way to do this is to ask the question: 'Are any prime numbers also even numbers?' If the answer is 'yes', a subtraction is required. If the answer is 'no', only addition is required.

Example 10.11

Find the probability of rolling a fair six-sided die and having a prime number or an even number show.

Solution

$$P(\text{prime or even}) = P(\text{prime}) + P(\text{even}) - P(\text{prime} \cap \text{even}) = \tfrac{3}{6} + \tfrac{3}{6} - \tfrac{1}{6} = \tfrac{5}{6}.$$

Example 10.12

Find the probability, when drawing a single card from a standard 52-card deck, that the card is either a 9 or a face card.

Solution

The word 'or' is used so that the question 'Are any 9's also face cards?' needs to be answered. There are not any 9's that are also face cards. The two events are mutually exclusive and hence only addition needs to be performed.

Therefore, $P(9 \text{ or face card}) = P(9) + P(\text{face card}) = \tfrac{4}{52} + \tfrac{12}{52} = \tfrac{16}{52} = \tfrac{4}{13}.$

Example 10.13

Find the probability, when drawing a single card from a standard 52-card deck, that the card is either a red or an ace.

Solution

The word 'or' is used so that the question 'Are any reds also aces?' needs to be answered. There are two cards that are both aces **and** red cards. The two events are **not** mutually exclusive and hence, after addition, subtraction needs to be performed.

Hence, P(red or ace) = P(red) + P(ace) − P(red ∩ ace)

$$= \frac{26}{52} + \frac{4}{52} - \frac{2}{52} = \frac{28}{52} = \frac{7}{13}.$$

Charts are often used to describe data so that the information is easy to understand.

Example 10.14

The following chart was used to show how many King High School juniors and seniors drive which kind of vehicle.

	Sports Car	Truck	Sedan
Junior	8	7	9
Senior	15	4	7

If a student is selected at random, find the following probabilities.
a) The student is a junior and drives a truck.
b) The student is not a junior.
c) The student drives a sports car or is a senior.
d) The student drives a sports car or a sedan.

Solution

	Sports car	Truck	Sedan	Totals
Junior	8	7	9	24
Senior	15	4	7	26
Totals	23	11	16	50

a) $P(\text{junior and truck}) = \dfrac{7}{50}.$

b) $P(\text{not a junior}) = \dfrac{15 + 4 + 7}{50} = \dfrac{26}{50} = \dfrac{13}{25}.$

c) P(sports car or senior) = P(sports car) + P(senior) − P(sports car and senior)

$$= \frac{23}{50} + \frac{26}{50} - \frac{15}{50} = \frac{34}{50} = \frac{17}{25}.$$

d) $P(\text{sports car or sedan}) = P(\text{sports car}) + P(\text{sedan}) = \dfrac{23}{50} + \dfrac{16}{50} = \dfrac{39}{50}.$

Exercise 10.4

1. Write the formula for the number of elements in sets for combined events. (Hint: $n(A \cup B) = n(A) + \ldots$)

2. Write the formula for the probability of combined events that are not mutually exclusive.

3. Draw a Venn diagram, for the general case, for combined events that are not mutually exclusive.

4. Determine if the following events are mutually exclusive or if they are combined events that are not mutually exclusive.
  a) On the draw of a single card, a red jack.
  b) On the draw of a single card, a black card or a 9.
  c) For the roll of two dice, 5 or the sum of 7.
  d) For the roll of two dice, a sum greater than 8 or doubles.
  e) On the draw of a letter from the alphabet, a vowel, or a consonant.
  f) On selecting a basketball player, male or over six feet tall.

5. Find each probability when drawing from a standard 52-card deck.
  a) P(drawing a red card)
  b) P(drawing a ten)
  c) P(drawing a red card and a ten)
  d) P(drawing a red card or a ten)

6. Find each probability when rolling two fair six-sided dice.
  a) P(rolling doubles)
  b) P(rolling a sum of ten)
  c) P(rolling doubles and a sum of ten)
  d) P(rolling doubles or a sum of ten)

7. If the probability that you will watch TV is 55%, the probability that you will talk on your cellphone is 65%, and the probability that you will watch TV and talk on your cellphone at the same time is 30%, find the probability that you will watch TV or talk on your cellphone.

8. The probability that you will do your homework is 75%. The probability that you will listen to your iPod is 40%. The probability that you will do your homework and listen to your iPod is 50%. Find the probability that you will do your homework or listen to your iPod.

9. During the random selection of a calendar year, it was found:
  - P(a person is born on a Tuesday) $= \frac{53}{366}$
  - P(a person is born on the 1st day of a month) $= \frac{12}{366}$
  - P(a person is born on a Tuesday, the 1st) $= \frac{3}{366}$.

  Find P(a person is born on a Tuesday or on the 1st day of a month).

10. During the random selection of a calendar year, it was found:
  - P(a person is born on a Wednesday) $= \frac{48}{365}$.
  - P(a person is born on the 29th of a month) $= \frac{11}{365}$.
  - P(a person is born on a Wednesday or on the 29th day of a month) $= \frac{57}{365}$.

  Find P(a person is born on a Wednesday, the 29th).

11. If $P(A) = 0.37$, $P(B) = 0.55$ and $P(A \cap B) = 0.13$, find:
  a) $P(A \cup B)$
  b) $P(A')$
  c) $P(B')$
  d) $P(A \cap B')$.   (Hint: Draw a Venn diagram with the given information and then use a version of the combined events formula.)

12. If P(A) = 0.75, P(B) = 0.29 and P(A ∪ B) = 0.88 and A and B are not mutually exclusive, find:

a) P(A ∩ B)

b) P(A′)

c) P(B′)

d) P(A′ ∩ B).

13. Using the sample space you made for Question 15, Exercise 10.1, find:

a) P(at most one head)

b) P(at least one tail)

c) P(at least two heads).

14. The probability a girl plays basketball is 14%. The probability that she plays volleyball is 17%, and that she plays both is 5%.

a) Draw a Venn diagram showing the percentages.

b) Find P(the girl does not play basketball).

c) Find P(the girl does not play volleyball).

d) Find P(the girl plays basketball or volleyball).

e) Find P(the girl plays basketball but does not play volleyball).

f) Find P(the girl plays neither basketball nor volleyball).

15. Data was collected for which gender reads which type of book. A book was selected at random. Find the probability that:

a) The book was read by a female or it was an action book.

b) The book was a historical book or it was read by a male.

c) The book was not read by a male and it was an action book.

	Action	Romance	Historical
Male	10	17	8
Female	3	4	8

16. The following data was collected in a parking lot. If a car is selected at random, find each probability.

	General Motors	Hyundai	Porsche
Silver	15	10	10
Green	25	20	5

a) The car was silver or a Porsche.

b) The car was not a Hyundai.

c) The car was not green and a General Motors.

10.5 Independent events

3.7 Probability of combined events, mutually exclusive events, independent events. Use of tree diagrams, Venn diagrams, sample space diagrams, and tables of outcomes. Probability using 'with replacement' and 'without replacement'. Conditional probability.

• **Examiner's hint:** The formula for independent events is found on page 4 of the Mathematical Studies SL Formula Booklet.

Two events are said to be independent if the probability of the first event does not influence the probability of the second event.

If two events, A and B, are **independent** then $P(A \text{ and } B) = P(A) \cdot P(B)$

Example 10.15

If the probability it will snow on Monday is 25% and on Tuesday is 40%, what is the probability that it will snow on both Monday and Tuesday?

Solution

According to the structure of the problem the probability that it will snow on Monday does not influence the probability it will snow on Tuesday. Therefore, the events are said to be independent.

Hence, $P(M \text{ and } T) = P(M) \cdot P(T) = (0.25)(0.40) = 0.10 = 10\%$.

Example 10.16

Find the probability of the toss of a coin landing heads up and the roll of a die showing a four.

Solution

Tossing the coin does not influence the roll of the die; therefore, the two events are independent.

Hence, $P(\text{head and four}) = P(\text{head}) \cdot P(\text{four}) = \frac{1}{2} \cdot \frac{1}{6} = \frac{1}{12}$.

A sample space supports the solution.

H1, H2, H3, **H4**, H5, H6, T1, T2, T3, T4, T5, T6

When the word 'and' is used as a connective between two events, or when the problem can be rephrased using the word 'and', the operation of multiplication is used.

In solving problems, when the word 'and' is used as the connective between two or more events, the set concept is intersection and for a probability answer, multiplication is used.

$$P(A \text{ and } B) = P(A) \cdot P(B)$$

The symbol ∩ is often substituted for the word 'and'. The formula is then:

$$P(A \cap B) = P(A) \cdot P(B)$$

Example 10.17

An urn contains 4 red marbles and 3 green marbles. Find the probability of drawing a red marble, replacing it and then drawing a green marble.

Solution

The probability of drawing a red marble then replacing it does not influence the probability of drawing a green marble. Therefore, the two events are independent. The word 'and' was used as a connective between the two events and so the operation of multiplication will be used.

$$\text{Hence, P(red and green)} = \text{P(red)} \cdot \text{P(green)} = \frac{4}{7} \cdot \frac{3}{7} = \frac{12}{49}.$$

Example 10.18

An urn contains 4 red marbles and 3 green marbles. Find the probability of drawing a red marble, replacing it, drawing a green marble, replacing it, and then drawing a red marble.

Solution

The word 'replacement' implies that the events are independent. The word 'and' is implied between the first and second draws; therefore, multiplication will be used.

Hence, P(red and green and red) $= \text{P(red)} \cdot \text{P(green)} \cdot \text{P(red)} = \frac{4}{7} \cdot \frac{3}{7} \cdot \frac{4}{7}$
$= \frac{48}{343}.$

Exercise 10.5

1. How can you tell if two events are independent?

2. State a method you could use to find the intersection of two events.

3. Which of the following events are independent?
   a) Tossing a coin three times.
   b) Drawing a card from a deck, shredding it, and then drawing another card from the same deck.
   c) Drawing a card from a deck, replacing it, and then drawing another card from the same deck.
   d) Having a high GPA and admission to a university.
   e) Studying for a geometry test and running a kilometre.
   f) Not exercising and gaining weight.

4. If there is a 25% chance that it will rain on any given day in Seattle, find the probability that it will rain for three consecutive days.

5. A coin is tossed and a six-sided die is rolled. Find the probability that a head and a 5 show.

6. An urn contains 4 red and 6 black marbles. Two marbles are drawn with replacement. Find each probability.
   a) A red is drawn first and then a black is drawn.
   b) A red and a black are drawn.

c) Two reds are drawn.

d) At least one red is drawn.

7. An urn has 3 red, 4 yellow, and 5 black marbles. Three marbles are drawn at random with replacement. Find each probability.

a) A red is drawn first, a yellow is drawn next, and a black is drawn last.

b) A red, yellow, and a black are drawn.

c) Two reds and a yellow are drawn.

d) At least one black is drawn.

8. Two cards are drawn from a standard 52-card deck with replacement. Find each probability.

a) Two spades are drawn.

b) A diamond is drawn and then a face card is drawn.

c) A queen is drawn first and then a king is drawn.

d) A king and a queen are drawn.

9. For a new class at a school, it is noted that 85% of the students have calculators. If four students are selected at random, find each probability.

a) The first student selected has a calculator and the other three do not.

b) Only one of the four students has a calculator.

c) Two of the students have calculators.

d) None of the students have calculators.

e) All of the students have calculators.

10. From a group of 20 athletes it is found that 13 played billiards, 12 played golf, and 5 played both billiards and golf.

a) Draw a Venn diagram that represents this information.

b) Find the probability that an athlete chosen at random

    (i) plays golf   (ii) does not play billiards

    (iii) plays billiards and golf   (v) plays billiards or golf.

c) State a reason as to whether or not the events are independent.

d) State a reason as to whether or not the events are mutually exclusive.

11. A ten-question True-False test is given. An unprepared student decides to guess at each question.

Find the probability that the student guessed all 10 questions correctly.

12. Find the probability that two people were born in the same month.

13. Three fair six-sided dice are tossed. Find the probability that

a) triple fives show

b) triple fives or triple sixes show

c) triples show.

14. Consider the Venn diagram below showing the probabilities of events A and B.

a) Find each probability.

    (i) $P(A)$

    (ii) $P(B)$

    (iii) $P(A \cup B)$

b) Are events A and B independent? Why or why not?

c) Are events A and B mutually exclusive? Why or why not?

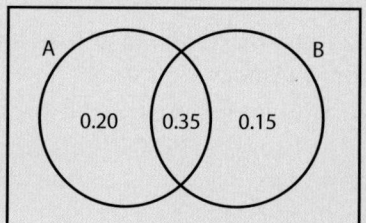

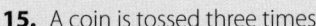

15. A coin is tossed three times.

   a) Write down all possible outcomes using H for heads and T for tails.

   b) Write as a fraction, the probability that

      (i) heads were the outcomes of the three tosses

      (ii) the outcomes of the first and second tosses were different.

10.6 Dependent events

3.7 Probability of combined events, mutually exclusive events, independent events. Use of tree diagrams, Venn diagrams, sample space diagrams, and tables of outcomes. Probability using 'with replacement' and 'without replacement'. Conditional probability.

Two events are said to be **dependent** if the probability of the first event occurring influences the probability of the second event occurring.

Example 10.19

An urn contains 4 red marbles and 3 green marbles. Find the probability of drawing a red marble, *not* replacing it, and then drawing a green marble.

Solution

Not replacing the red marble reduces the sample space from 7 to 6 marbles and thus influences the probability of drawing the green marble.

∴ P(red and green) = P(red) · P(green), given that the red marble was not replaced)

$$= \frac{4}{7} \cdot \frac{3}{6} = \frac{12}{42} = \frac{2}{7}.$$

 The notation $P(B|A)$ is read as, 'probability of B given that event A has already occurred' or simply 'probability of B given A.'

● **Examiner's hint:** The formula for conditional probability is found on page 4 of the Mathematical Studies SL Formula Booklet.

Dependent events are defined thus:

> **If two events are dependent, then P(A and B) = P(A) · P(B|A).**
> Using '∩' for 'and' the formula is: **P(A ∩ B) = P(A) · P(B|A).**

Example 10.20

When drawing two cards from a standard 52-card deck, find the probability of both cards being black.

Solution

Rephrase this problem as follows: When drawing two cards, one at a time, from a standard 52-card deck, find the probability that the first one is black, is not replaced, **and** the second card is black.

$$P(\text{black and black}) = P(B) \cdot P(B|B) = \frac{26}{52} \cdot \frac{25}{51} = \frac{25}{102}.$$

When solving probability problems like the one in Example 10.20, you should think about 'getting the black card you are aiming for', even though in reality you might not draw a black card.

Discuss the philosophy of drawing two black cards at one time versus drawing one black card and then drawing another black card. Is it possible to 'touch' two cards at the same time?

Suppose you tossed a fair coin 100 times and each time it landed heads up. Even though you know that the probability of the coin landing heads up on the next toss is one-half, discuss why you might 'emotionally feel' that the next toss is 'due' to land tails up.

Exercise 10.6

1. How can you tell if events are dependent?

2. State which of the following events are dependent.
   a) Two cards are drawn without replacement from a standard 52-card deck.
   b) Hours spent studying; GPA.
   c) Tossing a coin; rolling a die.
   d) Cholesterol level; eating saturated fat.
   e) Rolling two fair six-sided dice.

3. In an urn there are 4 white and 6 black marbles. Two marbles are selected. Find each probability.
   (Note: Unless it is explicitly stated, you must assume that there is **no** replacement.)
   a) Both marbles are white.
   b) The first marble selected is black and the next marble selected is white.
   c) A black and white marble are selected.

4. A box contains 5 red poker chips, 7 white poker chips, and 6 blue poker chips. Three chips are selected at random. Find each probability.
   a) All three chips are red.
   b) The first chip selected is red, the next chip selected is red, and the last chip selected is blue.
   c) Two red chips and one blue chip are selected.
   d) No white chips are selected.
   e) Neither blue nor white chips are selected.
   f) At least one red chip is selected.

5. From a standard 52-card deck two cards are drawn without replacement. Find each probability.
   a) Two jacks are selected.
   b) Two red cards are selected.
   c) An ace is selected first and then a five is selected.
   d) An ace and a five are selected.

6. Three coins are tossed. Find each probability.
   a) The first coin lands heads up, the second coin lands heads up, and the third coin lands tails up.
   b) Two of the coins land heads up and one of the coins lands tails up.
   c) At least one of the coins lands heads up.
   d) At most, two coins land tails up.

7. A group consists of 8 administrators, 10 teachers, and 12 students. A committee is formed by the random selection of four people. Find each probability.
   a) The first person is a student, the next two are teachers, and the last selected is an administrator.
   b) Two students and two teachers are selected.
   c) Three administrators and one student are selected.

8. A four-question multiple-choice test has five choices for each question. There is only one correct answer for each question. An unprepared student has decided to guess at each question. Find each probability.
   a) None were answered correctly.

b) All were answered correctly.

c) Three were answered correctly.

d) The first two were answered correctly and the last two were answered incorrectly.

e) Two were answered correctly.

9. There are three bags each containing 7, 8, and 9 balls respectively, as shown in the picture.

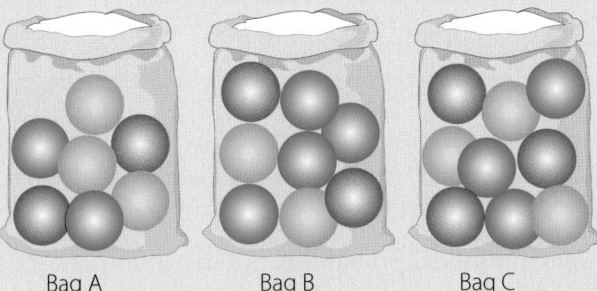

Bag A   Bag B   Bag C

a) Hannah takes a ball from bag A, notes the colour, and then places the ball on the table. Hannah then takes another ball from bag A and also places it on the table. Draw a tree diagram to show all probabilities.

b) Find the probability that a ball drawn at random from bag A will

   i) not be yellow

   ii) be neither red or yellow.

c) From bag B, balls are drawn randomly, one by one, and placed on the table. Find the probability that the first red ball drawn will appear at the *third* draw.

d) From bag C, three balls are drawn together at random. Find the probability that the three balls are all the same colour.

Conditional probability

3.7   Probability of combined events, mutually exclusive events, independent events. Use of tree diagrams, Venn diagrams, sample space diagrams, and tables of outcomes. Probability using 'with replacement' and 'without replacement'. Conditional probability.

It is fairly common to say something like, 'Given the fact that a student is a senior, what is the chance that she owns her own car?' or 'Given the fact that a card is a king, what is the probability it is the king of diamonds?'.

Answering questions like these involves a concept called **conditional probability**. It can be referred to as 'given probability', since the word 'given' is naturally used in the stated problems.

Example 10.21

Six identical marbles, numbered 1, 2, 3, 4, 5, and 6, are placed in an urn. A single draw is made. It is known only that an even-numbered marble is selected. What is the probability that the marble selected is a 4?

Solution

The sample space is 1, 2, 3, 4, 5, 6. When the question stated that only an even-numbered marble is selected, the sample space has been reduced in size.

The new sample space is 2, 4, 6. Solve this problem (as if it were the original):

Given that three marbles, numbered 2, 4, and 6, are placed in an urn, find the probability of drawing the marble numbered 4.

$$\text{Hence, P(a 4 is drawn)} = \tfrac{1}{3}.$$

A key word to look for is 'given'.

● **Examiner's hint:** The formula for conditional probability is found on page 4 of the Mathematical Studies SL Formula Booklet.

The above example and its solution illustrate the idea behind conditional probability. The 'condition' that an event is known reduces the sample space.

The formula for **conditional probability** is derived from the formula for the probability of dependent events.

$$P(A \cap B) = P(A) \cdot P(B|A)$$

$$\therefore \frac{P(A \cap B)}{P(A)} = P(B|A) \Rightarrow P(B|A) = \frac{P(A \cap B)}{P(A)}.$$

In words, $P(B|A) = \dfrac{P(A \cap B)}{P(A)}$ says: The probability of event B occurring given the fact that event A has already occurred is equal to the probability of the events A and B occurring at the same time (the intersection of A and B) divided by the probability of A occurring.

In the case of Example 10.21, P(a 4 is selected **given** that an even number had already been selected)

$$= P(4|\text{even}) = \frac{P(4 \cap \text{even})}{P(\text{even})} = \frac{1/6}{3/6} = \frac{1}{6} \cdot \frac{6}{3} = \frac{1}{3}.$$

Right is a Venn diagram depicting the information.

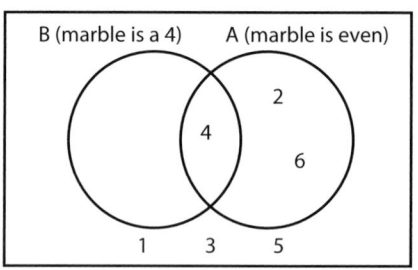

Example 10.22

Consider the data collected from King High School in Example 10.14:

	Sports car	Truck	Sedan	Totals
Junior	8	7	9	24
Senior	15	4	7	26
Totals	23	11	16	50

a) Find the probability that the kind of vehicle driven to school was a truck, given that the student was a senior.

b) Given that the vehicle driven to King High School was a sports car, find the probability that the student was a Junior.

Solution

Method I. The conditional probability formula.

a) P(truck given senior) $= \dfrac{P(\text{truck} \cap \text{senior})}{P(\text{senior})} = \dfrac{4/50}{26/50} = \dfrac{4}{50} \cdot \dfrac{50}{26} = \dfrac{4}{26} = \dfrac{2}{13}.$

b) P(junior given sports car) $= \dfrac{P(\text{junior} \cap \text{sports car})}{P(\text{sports car})} = \dfrac{8/50}{23/50}$

$$= \dfrac{8}{50} \cdot \dfrac{50}{23} = \dfrac{8}{23}.$$

Method II. Reduce the sample space.

a) Cover up all of the data in the 'Junior' row and the 'Totals' row.

Leave uncovered only the 'given' information in the 'Senior' row.

	Sports car	Truck	Sedan	Totals
Senior	15	4	7	26

Hence, P(truck given senior) $= \dfrac{4}{26} = \dfrac{2}{13}.$

b) Cover up all of the data in the 'Truck' and 'Sedan' columns and the 'Totals' column.

Leave uncovered only the 'given' information in the 'Sports car' column.

	Sports car	Truck	Sedan	Totals
Junior	8			
Senior	15			
Totals	23			

Hence, P(junior given sports car) $= \dfrac{8}{23}.$

1. Write the formula for conditional probability five times, each time using different letters for the two events.

2. Other than using the formula, what other method can be used to solve conditional probability problems?

3. Given that $P(B) = 0.8$ and $P(A \cap B) = 0.6$, find the $P(A|B)$.

4. Given that $P(C) = \frac{2}{3}$ and $P(C \cap D) = \frac{2}{5}$, find the $P(D|C)$.

5. Let $P(A) = 0.6$, $P(B) = 0.5$, and $P(A \cap B) = 0.2$.
   a) Draw a Venn diagram to represent the information.
   Find each of the following probabilities.
   b) $P(A \cup B)$ c) $P((A \cup B)')$ d) $P(A|B)$ e) $P(B|A)$

6. Let $P(C) = 45\%$, $P(D) = 27\%$, and $P(C \cap D) = 12\%$.
a) Draw a Venn diagram to represent the information.
Find each of the following probabilities.
b) $P(C \cup D)$ c) $P((C \cup D)')$ d) $P(C|D)$ e) $P(D|C)$

7. From a survey of 18 students it is found that 9 like language arts, 12 like mathematics, 5 like language arts and mathematics, and 2 like neither discipline.
a) Draw a Venn diagram to represent the information.
Find each of the following probabilities.
b) P(a student likes mathematics or language arts)
c) P(a student likes neither mathematics nor language arts)
b) P(a student likes mathematics given that he likes language arts)
b) P(a student likes language arts given that he likes mathematics)

8. From a standard 52-card deck, let event D be the selection of a diamond and event F be the selection of a face card. One card is selected at random,
a) Draw a Venn diagram to represent the information.
Find each of the following probabilities.
b) P(the card is a diamond or a face card)
c) P(the card is neither a diamond nor a face card)
d) P(a face card is selected given that a diamond has already been selected)
e) P(a diamond is selected given that a face card has already been selected)

9. Data was collected reporting seat belt usage and gender. The data was organized in the table below.

	Wears seat belt	Does not wear seat belt
Male	35	55
Female	45	15

If a person is selected at random, find each probability.
a) P(the person wears a seat belt)
b) P(the person was a male)
c) P(the person was male and wears a seat belt)
d) P(the person was male given that he wears a seat belt)
e) P(the person does not wear a seat belt given that the person was female)

10. A survey was given to students asking them their GPA and if they planned to attend a community college or a state university. The results are tabulated below.

	Below average	Average	Above average
Community college	20	25	5
State university	7	8	15

If a student was selected at random, find each probability.
a) P(the student had an average GPA)
b) P(the student went to a state university)
c) P(the student went to a state university and had an average GPA)
d) P(the student had an average GPA given that the student went to a state university)

e) P(the student went to a community college given that the student had an above average GPA)

11. The following table contains data collected from 200 randomly selected office workers.

Gender	Mode of transport to work			Total
	Car	Bus	Train	
Male	27	11	57	95
Female	16	34	55	105
Total	43	45	112	200

What is the probability of:
a) an office worker travelling by car given that the worker is female?
b) an office worker travelling by bus?
c) an office worker being male given that the office worker travels by train?
d) an office worker travelling by bus and not being male?

10.8 The probability of 'at least one'

3.7 Probability of combined events, mutually exclusive events, independent events. Use of tree diagrams, Venn diagrams, sample space diagrams, and tables of outcomes. Probability using 'with replacement' and 'without replacement'. Conditional probability.

 Rewriting the phrase 'at least' naturally yields the word 'or'.

Several examples of the phrase 'at least one' are given below.
- The phrase 'at least one' means one or two or three, etc.
- The phrase 'at least two' means two or three or four, etc.
- The phrase 'at least three' means three or four or five, etc.

Example 10.23

An urn contains 7 red marbles and 3 yellow marbles. When three marbles are chosen at random, find the probability that at least 1 red marble is chosen.

Solution

Method I. Using the meaning of the phrase 'at least one'.

At least one red marble means: one red **or** two red **or** three red marbles.

Let $y = P(\text{yellow})$ and $r = P(\text{red})$.

P(at least one red marble)

$= P(1 \text{ red and } 2 \text{ yellow})$ or $P(2 \text{ red and } 1 \text{ yellow})$ or $P(3 \text{ red and } 0 \text{ yellow})$

$= (\mathbf{ryy} \text{ or } \mathbf{yry} \text{ or } \mathbf{yyr}) \text{ or } (\mathbf{rry} \text{ or } \mathbf{ryr} \text{ or } \mathbf{yrr}) \text{ or } (\mathbf{rrr})$

This is relatively loose notation, but is shows the necessity of considering all permutations of the red and yellow marbles.

$$= (r \cdot y \cdot y \; + \; y \cdot r \cdot y \; + \; y \cdot y \cdot r) \; + \; (r \cdot r \cdot y \; + \; r \cdot y \cdot r \; + \; y \cdot r \cdot r) \; + \; (r \cdot r \cdot r)$$

$$= \frac{7}{10} \cdot \frac{3}{9} \cdot \frac{2}{8} + \frac{3}{10} \cdot \frac{7}{9} \cdot \frac{2}{8} + \frac{3}{10} \cdot \frac{2}{9} \cdot \frac{7}{8} + \frac{7}{10} \cdot \frac{6}{9} \cdot \frac{3}{8} + \frac{7}{10} \cdot \frac{3}{9} \cdot \frac{6}{8} + \frac{3}{10} \cdot \frac{7}{9} \cdot \frac{6}{8} + \frac{7}{10} \cdot \frac{6}{9} \cdot \frac{5}{8}$$

$$= \qquad 3 \cdot \frac{42}{720} \qquad + \qquad 3 \cdot \frac{126}{720} \qquad + \frac{210}{720}$$

$$= \frac{714}{720} = \frac{119}{120} = 0.992 \text{ to 3 s.f.}$$

Method II. Use the principle that **P(of at least one) = 1 − P(of none)**.

All of the permutations (the different ways) of selecting 3 marbles are given below.

$$\underbrace{yyy}_{\text{all 3 yellow}} \quad \text{or} \quad \underbrace{ryy}_{\text{1 red and 2 yellow}} \quad \text{or} \quad \underbrace{rry}_{\text{2 reds and 1 yellow}} \quad \text{or} \quad \underbrace{rrr}_{\text{all 3 red}}$$

If we think of each letter (y or r) as the probability of drawing that colour marble, the sum of the probabilities would be:

$$\frac{3}{10} \cdot \frac{2}{9} \cdot \frac{1}{8} + \frac{7}{10} \cdot \frac{3}{9} \cdot \frac{2}{8} + \frac{7}{10} \cdot \frac{6}{9} \cdot \frac{3}{8} + \frac{7}{10} \cdot \frac{6}{9} \cdot \frac{5}{8} = \frac{384}{720}$$

However, it is well known that the sum of all of the probabilities of the outcomes in a sample space equals 1.

Therefore, there must be some outcomes that have not been accounted for.

The outcomes that have **not** been accounted for are the permutations (the rearrangements of) of 1 red and 2 yellow marbles AND 2 red and 1 yellow marble.

The outcome **ryy** can be rearranged as: **yry** or **yyr**. All three arrangements satisfy the condition of drawing 1 red and 2 yellow marbles.

The outcome **ryy** only satisfies one condition: the red marble is drawn first and the next two draws are yellow marbles.

In the same way, **rry** can also be rearranged in two ways: **ryr** or **yrr**.

Therefore, the correct number of permutations for drawing 3 marbles is:

(**yyy**) or (**ryy** or **yry** or **yyr**) or (**rry** or **ryr** or **yrr**) or (**rrr**).

The sum of the probabilities is:

$$\left(\frac{3}{10} \cdot \frac{2}{9} \cdot \frac{1}{8} \right) + 3 \cdot \left(\frac{7}{10} \cdot \frac{3}{9} \cdot \frac{2}{8} \right) + 3 \cdot \left(\frac{7}{10} \cdot \frac{6}{9} \cdot \frac{3}{8} \right) + \left(\frac{7}{10} \cdot \frac{6}{9} \cdot \frac{5}{8} \right) = \frac{720}{720} = 1.$$

Now with some very loose notation, letting y = P(yellow) and r = P(red),

$$(yyy) + \underbrace{(ryy + yry + yyr) + (rry + ryr + yrr) + (rrr)}_{\text{These terms have 'at least one red marble'.}} = 1$$

$$\therefore \underbrace{(ryy + yry + yyr) + (rry + ryr + yrr) + (rrr)}_{\text{The probability of at least one red marble}} \underbrace{=}_{\text{equals}} \underbrace{1}_{\text{one}} \underbrace{-}_{\text{minus}} \underbrace{(yyy)}_{\substack{\text{the} \\ \text{probability} \\ \text{of no red} \\ \text{marbles.}}}$$

Therefore, P(at least one red marble)

$$= 1 - P(\text{of no red marbles}) = 1 - P(\text{all yellow marbles})$$

$$= 1 - \frac{3}{10} \cdot \frac{2}{9} \cdot \frac{1}{8} = 1 - \frac{6}{720} = \frac{714}{720} = \frac{119}{120} = 0.992 \text{ to 3 s.f.}$$

Did you know that in a room of 23 people there is a 50.7% chance that at least 2 of the people will have the same birthday? With 30 people there is a 70% chance and if the number is 50 then there is a 97% chance that 2 people will have the same birthday!

For more information on the birthday paradox problem, visit www.pearsonhotlinks.co.uk, enter the ISBN for this book and click on weblink 10.3.

For each of the following problems, use the method described in Example 10.23, i.e. P(at least one) = 1 − P(none).

1. Three coins are tossed. Find the probability that at least one shows heads up.

2. Four coins are tossed. Find the probability that at least one shows tails up.

3. An urn contains 7 blue and 3 red marbles. Two marbles are drawn at random without replacement. Find the probability at least one of them is a blue marble.

4. A box contains 5 black, 4 white, and 4 yellow chips. Three chips are selected at random. Find the probability that at least one chip is black.

5. A couple are planning on having five children. Find the probability that at least one of the children will be a girl.

6. Suppose the probability that it will rain on any given day is 15%. If three days are selected at random, find the probability that it will rain on at least one of those days.

10.9 Summarization of the laws of probability

3.6	Sample space; event A; complementary event, A'. Probability of an event. Probability of a complementary event. Expected value.
3.7	Probability of combined events, mutually exclusive events, independent events. Use of tree diagrams, Venn diagrams, sample space diagrams, and tables of outcomes. Probability using 'with replacement' and 'without replacement'. Conditional probability.

For questions and answers to some very interesting probability problems and concepts, visit www.pearsonhotlinks.co.uk, enter the ISBN for this book and click on weblinks 10.4 and 10.5.

1. Simple probability event: $P(A) = \dfrac{n(A)}{n(U)}$.

2. Complementary event: $P(A') = 1 - P(A)$.

3. Mutually exclusive events: $P(A \cup B) = P(A) + P(B)$, where $P(A \cap B) = 0$.
   a) Key word: 'or'
   b) Key phrase to ask: 'Are any of event A also event B?'
      (i) If 'no', then add only.
      (ii) If 'yes', then use combined events.

4. Combined events: $P(A \cup B) = P(A) + P(B) - P(A \cap B)$, where $P(A \cap B) \neq 0$.
   a) Key word: 'or'
   b) Key phrase to ask: 'Are any of event A also event B?'
      (i) If 'no', then use mutually exclusive events.
      (ii) If 'yes', then add and subtract.

5. Independent events: $P(A \cap B) = P(A) \cdot P(B)$.
   a) Key word: 'and'

b) Key phrase to ask: 'Will the probability of the first event influence the probability of the second event?'
    (i) If 'no', multiply the individual probabilities together.
    (ii) If 'yes', then use dependent events.

To discuss the Petersburg Paradox, visit www. pearsonhotlinks.co.uk, enter the ISBN for this book and click on weblink 10.6.

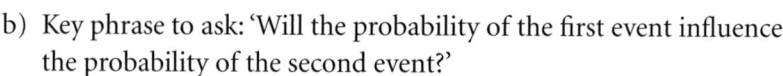

6. Dependent events: $P(A \cap B) = P(A) \cdot P(B|A)$.
   a) Key word: 'and'
   b) Key phrase to ask: 'Will the probability of the first event influence the probability of the second event?'
     (i) If 'yes', find the probability of the first event, find the probability of the second event by reducing the sample space, and multiply the two probabilities.
     (ii) If 'no', then use independent events.

7. Conditional events: $P(B|A) = \dfrac{P(A \cap B)}{P(A)}$.
   a) Key word: 'given'
   b) Key technique:
     (i) reduce the sample space by the 'listing' method.
     (ii) reduce the sample space by the 'cover up' method.

8. 'At least one' events: P(of at least one) = 1 − P(of none)
   a) Key phrase: 'at least one'
   b) Key technique: find the probability of none by multiplying and then subtract from 1.

Exercise 10.9

1. A bag contains 12 coloured balls: 7 red, 3 white, and 2 blue. One ball is selected at random and put into bag A.
   a) What is the probability that the ball selected is
     i) red?
     ii) not red?

   A second ball is selected at random and put into bag B.
   b) What is the probability that
     i) the ball in bag B is red given that the ball in bag A is red?
     ii) both balls are not red?
     iii) one bag contains a red ball and the other bag does not contain a red ball?

   A third ball is selected at random and put into bag C.
   c) What is the probability that
     i) all three bags contain red balls?
     ii) only one bag contains a red ball?
     iii) at least one bag contains a red ball?

2. Zuhair is planning an outdoor party in June. He is concerned that it may rain, so he has decided to collect some data on the weather in his city. He has found that in his city if it is sunny on one day, then the probability of it being sunny the next day is $\frac{3}{5}$. He has also found that the probability of a day being sunny in June is $\frac{4}{7}$.
   a) Construct a tree diagram to represent the data.
   b) What is the probability that only one day out of two days will be sunny?

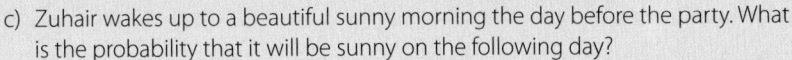

c) Zuhair wakes up to a beautiful sunny morning the day before the party. What is the probability that it will be sunny on the following day?

3. a) In a two-child family, the older child is a boy. What is the probability that the younger child is a girl?
   b) In a two-child family, one of the children is a boy. What is the probability that the other child is a girl?

4. Suppose that you have a deck of 60 cards, consisting of:
   • 20 black cards that are black on both sides
   • 20 white cards that are white on both sides
   • 20 mixed cards that are black on one side and white on the other side.

   The cards are shuffled. One is pulled out of the deck at random and placed on a table. The side facing up is black. You do not observe the other side. What is the probability that the other side is also black?

Connecting independent and mutually exclusive events

Independent and mutually exclusive events are often confused. Are the two concepts connected? Can you have one without the other or can they both occur at the same time?

In other words:

a) Can independent events also be mutually exclusive? Not mutually exclusive?
b) Can dependent events also be mutually exclusive? Not mutually exclusive?

The answer to all of the above questions is 'yes'.

The following chart shows an example for each situation.

	Independent	Dependent
Mutually exclusive	If $P(A) = 0$ and $P(B) = 0.5$ then $P(A \cap B) = 0 \cdot 0.5 = 0$.	If $P(\text{drawing an } 8) = \frac{4}{52}$ and, with replacement, $P(\text{drawing a } 7) = \frac{4}{52}$, then $P(8 \cap 7) = 0$. Also, $P(8\|7) = \frac{P(8 \cap 7)}{P(7)} = \frac{0}{4/52} = 0$.
Not mutually exclusive	If $P(\text{drawing a club}) = \frac{1}{4}$ and, with replacement, $P(\text{drawing a heart}) = \frac{1}{4}$, then $P(C \cap H) = \left(\frac{1}{4}\right)\left(\frac{1}{4}\right) = \frac{1}{16}$.	If $P(\text{drawing a diamond}) = \frac{13}{52}$ and, with replacement, $P(\text{drawing a face card}) = \frac{12}{52}$, then $P(D \cap F) = \frac{3}{52}$. Also, $P(F\|D) = \frac{P(F \cap D)}{P(D)} = \frac{3/52}{13/52} = \frac{3}{13}$.

Exercise 10.10

1. Using a standard 52-deck, fill in the chart below by making up examples similar to those in the example in section 10.10.

	Independent	Dependent
Mutually exclusive		
Not mutually exclusive		

2. Using any other sample space except a deck of cards, fill in the chart by making up examples which satisfy each condition.

	Independent	Dependent
Mutually exclusive		
Not mutually exclusive		

11 Descriptive Statistics

Assessment statements

Overview

By the end of this chapter you will be able to:
- classify data as discrete, continuous, nominal, ordinal, interval
- organize discrete data into frequency tables
- graph discrete data as a frequency polygon
- organize continuous data into frequency tables
- graph continuous data as histograms
- organize continuous data into cumulative frequency tables
- graph continuous data as cumulative frequency curves
- solve problems involving the measures of central tendency: mean, median and mode
- solve problems involving the measures of dispersion: range, interquartile range and standard deviation.

11.1 Classification of data

2.1 Classification of data as discrete or continuous.

When undertaking a research project requiring the collection of data, ideally we would like to gather data from the whole **population,** such as from each person or from measuring all objects being tested, but this is not always possible. For example, we may want to know if seatbelt usage for car

It is likely that the original purpose of statistics was for governmental purposes such as the taking of a census.

owners is independent of gender. Collecting the data from all car owners in one region would not be practical, or even possible.

Alternatively we could determine the owners' usage of seatbelts by gender by taking a **sample**. A sample is intended to be representative of the population.

A **random sample** is preferred as it is considered to be representative of the whole population. In the case of researching gender and usage of seatbelts we could select car owners by the last digit of their mobile telephone number, by the last digit on the number plate of their car, or by using a random number generator on your GDC.

However, when sampling we need to be aware of **biased sampling**, that is one that is not random. For example, in the case of researching gender and usage of seatbelts then selecting only female car owners, or a particular age group would be considered a biased sample which would not produce results that are representative of the population under investigation.

When an event is observed, it is common to record the information. The information that is recorded is called **data**. The information is often recorded with slash marks, called **tally** marks, in a table. For example, a manager recorded how a baseball player performed while batting during a five-game stretch. The data she observed are recorded in the table below.

For more information on the history of probability and statistics, visit www.pearsonhotlinks.co.uk, enter the ISBN for this book and click on weblink 11.1.

Game no.	At bat tallies	At bats	Hit tallies	Hits
1	////	4	/	1
2	7HL	5	//	2
3	////	4	//	2
4	///	3		0
5	////	4	//	2

There are several classifications of data.

- **Nominal** data cannot be ordered. An example would be political affiliation: Republican, Democrat, Green Party, etc. This type of data is called **qualitative**.

- **Ordinal** data can be ordered, but determining the differences between the rankings is not possible. An example would be places in a dance contest: 1st place, 2nd place, 3rd place, etc. This type of data is called **quantitative**.

- **Interval** data can be ordered and determining the differences between the rankings is possible. An example would be maths SAT scores: 780, 760, 700, 640, etc. This type of data is also called **quantitative.**

Is the number of grains of sand on a beach discrete?

Quantitative data can also be classified into two categories: **discrete and continuous**.

- Data is called **discrete** if you can count it. Examples include the number of students in a room, the number of goals a footballer makes, the number of steps walked each day, the number of text messages you make each day, etc.

- Data is called **continuous** if you can measure it. Examples include time, weight, and distance. Suppose a student was measured to be 1.6 m tall on January 1 and then six months later was measured to be 1.8 m tall. The student could not have been 1.6 m tall and then suddenly have been 1.8 m tall without having grown through all of the heights between 1.6 and 1.8 m. Therefore, height is considered to be a continuous variable.

What does it mean for time to be continuous? Has it always been continuous? Will it continue to be continuous? Was there a beginning point?

Example 11.1

Classify each as nominal, ordinal, or interval data.
a) 100-metre race times.
b) Colours used by maths teachers to write on their whiteboards.
c) The rating of trumpet solos as superior, excellent, or good.

Solution
a) This is an example of **interval** data since there is a specific difference between, for example, 11.5 and 11.6 seconds.
b) Colours cannot be ordered; therefore, this is an example of **nominal** data.
c) Excellent is better than good, but the difference cannot be exactly determined; therefore, this is an example of **ordinal** data.

Example 11.2

Classify each as discrete or continuous data.
a) Number of students in a high school.
b) The time it take to complete a test.

Solution
a) Since the number of students is countable, the data is **discrete**.
b) Since the time to complete the test could be recorded as any fraction of a minute, the data is **continuous.**

Exercise 11.1

1. Name the three main classifications of data.

2. Which type of data is qualitative?

3. Which two types of data are quantitative?

4. Name the two categories for classifying quantitative data.

5. Draw a tree diagram for the following types of statistics and classifications of data: qualitative, quantitative, discrete, continuous.

6. Draw a tree diagram for the following types of statistics and classifications of data: nominal, ordinal, interval.

7. Classify each as qualitative or quantitative data.
   a) Religious preference
   b) The number of seconds a download takes.
   c) The colours that a homeowners' association allows houses to be painted.
   d) The number of laps run by a track team member in one season.
   e) The currencies of the world.

8. Classify each as nominal, ordinal, or interval data.
   a) SAT scores
   b) Political parties
   c) Gender
   d) Grades of 1, 2, 3, 4, 5, 6, or 7 on the Mathematical Studies IB exam
   e) College Grade Point Average (GPA)

9. Determine if each type of data is discrete or continuous.
   a) The number of leaves on a tree.
   b) The time it would take to count the leaves on a tree.
   c) The volume of water in a swimming pool.
   d) The number of words in a book.
   e) The height of a child during her first 6 months of life.

11.2 Discrete data

| 2.2 | Simple discrete data: frequency tables. |

Once data has been collected, it should be organized in an appropriate manner, and then it should be displayed with an appropriate diagram.

Data is usually collected with a tally sheet and organized with a table or chart. For discrete data, there are several diagrams that are appropriate to use. Some of those are listed below.

- Number line graph. Data values are plotted above or below the appropriate natural numbers on a number line.
- Bar charts. (See Section 1.7.)
- Pie chart. (See Section 1.7.)
- Pictograph. (See Section 1.7.)
- Box-and-whisker diagram (box plot).

Number line graphs are not part of the Maths Studies curriculum. However, to find information concerning number line graphs, visit www.pearsonhotlinks.co.uk, enter the ISBN for this book and click on weblink 11.2.

Example 11.3

During a random day at a local mall, the following data were recorded when observing the colours of cars in the parking lot.

Colour	Number of cars
Red	卅卅 卅卅 卅卅 卅卅 卅卅
Blue	卅卅 卅卅 卅卅 卅卅 卅卅 卅卅 卅卅 卅卅 卅卅
Green	卅卅 卅卅 卅卅 卅卅 卅卅 卅卅 卅卅 卅卅
White	卅卅 卅卅 卅卅 卅卅 卅卅 卅卅 卅卅 卅卅 卅卅 卅卅 卅卅 卅卅 卅卅 卅卅
Black	卅卅 卅卅 卅卅 卅卅

a) Organize the data into a frequency table in alphabetical order.

b) Using the methods described in Section 1.7, construct each of the following graphs to display the data:

(i) bar chart   (ii) pie chart   (iii) pictogram

c) Name at least one advantage these graphs have over the list of raw data or the table of organized values.

Solution

a)

Colour	Number of cars
Black	20
Blue	45
Green	40
Red	25
White	70

b) (i)

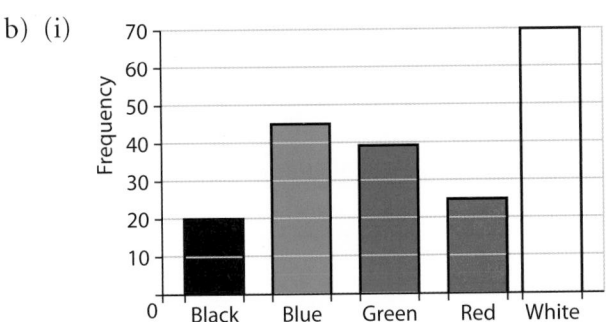

b) (ii) Black: $\dfrac{20}{200} \cdot 360° = 36°$

Blue: $\dfrac{45}{200} \cdot 360° = 81°$

Green: $\dfrac{40}{200} \cdot 360° = 72°$

Red: $\dfrac{25}{200} \cdot 360° = 45°$

White: $\dfrac{70}{200} \cdot 360° = 126°$

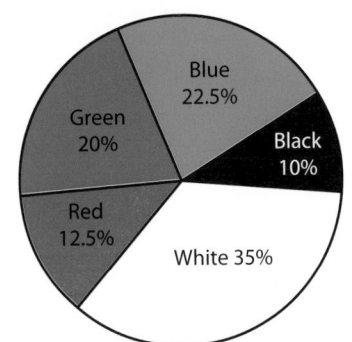

b) (iii)

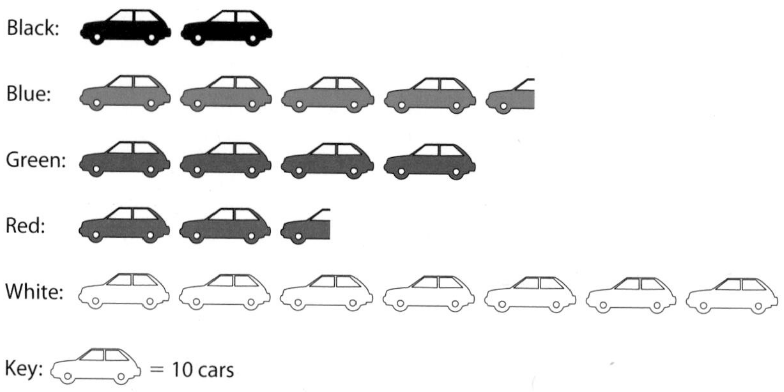

Black:

Blue:

Green:

Red:

White:

Key: = 10 cars

c) These graphs are very visual and easy to understand.

Exercise 11.2

1. Write one or two sentences describing the value of tally marks.

2. The number of absences, from a small high school, was collected during the fifth week of the school year. The data is shown below.

Grade level	Number of absences
9	10
10	30
11	18
12	42

Construct each of the following graphs to display the data:
a) bar graph
b) pie chart
c) pictogram

3. Ronnie collected spare change for the last 12 months. The type and the number of each coin is shown below.

Coin	Number
Penny	450
Nickel	200
Dime	150
Quarter	100

Construct each of the following graphs to display the data:
a) bar graph   b) pie chart   c) pictogram

4. The scores, out of a maximum of 20, on a recent mathematical studies test were:
17, 11, 9, 3, 19, 20, 15, 7, 11, 13, 12, 15, 18, 17, 6, 8, 14, 17, 19, 18, 5, 7, 11, 10, 10, 17.
Organize the data into a frequency table.

5. The ages of all 75 players in an NFL football team are

```
24  26  23  22  25  35  33  24  24  29  25  28  29  24  25
22  22  24  24  23  22  26  28  22  23  28  29  23  28  22
22  28  26  28  22  25  25  32  31  24  31  25  24  30  27
26  25  25  26  22  28  23  26  27  32  23  26  24  24  24
23  26  22  23  28  22  24  23  31  26  31  30  24  22  28
```

Organize the data into a frequency table.

6. Two dice are rolled and the numbers on the uppermost side of each die are added together. Use the following frequency table to determine the missing value *a*.

Sum of two uppermost sides	Frequency
2	3
3	3
4	4
5	3
6	5
7	6
8	7
9	*a*
10	5
11	4
12	4
Total	50

11.3 Continuous data

2.3 Grouped discrete or continuous data: frequency tables; mid-interval values; upper and lower boundaries.
Frequency histograms.

2.4 Cumulative frequency tables for grouped discrete data and for grouped continuous data; cumulative frequency curves, median, and quartiles.
Box-and-whisker diagram (box plots).

Constructing a frequency distribution table

When organizing data values, two situations could occur:

1) The number of different categories (**classes**) of data values is small enough to be able to list each one as its own row in an organizational chart (**frequency table**).

   - For example, suppose the data collected is the number of minutes spent on a treadmill during the days of a randomly selected month. The only number values collected (not counting the frequencies of each number) were: 25, 26, 27, 28, 30. A frequency table with only six rows would suffice. (A row for no minutes of 29 would still be needed.)

- An **ungrouped** frequency distribution table would be appropriate for this data.

Minutes	Frequency
25	6
26	7
27	5
28	4
29	0
30	8
Total	30

2) The number of different classes of data values is too large to be able to efficiently list each one as its own row in a frequency table.
- For example, suppose the data is the number of minutes spent talking on a cellphone. The numbers collected (not including the frequencies of each number) were: 2, 3, 6, 8, 15, 18, 25, 45. If the data were not grouped together, the frequency table would have to have 44 rows and would be much too cumbersome to deal with efficiently.
- A grouped frequency distribution would be appropriate for data of this type.

Continuous data as well as discrete data can be organized in grouped or ungrouped frequency distribution tables.

To construct a grouped frequency distribution table follow the steps below.

Step 1: Find the range of the data.

Step 2: Divide the range by the number of classes you desire (usually between 5 and 15 or so).

Step 3: Round the Step 2 value **up** to the nearest natural number. This number will be the **class width.**

Step 4: Construct a frequency table with that number of rows.

Step 5: Make sure that each row is a unique group of numbers with no overlap between the rows.

Step 6: An easy way to get the first row is to start with the lowest data value. Add the class width to the lowest value in order to get the lowest value of the next class.

Example 11.4

Data are collected for the number of minutes that 50 randomly selected middle school students studied for a maths test. The results are listed below.

8	52	38	48	42	9	15	36	36	53
10	8	46	46	9	11	12	24	49	34
10	11	9	11	45	25	25	37	14	16
20	22	12	43	36	23	23	26	27	16
21	29	29	38	30	47	34	39	48	46

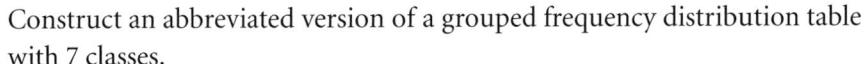

Construct an abbreviated version of a grouped frequency distribution table with 7 classes.

Solution

Following the above steps we have:

- Range $= 53 - 8 = 45$.
- $\frac{45}{7} \approx 6.4 = 7$ **rounded up** to the nearest natural number. This is the class width.
- Start the table with the lowest data value, 8.
- Add the class width, 7, to 8 to get the next class value, and so on.

Classes	Class boundaries	Tally	Frequency
8			
15			
22			
29			
36			
43			
50			

- Remember there can be no overlap between the classes. Hence, the first class must be from 8 to 14. The second class from 15 to 21, and so on.

Class	Class boundaries	Tally	Frequency
8 – 14	7.5 – 14.5		
15 – 21	14.5 – 21.5		
22 – 28	21.5 – 28.5		
29 – 35	28.5 – 35.5		
36 – 42	35.5 – 42.5		
43 – 49	42.5 – 49.5		
50 – 56	49.5 – 56.5		

- When data is continuous, there cannot be any gaps. Therefore, a 'Class boundaries' column is needed. To write the correct class boundary:
  - **decrease** the left class limit by one-half of the recorded data unit. This value is called the **lower class boundary.**
  - **increase** the right class limit by one-half of the recorded data unit. This value is called the **upper class boundary.**

Hence, for the class 8 – 14, the class boundaries would be 7.5 – 14.5. For the class 15 – 21, the class boundaries would be 14.5 – 21.5, etc.

- It does appear that there is 'overlap'. In fact there is not, since each class boundary can be expressed as an inequality that includes the lower boundary, but does not include the upper boundary.
  - For example, 7.5 – 14.5 means that $7.5 \leqslant$ a number of minutes (M) < 14.5.
  - 14.5 – 21.5 means, $14.5 \leqslant M < 21.5$, etc.
- Carefully tally the data values.
- Count the tallies and record the frequency as a natural number.
- Add the 'Frequency' column to double-check the total data values is 50.

Class	Class boundaries	Tally	Frequency
8 – 14	7.5 – 14.5	𝓣𝓗𝓛 𝓣𝓗𝓛 ///	13
15 – 21	14.5 – 21.5	𝓣𝓗𝓛	5
22 – 28	21.5 – 28.5	𝓣𝓗𝓛 ///	8
29 – 35	28.5 – 35.5	𝓣𝓗𝓛	5
36 – 42	35.5 – 42.5	𝓣𝓗𝓛 ///	8
43 – 49	42.5 – 49.5	𝓣𝓗𝓛 ////	9
50 – 56	49.5 – 56.5	//	2
			50

If the data were discrete, the *Class boundaries* column would not be needed.

There are two other columns that need to be added to make the frequency table complete. They are:

- mid-interval values (midpoints)
- cumulative frequency

There are times, when problem solving or when analyzing someone else's data, that the data set is not available to you. However, the frequency distribution table is available. When you do not know, or do not have available to you, the data set, then you can use the **midpoints** (mid-interval values) of each class as the best estimate of the true data values.

To find the midpoint (mid-interval value) find the mean average of the class limits:

$$MP = \frac{\textbf{lower boundary + upper boundary}}{\textbf{2}}.$$

For example, in the class 8 – 14, $MP = \dfrac{8 + 14}{2} = 11$.

For the 15 – 21 class, MP $= \dfrac{15 + 21}{2} = 18$.

In other words, instead of using the actual values of 15, 16, 16, 20 and 21, in the 15 – 21 class, you would use mid-interval values: 18, 18, 18, 18, 18, 18, 18.

Class	Mid–interval
8 – 14	11
15 – 21	18
22 – 28	25
29 – 35	32
36 – 42	39
43 – 49	46
50 – 56	53

Each new entry in the **cumulative frequency column** represents the sum of all of the frequencies up to and including the new entry.

Frequency	Cumulative frequency	cf
13	13	13
5	13 + 5	18
8	13 + 5 + 8	26
5	13 + 5 + 8 + 5	31
8	13 + 5 + 8 + 5 + 8	39
9	13 + 5 + 8 + 5 + 8 + 9	48
2	13 + 5 + 8 + 5 + 8 + 9 + 2	50
50		

● **Examiner's hint:** There is a short cut for finding the numbers in the 'cf' column. Start with 13 and to get the next cf, add the next frequency, i.e. 13 + 5 = 18, 18 + 8 = 26, 26 + 5 = 31, etc.

The entire frequency distribution table is shown below.

Class	Class boundaries	Tally	Freq.	Mid-interval	cf
8 – 14	7.5 – 14.5	7HL 7HL ///	13	11	13
15 – 21	14.5 – 21.5	7HL	5	18	18
22 – 28	21.5 – 28.5	7HL ///	8	25	26
29 – 35	28.5 – 35.5	7HL	5	32	31
36 – 42	35.5 – 42.5	7HL ///	8	39	39
43 – 49	42.5 – 49.5	7HL ////	9	46	48
50 – 56	49.5 – 56.5	//	2	53	50
			50		

Once the frequency distribution table is constructed, graphs can be drawn that will visually depict that data. A histogram is one such graph which will require the use of the upper and lower class boundaries and the frequencies to construct it as shown below.

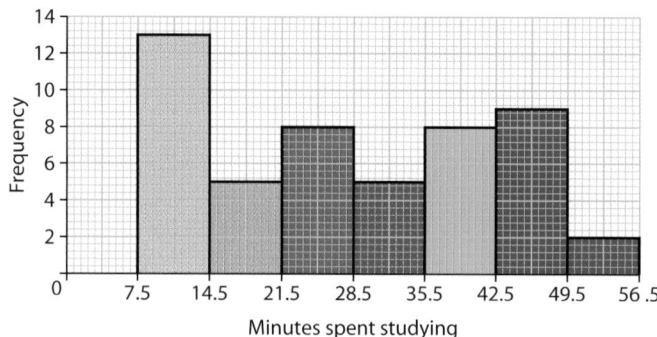

In general, to construct a frequency histogram:

- Use the *Class boundaries* and *Frequency* columns.
- Class boundaries are used on the *x*-axis and the frequency values are used on the *y*-axis.
- Let the lower and upper class boundaries lie on the cm marks.
- The graph looks like a bar graph, but there is no space between the bars.
- Use slash marks, or colours, to distinguish one bar from another.
- Use a squiggle, $\wedge\!\!\!\vee$, on the *x*-axis, next to the origin, to indicate starting number line values that are not necessary.

Exercise 11.3

1. Say whether each of the following is an example of discrete or continuous data.
   a) Selecting a random day in the month and counting the number of TV sitcom shows.
   b) Measuring the volume of liquid in a jar.
   c) Timing 10 randomly selected students in the 100-metre dash.
   d) Measuring the birth weights of Border collie canines.
   e) Counting the number of revolutions a car engine makes in one minute.

2. When would it be appropriate to construct a grouped frequency distribution for
   a) discrete data?
   b) continuous data?

3. When would it be appropriate to construct an ungrouped frequency distribution for
   a) discrete data?
   b) continuous data?

4. Given the following classes for grouped continuous data, fill in the missing cells.

a)

Class	Class boundaries
5 – 8	
9 – 12	
13 – 16	

b)

Class	Class boundaries
2.50 – 3.00	
3.10 – 3.60	

c)

Class	Class boundaries
0 – 7	

d)

Class	Class boundaries
5.40 – 5.80	

e)

Class	Class boundaries
3 –	
8 –	

5. Data are collected for length and rounded to the nearest metre. Find the class boundaries for the following.

a) 100 m   b) 25 m   c) 7 m

6. The number of text messages sent by a group of students in one day was

```
26  14  28  13   28  29  15  13  23   20  17  16
14  27  25  17   19  26  16  13  22   13  18  24
23  21  25  18   24  22  14  29  22   19  28  17
26  26  19  27   19  27  27  24  28   23  15  28
20  29  15  16   24  13  14  19  20   17  15  28
24  20  15  29   27  21  20  14  20   19  28  14
16  24  24  28   20  20  28  14
```

a) Organize the data into a grouped frequency table.

b) Draw a histogram to represent this information.

7. The time taken for students to travel to school is shown in the following table.

Time taken (minutes)	Number of students
$10 \leqslant t < 20$	5
$20 \leqslant t < 30$	16
$30 \leqslant t < 40$	20
$40 \leqslant t < 50$	32
$50 \leqslant t < 60$	26
$60 \leqslant t < 70$	7
$70 \leqslant t < 80$	3

a) Write down the lower and upper class boundaries for the 4th class.

b) Draw a histogram to represent the data.

8. A national park visitors' centre recorded the number of persons who looked at an exhibition on the local flora, each half hour and recorded their numbers in the table below.

Number of visitors per half hour	Frequency
$10 \leqslant t \leqslant 15$	6
$16 \leqslant t \leqslant 20$	16
$21 \leqslant t \leqslant 25$	22
$26 \leqslant t \leqslant 30$	33
$31 \leqslant t \leqslant 35$	35
$36 \leqslant t \leqslant 40$	15
$41 \leqslant t \leqslant 46$	9

Draw a histogram to represent the data.

9. A mathematical studies student is investigating the number of persons visiting a museum as part of their project. The data were collected and compiled into a frequency table as shown below.

Time	Number of visitors
$10{:}00 \leqslant t < 11{:}00$	29
$11{:}00 \leqslant t < 12{:}00$	38
$12{:}00 \leqslant t < 13{:}00$	21
$13{:}00 \leqslant t < 14{:}00$	15
$14{:}00 \leqslant t < 15{:}00$	11
$15{:}00 \leqslant t < 16{:}00$	75
$16{:}00 \leqslant t < 17{:}00$	83
$17{:}00 \leqslant t < 18{:}00$	53
$18{:}00 \leqslant t < 19{:}00$	17

Draw a histogram to represent the data.

Measures of central tendency

11.4

| 2.5 | Measures of central tendency. For simple discrete data: mean; median; mode. For grouped discrete data: estimate of a mean; modal class. |

In everyday conversation, the word *average* is often used. What is the *average* wage the employees in a company make? What is the *average* selling price of a house? What is the *average* number of minutes teenagers talk on their cellphones per day? What is the *average* dress size a clothing buyer should order?

It is clear that the use of the word *average* is ambiguous at best. If a prospective employee enquires about the average wage a company employee makes, the personnel manager might respond with $37 500, when, in fact, most of the employees make $32 000. A student might report to his mother that the average test score on his maths test was 72% and he received 77%. But, in fact, 50% of the scores were less than 80%.

This ambiguity leads to a need for a more thorough discussion of the word average. There are three types of averages we will study. They are **mean**, **median**, and **mode**.

The **mean average**, or simply the **mean**, is defined as the sum of all the data values divided by the number of data values. The symbol for the mean is $\bar{x}$ (pronounced 'x-bar').

The IB formula is: $\bar{x} = \dfrac{\sum\limits_{i=1}^{k} f_i x_i}{n}$, where $n = \sum\limits_{i=1}^{k} f_i$.

Did you hear about the politician who promised that if he were elected he'd make certain that everybody would get an above-average income?

For more jokes, visit www.pearsonhotlinks.co.uk, enter the ISBN for this book and click on weblink 11.3.

● **Examiner's hint:** The formula for the mean average is given on page 3 of the Mathematical Studies SL Formula Booklet (MSSLFB).

Example 11.5

For the following data set, find the mean, $\bar{x}$.

2, 7, 7, 10, 10, 10, 10, 12, 12, 18, 18, 18, 32, 44

Solution
Method I. Use the IB formula.
Organize the array into a (limited) discrete frequency distribution.

Class	Frequency
2	1
7	2
10	4
12	2
18	3
32	1
44	1

The total number of data values is:

$$n = \sum_{i=1}^{6} f_i = 1 + 2 + 4 + 2 + 3 + 1 + 1 = 14.$$

Hence, $\bar{x} = \dfrac{\displaystyle\sum_{i=1}^{6} f_i x_i}{n}$

$$= \frac{1 \cdot 2 + 2 \cdot 7 + 4 \cdot 10 + 2 \cdot 12 + 18 \cdot 3 + 1 \cdot 32 + 1 \cdot 44}{14} = 15.$$

Method II. Use your TI calculator.

Press the following keystrokes.

> STAT, > (CALC), 1:1 Var Stats (ENTER), 2ND L1,V,(leave FreqList Blank), V(Calculate), ENTER (as shown below)

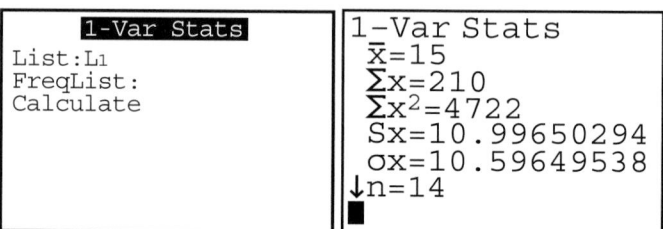

```
1-Var Stats
List:L1
FreqList:
Calculate
```

```
1-Var Stats
x̄=15
Σx=210
Σx²=4722
Sx=10.99650294
σx=10.59649538
↓n=14
■
```

Method III. This method is very useful when the number of data values is large and the number of classes is relatively small.
Enter class values (the individual data values with no duplication) into List 1 data values and enter the corresponding frequencies into List 2.

Now press the following keys:

> STAT, >(CALC), 1: 1 Var Stats (ENTER), 2ND L1,V, 2ND L2, V(Calculate), ENTER (as shown below). Therefore, $\bar{x} = 15$.

L1	L2	L3	1
	1	......	
7	2		
10	4		
12	2		
18	3		
32	1		
44	1		
L1(1)=2			

```
1-Var Stats
List:L1
FreqList:L2
Calculate
```

```
1-Var Stats
x̄=15
Σx=210
Σx²=4722
Sx=10.99650294
σx=10.59649538
↓n=14
■
```

Often the fastest way to find the mean is to simply add the data values and divide by the total number. This is the fastest when the number of data values is small.

Example 11.6

Find the value of x if the mean of the following data set is 12.

26, 6, 6, x, 3, 13, 8, 15, 16, 18, 2

Solution

$$\frac{26 + 6 + 6 + x + 3 + 13 + 8 + 15 + 16 + 18 + 2}{11} = 12$$

$$x + 113 = 132$$

$$\therefore \qquad x = 19.$$

Example 11.7

Given the following grouped discrete frequency distribution, find the mean of the data set.

Class	Frequency
4 – 8	3
9 – 13	5
14 – 18	4
19 – 23	7

Solution

The original data set is unknown. Therefore, the best approximation for the data set will be the mid-interval values (midpoints). This also implies that the mean we find will only be an approximation of the true mean.

Class	Frequency	Mid-interval
4 – 8	3	6
9 – 13	5	11
14 – 18	4	16
19 – 23	7	23

Therefore, the data set can be represented as:

6, 6, 6, 11, 11, 11, 11, 11, 16, 16, 16, 16, 23, 23, 23, 23, 23, 23, 23

The mean can now be found using any one of the previous methods. The method that lends itself best to the solution is Method III. Enter the mid-interval values into List 1 (as if they were the original data values) and enter the frequencies into List 2. Now press the following keys

STAT, CALC, ENTER, L_1, L_2, ENTER.

Therefore, $\bar{x} = 15.7$ to 3 significant figures.

The **median** of a data set represents the number in the middle of a data array. If the data set has an odd number of values, the median will be a number in the data set. If the data set has an even number of data values, the median will be the midpoint between the two middle values.

The **mode** of a data set represents the number that occurs the *most*. In a data set, there can be no modes, one mode, two modes (bimodal) or even three modes (trimodal).

Example 11.8

Find the median and the mode for the following discrete data set.

11, 12, 7, 4, 5, 5, 10, 11, 25, 3, 12, 12, 5

Solution

Arrange the data in an array: 3, 4, 5, 5, 5, 7, 10, 11, 11, 12, 12, 12, 25

There are an odd number of data values; therefore, the median is the middle number, in this case, 10. There are three 5's and three 12's. Therefore, there are two modes, 5 and 12.

Example 11.9

Given the following (partial) frequency distribution for continuous data, find the class in which the median occurs and the modal class (the class in which most data values occur).

Class boundaries	Frequency
16.5 – 19.5	9
19.5 – 22.5	3
22.5 – 25.5	7
25.5 – 28.5	2

Solution

The total number of data values is 21. Therefore, the median is the 11th value (if the original data were known). The 11th value, hence the median, must occur in the 19.5 – 22.5 class.

The modal class is 16.5 – 19.5, since it has the most frequencies.

Exercise 11.4

1. Name the three measures of central tendency.

2. Using the IB formula, $\bar{x} = \dfrac{\sum\limits_{i=1}^{k} f_i x_i}{n}$, where $n = \sum\limits_{i=1}^{k} f_i$, find the mean of the following data sets.

   a) 7, 7, 5, 3, 2, 5, 7, 3, 5, 7, 4

   b) 40, 37, 31, 31, 25, 25, 20, 37, 20, 31, 20, 31

3. Using your GDC and Method II, find the mean of each data set.

   a) 3, 5, 12, 14, 6, 5, 7, 8, 8, 8, 11, 10

   b) 1, 4, 12, 5, 6, 4, 13, 17, 18, 6, 4, 3, 2, 1

4. Use the technique in Method III to find the mean of the data sets that are organized as follows.

   a)

Class	Frequency
5	14
8	10
15	16
19	9
26	36

b)

Class	Frequency
11.4	8
11.8	6
12.1	7
12.5	9
12.9	4

5. a) Find the median and the mode for the frequency distribution in 4a) above.

    b) Find the median and the mode for the frequency distribution in 4b) above.

6. Given the grouped discrete frequency distribution below, find the:

    a) mean   b) median class   c) modal class.

Class	Frequency
1 – 5	2
6 – 10	7
11 – 15	10
16 – 20	5
21 – 25	1

7. Given the grouped discrete frequency distribution below, find the:

    a) mean   b) median class   c) modal class.

Class	Frequency
12 – 20	5
21 – 29	3
30 – 38	2
39 – 47	4

8. Given the following frequency distribution for continuous data, find the:

    a) mean   b) median class   c) modal class.

Class boundaries	Frequency
3.5 – 6.5	8
6.5 – 9.5	9
9.5 – 12.5	3
12.5 – 15.5	7
15.5 – 18.5	9

9. Given the following frequency distribution for continuous data, find the:
   a) mean
   b) median class
   c) modal class.

Class boundaries	Frequency
25.7 – 28.3	2
28.3 – 31.3	3
31.3 – 34.3	5
34.3 – 37.3	12
37.3 – 40.3	15

11.5 Graphs of continuous data

> **2.4** Cumulative frequency tables for grouped discrete data and for grouped continuous data; cumulative frequency curves, median, and quartiles. Box-and-whisker diagram (box plots).

For continuous data, there are several appropriate diagrams. Some of these are listed below.

- Histogram
- Cumulative frequency graph
- Scatter plot (see Chapter 12)
- Box-and-whisker diagrams (though most often used with discrete data)

To construct a cumulative frequency graph.

- Use the *Class boundaries* and the *Cumulative frequency* (*cf*) columns.
- Plot the ordered pairs (upper class boundary, cf).
- Let the upper class boundaries lie on the cm marks.
- Plot a point on the *x*-axis one class width to the left of the leftmost upper boundary.

- Connect the ordered pairs with straight line segments or with a smooth 'S' style curve.
- **Do not** connect the right side of the graph to the *x*-axis.
- A good *y*-axis scale is 1 cm = 5 divisions or 1 cm = 10 divisions.
- Do not use a squiggle on the *x*-axis unless absolutely necessary.

Cumulative frequency graphs are used to find how many values are below a certain value.

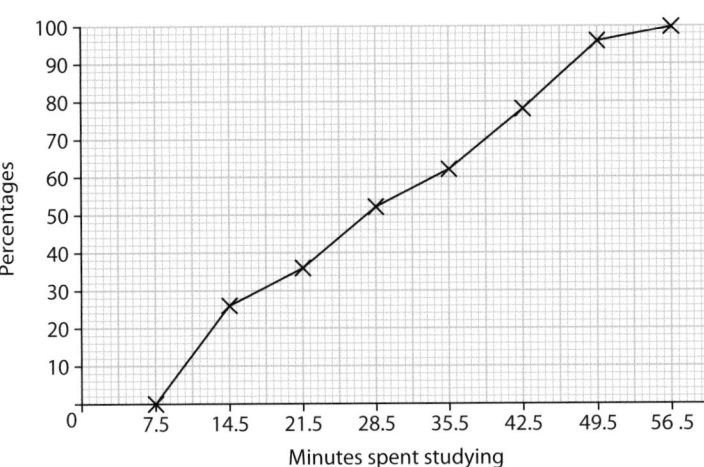

Example 11.10

Using the data organized in the frequency distribution table on page 297, construct each of the following.

a) Histogram
b) Cumulative frequency graph

• **Examiner's hint:** When reading answers from a graph, the IB examiners will always provide some leeway and will sometimes award 'follow through' (ft) marks, depending on the working you show.

Solution

a) Histogram

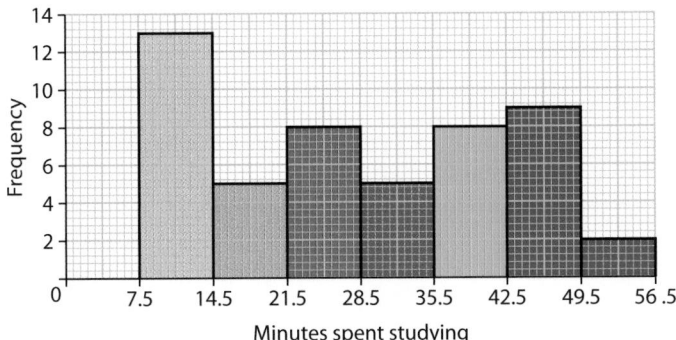

b) Cumulative frequency graph

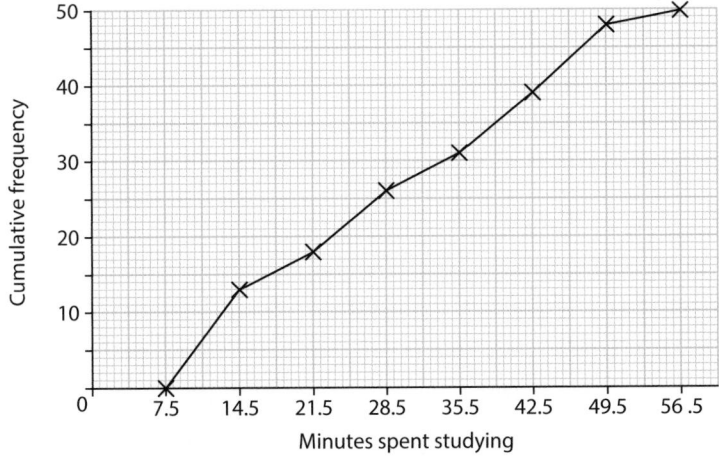

• **Examiner's hint:** Based on how you constructed your graph and how you drew your lines, there will be some leeway in the answer accepted. Answers accepted will most likely be 28 ± 2 minutes.

A cumulative frequency graph can be used to find percentiles or quartiles. Percentiles are used to separate data into $\frac{1}{100}$ or 1% whereas quartiles separate the data into quarters.

Exercise 11.5

1. Data have been collected and a grouped frequency distribution for continuous data has been constructed and partially filled in.
   a) Copy and complete the frequency distribution chart.
   b) Construct each of the following graphs.
      (i) Histogram
      (ii) Cumulative frequency graph
   c) Use your graph to estimate the median score:

Class	Class boundaries	Tally	Freq.	Mid-interval	cf
2 – 4		//			
5 – 7		////			
8 – 10		7HL //			
11 – 13		7HL			
14 – 16		//			

2. Data have been collected and a grouped frequency distribution for continuous data has been constructed and partially filled in.
   a) Copy and complete the frequency distribution chart.
   b) Construct each of the following graphs:
      (i) Histogram
      (ii) Cumulative frequency graph

Class	Class boundaries	Tally	Freq.	Mid-interval	cf
10.5 – 10.9		//			
11.0 – 11.4		7HL			
11.5 – 11.9		7HL 7HL /			
12.0 – 12.4		7HL 7HL 7HL /			
12.5 – 12.9		7HL ////			
13.0 – 13.4		7HL //			

3. Data were collected for the amount of time, to the nearest minute, for teenagers putting a puzzle together.

a) Complete the following table.

Time taken (minutes)	Number of students	Mid interval	Cumulative freq.
$10 \leqslant t < 20$	5		
$20 \leqslant t < 30$	6		
$30 \leqslant t < 40$	10		
$40 \leqslant t < 50$	12		
$50 \leqslant t < 60$	6		
$60 \leqslant t < 70$	5		
$70 \leqslant t < 80$	6		

b) Construct each of the following graphs.

(i) Histogram

(ii) Cumulative frequency curve

c) By drawing lines on your cumulative frequency graph, find:

(i) the total number of teenagers who could put the puzzle together in less than 40 minutes;

(ii) the percentage of teenagers that could put the puzzle together in less than 20 minutes;

(iii) the average number of minutes for 50 percent of the students to solve the puzzle.

4. Data on the number of hours spent by IB students on their homework per week were collected and summarized in the table below.

Time taken (hours)	Number of students	Cumulative freq.
10	15	15
11	11	26
12	5	31
13	7	a
14	25	63
15	11	74
16	19	93
17	17	110
18	13	123
>18	5	b

a) Write down the values of a and b.

b) How many students were surveyed?

c) How many students spent less than 14 hours doing homework?

d) What percentage of students spent more than 16 hours doing homework?

5. The weights of all players on an American football team were compiled and written in the frequency table below.

Weight of players in kg	Frequency	Cumulative frequency
$70 \leqslant w < 80$	1	1
$80 \leqslant w < 90$	14	15
$90 \leqslant w < 100$	14	29
$100 \leqslant w < 110$	12	36
$110 \leqslant w < 120$	16	a
$120 \leqslant w < 130$	3	55
$130 \leqslant w < 140$	11	66
$140 \leqslant w < 150$	7	73
$150 \leqslant w < 160$	1	74

a) (i) Write down the value of a.

   (ii) On graph paper, draw the cumulative frequency curve for these data.
         Use a scale of 1 cm to represent 10 kg on the horizontal axis and 1 cm to
         represent 5 units on the vertical axis. Label the axes clearly.

b) Use the graph to find the median weight of the players.

c) Determine the number of players whose weights are within 20% of the
    median weight.

11.6 Box-and-whisker diagram

2.4	Cumulative frequency tables for grouped discrete data and for grouped continuous data; cumulative frequency curves, median, and quartiles. Box-and-whisker diagram (box plots).

To construct a box-and-whisker diagram we need five pieces of
information,

- lower quartile
- upper quartile
- median
- lowest (minimum) value
- highest (maximum) value

When raw data is arranged
in increasing (or decreasing)
order, the listing is called an
array.

A quartile is a number that sections off 25% of the data. Phrases such
as 'the lower quartile' or 'upper quartile' are often used when verbally
discussing data. When discussing a data set in terms of quartiles, there
are three numbers to consider, since three numbers will separate an array
into four parts. When dividing the array into four equal parts, the three
numbers are called **quartiles**. The numbers that are used as the dividing
lines between the sections are denoted as:

Q_1, Q_2, Q_3, where Q_1 is the upper boundary of the first 25% of the data,

Q_2 is the upper boundary of the second 25% of the data, and

Q_3 is the upper boundary of the third 25% of the data.

An array will either have an odd or an even number of data values. For example, consider the array 2, 5, 11, 15, 17, 22, 30. This array has an odd number of data values.

- The lowest value is 2.
- The middle value is 15. The middle value is called the **median** and it represents Q_2.
- The largest value is 30.

To find the first quartile, Q_1, proceed as follows:

Step 1: List the data in order from lowest to highest as shown above.

Step 2: Find the median. In this example, the median is 15.

Step 3: Find the number that is the median (the middle number) of the data values *below* the median of the entire data array. In this case, the median of the numbers 2, 5 and 11 is 5.

Step 4: 5 is called the **first quartile**. It is denoted Q_1.

Step 5: Find the third quartile by finding the median of the data values that are *above* the median of the entire array. The median of the numbers 17, 22, and 30 is 22. Therefore, 22 is called the **third quartile** and is denoted Q_3.

2	5	11	15	17	22	30
↑	↑		↑		↑	↑
Lowest value	Q_1		Q_2 Median		Q_3	Highest value

The diagram above is often referred to as the 5-number summary of the data array.

When the array has an even number of data values, the procedure for finding the 5-number summary stays the same, except that an even number of data values does not have a data value that represents the median. There are two 'middle numbers'. Consider the array:

1, 4, 7, 13, 16, 20

The lowest (minimum) number is 1.

The highest (maximum) number is 20.

7 and 13 are the two 'middle numbers'. To find the median of the entire data array, find the (mean) average of 7 and 13.

Therefore, **median** $= Q_2 = \dfrac{7 + 13}{2} = 10$.

The data values below 10 are 1, 4 and 7. Therefore, $Q_1 = 4$.

The data values above 10 are 13, 16 and 20. Therefore, $Q_3 = 16$.

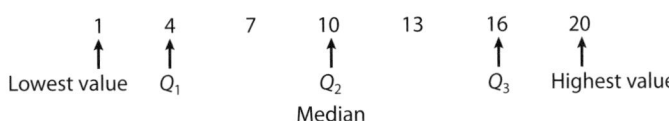

1	4	7	10	13	16	20
↑	↑		↑		↑	↑
Lowest value	Q_1		Q_2 Median		Q_3	Highest value

Example 11.11

Data are collected for the number of text messages that 150 randomly selected college students sent during one day. The results are presented in the frequency table below.

Find the values in the 5-number summary.

Number of text messages	Frequency	Cumulative frequency
8	2	2
9	5	7
10	3	10
11	13	23
12	8	31
13	23	54
14	9	63
15	33	96
16	21	117
17	22	139
18	5	144
19	2	146
20	4	150

Solution

a) The lowest number of text messages (lowest value) is 8.

b) The highest value is 20.

c) The median is 15.

d) Q_1 is halfway between the lowest value and the median. Therefore $Q_1 = 13$.

e) Q_3 is halfway between the median and the highest value. Therefore $Q_3 = 16$.

The diagram that is used to depict the 5-number summary is called a **box-and-whisker diagram** or simply a **box plot**. The rules for constructing a box plot are given below.

Step 1: Draw a number line, preferably using the 2-mm IB graph paper or a number line only, scaled in cm.

Step 2: Scale the number line so that the lowest and highest values are near each end.

Step 3: Locate the lowest value, Q_1, median, Q_3, and the highest value at the appropriate places about 1 cm above the number line.

Step 4: Draw a rectangular box between the first and third quartiles.

Step 5: Draw a vertical line segment in the box through the median.

Step 6: Draw line segments (the whiskers) from the sides of the box to the lowest and highest values.

Example 11.12

Draw a box plot that shows the 5-number summary from Example 11.11.

Solution

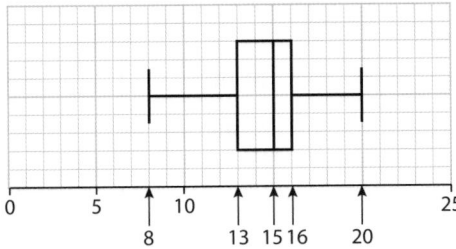

At a glance, the data represented by the box plot can be interpreted in several ways.

- The **range** of the entire data array is $20 - 8 = 12$.
- The range of the lowest 25% of the data is $15 - 8 = 7$. It is relatively small compared to the range of the upper 25% of the data whose range is $20 - 16 = 4$.
- 50% of the data lies between 13 and 16. Since the box is relatively long, $(39 - 14 = 25)$, the data is spread out in that range.
- The second and third 25% sections of data are about the same width, which means that the data is fairly evenly spread out in those ranges.

Data can be entered into your TI calculator using the STAT menu. The following steps and keystrokes will help you learn to use your calculator to enter data and do statistical analysis.

- STAT   This keystroke begins the process of entering data for analysis.

- 5: SetUpEditor, ENTER   These keystrokes reset the EDIT window so that Lists 1 through 6 show. List 1 is denoted as L_1.

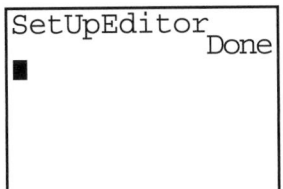

- STAT, Edit (ENTER)   These keystrokes access the window that will allow you to enter data.

- ∧   This keystroke will highlight 'L_1'.

- CLEAR, ∨   These keystrokes will clear all of the previous entries in List 1 (L_1).

- STAT, ENTER, ∧, DEL   These keystrokes will delete not the entries in List 1, but the list itself. If this happens, press STAT, 5, ENTER to reset the lists.

- STAT, ENTER, 8, ENTER

  These keystrokes will enter 8 as a data value into List 1.

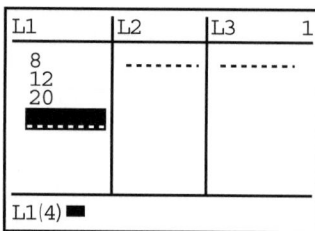

L1	L2	L3	1
8			
12			
20			

L1(4)■

```
EDIT CALC TESTS
1:1-Var Stats
2:2-Var Stats
3:Med-Med
4:LinReg(ax+b)
5:QuadReg
6:CubicReg
7↓QuartReg
```

```
1-Var Stats L1■
```

- 12, ENTER, 20, ENTER These keystrokes continue to enter the data values 12 and 20 into List 1.

- STAT, > (CALC), 1: 1 − Var Stats, ENTER, 2ND, 1 (L1)

  These keystrokes will do the statistical analysis on the data that was entered in L_1. ('1 − Var Stats' means One Variable Statistics).

```
1-Var Stats
 x̄=13.33333333
 Σx=40
 Σx²=608
 Sx=6.110100927
 σx=4.988876516
↓n=3
■
```

STAT, > (CALC), 1: 1 − Var Stats, ENTER, ENTER, will also do the statistical analysis on L_1. When not designated, the TI defaults to List 1 or to L_1, L_2 when accessing 2: 2 − Var Stats.

```
1-Var Stats
↑n=3
 minX=8
 Q₁=8
 Med=12
 Q₃=20
 maxX=20
■
```

It is also possible to graph a box plot with the TI calculator. Using the data from Example 11.11, press the following keystrokes to see the graph.

Instead of clearing each function, you could turn each function **off** by placing the cursor on the '=' sign and pressing ENTER.

- Y = CLEAR (Clear all functions, Y_1, Y_2, etc.)
- STAT, ENTER (Enter the number of text messages data into L_1.)
- STAT, ENTER (Enter the frequency data into L_2) as shown below.

L1	L2	L3	2
8	2		
9	5		
10	3		
11	13		
12	8		
13	23		
14	9		

L2(4)=13

- 2ND, Y = (STAT PLOT)
- ENTER (The above keystrokes access the Plot window above.)
- ENTER (On) (This keystroke turns Plot1 on.)

- $>, >, >, >$, ENTER (These keystrokes select the box plot w/o outliers.)
- , Xlist: L_1 (These keystrokes select List 1 as the data list.)
- $\vee$, Freq: L_2
- ZOOM, 9 (9: ZoomStat)
- TRACE, $<, <, >, >, >, >$ (These keystrokes will show the 5-number summary.)

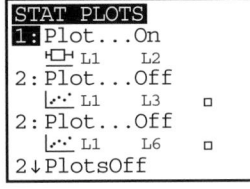  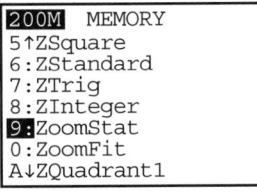 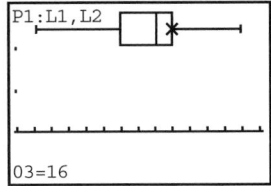

- The graph below shows an example of an outlier as an individual point.

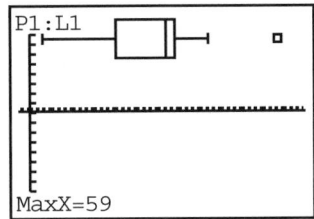

Press Y =, and make sure that 'Plot1' is **not** highlighted when you are ready to graph a function.

Notes on outliers

- In general, an outlier is a data value that seems to be far removed from the rest of the data values.
- A graph of all of the data values, either on a number line or on a coordinate plane, will yield visual clues.
- The statistician may choose to ignore outliers or use them as his/her study might dictate.
- There is no real consensus on how to determine outliers. Two methods are described below.
  - Below or above 1.5(IQR) from Q_1 and Q_3.
  - Greater than 3 standard deviations from the mean. (See Section 11.7).

• **Examiner's hint:** The formula for the Interquartile Range (IQR) is found in Topic 2, page 3 of the Mathematical Studies SL Formula Booklet.

• **Examiner's hint:** The treatment if **outliers** will not be examined but may be incorporated in projects.

Exercise 11.6

1. Use the following data set: 17, 23, 52, 75, 2, 39, 17, 38, 28
   a) Find, without the use of a GDC, the
      (i) 5-number summary
      (ii) interquartile range
   b) Draw the box plot.
   c) Use a GDC to check your answers.

2. Use the following data set: 1, 15, 37, 37, 35, 18, 17, 31, 40, 37, 17, 16
   a) Find, without the use of a GDC, the
      (i) 5-number summary
      (ii) interquartile range
   b) Draw the box plot.
   c) Use a GDC to check your answers.

3. Data were collected showing how many cellphone calls were made each day during the month of January.

a) Construct a box-and-whisker diagram.

```
  6  40  21  25   8  18  13  22  29  29
 38   7  12  30  34  34  42  14  15  18
 27  26  44  37  27  21  43  34  32  27
```

b) Use a GDC to check your answers.

Use the box plot obtained above.

c) Find, without using your GDC, the
- (i)   median
- (ii) mode
- (iii) range
- (iv) interquartile range
- (v)   first quartile
- (vi) third quartile.

4. Data were collected from 40 high school students on how many classroom tests they took in their junior year.

```
 70  81  81  90  72  73  89  86  94  92
 82  75  88  88  95  70  71  74  80  91
 89  75  72  81  74  76  77  81  85  95
 83  82  82  71  77  78  89  91  86  71
```

a) Construct a box-and-whisker diagram using your GDC.

b) Hence, write down
- (i)   median
- (ii) range
- (iii) first quartile
- (iv) third quartile
- (v)   interquartile range.

5. The following discrete frequency distribution reports scores for a high school calculus class. Find the 5-number summary, and the IQR.

Score	Frequency
1	1
2	4
3	7
4	10
5	8

6. The following discrete frequency distribution reports the number of strokes a golfer made during the last 25 rounds of golf she played. Find the 5-number summary, and the IQR.

Score	Frequency
78	2
79	3
80	5
81	7
82	5
84	2
87	1

7. The number of songs found on the newest digital recorder/phone device from a randomly selected group of students is given below.

$$52 \quad 71 \quad 73 \quad 77 \quad 82 \quad 85 \quad 85 \quad 85 \quad 90 \quad 91 \quad 91 \quad 94$$
$$96 \quad 97 \quad 101 \quad 103 \quad 105 \quad 111 \quad 112 \quad 114 \quad 117$$

Find each of the following:
a) mean   b) median   c) mode
d) range   e) first quartile   f) third quartile
g) interquartile range

11.7 Measures of dispersion

2.6 | Measures of dispersion: range, interquartile range, standard deviation.

The basic measures of how data is spread or dispersed are:
- range
- interquartile range
- variance
- standard deviation.

The **range** of a data set is the highest value minus the lowest value.

The **interquartile range** is the difference between the third and first quartile (discussed in Section 11.6).

The **variance** of a data set is the average distance the square of each data value is from the mean.

The **standard deviation** is defined as the square root of the variance. In laymen's terms, a standard deviation is a number that tells you how far from the mean a data value is, with respect to how far the other data values are from the mean.

The standard deviation of a data set can be produced from a process called first principles.

- Find the mean of the data.
- Subtract the mean from each data value. (This step finds the distance each data value is from the mean.)
- Square the differences. (This step exaggerates the distances the data values are from the mean.)
- Find the mean average of the squares. This average is called the variance.
- Take the square root of the variance. This number is called the standard deviation.

● **Examiner's hint:** Even though variance is not a syllabus content item, it is necessary to study this concept prior to studying the concept of standard deviation.

Example 11.13

Consider the following data array: 15, 20, 30, 35. Find the standard deviation using first principles.

Solution

- $\bar{x} = \dfrac{15 + 20 + 30 + 35}{4} = 25$

- $15 - 25 = -10 \qquad (-10)^2 = 100$
  $20 - 25 = -5 \qquad (-5)^2 = 25$
  $30 - 25 = 5 \qquad\qquad 5^2 = 25$
  $35 - 25 = 10 \qquad\quad 10^2 = 100$

- $\dfrac{100 + 25 + 25 + 100}{4} = 62.5$. This is the variance.

- The standard deviation $= \sqrt{62.5} = 7.91$ to 3 significant figures.

7.91 represents the average distance that a data value is from the mean of 25, with respect to the distances the other data values are from 25.

One reason for taking the square root of the variance is that the variance is in terms of 'square units'. For example, if the data values represent ages of trees in years, the variance would be '62.5 square years'! This is clearly a meaningless number. Taking the square root will yield 7.91 years, which is very understandable.

Example 11.14

Using your TI calculator, find the standard deviation of the following data array:

$$5, 10, 10, 15, 15, 20, 30, 35, 35, 40, 40, 45$$

Solution

Press the keys: STAT, Edit (ENTER)

Clear List 1 and enter the data values into L_1.

STAT, > (CALC), 1: 1 Var Stats (ENTER), 2ND L1,∨, (leave FreqList Blank),∨(Calculate),

ENTER (as shown below)

• **Examiner's hint:** Students are expected to be able to find the standard deviation via a GDC.

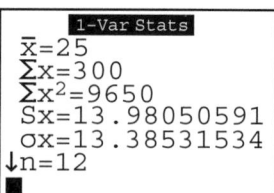

Consider the data sets in Examples 11.13 and 11.14.
- Construct a number line graph for both data sets.

  - Example 11.13 data set:

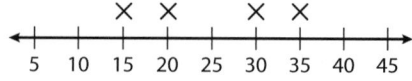

- Example 11.14 data set:

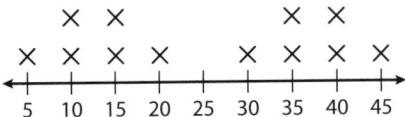

- Notice that the mean, 25, is the same for both data sets.

- The data set for Example 11.13 is grouped about the mean more closely than the data set for Example 11.14.
  - The standard deviation is 7.91 for the data set for Example 11.13.
  - The standard deviation is 13.4 for the data set for Example 11.14.

- The larger the standard deviation, the more the data is spread (dispersed) from the mean. The smaller the standard deviation, the more the data is grouped about the mean.

Notations for standard deviation: s_x vs σ_x

What, if any, are the inherent dangers in collecting and analyzing the statistics of social, political and religious data?

Statistics involves gathering data, analyzing the data, and graphing the results of that analysis.

When performing inferential statistical analysis (see Chapter 12), the statistician will gather data from a small sample of population under study and then based on the analysis of that data he/she will make an inference (a statement) about the population under study.

When data is being collected from a sample, it is customary to use regular, italicized, alphabet letters to indicate the statistic. For example, $\bar{x}$ represents the mean of the sample.

When a number is used to represent the entire population, then the number is called a **parameter**. Greek letters are used to describe a parameter.

There are two symbols that are used to represent the standard deviation.
- The symbol, s, is customarily used to represent the standard deviation of the **sample**.
- The symbol, σ, is customarily used to represent the standard deviation of the **population**.

Using the TI calculator, once the data have been entered in to List 1 (L_1) and the frequencies in List 2 (L_2) then using the TI calculator STAT, CALC, 1, 2nd, 1, ENTER, 2nd, 2, ENTER, σ, ENTER.

- **Examiner's hint:** For example, the mean (μ) of the (entire) population can be represented as $\mu = 56$. μ is the lower case Greek letter 'mu'. It is pronounced 'mew'.

- **Examiner's hint:** Be aware that the population standard deviation is generally unknown, and that the sample standard deviation is an estimate only.

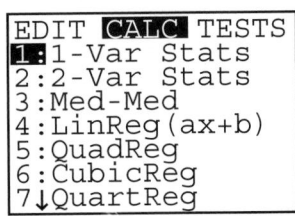

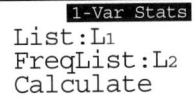

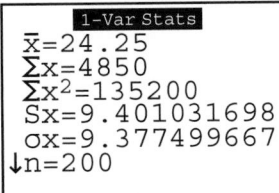

- **Examiner's hint:** In the IB examination questions when using your TI GDC to find the standard deviation choose σ_x from the 1: 1 – Var Stats window.

As you can see, $s_x = 9.401$ and $\sigma x = 9.377$ are both listed as standard deviations. The one that corresponds correctly with the Mathematical Studies course is the population standard deviation, σ.

1. What do each of the following symbols represent in the study of statistics?
   a) $\bar{x}$ b) μ c) s^2
   d) s e) σ^2 f) σ

2. Describe what a standard deviation is.

3. What is another name for the standard deviation squared?

4. Name the three measures of dispersion.

5. Data set A has a mean of 25 and a standard deviation of 7, and data set B has a mean of 25 and a standard deviation of 3. What general statement can you make about the data sets?

6. Use your GDC to find the standard deviation, σ_x, for:
   a) 4, 9, 12, 17, 18
   b) 3, 7, 10, 16

7. Using your GDC find the standard deviation, σ_x by entering the data into List 1 and the corresponding frequencies into List 2 for
   a) 2, 7, 7, 10, 10, 10, 10, 12, 12, 18, 18, 18, 32, 44
   b) 3, 6, 6, 6, 6, 7, 7, 7, 9, 9, 9, 9, 15, 15, 15, 15

8. Ten students were given two tests worth a maximum of 50 points each, one on mathematics and one on English. The table shows the results of the tests for each of the ten students.

Student	A	B	C	D	E	F	G	H	I	J
Mathematics	15	38	47	43	50	42	12	45	42	27
English	28	40	30	48	18	23	46	30	36	50

   a) (i)   Find the mean score for the mathematics test.
      (ii) Find the median score for the mathematics test.
      (iii) Find the standard deviation for the mathematics test.
   b) (i)   Find the mean score for the English test.
      (ii) Find the median score for the English test.
      (iii) Find the standard deviation for the English test.
   c) Comment on your findings.

9. A high school in Australia has students in grades 7 to 12. A random sample of 50 students was asked their age. The raw data were organized into the following frequency table.

Age	13	14	15	16	17	18	19	20
Frequency	1	5	n	12	11	7	3	1

   a) Write down the value of n.
   b) Calculate the mean age of students.
   c) What percentage of the students is older than 17?
   d) Write down the modal age.
   e) Write down the standard deviation.

12 Statistical Applications

Assessment statements

4.1 The normal distribution. The concept of a random variable; of the parameters μ and σ; of the bell shape; the symmetry about $x = \mu$. Diagrammatic representation. Normal probability calculations. Expected value. Inverse normal calculations.

4.2 Bivariate data: the concept of correlation. Scatter diagrams; line of best fit, by eye, passing through the mean point. Pearson's product-moment correlation coefficient, r. Interpretation of positive, zero, and negative, strong or weak correlations.

4.3 The regression line for y on x. Use of the regression line for prediction purposes.

4.4 The χ^2 test for independence: formulation of null and alternative hypotheses; significance levels; contingency tables; expected frequencies; degrees of freedom; p-values.

Overview

By the end of this chapter you will be able to:
- sketch a normal distribution
- find a probability for a normal distribution
- find the expected value for a normal distribution
- understand the difference between descriptive statistics and inferential statistics
- understand the concept of correlation
- interpret the correlation as strong, weak or none
- draw a scatter plot
- draw a line of best fit by eye
- draw a line of best fit through the mean point
- predict values using the graph and the regression equation
- understand the basic concepts of hypothesis testing
- write the null and alternative hypotheses
- find expected frequencies, degrees of freedom
- determine if two variables are independent of each other by performing the chi-square test for independence.

12.1 The normal distribution

4.1 The normal distribution. The concept of a random variable; of the parameters μ and σ; of the bell shape; the symmetry about $x = \mu$. Diagrammatic representation. Normal probability calculations. Expected value. Inverse normal calculations.

Often we collect continuous data, such as personal data including height and weight, data related to time such as travelling time to school. Such data can be graphed in the form of histograms as shown in the following example. Data were collected on the heights of students. The following histogram was drawn using data from the Australian Bureau of Statistics Census at school site (http://www.abs.gov.au/censusatschool)

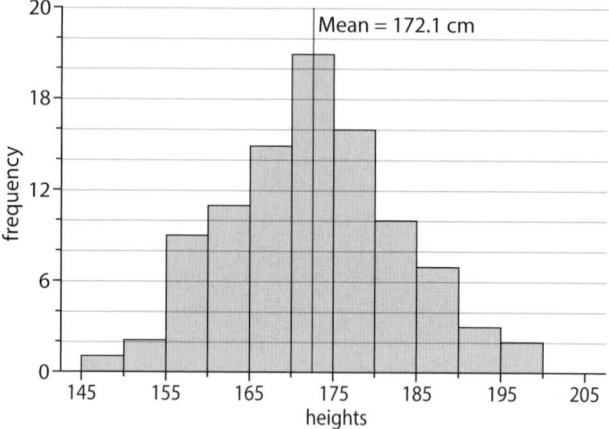

Notice how the data is approximately symmetrical about the mean of 172.1 cm which is indicated by the red line. When the data is distributed in such a way it can be modelled by a normal distribution. The normal distribution is a model for real data and is ideally suited to situations where the data is evenly distributed about the mean (μ) as is the case here. The normal distribution can be overlaid on a histogram as shown below.

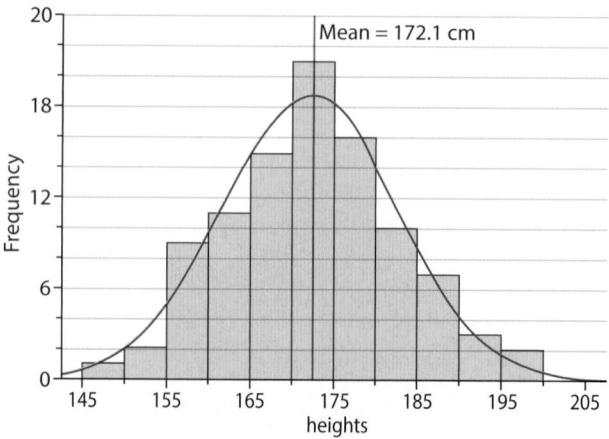

Note how the blue line looks like a bell (which is why the normal distribution is often called the bell shape curve) and that the highest point on the curve coincides with the mean, whereas the ends of the curve are asymptotes to the *x*-axis.

Properties of the normal distribution.

The area under the normal distribution curve is the sum of all the probabilities and thus the total area is 1 or 100%. The curve is symmetrical about the mean and thus the area to the right of the mean (μ) and the area to the left are both equal to 0.5 (50%).

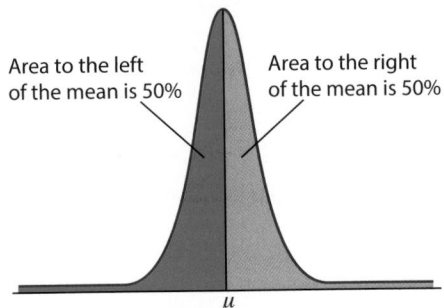

The shape (or flatness) of the curve is determined by the value of σ the standard deviation. If the standard deviation is small then the curve will be very narrow with steep sides, but as σ becomes larger then the curve flattens out as shown in the figure below.

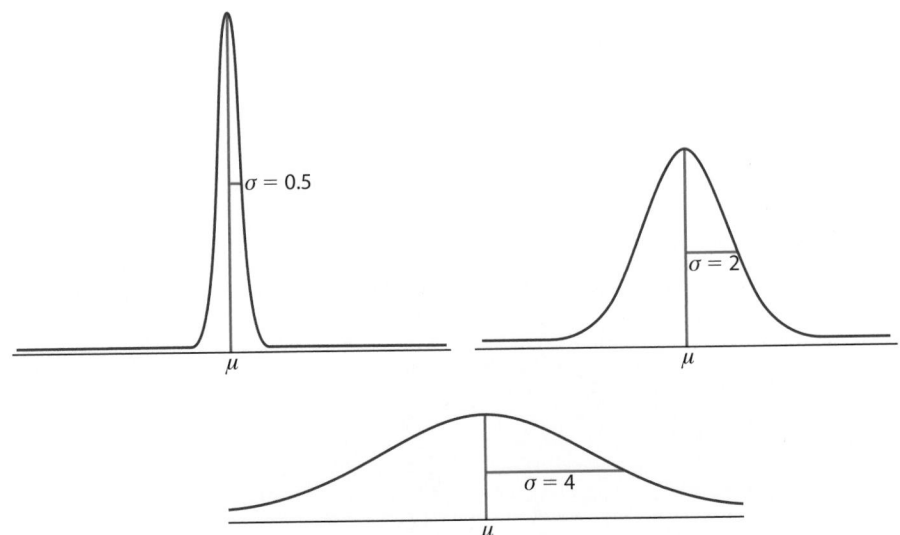

As we have already indicated the standard deviation (σ) plays an important role in determining the flatness of the bell shaped curve. Because the curve is symmetrical the standard deviation is the same on both sides of the mean as shown on the right.

For the normal distribution with mean (μ) and standard deviation (σ):

approximately 68% of the observations fall within 1 standard deviation, σ, of the mean, μ:

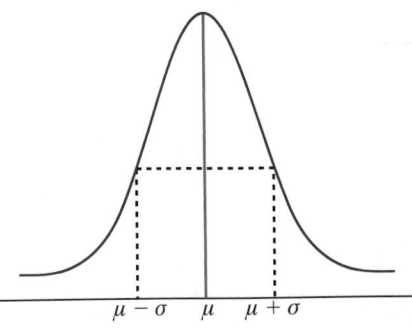

approximately 95% of the observations fall within 2 standard deviations, σ, of the mean, μ:

approximately 99% of the observations fall within 3 standard deviations, σ, of the mean, μ.

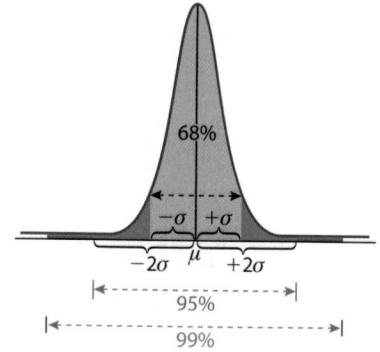

● **Examiner's hint:** Students will be expected to use the GDC when calculating probabilities.

We can use the normal distribution to calculate probabilities of events that can be modelled by the normal distribution.

Example 12.1

The maximum daily temperatures in June in any year approximate to a normally distributed random variable with a mean of 18 °C and a standard deviation of 2 °C.

a)   During what percentage of days in June will the temperature range between 14 °C and 22 °C?

b)   What percentage of days in June will have a temperature less than 20 °C?

c)   What percentage of days in June will have a temperature greater than 21 °C?

Solution

a)   Sketch and shade the region on a normal curve.

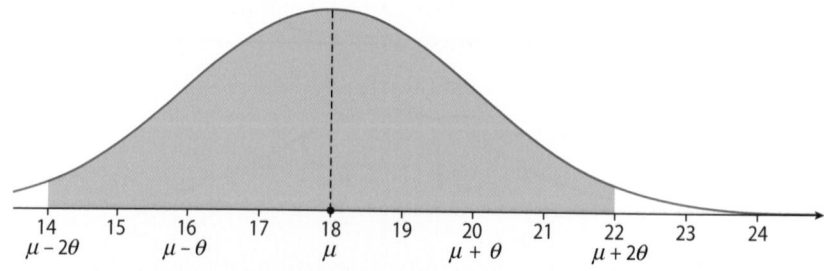

The temperatures between 14 °C and 22 °C lie within two standard deviations of the mean $(14 = 18 - 2 \times 2$ and $22 = 18 + 2 \times 2)$.

So 95% of the daily maximum temperatures will lie between 14 °C and 22 °C.

b)   For this question we are interested in temperatures less than 20 °C. Graphically this is shown below.

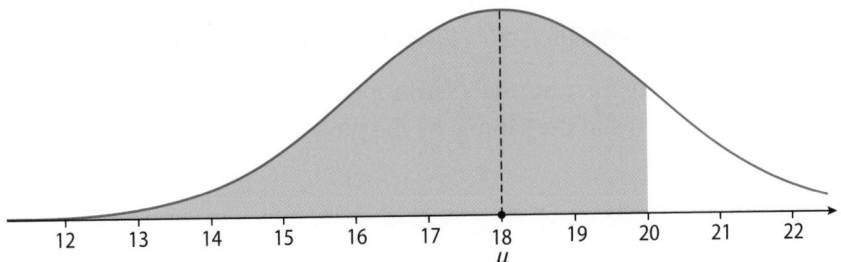

As 20 °C is 1 standard deviation above the mean, then we can calculate the percentage by recognizing that:

50% of all observations are below the mean; between 18 and 20 is half of one standard deviation, which is $\frac{1}{2} \times 68\% = 34\%$; the percentage of days where the temperature is below 20 °C is 50% + 34% = 84%.

c) Graphically we are interested in the area for temperatures greater than 23 °C.

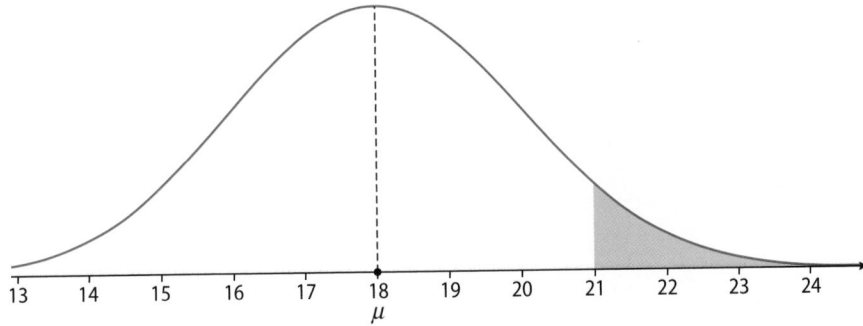

As 21 °C is not a whole multiple of the standard deviation away from the mean, that is 21 − 18 = 3 and the standard deviation is 2, then we can use the GDC as follows.

Step 1. Press 2nd, VARS (DISTR), > (DRAW), 1:ShadeNorm(, ENTER, lower: 21, upper:1E99 (1, 2nd, , , 99), μ = 18, σ = 2, ∨, ENTER

Notice here that we put 1E99 as the upper limit. You can put a number as an upper limit far enough from the mean to make sure you are receiving the correct cumulative distribution, 1E99 (and −1E99) are very large (or very small numbers) and can be used for all calculations.

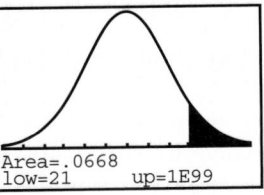

The resulting normal curve is displayed with an area of 0.0668. In other words, approximately 6.68% of days in June will have a maximum temperature greater than 21 °C.

Expected value

The expected value for a normal distribution is calculated by multiplying the number of items in the sample by the probability.

Example 12.2

The distribution of delivery times for a speedy courier service in a major city is approximately normally distributed with a mean of 45 minutes and standard deviation of 10 minutes.

a) What percentage of delivery times will be between 30 and 70 minutes?

b) The number of deliveries in any day is 300. Calculate the number of deliveries that will have a time between 30 and 70 minutes.

Solution

a) As both 30 and 70 cannot be written as the mean ± a whole multiple of the standard deviation we will need to use the GDC. Firstly, sketch the normal curve.

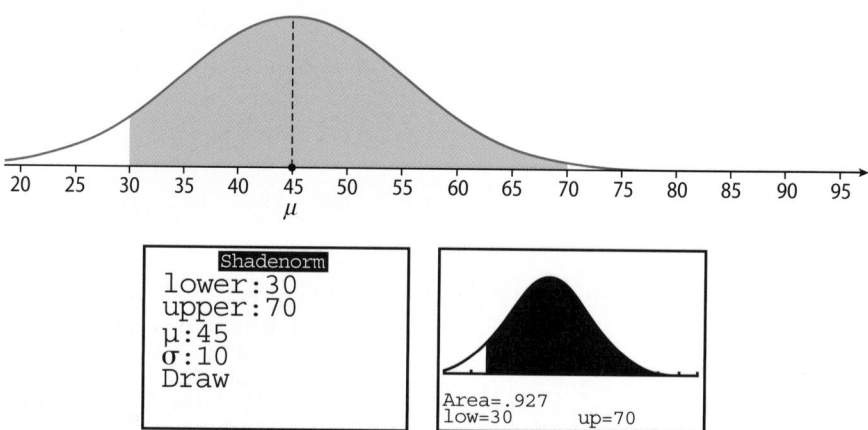

92.7% of all deliveries will occur between 30 and 70 minutes.

b) The expected number of deliveries made between 30 and 70 minutes is $0.927 \times 300 = 278.1$.

The expected number of deliveries is 278.

The inverse normal distribution

Sometimes you are aware of the percentage area (or probability) and are required to determine the value in the data that has this cumulative probability.

Example 12.3

A factory produces packets of coffee. The weights of the packets of coffee are normally distributed with a mean of 500 g and a standard deviation of 3 g. It is known that 5% of the packets contain less than x g. Find the value of x.

Solution

Step 1. Sketch diagram

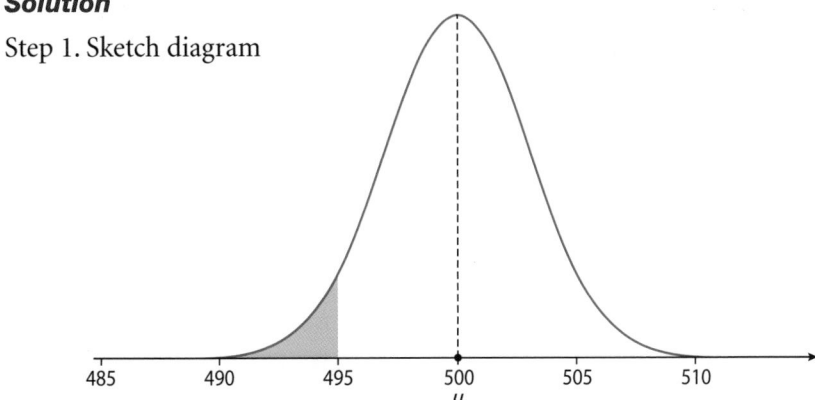

Step 2: Press 2nd, VARS (DISTR), ∨, 3:invNorm(0.05 (5%), $\mu = 500$, $\sigma = 500$, ∨, ENTER, ENTER

$x = 495$ (3 sf)

In other words, 5% of packets of coffee weigh less than 495 grams.

Exercise 12.1

1. A random variable X is normally distributed with a mean of 185 and a standard deviation of 15.

   a)   Calculate the probability that X is less than 175.

   b)   Calculate the probability that X is greater than 190.

   c)   Calculate the probability that X is between 175 and 198.

2. A random variable X is normally distributed with a mean of 16 and a standard deviation of 3.

   a)   Sketch a normal distribution diagram to represent this information, indicating the mean, and the location of 1, 2, and 3 standard deviations from the mean.

   b)   Find the probability that X is less than 19.

   c)   Calculate the probability that X is greater than 10.

   d)   Calculate the probability that X is between 10 and 19.

3. A random variable X is normally distributed with a mean of 37 and a standard deviation of 4.

a) Sketch a normal distribution diagram illustrating the information and shade the area represented by $P(X > 41)$.

b) Write down $P(X > 41)$.

c) Sketch a normal distribution diagram illustrating the information and shade the area represented by $P(32 < X < 43)$.

d) Calculate the probability $P(32 < X < 43)$.

4. The heights of 200 sixteen-year-old female students are normally distributed with a mean of 168 cm and a standard deviation of 10 cm.

a) Sketch a normal distribution diagram illustrating the information and shade the area represented by $P(155 < X < 175)$.

b) Calculate the percentage of females who have heights between 155 cm and 175 cm.

c) The school's female basketball team is looking for students who are taller than 175 cm. What is the expected number of students who are greater than 175 cm?

5. The working life of the battery in a mobile phone is normally distributed with mean 12 000 hours and standard deviation of 1500 hours.

a) The average mobile phone operates 24 hours a day, 365 days a year. What percentage of batteries will need to be replaced at the end of the first year?

b) It is known that 5% of the batteries have a lifetime of n hours (ie they cannot be recharged after n hours). Find the value of n.

6. An urban highway has a speed limit of 50 kmh^{-1}. It is known that the speeds of vehicles travelling on the highway are normally distributed, with a mean of 45 kmh^{-1} and a standard deviation of l0 kmh^{-1}.

a) Draw a normal diagram to represent this information and shade area for vehicles exceeding the speed limit, i.e. speeds greater than 50 kmh^{-1}.

b) What percentage of vehicles are exceeding the speed limit?

c) The authorities are also concerned about slow-moving vehicles. It has been found that 10% of vehicles are driving too slowly. At what speeds are these vehicles travelling?

7. Bags of sugar are labelled 1 kg. The bags are filled by machine and the actual weights are normally distributed with mean 1.05 kg and standard deviation 0.04 kg.

a) What is the probability a bag selected at random will weigh less than 1 kg?

In order to reduce the number of underweight bags (bags weighing less than 1 kg) the mean was increased to 1.1 kg, without changing the standard deviation.

b) What is the new percentage of bags that weigh less than1 kg?

c) If a shop purchases 1000 bags of sugar after the machine has been adjusted, how many bags will weigh under 1 kg?

8. The reaction times to catch a falling object for a group of male students were measured and the data were found to be normally distributed with a mean of 0.334 seconds and a standard deviation of 0.084 seconds.

a) Draw a normal distribution diagram to illustrate this information.

b) A researcher has suggested that persons with reaction times of less than 0.20 seconds have the potential to be successful at computer games. Show this information on the diagram, in part a).

c) What percentage of students have reaction times less than 0.20 seconds.

9. The mass of packets of a breakfast cereal is normally distributed with a mean of 500 g and standard deviation of 15 g.

a) Find the probability that a packet chosen at random has a mass:

    i) less than 490 g;

    ii) more than 520 g;

    iii) between 490 g and 520 g.

b) If the mass of 5% of the packets is less than w g, then a batch is recalled. Find the value of w.

10. A ballpoint pen manufacturer suggests that the distance their pens will write is normally distributed with a mean of 2500 metres and a standard deviation of 200 metres.

a) What proportion of the company's pens last for:

    i) less than 2100 metres?

    ii) more than 2300 metres?

    iii) between 2000 and 3000 metres?

b) The manager of the company wishes to claim that 90% of the pens can write more than x metres. What should the value x be?

c) If 1000 pens are sold by a local store, what is the expected number of pens that will last less than 2000 metres?

12.2 Line of best fit (the regression line)

4.2	Bivariate data: the concept of correlation. Scatter diagrams; line of best fit, by eye, passing through the mean point. Pearson's product-moment correlation coefficient, r. Interpretation of positive, zero, and negative, strong or weak correlations.
4.3	The regression line for y on x. Use of the regression line for prediction purposes.

Chapter 11 explored the branch of statistics called descriptive statistics. This chapter will study the branch called inferential statistics. As the word **inferential** implies, conclusions about data that have *not* been collected are *inferred* from data that have been collected. In this branch of statistics, conjectures (statements thought to be true) *cannot be proved*. Conjectures can be accepted to a high degree of probability, but never as a certainty, since not all data has been collected.

Suppose that a manufacturer of youth football equipment needs to know the average weight of all 12-year-old boys in Italy. It would be impractical, if not impossible, to weigh all of the 12-year-old boys. The manufacturer can still find the answer to his question by selecting a sample of 12-year-old boys from across Italy and inferring the weight of all 12-year-old boys.

As another example, suppose a statistician wants to know if there is a correlation between PSAT maths scores and the Maths Studies IB exam scores. If there is a correlation, then he/she will be able to set up a probability model that will determine student improvement by teacher. If there is no correlation, then he/she will have to look elsewhere to determine student improvement. It would, again, be impractical to collect and use all of the PSAT and Maths Studies exam scores that have ever been given. However, the statistician can still be reasonably certain of the answer by collecting and analyzing a sample of those scores.

To compare the correlation between PSAT maths scores and Maths Studies IB scores we can use a type of inferential statistics called **regression analysis**. There is much theory which underlies the concept, most of which is beyond the scope of this course. The following example will show how the process works and the discussion that follows will refine the concept specific to the IB syllabus content.

When we graph a data set, such as the heights and weights of Olympic athletes at the 2012 Olympic games in London (data ref http://www.bbc.co.uk/news/uk–19050139), we will obtain a scatter plot of the data, such as the one below.

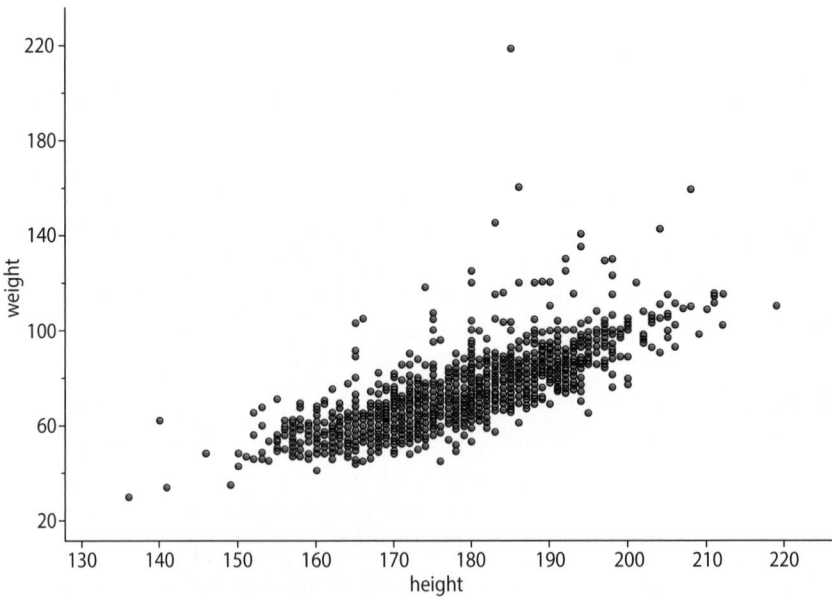

It is evident when looking at the data that many of the data points are clustered together and that the data points approximate to a straight line. That is, the weights of the athletes are dependent on their heights, where height is the independent variable and weight is the dependent variable.

The closer the points in a scatter plot fit a straight line the stronger the relationship between the dependent and independent variables.

It is possible to get a feel for the strength of the relationship between the dependent and independent variable by looking at the graphs. However, to obtain a more accurate measure of the correlation between the variables we need to calculate the correlation coefficient (also known as Pearson's product moment correlation coefficient, or r value). Later in the chapter we will calculate the r value but for now we will look at some scatter plots to see the different types of correlation.

Positive or negative correlation

In the same way that we can have a positive or negative straight line graph we can also have positive or negative correlation as shown in the following diagrams.

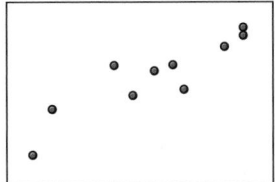

Positive correlation   Negative correlation

When looking at scatter plots there can be no correlation, weak, moderate, or strong correlation. Examples of this appear in the following diagrams. Note that both positive and negative correlations can occur.

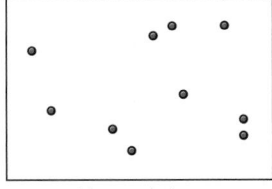 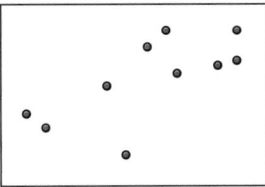 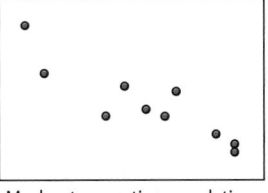 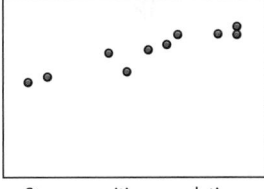

No correlation   Weak positive correlation   Moderate negative correlation   Strong positive correlation

The graphing calculator can allow you to determine a linear equation for the set of data and will also provide you with a measure of the strength of the correlation between the variables in the form of the r value. The value r is known as Pearson's product-moment correlation coefficient and can be used to measure the *strength* of the *linear* relationship between two variables. The r-values lay between -1 and $+1$.

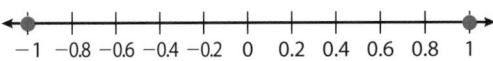

The closer to $+1$ the r-value is, the stronger the (positive) correlation between the variables.

The closer to 0 the r-value is, the weaker the correlation is.

The closer to -1 the r-value is, the stronger the (negative) correlation between the variables.

Strength of the correlation and *r* values

In summary, there exists a:

- strong positive relationship, when *r* is between 0.75 and 1
- moderate positive relationship, when *r* is between 0.50 and 0.74
- weak positive relationship, when *r* is between 0.25 and 0.49
- a very weak or no relationship, when *r* is between –0.24 and 0.24
- weak negative relationship, when *r* is between –0.25 and –0.49
- moderate negative relationship, when *r* is between –0.50 and –0.74
- strong negative relationship, when *r* is between –0.75 and –1

Correlation between two variables means that there is some association between them. However, correlation does not imply causation. In other words correlation indicates that two variables are related but if one variable changes it does not mean that the other changes in the same manner. For example, it may be found that there is a strong correlation between the number of flowers in the garden and the number of birds in a garden. But it does not necessarily mean that a decrease in flowers in a garden will cause a decrease in the number of birds in a garden. However, causal relationships can exist, such as the number of portable tablet computers sold and the average household income in a particular country. Proving causation is beyond the scope of this course.

Linear regression

The following example will demonstrate a process called **linear regression**. It will show how to draw the line (the line of best fit) that best represents a set of points. It will show how to determine if there is a correlation between two variables, and it will then demonstrate how to make inferences (called predictions) about data that was not collected.

To view different correlations and their respective coefficients, visit www. pearsonhotlinks.co.uk, enter the ISBN for this book and click on weblink 12.1.

Example 12.4

A statistician wants to know if there is a correlation between PSAT math scores and the Math Studies IB exam scores. She collected the following data from 10 randomly selected students.

Test	Student selected									
	1	2	3	4	5	6	7	8	9	10
PSAT	52	65	74	72	53	61	66	75	58	52
IB	5	5	6	7	4	4	6	7	5	2

a) Is there a correlation between PSAT and IB math test scores?
b) What kind of correlation is it?
c) Draw the scatter plot showing the data collected.
d) Draw the line of best fit.
e) If a PSAT score is 57, predict the corresponding IB score.
f) If an IB score was a 2, predict the corresponding PSAT score.

Solution

Before being able to answer the questions we need to enter the data into the GDC using the following steps.

Step 1: Press the Mode button on your GDC. Scroll down to the second page and ensure that Stats Diagnostics is ON by moving the cursor to it and press ENTER.

Step 2: Enter the data for the scores in the PSAT row into List 1 (L_1) in your GDC.

Step 3: Enter the data for the scores in the IB row into List 2 (L_2) in your GDC.

Step 4: Press the keys: STAT, > (CALC), 4:LinReg(ax + b), ENTER

Step 5: Ensure that L_1 and L_2 are entered in Xlist and Ylist respectively and scroll down to StoreRegeq, then press APLHA TRACE to select where the regression equation is stored, usually Y1.

Step 6: Scroll down to highlight CALCULATE and press ENTER

Step 7: The resulting linear equation is given along with the r and r^2 values.

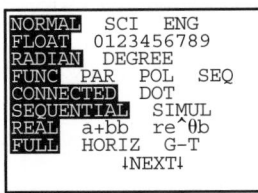

a) Having entered the data and found that the r value is 0.8195301899, we can say that the r value is close to one and thus there is a correlation between PSAT and IB scores.

b) Since the r-value, 0.8195, is fairly close to 1, we can say that there is a **strong positive correlation** between the PSAT and IB math scores. In other words, the higher the PSAT score the higher the IB score, and vice versa.

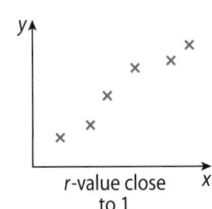

r-value close to 1

c) To draw the scatter plot, plot the ordered pairs (PSAT, IB) on graph paper.
- As has been suggested before, use of the 2-mm IB graph paper is recommended.
- Use a scale of 1 cm = 1 unit whenever possible.
  - In this example, 1 cm = 1 unit is not possible on the x-axis, since the PSAT scores range from 52 to 75.

- **Examiner's hint:** Terms such as 'strong positive' or 'moderately strong negative' are expected to be used.

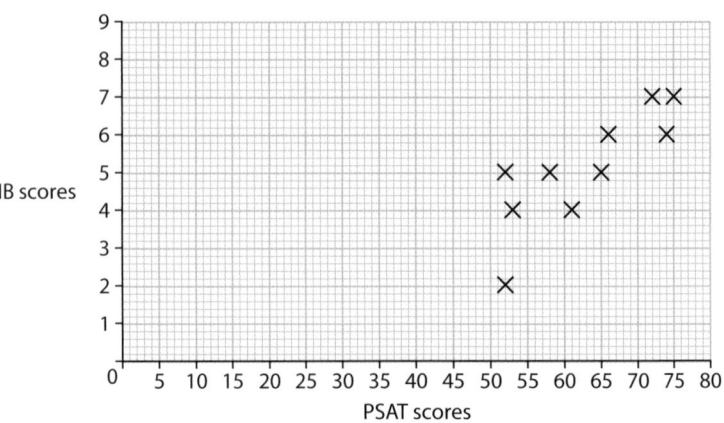

The line of best fit is also called the **regression line**, or the **regression line for *y* on *x***.

d) The line of best fit is a straight line that best represents the ordered pairs plotted.

- The less scattered the points are, the closer to the line of best fit they will be and the stronger the *r*-value (towards 1) will be.
- The more scattered the points are, the farther away from the line of best fit they will be and the weaker the *r*-value (towards 0) will be.
- If the points are scattered in a circular grouping (like darts, missing the bull's eye, but scattered all around it), then it will be not be possible to draw a line of best fit. The *r*-value will be very close to 0.

Since the *r*-value (0.8195) is close to 1, we can correctly draw a line of best fit. The following steps describe the procedure.

Step1: Plot $(\bar{x}, \bar{y})$. The line of best fit *must* pass through this mean point!

- Press the keys: STAT, > (CALC), 2: 2 – Var Stats. Ensure that correct lists are entered for XList and YList. In this example the data is in L1 and L2. Scroll down to Calculate and press ENTER.
- Read $\bar{x}$ = 62.8. This is the mean of the PSAT scores.
- Press ∨, and find $\bar{y}$ = 5.1. This is the mean of the IB scores.
- Plot and label $(\bar{x}, \bar{y}) = (62.8, 5.1)$.

```
EDIT CALC TESTS
1:1-Var Stats
2:2-Var Stats
3:Med-Med
4:LinReg(ax+b)
5:QuadReg
6:CubicReg
7↓QuartReg
```

```
2-Var Stats
Xlist:L₁
Ylist:L₂
FreqList:
Calculate
```

```
2-Var Stats
x̄=62.8
Σx=628
Σx²=40168
Sx=9.003702942
σx=8.541662602
↓n=10
```

```
2-Var Stats
↑ȳ=5.1
Σy=51
Σy²=281
Sy=1.523883927
σx=1.445683229
↓Σxy=3304
```

Step 2: Find another point that lies on the line of best fit.

- As we have saved the regression equation to Y1 then we can use the table of values to find another point.
- Press 2nd TABLE, and there will be a table of values for the equation in Y1. But, the interval between the values may not be appropriate so PRESS the + button and the symbol ΔTbl will appear. Choose a more appropriate step for the function, in this example, 5 may be more appropriate. Type 5 and press

ENTER, and a new table will appear.

X	Y1	
0	-3.611	
1	-3.472	
2	-3.333	
3	-3.195	
4	-3.056	
5	-2.917	
6	-2.779	
Press + for △Tbl		

X	Y1	
0	-3.611	
1	-3.472	
2	-3.333	
3	-3.195	
4	-3.056	
5	-2.917	
6	-2.779	
△Tbl=1		

X	Y1	
0	-3.611	
5	-2.917	
10	-2.224	
15	-1.53	
20	-.8366	
25	-.1431	
30	.55044	
X=0		

X	Y1	
15	-1.53	
20	-.8366	
25	-.1431	
30	.55044	
35	1.244	
40	1.9375	
45	2.631	
X=45		

- Scroll down to a value for X (PSAT scores) that is well away from $\bar{x} = 62.8$ e.g 45 and write down the ordered pair, (45, 2.631).
- Plot the ordered pair (45, 2.6).

Connect (62.8, 5.1) and that point (45, 2.6).

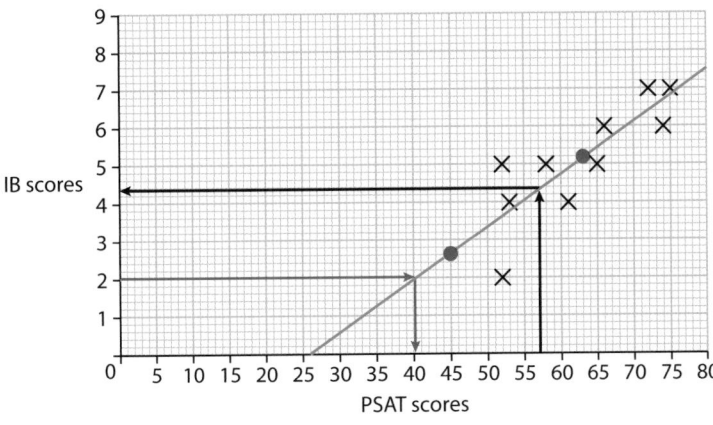

◀ **Figure 12.1** This is the only correct line of best fit.

e) Draw a line from the PSAT score of 57 to the line of best fit and then to IB scores axis.
The IB score is about 4.2. However, IB scores are discrete and therefore the closest IB score is 4.

f) Draw a line from the IB score of 2 to the line of best fit and then to the PSAT axis.
The PSAT score appears to be closest to 40.

 Mathematical work on regression analysis is credited to the French mathematician Adrien-Marie Legendre (1805) and the German mathematician Carl Friedrich Gauss.

Using your GDC to check your work

The TI calculator can graph both the scatter plot and the line of best fit. It is important to be able to draw the scatter plot and the line of best fit on graph paper, but it is also important to be able to use your GDC to produce the results. Using the TRACE key will enable you to check your work, or perhaps determine an answer when time is running short during your IB examination.

Follow the steps below and review the screenshots to gain further understanding.

Step 1: Enter the data into List 1 and List 2 as described before.

Step 2: Find the a and b values as well as storing the regression equation as described before into Y_1.

Step 3: Now press the following keys:
- 2ND, Y=
- ENTER (1: Plot1…Off)
  Pressing ENTER accesses the statistical plot window.
- ENTER (This turns Plot 1 on.)

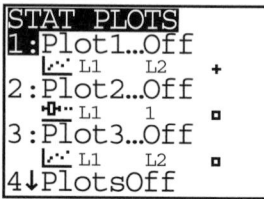
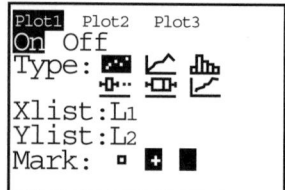

- ∨, ENTER (This selects the scatter plot graph.)
- Xlist: L_1 (Make sure that L_1 is displayed, since the data for the PSAT scores was entered in List 1.)
- Ylist: L_2 (Make sure that L_2 is displayed, since the data for the IB scores was entered in List 2.)

∨, >, ENTER (This selects the '+' as the mark used to show the points on the scatter plot.)

- ZOOM, 9 (for 9: ZoomStat, this will graph the scatter plot and the correct line of best fit.)

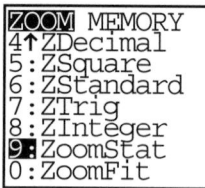
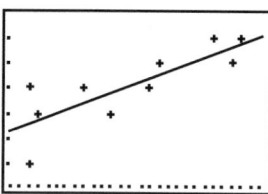
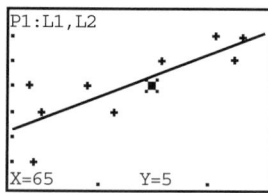

- TRACE, >, >, >, … (This causes the cursor to cycle through each ordered pair in the scatter plot in the order entered in L_1 and L_2. When you get to the 10th ordered pair, you will have to press <, < , … in order to cycle back through the set of ordered pairs.)
- Press either the ∧ or ∨ to alternate between the scatter plots and the line of best fit.
- When the cursor is on the line of best fit, press the < or > to make the cursor move along the line of best fit.
- Once the cursor is on the line of best fit type 57 then press ENTER and the cursor will move to the value $x = 57$ enabling you to verify that the y-value (approximately 4.3) is the same value as obtained from the graph in Figure 12.1.

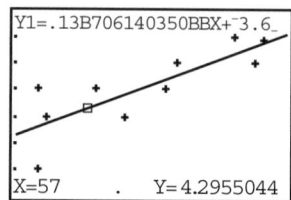

- In order to verify the x-value, of approximately 40, you must change the window so that the cursor can move left of 49.7.
- Press WINDOW, and change the values as shown below.

- Press GRAPH
- Move the cursor so that $y \approx 2$ and verify that the x-value, 40, is the same value as obtained from the graph in Figure 12.1.

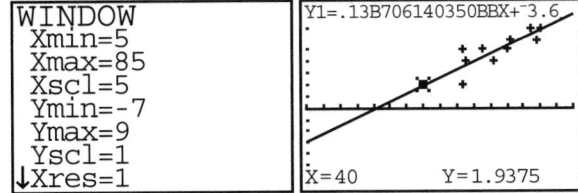

```
WINDOW
 Xmin=5
 Xmax=85
 Xscl=5
 Ymin=-7
 Ymax=9
 Yscl=1
↓Xres=1
```

```
Y1=.13B706140350BBX+-3.6
                    +  +  +
              +  +  +  + +
          +   +
      +  +
  *       +
X=40          Y=1.9375
```

Additional notes tying together the concepts and techniques

Listed below are a few notes that will help tie together some loose ends and round out the discussion about the concepts surrounding the line of best fit.

- The graph of the line of best fit is usually only drawn in the first quadrant, since it doesn't make any sense to talk about a negative PSAT or IB score.
- One of the easiest ways to draw the correct line of best fit is to plot $(\bar{x}, \bar{y})$ and the y-intercept, i.e. $(62.8, 5.1)$ and $(0, -3.6)$. Although this is faster, sometimes the y-intercept might be too large to fit on the graph.
- Predicting values that lie **inside** the data range is called **interpolation**. Predicting values is safest in this range since there are other data values on either side.
- Predicting values that lie **outside** the data range is called **extrapolation**. The farther the number, for which you want a predicted value, is from the minimum or maximum data value, the greater the chance that your prediction will be meaningless.
  For example, there is a well-known correlation between a person's age and his/her height. It is a strong positive correlation; the older you get, the taller you are. This correlation holds true through middle to late teenage years (for most people). If you were to try and predict a person's height when he/she was 35 years old, the prediction model would clearly show absurd information (i.e. 10 feet tall!).
- If there is no correlation, a line of best fit *cannot* be constructed.
- If the r-value is questionable, 0.5 or so, then it takes a hypothesis test in order to determine if there is a correlation. That type of hypothesis testing is beyond the scope of this course.

Example 12.5

The mean point $(\bar{x}, \bar{y})$ of a set of data is $(14.4, 35.2)$ and the gradient of the regression line is 0.771. Write the equation of the regression line in $y = ax + b$ form where $a, b \in \mathbb{R}$.

Solution

Gradient $a = 0.771 = \dfrac{y - 35.2}{x - 14.4}$

$$y - 35.2 = 0.771(x - 14.4)$$
$$y - 35.2 = 0.771x - 11.1024$$
$$y = 0.771x - 11.1024 + 35.2$$
$$y = 0.771x + 24.1$$

Examples of scatter plot visuals

Example 12.6

Find each r-value and interpret the correlation using words such as positive, negative, none, strong, moderate, and/or weak.

a)
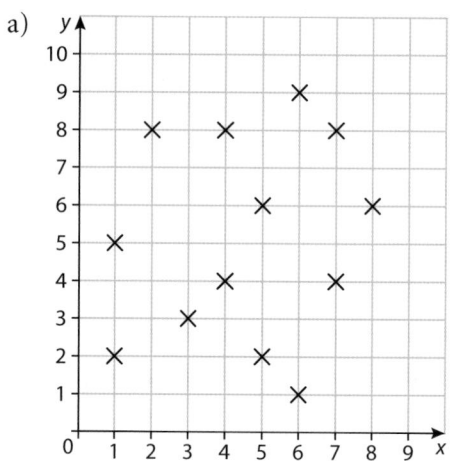

b)
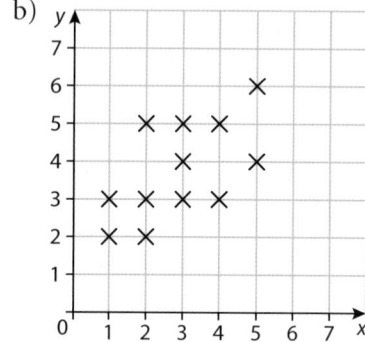

c)
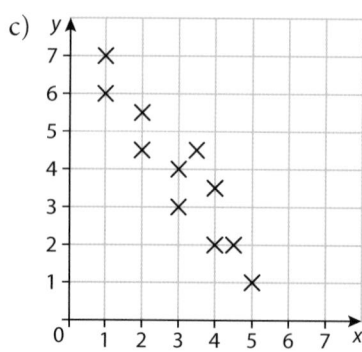

Solution

For each scatter plot, enter all of the ordered pairs into List 1 and List 2. Use LinReg from the STAT, CALC window and note each respective r-value.

a) $r = 0.170$ to 3 s.f. This r-value is very close to 0; therefore, there is **no (linear) correlation** between the variables. Note that the points are scattered in an almost circular pattern.

b) $r = 0.601$ to 3 s.f. This r-value is a **moderate positive correlation** between the variables. Note that the points are scattered in more of a linear pattern.

c) $r = -0.928$ to 3 s.f. This r-value is a **strong negative correlation** between the variables. Note that the points are very nearly in a straight line whose gradient is negative.

A summary

A summary of the steps needed to do a complete linear regression problem is given below.

Step 1: Enter the data into List 1 and List 2.

Step 2: Determine the r-value by accessing LinReg $(ax + b)$.

a) If the r-value is approximately 0.50 or greater (or between -0.5 and -1), there is (probably) a linear correlation between the variables and a line of best fit can be constructed.

b) If the r-value is between -0.5 and $+0.5$, there is (probably) no linear correlation between the variables and a line of best fit **cannot** be constructed.

c) The following steps assume that the r-value is large enough that a line of best fit can be constructed.

Step 1: Find and plot $(\bar{x}, \bar{y})$ by accessing 2 – Var Stats.

Step 2: Find the equation of the line of best fit by accessing LinReg $(ax + b)$.

Step 3: Ensure that the equation for the line of best fit is stored in Y_1.

Step 4: Find and plot another ordered pair that lies on the line of best fit by using the TABLE window.

Step 5: Plot that point found in Step 4.

Step 6: Draw the line of best fit through $(\bar{x}, \bar{y})$ and the point found in Step 4.

Step 7: Draw lines on your graph, from either axis to the line of best fit, to predict values.

Step 8: Use your GDC, the STAT PLOT window and ZOOM 9, to check your work.

The r-value (0.50) given in a) and b) right is only an approximation used to determine whether there is linear correlation or not. A hypothesis test must be run in order to see if there is a linear correlation between the variables. That test is not in the IB curriculum. However, the concept will be touched on in Section 12.3.

• **Examiner's hint:** In regression line problems set in the IB examinations, the r-values will be unambiguous.

Example 12.7

A Mathematical Studies student collected data to determine if there is a correlation between the age of a high school student and the number of hours of homework he/she did per week.

The results from 10 randomly selected students are shown below.

Age (x)	13	16	18	14	17	18	16	17	14	14
Hours (y)	14	12	4	9	9	9	7	6	13	10

1. Draw a scatter plot.

2. Determine the *r*-value.

3. Interpret the *r*-value as it relates to the relationship between the variables.

4. Determine the equation of the regression line.

5. Construct a regression line if the *r*-value suggests a correlation between the variables.

6. By drawing lines on your graph predict
   a) the number of hours, to the nearest hour, a 15-year-old student will study each week
   b) the age, to the nearest half-year, of a student who studies 2 hours per week. Comment on this answer.

7. Use the equation of the regression line and predict
   a) the number of hours, to the nearest hour, a 13.5-year-old student will study each week
   b) the age, to the nearest half-year, of a student who studies 16 hours per week. Comment on this answer.

Solution

1.

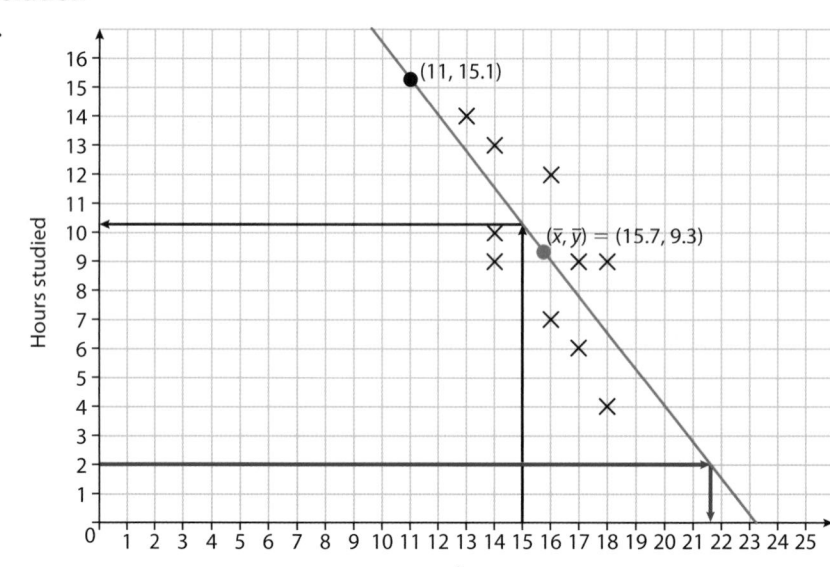

2. STAT, EDIT, now enter the values into List 1 and List 2, STAT, CALC, LinReg, ENTER.

∴ $r = -0.720$ to 3 s.f.

3. There is a moderately strong negative correlation between the age of a high school student and the number of hours he/she studies per week.

4. STAT, CALC, LinReg, L1, L2, Y1, CALCULATE.

∴ $y = -1.232\,56x + 28.6512$

Hence, $y = -1.23x + 28.7$ to 3 s.f.

5. See graph in 1.

6. a) See the graph. Therefore, the number of hours equals 10 to the nearest hour.

   b) See the graph. Therefore, the age of the student equals 22 to the nearest year. Since 2 hours of studying per week is outside of the range of data values, we are attempting to extrapolate an answer. It is clear that, in this case, the age of a high school student cannot be 22 years. The prediction model fails at this point.

7. a)
$$y = -1.232\,56\,x + 28.6512$$
$$y = -1.232\,56(13.5) + 28.6512$$
∴ $y = 12.011\,64 = 12.0$ hours to the nearest half-hour.

   b)
$$y = -1.232\,56\,x + 28.6512$$
$$16 = -1.232\,56\,x + 28.6512$$
$$-12.6512 = -1.232\,56\,x$$
∴ $\dfrac{-12.6512}{-1.232\,56} = x$

Hence, $x = 10.264 = 10$ years to the nearest half-year.

Even though the mathematics of the regression equation says the answer is that the student must be 10 years old, there are no 10-year-old students who are in high school. This is another example of what can happen when extrapolating. Although not all extrapolations will be incorrect, caution must be used when extrapolating.

Drawing a regression line by eye

When drawing a regression line after collecting data, you have all of the necessary information to draw the one and only correct line of best fit as explained above. However, during a testing situation, you might only be given the scatter plot and not the raw data. The question might still ask for a prediction and therefore a regression line would be necessary. Such a line would be called a regression line drawn by eye.

The following points might help in drawing such a line.

● If possible, approximate the ordered pairs and enter them into List 1 and List 2. This will enable you to find all the necessary information to draw the regression line.

- When it is not possible to approximate the scatter points either due to time constraints, the number of data points, or the lack of clear scales on the *x*- and *y*-axes, then consider the following.
  - Draw a line that tends to go through the middle of the ordered pairs so that roughly as many points are above the line as are below it.
  - One technique is to use your pencil to approximate the line.
  - Although a regression line could pass through the origin, most problems are designed so that does not occur.
  - Although scatter points may possibly lie on the regression line, do not connect, purposely, two of the scatter points in order to draw the line.

Example 12.8

Draw a line of best fit by eye for the following scatter plot.

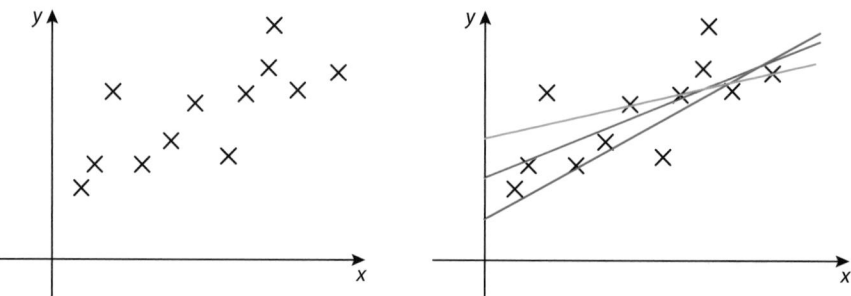

Solution
Two of the possible 'lines' that might be drawn by 'eye', and would not be acceptable, are drawn in blue and green. The one and only correct line is drawn in red.

Exercise 12.2

1. Explain, in layman's terms, what the *r*-value tells you about a scatter plot.

2. Write the range of the *r*-value in
   a) inequality form
   b) open interval form.

3. What ordered pair must the regression line pass through?

4. Will the line of best fit always pass through the origin? Could it?

5. Explain why no line of best fit may be drawn if *r* = 0.10.

6. Is it possible to draw the regression line knowing only the *r*-value? Why or why not?

7. What is the minimum amount of information needed to draw the correct line of best fit?

8. Name two variables that do not have any linear correlation.

9. Name two variables that (probably) have a strong positive linear correlation.

10. Name two variables that (probably) have a strong negative correlation.

11. Interpret the following *r*-values using combinations of the following words: positive, negative, strong, moderate, weak, none. Make sure to always use the words 'correlation' and 'variables'.

a) 0.901   b) 0.113   c) −0.626

d) 0.595   e) −0.843

12. Sketch a scatter plot of about 10 to 15 points that could represent each of the following *r*-values.

a) 0.952   b) −0.845   c) 0.322   d) −0.172

13. Name three ways in which the gradient-intercept form of a linear function, $y = mx + c$, differs from the linear regression form, $y = ax + b$.

14. Given the mean point $(\bar{x}, \bar{y})$ and the gradient of the regression line, write the equation of the regression line in

$y = ax + b$ form, where $a, b \in \mathbb{R}$.

a) $(13, 25);\ 2.5$

b) $(23.4, 67.8);\ -5.43$

c) $(10.75, 15.25);\ 8.825$

d) $(9, 20);\ -0.623$

15. Draw the graph of each regression line in Question 14.

16.

x	2	1	2	3	3	4	5	5	6	6
y	3	3	5	3	5	6	5	4	7	5

a) Predict y given that $x = 5.5$.

b) Predict x given that $y = 7$.

17.

x	10	18	20	22	27	33	38	41	42	50	52	57
y	70	90	60	86	32	57	42	22	33	12	41	21

a) Predict y given that $x = 60$.

b) Predict x given that $y = 28$.

18. Given that $(\bar{x}, \bar{y}) = (4.5, 7)$ and the scatter plot below,

a) plot the mean point

b) draw the line of best fit by eye

c) approximate the
     r-value

d) using lines on your
     graph, predict
     (i) y when x is 3
     (ii) x when y is 11

e) comment on your
     predictions.

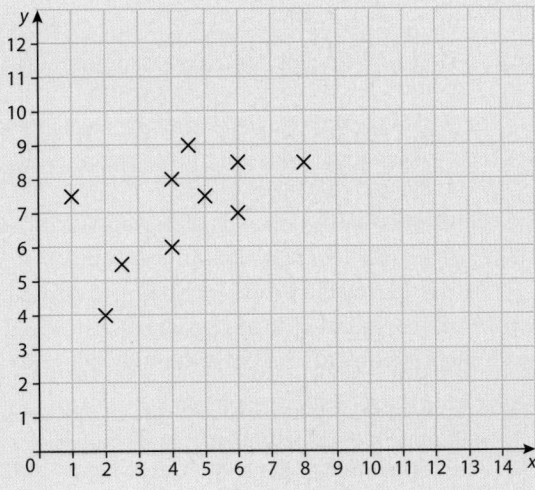

19. Given that $(\bar{x}, \bar{y}) = (60, 65)$ and the scatter plot below,
  a) plot the mean point
  b) draw the line of best fit by eye
  c) approximate the *r*-value
  d) using lines on your graph, predict
    (i) *y* when *x* is 90
    (ii) *x* when *y* is 80
  e) comment on your predictions.

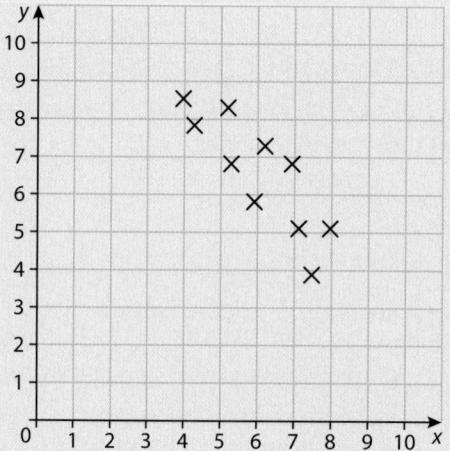

20. For each scatter plot, draw the regression line by eye.

a)   b)

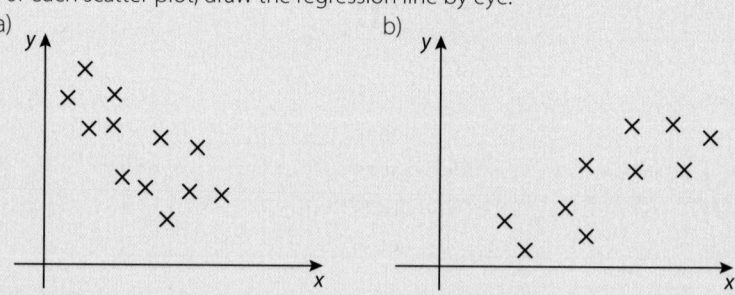

21. The heights and weights of 10 athletes were measured and the results are given in the table below.

Height	168	170	172	173	175	178	179	179	180	184
Weight	67	69	69	68	67	78	81	73	79	80

  a) On graph paper, draw a scatter diagram to illustrate these data. (Use a scale of 1 cm on the horizontal axis to represent 10 cm in height and 1 cm on the vertical axis to represent 10 kilograms.)
  b) Write down the coordinates of the mean point $(\bar{x}, \bar{y})$.
  c) Write down the value of **r**, the Pearson's product-moment correlation coefficient for this set of data.
  d) Write down the equation of the regression line.
  e) Draw your regression line on the graph
  f) Use your diagram, showing your methods clearly, to estimate
    (i) the height of an athlete whose weight is 70 kilograms
    (ii) the weight of an athlete whose height is 174 cm.

22. A Mathematical Studies student was investigating the number of people visiting a museum and the relationship to the number of hours of sunlight there was in a particular day.

The student constructed a table showing the number of visitors entering the museum and the number of hours of sunlight for that particular day.

Hours of sunlight	2	3	4	7	6	7	8	9	10	11	12	13	14
Number of visitors	7	8	7	7	7	7	6	6	6	5	6	4	4

a) On graph paper, draw a scatter diagram to illustrate these data. (Use a scale of 1 cm on the horizontal axis to represent 1 hour of sunlight and 1 cm on the vertical axis to represent 1 visitor.)

b) Write down the product-moment correlation coefficient (r) for this data.

c) What is the nature of the relationship between the number of hours of sunshine and the number of visitors?

d) Write down the coordinates of the mean point $(\bar{x}, \bar{y})$ and draw the point on your graph.

e) (i) Determine the equation of the linear regression line for y on x.
   (ii) Draw the regression line on the graph drawn in part a).
   (iii) Find the expected number of visitors when there are 12 hours of sunlight. Show how you arrived at your answer.

12.3 Hypothesis testing

Hypothesis testing is another area of inferential statistical mathematics. The concepts usually take several chapters in a book that is entirely devoted to the study of statistics. Several courses (e.g. biology and psychology) use hypothesis testing to perform the mathematics necessary to complete their Internal Assessment (IA) component.

The only hypothesis test that is required for Mathematical Studies is the chi-square test. However, so that you will more fully appreciate chi-square, a short introduction into the general concepts and methodologies of hypothesis testing is given on the facing page.

In general, hypothesis testing involves making a conjecture about some facet of our world, collecting data from a sample, performing mathematics according to a specific algorithm, and then deciding whether or not the conjecture is true for the entire population.

The null hypothesis, H_0, states that there is no significant difference between two (population) parameters (i.e. two numbers are the same). The alternative hypothesis, H_1, states that there *is* a significant difference between two (population) parameters (i.e. two numbers are different).

Ronald Fisher (English statistician, 1890-1962), Jerzy Neyman (Polish-American statistician, 1894-1981), Karl Pearson (British statistician, 1857-1936) and Egon Pearson (son of Karl, British statistician, 1895-1980) are largely responsible for the development of hypothesis testing.

Two specific examples of hypothesis testing

Example 12.9

Suppose that you see a statement in the newspaper that says the average price of regular gasoline is $3.21 per gallon. However, you have noticed

● **Examiner's hint:** Questions such as the next two examples will *not* be set on the IB examinations. The only hypothesis test that will be tested is chi-square. Chi-square will be discussed in the next section.

The following algorithm will look difficult and may seem confusing. This is only because it is a brand new procedure. After practice, you will be comfortable with using the algorithm.

that locally most of the prices are about \$3.17 per gallon. You want to see if the statement is correct or not.

It should be clear on two accounts, cost and time, that it will be impractical, if not impossible, to collect all of the gas prices from all of the fuel stations in the large city in which you reside.

However, the following algorithm will provide the answer that is both cost- and time-effective.

Solution

Step 1: State the conjecture (hypothesis) in two parts, the null H_0 and the alternative H_1.

H_0: The average price for regular gasoline is \$3.21. This is the null hypothesis (i.e. $\mu = 3.21$, where μ is the hypothesized mean average of the population of all gas stations).

H_1: The average price for regular gasoline is *not* \$3.21. This is the alternative hypothesis (i.e. $\mu \neq 3.21$).

   a) By definition, the null hypothesis says that two numbers are not significantly different and therefore must always contain the '=' sign.

   b) Notice that H_0 and H_1 are opposites (or complements of each other).

Now you must gather data. Let's say that in one day you randomly selected 10 gas stations and recorded the following price per gallon for regular fuel from each.

$$3.18, 3.16, 3.17, 3.22, 3.21, 3.18, 3.16, 3.17, 3.17, 3.18$$

Step 2: Find the *p*-value.

Enter the data into List 1 (L_1). Now press the keys:

● STAT

● >, >, (the TESTS window)

● ∨, ENTER, (2: T-Test…)

● Input: Data

● μ_0: 3.21 (This is the hypothesized mean.)

● List: L_1

● Freq:1

● $\mu: \neq \mu_0$ (This refers to the alternative hypothesis.)

● ∨, Calculate, ENTER

L1	L2	L3	1
3.18	------	------	
3.16			
3.17			
3.22			
3.21			
3.18			
3.16			
L1(1)=3.18			

L1	L2	L3	1
3.21			
3.18			
3.16			
3.17			
3.17			
3.18			

L1(10)=3.18			

EDIT CALC **TESTS**
1:Z-Test…
2:T-Test…
3:2-SampZTest…
4:2-SampTTest…
5:1-PropZTest…
6:2-PropZTest…
7↓ZInterval…

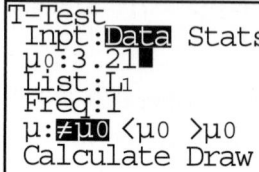

```
T-Test
Inpt:Data Stats
μ0:3.21
List:L1
Freq:1
μ:≠μ0 <μ0 >μ0
Calculate Draw
```

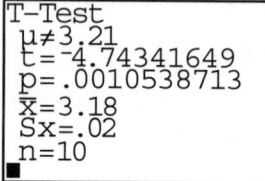

```
T-Test
μ≠3.21
t=-4.74341649
p=.0010538713
x̄=3.18
Sx=.02
n=10
```

Now closely examine the T-Test window. The window has a lot of information.

- $\mu \neq 3.21$ This is the alternative hypothesis.
- $t = -4.743\,416\,49$.   This is called the test value. In this case it is referred to as 't'.
- $p = 0.001\,053\,8713$.   This is the probability value. It is the probability of evidence **against** the null hypothesis. In other words, the smaller this number the more chance that the two numbers in question, 3.21 and 3.18, really are significantly different.
- $\bar{x} = 3.18$.   This is the mean of the sample data collected.
- $s_x = 0.02$ This is the standard deviation of the sample data collected.
- $n = 10$ This is the number of data collected.

Step 3: Select an alpha level.

An alpha level is a percentage referring to the chance that your answer might be wrong. Since you plan on inferring about the gas prices for the population based on a sample of only 10 gas prices, your results might be incorrect. It just might be the case that you happened to select 10 stations whose prices were below (or above) the 'true' mean average of all gas station prices. Common alpha (α) levels are 1%, 5% or 10%. For this problem, select 5% ($\alpha = 0.05$).

Step 4: Compare the p-value to the alpha level: p-value $<$ alpha level $(0.001\,05 < 0.05)$.

Step 5: Interpret the comparison between the p-value and the alpha level. There are two basic ways to decide whether \$3.21 and \$3.18 are **significantly** different or not.

- Comparing the test value ($t = -4.743\,416\,49$) with the critical value (unknown at this time). This method will be used in the next section when studying chi-square. You will learn how to find the critical value for chi-square.
- Comparing the p-value to the alpha level. This is perhaps the easiest way to determine the answer. It is also a required method for the IB examinations.
    - The p-value that was calculated was $0.001\,05$ to 3 s.f. The alpha level that was chosen was 0.05.
    - Since $0.001\,05 < 0.05$, we can say that there is enough evidence (the mean average of the sample data collected) *against* the null hypothesis (which says that two numbers are *not* significantly different). So what the newspaper *claimed* to be true is not true. In other words, the newspaper incorrectly reported that the average price of regular gasoline was \$3.21.

A few notes are in order.

1 There is a chance that a mistake was made and it is possible that the newspaper was correct in their report that the mean price was $3.21. The chance you made the mistake (of saying that the newspaper's assertion was wrong) was exactly the same as the alpha level, 5%.

2 Suppose you wanted to decrease the chance that you would make a mistake (and therefore not perhaps embarrass yourself when they called!). You choose $\alpha = 0.001$ even though it is not one of the common levels.

- Now the p-value is greater than the alpha level ($0.001\ 05 > 0.001$).
- This changes the answer. Now there is **not** enough evidence (the mean average of the sample data collected) **against** the null hypothesis.
- Therefore, the two numbers are **not** significantly different! In other words, they are statistically the same!

Example 12.10

A researcher collects data to test the hypothesis (the conjecture or the claim) that the average score on the SAT mathematics section is different for boys and girls. The results for ten randomly selected boys and eight randomly selected girls are shown below.

Boys' scores	740	580	640	760	660	680	600	600	780	640
Girls' scores	720	520	520	560	700	680	580	520		

Solution

Step 1: Write the null and alternative hypotheses.

H₀: The average scores on the mathematics section of the SAT are **not** significantly **different** for boys and girls. In symbols, $\mu_1 = \mu_2$, where μ_1 is the mean of the population of boys' scores and μ_2 is the mean of the population of girls' scores.

H₁: The average scores on the mathematics section of the SAT **are** significantly **different** for boys and girls. In symbols, $\mu_1 \neq \mu_2$.

Step 2: Find the p-value.

- Enter the boys' scores into List 1 and the girls' scores into List 2. Press the following keys to find the p-value (see screenshots below):
- STAT
- >, > (TESTS)
- 4: 2 – SampTTest
  - Use this test because you want to compare two (sample) averages against each other.
  - Use this test instead of the 2 – SampZTest since the size of the sample < 30.
- Set the window as shown in the screenshot.
- ∨, ∨, …(to Calculate)
- Calculate
- ENTER

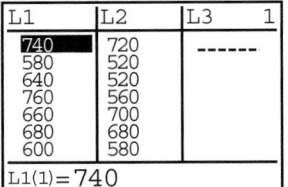

L1	L2	L3	1
740	720	------	
580	520		
640	520		
760	560		
660	700		
680	680		
600	580		

L1(1)=740

L1	L2	L3	1
660	700		
680	680		
600	580		
600	520		
780	------		
640			
▬			

L1(11)=

```
EDIT CALC TESTS
1:Z-Test…
2:T-Test…
3:2-SampZTest…
4:2-SampTTest…
5:1-PropZTest…
6:2-PropZTest…
7↓ZInterval…
```

```
2-SampTTest
 Inpt:Data Stats
 List1:L1
 List1:L2
 Freq1:1
 Freq2:1
 μ1:≠μ2 <μ2 >μ2
↓Pooled:No Yes
```

The 2 – SampTTest calculation window shows a lot of information. Find the *p*-value.

- *p*-value = 0.0943 to 3 s.f.

```
2-SampTTest
↑List1:L1
 List2:L2
 Freq1:1
 Freq2:1
 μ1:≠μ2 <μ2 >μ2
 Pooled:No Yes
 Calculate Draw
```

```
2-SampTTest
 μ1≠μ2
 t=1.799390819
 p=.0943140279
 df=13.51532195
 x̄1=668
↓x̄2=600
```

```
2-SampTTest
 μ1≠μ2
↑x̄2=600
 Sx1=70.6792442
 Sx2=86.1891607
 n1=10
 n2=8
```

Step 3: Select an alpha level. Select $\alpha = 0.05$.

Step 4: Compare the *p*-value to the alpha level.

- *p*-value > alpha level, i.e. $0.0943 > 0.05$.

Step 5: Interpret the comparison.

- Since the *p*-value is greater than the alpha level, there is *not* enough evidence against the null hypothesis to suggest that the boys and girls scores are significantly *different*.
- In other words, the boys' average of 668 and the girls' average of 600 are (statistically) the same!

Again, it should be noted that if the alpha level had originally been chosen as 0.10, then the results would be different! If that were the case then there would have been enough evidence to support the alternative hypothesis that the scores were different!

There are many hypothesis tests. They can be used with nominal data, ordinal data and, as we have seen, with interval data. They answer conjectures that are posed about averages, proportions, standard deviations, variances, and correlations, to mention a few.

Posing a question in terms of a conjecture, as in the above examples, is an excellent way to begin the Mathematical Studies project. The conjecture, gathering of the data, running the hypothesis test, and drawing a conclusion satisfy the main components of the project.

Exercise 12.3

Using Example 12.9 as a guide, solve Questions 1 and 2. Make sure to structure your solution as follows:

Step 1: Write the null and alternative hypotheses.

Step 2: Find the *p*-value via your GDC.

Step 3: Select an α-level. For these problems, select $\alpha = 0.01, 0.05$ and 0.10.

Step 4: Compare the *p*-value to the alpha level.

Step 5. Interpret the comparison between the *p*-value and the alpha level.

1. Sean conjectures that the average price of a new laptop is €1300. He randomly collects data by making a few phone calls and looking on the internet. His results are shown below.

    1395, 1290, 1400, 1490, 1100, 1535, 1370, 1480, 1270, 1430

Determine whether or not there is enough evidence to support his conjecture

a) for $\alpha = 0.01$ b) for $\alpha = 0.05$ c) for $\alpha = 0.10$.

2. Debbie hypothesizes that the average cost of eating out for a family of four is $83.85. She uses a convenience sample and surveys her friends. The data she collected are reported below.

79.23, 90.25, 67.95, 81.15, 64.77, 80.88, 59.95, 88.75, 92.21, 73.44, 78.23, 56.80

Determine whether or not there is enough evidence to support her hypothesis

a) for $\alpha = 0.01$ b) for $\alpha = 0.05$ c) for $\alpha = 0.10$.

Using Example 12.10 as a guide, solve Questions 3 and 4.

3. After some sporadic observations, Doug conjectured that the average number of minutes *over* the allotted lunch hour for men was different than that for women. During a two-month span, he collected data and then randomly chose 10 men and 10 women. His results are shown below.

	Minutes over 1 hour									
Men	5	9	5	9	8	10	5	10	11	1
Women	0	10	4	3	2	3	7	4	5	3

Determine whether or not there is enough evidence to support his conjecture

a) for $\alpha = 0.01$ b) for $\alpha = 0.05$ c) for $\alpha = 0.10$.

4. As the manager of the city's water department, Grace hypothesized that single-family dwellings used, on average, a different number of gallons of water each month than those who resided in apartments. She randomly selected 8 single-family dwellings and 8 apartments from the city's residence rolls and recorded the water usage for the month of November. The results are shown below.

	Water usage in gallons							
Single family	1700	1800	1650	2250	2000	1380	2500	1600
Apartment	1650	1500	1320	1250	1780	1800	1750	1480

Determine whether or not there is enough evidence to support her hypothesis

a) for $\alpha = 0.01$ b) for $\alpha = 0.05$ c) for $\alpha = 0.10$.

Using methods similar to Examples 12.9 and 12.10, solve Questions 5 and 6.

5. Bart conjectured 70% of the greenware pieces that are poured from ceramic moulds are dinnerware pieces. He randomly selected several ceramic shops and recorded the following data for 500 pieces of greenware. (Hint: 1–PropZTest …)

Dinnerware: 325
Non-dinnerware: 175

Determine whether or not there is enough evidence to support his conjecture

a) for $\alpha = 0.01$
b) for $\alpha = 0.05$
c) for $\alpha = 0.10$.

6. When trying to decide if the *r*-value is large enough to say that there is enough of a correlation to construct a regression line, and thus make predictions by extrapolating (outside the data range) or interpolating (inside the data range), you must do a hypothesis test. The TI calculator has such a test in the STAT, TEST window.

Run the LinRegTTest... and determine under what conditions (what α-level(s)) there would be enough evidence to support the conjecture that there is a significant enough correlation between the variables to construct a regression line for the following data:

x	2	1	2	3	3	4	5	5	6	6
y	3	3	5	3	5	6	5	4	7	5

12.4 The chi-square hypothesis test

4.4 The χ^2 test for independence: formulation of null and alternative hypotheses; significance levels; contingency tables; expected frequencies; degrees of freedom; p-values.

Chi-square (χ^2) is the name for the hypothesis test that determines whether or not two variables are related. The procedure for this hypothesis test is the same as for all other hypothesis tests: make a conjecture, write the null and alternative hypotheses, determine the p-value (via GDC), select an alpha level, make a comparison between the p-value and the alpha level (or between the χ^2 test statistic and the critical value), and interpret that comparison.

• **Examiner's hint:** Chi-square is the only hypothesis test you need to know.

Chi-square using a GDC

The following example will show the entire procedure via the TI calculator. After the example, a more detailed discussion will be presented, and procedures will be given for showing the procedure with all of the working that could be required for the IB examination.

Example 12.11

A researcher conjectures that seat belt usage, for drivers, is related to gender. She gathers data by recording seat belt usage at several randomly selected intersections. The data has been recorded and organized in the table right. Construct a chi-square hypothesis test to determine if there is enough evidence to support the researcher's conjecture.

	Seat belt usage	
Gender	Yes	No
Female	50	25
Male	40	45

Solution

Step 1: Write the null and alternative hypotheses.

   H_0: Seat belt usage *is not related* to gender.
   - This is in agreement with the definition of the null hypothesis that states there is no significant difference between two parameters (numbers about a population).

This table is called a **contingency table**.

- 'Independent of' is often used in place of 'not related to'.

H_1: Seat belt usage *is related* to gender.

- 'Related to' means that there is a relationship between seat belt usage and gender. If there is a relationship, one of the variables is dependent upon the other.
- 'Dependent on' is often used in place of is 'related to'.

Step 2: Find the *p*-value.

Enter the data into the calculator. Unlike the previous two examples, this data will be entered in a matrix array. Press the following keys: 2ND, x^{-1} (MATRIX)

- >, > (EDIT)
- ENTER
- 2, ENTER, 2, ENTER
- 50, ENTER, 25, ENTER, 40, ENTER, 45, ENTER
  - This step enters the data in the contingency table into MATRIX [A].
  - See the screenshots below.

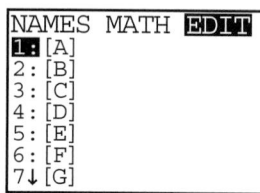
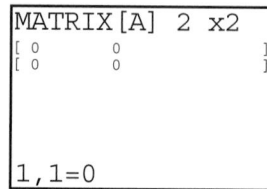
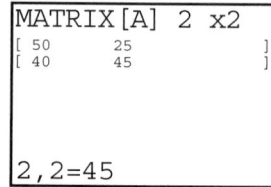

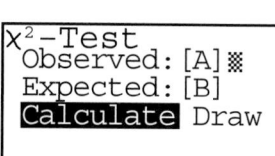

Now press the keys to find the *p*-value:

- STAT
- >, > (TESTS)
- ∨, ∨, …(until C: χ^2 – Test …)
- ENTER
- ∨, ∨ (Calculate)
- ENTER
- See the screenshots right.
- The χ^2 – Test window shows several lines of information:
- χ^2 = 6.2247 to 5 s.f.
  p = 0.0126 to 3 s.f.
  df = 1

Step 3: Select an alpha level. Select $\alpha = 0.01$

- Remember, the alpha level represents the chance of making a mistake, the mistake being that you reject the null hypothesis when it is actually true.
- In technical terms, this type of a mistake is called a Type I error.

Step 4: Compare the *p*-value to the alpha level.

- *p*-value > alpha level (0.0126 > 0.01).

Step 5: Interpret the comparison.

- Since the *p*-value is greater than the alpha level, there is **not** enough evidence against the null hypothesis to suggest that seat belt usage is related to gender.

- In other words, we do not reject the null hypothesis.
  - Another way to say 'do not reject' is to say 'accept'.
- Therefore, we accept the null hypothesis that there is no relationship between seat belt usage and gender.
- Hence, there is **not** enough evidence to support the researcher's conjecture that seat belt usage and gender are related.

Again, it should be noted that if the alpha level had originally been chosen as 0.05, then the results would be different since p-value $<$ alpha level ($0.0126 < 0.05$). If that were the case then there would have been enough evidence *against* the null hypothesis to suggest that seat belt usage is related to gender!

It is not technically correct to say 'accept', just as in a court of law the jury will not proclaim the defendant 'innocent'. In the same way that a jury will proclaim a defendant 'not guilty', it is technically correct to say 'do not reject' the null hypothesis.

Exercise 12.4

For Questions 1–4, use your GDC and the method in Example 12.11 to answer parts a), b) and c).

Make sure to structure your solution as follows:

Step 1: Write the null and alternative hypotheses.
Step 2: Find the p-value via your GDC.
Step 3: Select an α-level. For these problems, select $\alpha = 0.01, 0.05,$ and 0.10.
Step 4: Compare the p-value to the alpha level.
Step 5: Interpret the comparison between the p-value and the alpha level.

• **Examiner's hint:** 'Accept' the null hypothesis (if appropriate) is considered correct for the conclusion of a chi-square hypothesis test.

1. Nikki hypothesized that automobile drivers making a complete stop at a stop sign was independent of gender. She collected data at randomly selected intersections and organized it in the table shown below.

	Male	Female
Stopped	14	23
Did not stop	27	17

Determine whether there is enough evidence to accept or reject the null hypothesis

a) for $\alpha = 0.01$ b) for $\alpha = 0.05$ c) for $\alpha = 0.10$.

2. Mike conjectured that the number of cups of coffee consumed per day was not related to the age of the individual. He collected data by randomly selecting 119 people. He recorded and organized the data in the table below.

		Ages					
		15–25	26–36	37–47	48–58	59–69	70–80
Cups per day	0–1	8	13	4	7	5	6
	2–3	8	10	7	4	9	6
	4–5	2	7	13	10	3	5

Determine whether there is enough evidence to accept or reject the null hypothesis

a) for $\alpha = 0.01$ b) for $\alpha = 0.05$ c) for $\alpha = 0.10$.

3. Charles noticed there seemed to be a connection between the kinds of movies high school students enjoy and the types of extra-curricular activities they participate in. He randomly collected data to test the hypothesis that the extra-curricular activities a student participated in is dependent on movie genre. The data was tabulated and organized as shown below.

		Extra-curricular activity			
		Visual arts	Sports	Performing arts	Community service
Movie	Comedy	10	21	11	15
	Romance	17	2	17	3
	Action	3	15	2	6

Determine whether there is enough evidence to accept or reject the null hypothesis

a) for $\alpha = 0.01$ b) for $\alpha = 0.05$ c) for $\alpha = 0.10$.

4. From what Lauren has observed, she believes that the number of hours exercised per week is dependent on gender. She collected data randomly and organized the results in the table below.

	Hours exercised per week		
	Low	Average	High
Male	5	10	12
Female	9	8	4

Determine whether there is enough evidence to accept or reject the null hypothesis

a) for $\alpha = 0.01$ b) for $\alpha = 0.05$ c) for $\alpha = 0.10$.

How do you think that the often heard quote 'There are three kinds of lies – lies, damned lies and statistics' relates to hypothesis testing?

12.5 The chi-square test – manual working

> **4.4** The χ^2 test for independence: formulation of null and alternative hypotheses; significance levels; contingency tables; expected frequencies; degrees of freedom; p-values.

The following example will use the conjecture and data from Example 12.11. The difference will be that all working, along with annotated notes, will be shown.

Example 12.12

Melissa conjectures that seat belt usage, for drivers, is related to gender. The data is shown in the contingency table below. Test Melissa's conjecture using a chi-square hypothesis test at $\alpha = 0.01$.

	Seat belt usage	
Gender	Yes	No
Female	50	25
Male	40	45

Solution

Step 1: This step is exactly the same.

H_0: Seat belt usage *is not related* to gender.

H_1: Seat belt usage *is related* to gender.

The next two steps will be much longer. In essence, we will compute two values.

- The chi-square test statistic (χ^2).
- A value called the critical value (CV).

We will then compare those two values in much the same way that we compared the *p*-value and the alpha level.

Step 2: Calculating the chi-square (χ^2) test statistic.

- Expand the contingency table to include an extra row and column.
- Find each row and column total.
- The cell where the row total and column total meet is the total sum (the sum of all of the data values). Find the total sum.

	Seat belt usage		
Gender	Yes	No	Row totals
Female	50	25	**75**
Male	40	45	**85**
Column totals	**90**	**70**	**160**

Total sum

- The data is placed in four cells in the table. Each cell can be described by its row and column designation.
  - The number 50 is in cell '11': the 1st row and 1st column.
  - The number 25 is in cell '12': the 1st row and 2nd column.
  - The number 40 is in cell '21': the 2nd row and 1st column.
  - The number 45 is in cell '22': the 2nd row and 2nd column.

- The numbers 50, 25, 40, 45 are called the observed values.

- However, even though we observed (collected) those values, they are not what we would 'expect' as values.
  - We would intuitively 'expect' 25% of the data, 40, in each cell. 25% is used since there are four cells.

- Mathematically we will 'expect' a different number in each cell based on the observed values.
  - The probability that a person wears a seat belt is $\frac{90}{160}$. There are a total of 75 females, so the mathematical expected value for cell 11 is $\frac{90}{160} \cdot 75 = \frac{90 \cdot 75}{160} = 42.1875$.
  - An easy way to remember this is to memorize the formula:

Expected value for a cell

$$= \frac{\text{column total (for that cell)} \cdot \text{row total (for that cell)}}{\text{total sum (of all the data values)}}$$

$$= EV_{rc} = \frac{CT \cdot RT}{TS}$$

- $EV_{11} \quad = \frac{90 \cdot 75}{160} = 42.1875$

- $EV_{12} \quad = \frac{70 \cdot 75}{160} = 32.8125$

- $EV_{21} \quad = \frac{90 \cdot 85}{160} = 47.8125$

- $EV_{22} \quad = \frac{70 \cdot 85}{160} = 37.1875$

Expected values	Seat belt usage	
Gender	Yes	No
Female	42.1875	32.8125
Male	47.8125	37.1875

- Once the χ^2 has been completed on the GDC the expected values are easily checked by pressing:
  - MATRIX
  - 2: [B]
  - ENTER
  - $>, >, >$ (to see cells 12 and 22)

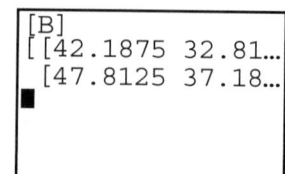

- It is convenient to list the observed values and expected values side by side and complete the chart below.

Cell	f_o	f_e	$f_o - f_e$	$(f_o - f_e)^2$	$\dfrac{(f_o - f_e)^2}{f_e}$
11	50	42.1875	7.8125	61.035	1.4468
12	25	32.8125	−7.8125	61.035	1.8601
21	40	47.8125	−7.8125	61.035	1.2765
22	45	37.1875	7.8125	61.035	1.6413
					Sum = 6.2247

- The notations in the top row are IB notations.
  - f_o = frequencies of the observed values.
  - f_e = the frequencies of the expected values.
- The 'Sum' cell is the chi-square (χ^2) test statistic.

The chi-square test statistic $= \chi^2_{calc} = \sum \dfrac{(f_o - f_e)^2}{f_e}$.

The chi-square formula says in summation form what the chart shows how to do.

$$\chi^2_{calc} = \sum \frac{(f_o - f_e)^2}{f_e} = \frac{(50 - 42.1875)^2}{42.1875} + \frac{(25 - 32.8125)^2}{32.8125}$$
$$+ \frac{(40 - 47.8125)^2}{47.8125} + \frac{(45 - 37.1875)^2}{37.1875} = 6.2247$$

$\therefore \ \chi^2_{calc} = 6.22$ to 3 s.f.

Step 3: Calculating the critical value.

Suppose you had one million dollars, in one-dollar bills, in your book bag. You give one dollar away and then your friend asks you how much money you have. You are likely to say 'one million dollars', even though you actually have one dollar less. There is just not a significant enough difference to say 'nine hundred and ninety -nine thousand, nine hundred and ninety-nine dollars'. However, at some point after you gave away enough money, there would be a significant enough difference for you to say something other than 'one million dollars'. A critical value is just that, a number that is the boundary for saying whether a statistic is significant or not.

- A critical value is a number that separates the 'reject the null hypothesis' statement from the 'accept the null hypothesis' statement.
- If the test statistic (χ^2_{calc}) falls to the left of the critical value (CV), the answer is 'do not reject the null hypothesis' or 'accept the null hypothesis'.
- If the test statistic (χ^2_{calc}) falls to the right of the CV, the answer is, 'reject the null hypothesis' or 'accept the alternative hypothesis'.

In examinations the critical value (CV) will always be given and in this case the CV = 6.635.

Step 4: Compare the χ^2_{calc} test statistic to the critical value.
- $6.2247 < 6.635$

Step 5: Interpret the comparison between the χ^2_{calc} test statistic and the critical value.
- Since $\chi^2_{calc} < CV$, do not reject the null hypothesis (accept the null hypothesis).
- Therefore, there is not enough evidence to suggest that seat belt usage is related to gender.
- Another way to express this is to say that seat belt usage is not related to gender.

Again, it should be noted that if the alpha level had originally been chosen as 0.05, then the results would be different since $\chi^2_{calc} > CV$ (i.e. $6.2247 > 3.841$). If that had been the case then there *would have been* enough evidence to suggest that seat belt usage *is related* to gender!

In summary: 1   a)   Write the null hypothesis using 'is not related to' or 'independent'.
        b)   Write the alternative hypothesis using 'is related to' or 'dependent'.
    2   Calculate the chi-square test statistic
        a)   using your GDC (in examinations), or
        b)   using the χ^2_{calc} formula (not required for examinations).
    3   Determine
        a)   the p-value by using your GDC, or
        b)   the critical value (this will be given in examinations)
    4   Compare
        a)   the p-value against the alpha level, or
        b)   the χ^2_{calc} against the critical value.
    5   Interpret the comparison.
        a)   If the p-value $>$ alpha level or $\chi^2_{calc} < CV$, **do not** reject the null hypothesis.
        b)   If the p-value $<$ alpha level or $\chi^2_{calc} > CV$, **reject** the null hypothesis.

> It is helpful to think of the p-values and alpha values as **area** numbers and the critical values and χ^2_{calc} values as **number line** numbers.

Example 12.13

Given the p-value/alpha level or the critical value/χ^2_{calc}, state whether to reject H_0 or not.
a) p-value $= 0.0123$, $\alpha = 0.05$
b) $CV = 7.815$, $\chi^2_{calc} = 7.25$
c) p-value $= 0.129$, $\alpha = 0.10$
d) $CV = 9.488$, $\chi^2_{calc} = 15.23$

Solution
a)   Reject H_0 since $0.0123 < 0.05$.

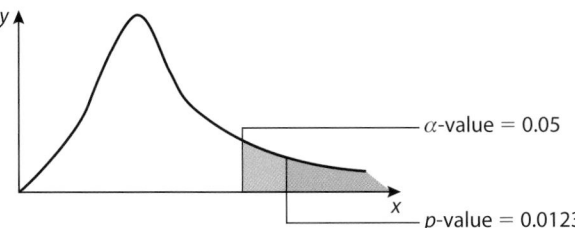

● **Examiner's hint:** Think of the p-value and the alpha level as area values. $\alpha = 0.05$ encompasses more area under the χ^2 curve than the p-value (0.0123).

b) Do not reject H_0 since $7.25 < 7.815$.

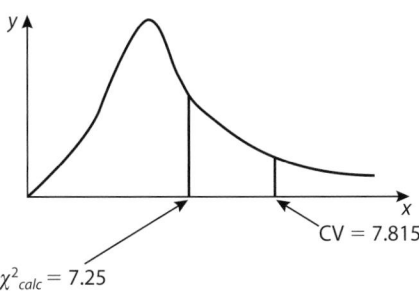

● **Examiner's hint:** Think of the χ^2_{calc} and CV values as number line values so that 7.25 is to the left of 7.815.

c) Do not reject H_0 since $0.129 > 0.10$

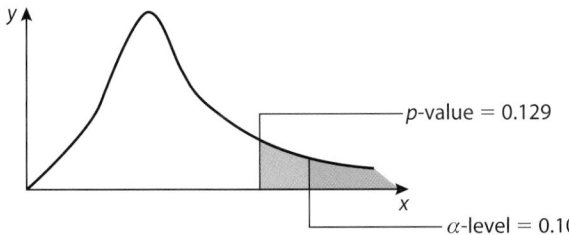

● **Examiner's hint:** The area 0.129 is greater than the area 0.10 therefore do not reject H_0.

d) Reject H_0 since $15.23 > 9.488$.

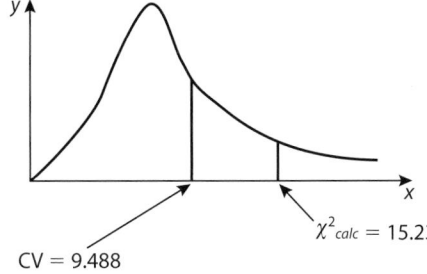

● **Examiner's hint:** 15.23 is to the right of 7.815 on the number line, therefore reject H_0.

One final example for χ^2

We will solve the problem in this final example using the GDC in Solution I, and without the GDC in Solution II.

Example 12.14

A survey is taken to determine if music preference is independent of a teenager's age. The data was collected and organized in the contingency table below. At $\alpha = 0.05$, determine if the conjecture can be supported.

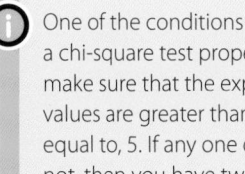

One of the conditions for using a chi-square test properly is to make sure that the expected values are greater than, or equal to, 5. If any one of them is not, then you have two choices before performing the test: gather more data or combine categories.

● **Examiner's hint:** The largest contingency table that will be tested will be 4 × 4.

Age	Music preference		
	Country	Rap/blues	Rock
13–15	8	17	9
16–19	15	8	13

Solution I: Using a GDC.

Step 1: Write the null and alternative hypotheses.

H_0: Music preference is independent of teenagers' age group.

H_1: Music preference is dependent on teenagers' age group.

```
MATRIX[A]  2 ×3
[ 8      17     9    ]
[ 15      8     13   ]


2,3=13
```

```
EDIT CALC TESTS
0↑2-SampTInt…
A:1-PropZInt…
B:2-PropZInt…
C:X²-Test…
D:2-SampFTest…
E:LinRegTTest…
F:ANOVA(
```

```
X²-Test
 Observed: [A]※
 Expected: [B]
 Calculate Draw

```

```
X²-Test
 x²=6.045499755
 p=.0486672053
 df=2

■
```

Step 2: Find the p-value and compare it with the α-level.

• Enter the data in the contingency table into MATRIX [A] in a 2 × 3 matrix.

• Press: STAT, >,> (TESTS), ∨, ∨, … (to χ^2 – Test …), ENTER, ∨, ∨, (Calculate), ENTER (see screenshots right).

• p-value = 0.0487.

Step:3 Select an α-level: $\alpha = 0.05$.

Step 4: Compare the p-value to the alpha level: p-value < alpha level.

Step 5: Interpret the comparison.

Reject H_0 since $0.0487 < 0.05$.

• There is enough evidence against the null hypothesis to say we can accept that music preference is dependent on teenagers' age group.

As before, if $\alpha = 0.01$, then $0.0487 > 0.01$, and there would *not* have been enough evidence against the null hypothesis in order to reject it. Therefore, we would have accepted the null hypothesis that music preference is independent of teenagers' age group!

Solution II: Without a GDC.

Step 1: Write the null and alternative hypotheses.

H_0: Music preference is independent of teenagers' age group.

H_1: Music preference dependent on teenagers' age group.

Step 2: Find the χ^2 test statistic.

a) Find the expected values.

Age	Music preference			Row totals
	Country	Rap/blues	Rock	
13–15	8	17	9	**34**
16–19	15	8	13	**36**
Column totals	**23**	**25**	**22**	**70**

$$E_{11} = \frac{23 \cdot 34}{70} = 11.1714 \quad E_{12} = \frac{25 \cdot 34}{70} = 12.1429 \quad E_{13} = \frac{22 \cdot 34}{70} = 10.6857$$

$$E_{21} = \frac{23 \cdot 36}{70} = 11.8286 \quad E_{22} = \frac{25 \cdot 36}{70} = 12.8571 \quad E_{23} = \frac{22 \cdot 36}{70} = 11.3143$$

● **Examiner's hint:** The row and column totals for the expected values will be the same as the respective row and column totals for the observed values, e.g. 11.171 + 11.829 = 23.

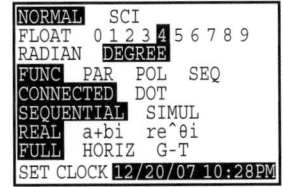

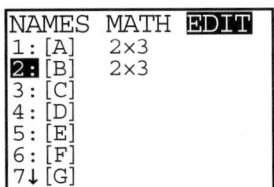

● **Examiner's hint:** MODE, FLOAT, 4 will round all entries/calculations to 4 decimal places.

```
[B]
[[11.1714 12.14…
 [11.8286 12.85…
■
```

```
[B]
…12.1429 10.685…
…12.8571 11.314…
■
```

```
[B]
….1429 10.6857]
….8571 11.3143]]
■
```

b) Now, calculate the χ^2 test statistic using the IB formula.

$$\chi^2_{calc} = \sum \frac{(f_o - f_e)^2}{f_e} = \frac{(8 - 11.1714)^2}{11.1714} + \frac{(17 - 12.1429)^2}{12.1429} + \frac{(9 - 10.6857)^2}{10.6857} +$$

$$\frac{(15 - 11.8286)^2}{11.8286} + \frac{(8 - 12.8571)^2}{12.8571} + \frac{(13 - 11.3143)^2}{11.3143}$$

$$= 6.05 \text{ to 3 s.f.}$$

 Access matrix [B] to check your expected value.

c) Use GDC to find the χ^2 test statistic

```
      χ²-Test
X²=6.045499755
p=.0486672053
df=2
```

$\chi^2 = 6.05$ 3 s.f.

Step 3: Critical Value (CV) = 5.991 (given in the examinations).

Step 4: Compare the χ^2 test statistic to the critical value:
χ^2 test statistic > CV i.e. 6.05 > 5.991.

Step 5: Interpret the results. Reject the null hypothesis since 6.05 > 5.991.

- Therefore, there is not enough evidence to suggest that music preference is independent of a teenager's age group.

This is an example of a Type I error, 'rejecting the null hypothesis when it is actually true'.

The chance of making a Type I error is the alpha level chosen. In this case, there was a 5% chance of making a mistake by rejecting the null hypothesis.

If, however, the alpha level had been chosen as 0.01, then the CV would have been 9.210 and the result would have been to accept (do not reject) the null hypothesis. A Type II error would have occurred in this scenario – 'not rejecting the null hypothesis when it was actually false'.

When performing inferential statistics, there will *always* be a chance that a mistake has been made. This is because you are making a statement about a large group by observing and taking statistics from a small group.

Exercise 12.5

1. Describe in one or two sentences the concept of the null hypothesis.

2. Describe in one or two sentences the concept of the alternative hypothesis.

3. The alpha level is the percentage chance of making a _____ _____ error.

4. A Type ___ error is the chance of rejecting _____ when it is actually _____.

5. The alpha level is the _____ under the curve to the right of the _____ value.

6. The _____ value separates the _____ from the _____.

7. The p-value is the _____ of evidence _____ the _____ hypothesis.

8. If a contingency table has 3 rows and 4 columns, the number of degrees of freedom is _____.

9. In order to decide whether or not to reject H_0, two pairs of values could be compared. Name them.

10. Another way of saying 'accept the null hypothesis' is _____.

11. What are the two common phrases that can be used to write the null hypothesis?

12. What are the two common phrases that can be used to write the alternative hypothesis?

13. Write the meaning of each of the following symbols.
    a) H_0 b) H_1 c) μ d) df
    e) CV   f) χ^2 g) p-value   h) α-level

14. It can be helpful to think about the χ^2 test statistic and the critical value as _____ line values.

15. It can be helpful to think about _____ and _____ as area values.

16. What is the main difference between descriptive statistics and inferential statistics?

17. Why is it not possible to 'prove' a statement using inferential statistics?

18. Given the number of degrees of freedom and the alpha level, find the critical value. (Hint: Use your GDC Equation solver and X²cdf under the Distribution Menu)
    a) df = 1; $\alpha = 0.1$ b) df = 3; $\alpha = 0.05$ c) df = 3; $\alpha = 0.10$
    d) df = 4; $\alpha = 0.01$ e) df = 6; $\alpha = 0.05$ f) df = 9; $\alpha = 0.10$

19. Say whether to 'reject' or to 'accept' ('do not reject') H_0.

    a) p-value = 0.0832; $\alpha = 0.05$ b) p-value = 0.1567; $\alpha = 0.10$

    c) p-value = 0.0245; $\alpha = 0.05$ d) p-value = 4.53×10^{-4}; $\alpha = 0.01$

    e) $\chi^2_{calc} = 10.23$;   CV = 5.99   f) $\chi^2_{calc} = 7.881$;   CV = 9.488

    g) $\chi^2_{calc} = 6.630$;   CV = 6.635   h) $\chi^2_{calc} = 11.5$;   CV = 7.779

20. Using the general form of the chi-square curve, draw a diagram that depicts the relationship between the p-value and the α-level for Question 19 parts a), b), c), and d).

21. Using the general form of the chi-square curve, draw a diagram that depicts the relationship between the χ^2 test statistic and the critical value for Question 19 parts e), f), g), and h).

For Questions 22 and 23, perform a chi-square hypothesis test. Show all working while performing the five steps demonstrated in Example 12.11.

22. Evelyn hypothesizes that political affiliation, in the United States, is dependent on gender. She collected data by randomly selecting 80 names from the phone directory. Evelyn organized the data in the contingency table below.

Gender	Political affiliation		
	Democrat	Independent	Republican
Male	14	3	23
Female	21	6	13

    a) Given that the critical value at the 5% level is 5.991, should the null hypothesis be rejected or accepted?

    b) Is there enough evidence to support Evelyn's hypothesis at the 5% level?

    c) Given the critical value at the 10% level is 4.605, should the null hypothesis be rejected or accepted?

    d) Is there enough evidence to support Evelyn's hypothesis at the 10% level.

23. Amy conducted a survey to determine whether transmission type of car (automatic or manual) was independent of age of owner. The ages were separated into two groups: 30 or younger and over 30. The results were tallied and organized in the table below.

Age groups	Transmission type	
	Automatic	Manual
$\leqslant 30$	35	51
> 30	54	40

    a) At a 1% significance level should the null hypothesis be rejected or accepted?

    b) Is there enough evidence to support Amy's hypothesis at a 1% level?

    c) At a 5% significance level should the null hypothesis be rejected or accepted?

    d) Is there enough evidence to support Amy's hypothesis at a 5% level?

● **Examiner's hint:** You are expected to be able to calculate the χ^2 test statistic with your GDC when raw data is given. You should also be able to perform an entire χ^2 hypothesis test without your GDC. You are advised to check the expected values by accessing MATRIX [B].

24. A survey was conducted at a high school to determine which was the most popular coloured M&M's candy. The observed data is given in the table below.

Observed	Blue	Green	Yellow	Red	Totals
Males	31	27	20	25	103
Females	51	32	21	18	122
All	82	59	41	43	225

Expected	Blue	Green	Yellow	Red
Males	p	q	18.8	19.7
Females	r	s	22.2	23.3

a) (i) Show that the expected number of males choosing blue (p) is 37.5.

   (ii) Hence complete the table for q, r, s.

b) (i) Write down the null hypothesis H_0.

   (ii) Write down the alternative hypothesis H_1.

   Perform a χ^2 test at the 5% significance level.

c) (i) Calculate the number of degrees of freedom.

   (ii) Write down the chi-squared value.

   The critical value is 7.815.

d) What can you say about gender and the preferred colour of M & M's?

13 Calculus

Assessment statements

Overview

By the end of this chapter you will be able to:

- understand the relationship between the gradient of a function and its rate of increase
- define the derivative of a function and show how to calculate it for a variety of different functions
- analyze the graph of a function using its derivative to determine the properties of the function
- apply differential calculus to real-life problems.

13.1 The definition of the derivative

7.1   Concept of the derivative as a rate of change. Tangent to a curve.

When travelling in a car, we often wonder how fast we are going. If our
view of the speedometer is blocked from the back seat, can we calculate our
mean speed from inside the car?

We know that speed $= \dfrac{\text{distance}}{\text{time}}$.

So *if* we can measure the distance and time, and *if* we assume that we are
going to travel at the same speed for the entire duration of our journey, we
can calculate an average speed.

If we pass 10 telephone poles, spaced 100 m apart, in 50 seconds, we know that our mean speed over that distance is:

$$\frac{\text{distance}}{\text{time}} = \frac{100 \times 10}{50} = 20\,\text{ms}^{-1}, \text{which we can convert to } 72\,\text{km/hr.}$$

For a more accurate measurement of our mean speed, we count more and more telephone poles over a longer period of time.

If we want to find out how fast we are travelling in *an instant*, we use a different procedure. Instead of increasing the length of time as we record our distances, we keep decreasing the time intervals, and taking more accurate measurements of the distance.

For a function, its average rate of change between two points is the same as its slope between those two points. The line between these two points is called the **secant line**. The line through a point on a function with the same rate of change of the function at that point is called the **tangent line**.

Figure 13.1 The relationship between the secant line between two points and the tangent line at a point.

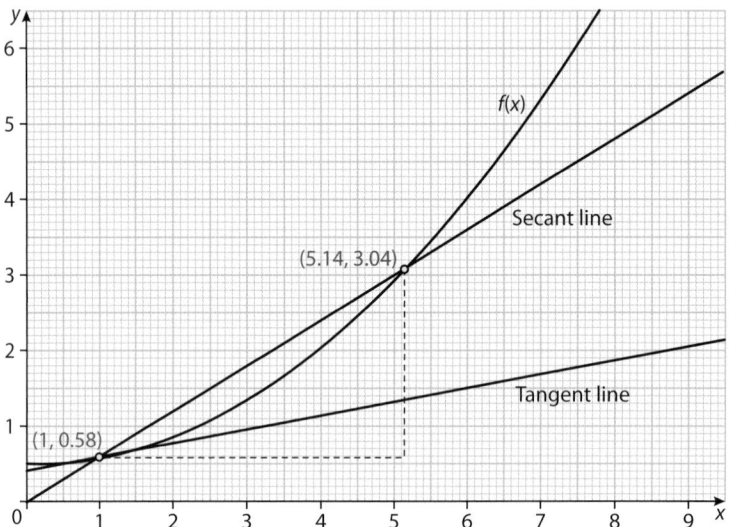

Example 13.1

$$f(x) = x^2$$
$$A\,(2, 4)$$

a) Find the slope of the secant line between A and each point in the table below.

x	$f(x) = x^2$	$B\,(x, f(x))$	Slope of secant line
2.5			
2.1			
2.01			
2.001			
2.0001			

b) Estimate the slope of the tangent line at point A.

Solution

a) First we fill in the table from left to right

x	$f(x) = x^2$	$B(x, f(x))$	Slope of secant line
2.5	$(2.5)^2 = 6.25$	$(2.5, 6.25)$	$\dfrac{6.25 - 4}{2.5 - 2} = 4.5$
2.1	$(2.1)^2 = 4.41$	$(2.1, 4.41)$	$\dfrac{4.41 - 4}{2.1 - 2} = 4.1$
2.01	$(2.01)^2 = 4.0401$	$(2.01, 4.0401)$	$\dfrac{4.0401 - 4}{2.01 - 2} = 4.01$
2.001	$(2.001)^2 = 4.004\,001$	$(2.001, 4.004\,001)$	$\dfrac{4.004\,001 - 4}{2.001 - 2} = 4.001$
2.0001	$(2.0001)^2 = 4.000\,400\,01$	$(2.0001, 4.000\,400\,01)$	$\dfrac{4.000\,400\,01 - 4}{2.0001 - 2} = 4.0001$

b) The slopes of the secant lines, which are getting closer and closer to the tangent line, are 4.5, 4.1, 4.01, 4.001, 4.0001. This suggests that the slope of the tangent line will be 4 at point A.

 How can a mathematical process, which is defined as the end result of division of ever smaller and smaller quantities, result in an integer?

We can use the GDC to investigate gradients of tangents at points on the curve.

Returning to Example 13.1

When $f(x) = x^2$ we found that at the point $x = 2$ the gradient is 4.

On the GDC

Step 1. Enter $y = x^2$ into the GDC. Y = , Y1 = x^2

Step 2. Set the window as shown, then press GRAPH.

```
WINDOW
 Xmin=-6
 Xmax=6
 Xscl=1
 Ymin=-2
 Ymax=10
 Yscl=1
↓Xres=1
```

Step 3. Draw a tangent at $x = 2$ using the following steps.

2nd, PRGM (Draw), 5: Tangent(, ENTER, 2. A tangent will be drawn at the point $x = 2$ and the equation is given which in this case is $y = 4x - 4$. Hence, the gradient of the tangent is 4.

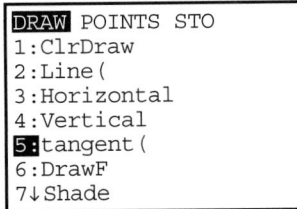
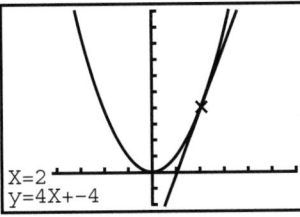

Now let's try the same steps for $x = -2$ using the following steps.

2nd, PRGM (Draw), 5: Tangent(, ENTER, −2. A tangent will be drawn at the point $x = 2$ and the equation is given which in this case is $y = -4x - 4$.

Here, the gradient of the tangent is − 4.

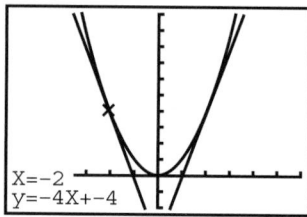

Repeating the process for $x = -1$, the gradient is −2.

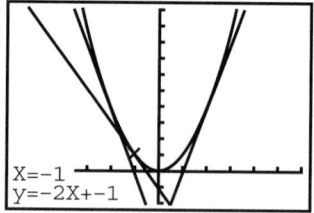

Repeating the process for $x = 1$, the gradient is 2.

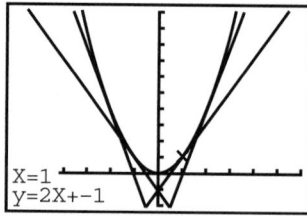

Repeating the process for $x = 0$, the gradient is 0. This is a special case and we will return to a discussion of the case where the gradient of a curve is zero later in the chapter.

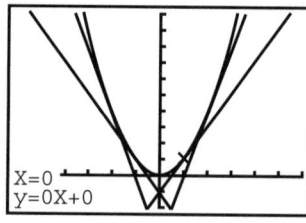

We can now construct a table of values for the gradients and the different values of x when $y = x^2$

Table 13.1

x-coordinate	−2	−1	0	1	2
Gradient of tangent	−4	−2	0	2	4

Notice that there appears to be a pattern. That is, the gradient is twice the value of x.

If we plot these values on a graph then we get the following straight line graph which is the graph of the gradient of the function $y = x^2$. The equation for the line is given as $y = 2x$.

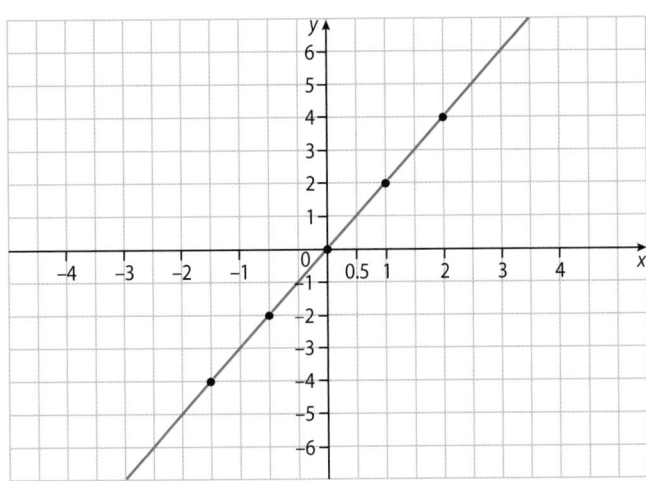

Let's take a look at $y = x^3$.

Using the same GDC steps as before, a table of gradients for $y = x^3$ follows

◄ Table 13.2

x-coordinate	−2	−1	0	1	2
Gradient of tangent	12	3	0	3	12

If we plot the points on a graph and connect the points we can see that this time the graph is a curve whose equation is given by $y = 3x^2$.

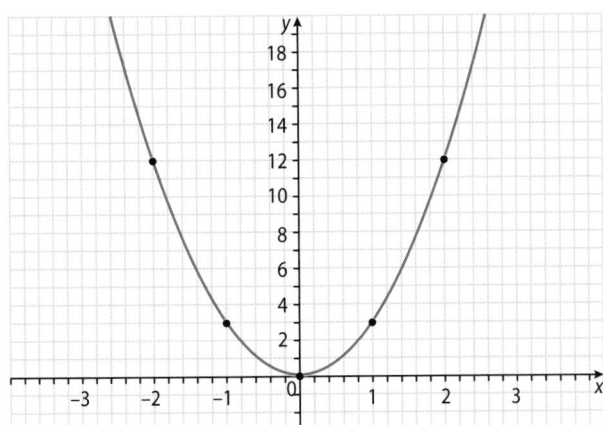

If we repeat the activity for other powers of x we can determine the gradient function which is the equation that describes the line passing through the tangent points. The following table summarizes the list of gradient functions for equations involving powers of x.

◄ Table 13.3

Function	Gradient function
$y = x^2$	$y = 2x$
$y = x^3$	$y = 3x^2$
$y = x^4$	$y = 4x^3$
$y = x^5$	$y = 5x^4$

There is another method for determining the gradient function, which is also known as the **derivative** (written as $f'(x)$ or y' or $\dfrac{dy}{dx}$).

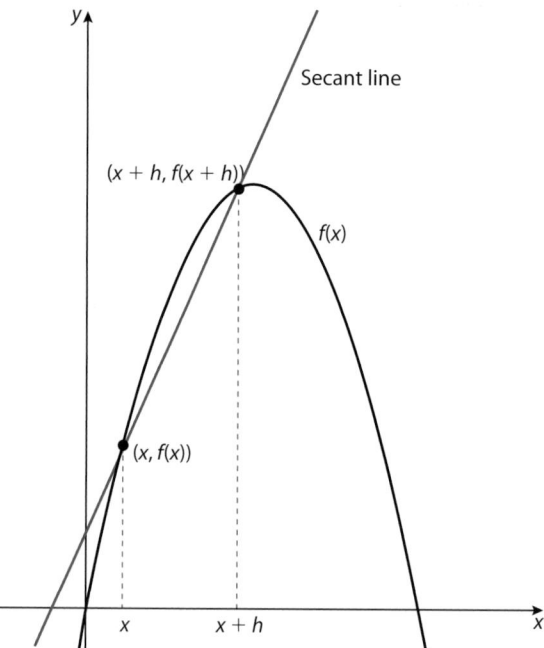

The definition of the **derivative** $\left(\text{written } f'(x) \text{ or } y' \text{ or } \dfrac{dy}{dx}\right)$ at point $(x, f(x))$ is:

$$f'(x) = \lim_{h \to 0} \left(\frac{f(x + h) - f(x)}{h} \right)$$

The 'lim' represents the process we used in our example above, continued forever. The expression $\left(\dfrac{f(x + h) - f(x)}{h} \right)$ represents the slope of each secant line when we notice that $(x + h) - x = h$.

Example 13.2

Use the definition of the derivative to find the value of the derivative of $f(x) = x^2$ when $x = 1$.

Solution

First we write down the definition of the derivative:

$$f'(x) = \lim_{h \to 0} \left(\frac{f(x + h) - f(x)}{h} \right)$$

We then substitute our value of x and $f(x)$ into the definition and simplify:

$$f'(x) = \lim_{h \to 0} \left(\frac{(x + h)^2 - x^2}{h} \right)$$

$$f'(x) = \lim_{h \to 0} \left(\frac{(1 + h)^2 - 1^2}{h} \right)$$

$$f'(x) = \lim_{h \to 0} \left(\frac{1 + 2h + h^2 - 1^2}{h} \right)$$

$$f'(x) = \lim_{h \to 0} \left(\frac{2h + h^2}{h} \right)$$

$$f'(x) = \lim_{h \to 0} \left(\frac{h(2 + h)}{h} \right)$$

$$f'(x) = \lim_{h \to 0} (2 + h)$$

We notice that $2 + h$ is equal to 2 when $h = 0$ and we substitute for h (since it will now not result in an undefined answer). Therefore, $f'(x) = 2$.

 The difference in notation we see in the definition of the derivative is because of the influence of the two great men who invented calculus, Sir Isaac Newton from Britain and Gottfried Leibniz from Germany.

Exercise 13.1

1. Find the slope of the secant line for the function $f(x) = x^2$ between (1, 1) and:

a) $(2, 4)$ b) $(1.5, 2.25)$ c) $(1.1, 1.21)$ d) $(1.01, 1.0201)$

2. Given: $f(x) = \frac{1}{3}x^3$, $A = \left(1, \frac{1}{3} \right)$

a) Fill in the table below, giving your answers to the nearest ten-thousandths place.

 The tangent line has the same slope as a function at a point and passes through the function at that point. The derivative at x is equal to the slope of the function at $(x, f(x))$.

x	$f(x) = \frac{1}{3}x^3$	$B(x, f(x))$	Slope of secant line between A and B
1.5			
1.1			
1.01			
1.001			
1.0001			

b) Estimate the slope of the tangent line at point A.

3. Given: $f(x) = \frac{4x - 3}{2}$, $A = \left(0, -\frac{3}{2} \right)$

a) Fill in the table below, giving your answers to the nearest ten-thousandths place.

x	$f(x) = \frac{4x - 3}{2}$	$B(x, f(x))$	Slope of secant line between A and B
0.5			
0.1			
0.01			
0.001			
0.0001			

b) Estimate the slope of the tangent line at point A.

4. Use the definition of the derivative to find $f'(x)$ for $f(x) = 3x + 4$ at $x = 1$.

5. Use the definition of the derivative to find $f'(x)$ for $f(x) = 2x^2$ at $x = -1$.

6. $f(x) = x^2 - x$

a) Find the equation of the secant line through A $(1, 0)$ and B $(2, 2)$.

b) Fill in the table of values below, to a sensible degree of accuracy.

x	$f(x) = x^2 - x$	B$(x, f(x))$	Slope of secant line between A and B
2			
1.5			
1.1			
1.01			
1.001			

c) Use the definition of the derivative to find the value of $f'(1)$.

7. $f(x) = x^3$

a) Find the equation of the secant line through A $(1, 1)$ and B $(2, 8)$.

b) Fill in the table of values below, rounding your answers to the nearest thousandth.

x	$f(x) = x^3$	B$(x, f(x))$	Slope of secant line between A and B
2			
1.5			
1.1			
1.01			
1.001			

c) Use the definition of the derivative to find the value of $f'(1)$.

8. $f(x) = x^2 - 3x$

a) Draw the graph of $f(x)$ for $0 \leqslant x \leqslant 4$.

b) (i) Find the equation of the secant line through $(1.5, -2.25)$ and $(3, 0)$.

(ii) Graph your answer to part b) on the same axis you graphed $f(x)$.

c) (i) Use the definition of the derivative to find $f'(3)$.

(ii) Graph the tangent line to $f(x)$ at $x = 3$.

13.2 The power rule

7.2 The principle that
$$f(x) = ax^n \Rightarrow f'(x) = anx^{n-1}$$
The derivative of functions of the form
$$f(x) = ax^n + bx^{n-1} + ..., \text{ where all exponents are integers.}$$

We noticed in Table 13.3 that there appeared to be a pattern in the gradient functions of the powers of x.

For example for $f(x) = x^2$ the **gradient function** $f'(x) = 2x$

and $f(x) = x^3$, $f'(x) = 3x^2$

The general rule for finding the derivative is

$$f(x) = ax^n \Rightarrow f'(x) = anx^{n-1}$$

This is known as the power rule.

The **power rule** can be worked out from the definition of the first derivative. We use it to differentiate functions like $f(x) = 5x^3 - 2x^2 + 7x + 4$ and other polynomial functions. We take the power of x and multiply it by the coefficient of the polynomial *and* reduce the power of x by 1, repeating this for each term of the polynomial. We notice that this means the derivative of a polynomial will have degree of 1 less than the original polynomial.

Example 13.3

Differentiate $f(x) = 6x^3 + 10x^2 - 3x + 7$.

Solution

The first term of $f(x) = 6x^3 + 10x^2 - 3x + 7$ is $6x^3$. The derivative of $6x^3$ is $6 \times 3 \times x^2 = 18x^2$. Similarly the derivative of $10x^2$ is $20x$ and the derivative of $-3x$ is -3. The derivative of 7 is 0, since the power of x in this case is 0 (recall: $x^0 = 1$). Recombining these individual derivatives we find that $f'(x) = 18x^2 + 20x - 3$.

 The derivative of the formula for the volume of a sphere is equal to the formula for the surface area of a sphere. See if you can prove this yourself.

When we have a polynomial with a negative power, we use the same procedure. We notice that, in this case, the size of the absolute value of the power will actually *increase* when we take the derivative.

Example 13.4

Find the derivative of $f(x) = 3x^2 - 4x^{-3} + 1$.

Solution

We split $f(x) = 3x^2 - 4x^{-3} + 1$ into its individual terms, and find the derivative of each of them. Hence, the derivative of $3x^2$ is $6x$, the derivative of $-4x^{-3}$ is $12x^{-4}$ and the derivative of 1 is 0. We recombine these to find that $f'(x) = 6x^2 + 12x^{-4}$.

● **Examiner's hint:** Always convert powers of x that are in the denominator of a fraction to negative powers of x in the numerator. This will cut down on errors when finding the derivative of a function.

If we are asked to find the value of the derivative at a particular point, we first find the derivative function, $f'(x)$, and then substitute the value of x into this function.

Example 13.5

Find the value of the derivative of $f(x) = 6x^2 + 6$ at $x = -2$

   (i)   Using the power rule

   (ii) Using the GDC.

Solution

   (i)   First find $f'(x) = 12x$. We then substitute $x = -2$ into $f'(x)$ and find that $f'(-2) = 12(-2) = -24$.

(ii) On your GDC, press MATH,∨,∨,∨,... 8:nDeriv(, ENTER. The following screen will appear:

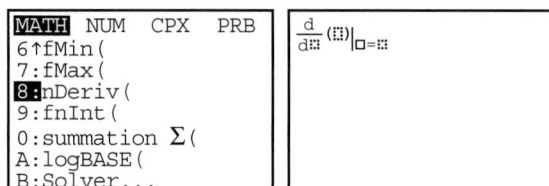

Now enter $x, > 6x^2 + 6, > -2$ as shown below and press ENTER.

```
d
── (6X²+6)|
dx         x=-2
                        -24
```

Note that the variable used when differentiating is unimportant to the process. For example, if $g(t) = 4t^3 - 10t$ then $g'(t) = 12t^2 - 10$.

We can also find the second derivative of $f(x)$, which we write as $f''(x)$ or $\dfrac{d^2y}{dx^2}$. The second derivative of a function is defined as the derivative of the derivative of the function. For example, if the function is $y = 4x^3 + 2x$, then $\dfrac{dy}{dx} = 12x^2 + 2$ and $\dfrac{d^2y}{dx^2} = 24x$.

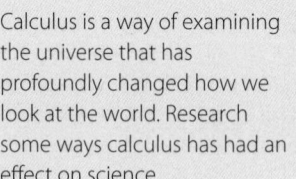

Calculus is a way of examining the universe that has profoundly changed how we look at the world. Research some ways calculus has had an effect on science.

Example 13.6

Find the value of $f''(-1)$ if $f(x) = 10x - \dfrac{4}{x^2} + 2$.

Solution

First we rewrite $f(x) = 10x - 4x^{-2} + 2$, remembering our rules of exponents. $f'(x) = 10 + 8x^{-3}$ and $f''(x) = -24x^{-4}$, which we would normally write as $f''(x) = \dfrac{-24}{x^4}$.

The power rule lets us determine the derivative of polynomial functions without using the definition of the derivative. The derivative of the derivative is the second derivative.

Finally $f''(-1) = \dfrac{-24}{(-1)^4} = -24$.

Exercise 13.2

1. Find the derivative of each of the following.
    a) $f(x) = 6x^2 - 5x + 2$ b) $f(x) = 4x^3 + 10x - 7$
    c) $f(x) = \frac{2}{3}x^3 + \frac{9}{2}x^2 + 3$ d) $f(x) = -3x^7 + 5x^4 - 3x^2 + \frac{3}{5}$

2. Find the derivative of each of the following.
    a) $f(x) = 2x^{-3} + \frac{3}{4}x$ b) $f(x) = \frac{4}{x^4} + \frac{2}{x}$
    c) $f(x) = \frac{2+x}{x^3}$ d) $f(x) = \frac{3-x^2}{4x^4}$

3. Find the value of $f'(1)$ for each of the following.
    a) $f(x) = 2x^3 + 3x + 1$ b) $f(x) = \frac{1}{3}x^6 + \frac{2}{x} - 2$
    c) $f(x) = \frac{4}{x^2} + \frac{2}{5}x$ d) $f(x) = 100 - 6x - 3x^4$

4. Find $f'(-1)$ for each of the following functions.
    a) $f(x) = 2x^3 - 8x + 11$ b) $f(x) = \frac{1}{4}x^4 + \frac{1}{3}x^3 + \frac{1}{2}x^2$

c) $f(x) = \dfrac{18x^3 + 12x}{3}$

d) $f(x) = 100x(200 - x)$

5. Find the second derivative of each of the following functions.
   a) $y = x^2 + 4x + 3$
   b) $f(x) = x^7 + x^5$
   c) $D(t) = 40t - 10t^2$
   d) $y = 4x + 1$

6. Find the value of the second derivative at $x = 1$.
   a) $f(x) = 3x^3 + 2x$
   b) $f(x) = x^4 + x^3$
   c) $H(x) = \dfrac{3}{x^2} + 3x^2$
   d) $g(x) = x^{-1} + x^{-2}$

7. The number of balloons sold each year by the 123 Balloon company can be approximated by the formula:
$$B(x) = 0.0104x^5 - 0.3058x^4 + 3.099x^3 - 12.299x^2 + 16.983x$$
   where $B(x)$ is the number of balloons in the millions and x is the number of years since 1996.
   a) How many balloons did the company sell in 2007?
   b) (i) Find $B'(x)$.
      (ii) Hence, what was the rate of change in the number of balloons being sold in the year 2007 to the nearest hundred thousand?

8. The number of users of a particular social networking website can be modelled using the formula:
$$N(x) = 0.074x^3 - 1.46x^2 + 8.58x$$
   where $N(x)$ is the number of users in the thousands per day and x is the number of months the website has been operational.
   a) (i) Fill in the table of values below.

x	0	2	4	6	8	10	12
$f(x)$							

   (ii) Graph $N(x)$ from $0 \leqslant x \leqslant 12$.
   b) (i) Find $N'(x)$.
      (ii) Find $N'(2)$ and hence the rate at which the number of users was increasing during the second month.

9. $D(t) = 100 - 5t^2$ gives an expression relating $D(t)$ in metres, the distance a ball is from the ground, and t, the time the ball is in the air.
   a) Find the initial position of the ball, $D(0)$.
   b) Solve for the amount of time the ball takes to hit the ground, $D(t) = 0$.
   c) (i) Find the derivative of $D(t)$.
      (ii) Find the value of $D'(t)$ at the time the ball hits the ground, and hence the speed with which the ball hits the ground.
      (iii) What does the sign of your answer to c)(ii) mean?
   d) Find the second derivative of $D(t)$ and hence the acceleration of the ball at time t.

10. $f(x) = 2x^3 + 3x^2 - 12x$
    a) (i) Find the derivative of $f(x)$.
       (ii) Hence, find $f'(-1)$.
    b) (i) Fill in the table of values below.

x	-3	-2	-1	0	1	2
$f(x)$						

    (ii) Graph $f(x)$ for $-3 \leqslant x \leqslant 2$ and $-10 \leqslant y \leqslant 25$.
    c) On your graph, plot the tangent line to $f(x)$ at $x = -1$.
    d) (i) Find $f''(x)$.
       (ii) Hence, find $f''(-1)$.

Gradients of curves and equations of tangents and normals

| 7.3 | Gradients of curves for given values of x. Values of x where $f'(x)$ is given. Equation of the tangent at a given point. Equation of the line perpendicular to the tangent at a given point (normal). |

The gradient of a curve at any point on a curve can be found using the power rule.

Example 13.7

The curve of $f(x) = x^3 - 3x^2 - 9x + 13$ is shown below.

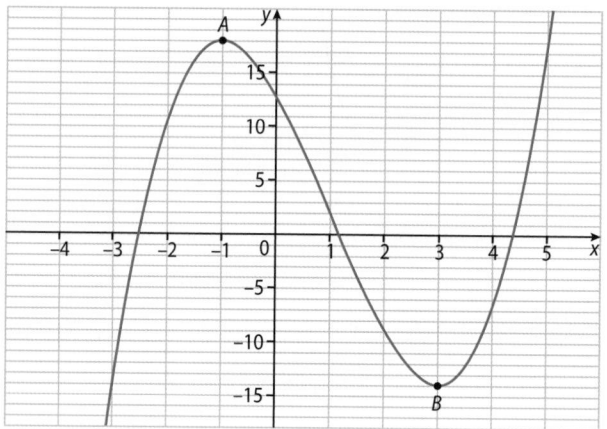

a) Write down the sign of the gradient
   (i)   Between $x = -3$ and the point A
   (ii) Between the points A and B
   (iii) Between the point B and $x = 5$.

b) Complete the table of the gradients for the given values of x.

x	−3	−2	−1	0	1	2	3	4	5
$f'(x)$									

Solution

a) (i) Between $x = -3$ and the point A, the gradient is positive, i.e.
        $f'(x) > 0$

   (ii) Between the points A and B, the gradient is negative, i.e. $f'(x) < 0$

   (iii) Between the point B and $x = 5$, the gradient is positive, i.e. $f'(x) > 0$

b) We can find the gradient at any value for x, by firstly differentiating and then substituting for x.

   $f(x) = x^3 - 3x^2 - 9x + 13$

   Then $f(x) = 3x^2 - 6x - 9$

   To find the gradient when $x = 1$ we want $f'(x) = 3(1)^2 - 6(1) - 9 = -12$.
   The gradient at $x = 1$ is -12 or $f'(1) = -12$.

Alternatively, we can use the GDC to determine the gradients at points.

Enter the equation in Y1 then, set window, as shown

Press GRAPH, 2nd, TRACE (CALC), ∨ …, 6:dy/dx, ENTER, choose a value for x e.g. 5

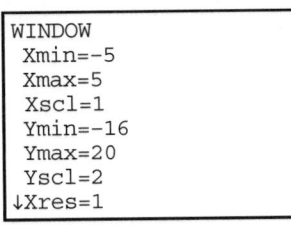
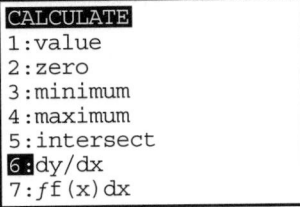

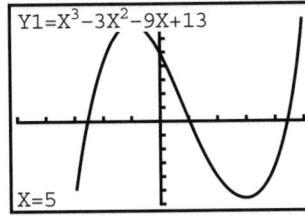
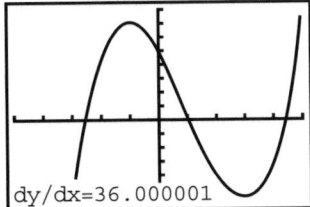

The derivative, or gradient of the curve at $x = 5$ is $\dfrac{dy}{dx} = 36$. Note that since the GDC can only approximate to the exact solution there may be a small variation between the exact solution and the GDC solution.

The gradients between $-5 \leqslant x \leqslant 5$ have been calculated and are given below.

x	−3	−2	−1	0	1	2	3	4	5
$f'(x)$	36	15	0	−9	−12	−9	0	15	36

Notice the two special cases where the gradients are 0, that is at $x = -1$ and $x = 3$. We will discuss this in more detail in section 13.4.

When determining gradients at any point on a curve we can also determine the equation of a tangent at that point.

At every point of the functions we discuss on this course, there is a tangent line through that point with the same gradient as the function at that point. If we know what the gradient of the tangent line is, we can determine a lot of information about the original function. For instance, if the gradient of the tangent line is positive, we know the function is increasing at that point.

Example 13.8

Find the gradient of the tangent line of $f(x) = 2x^2 - 4x + 1$ at:

a) $x = 0$

b) $x = 1$

c) $x = 2$

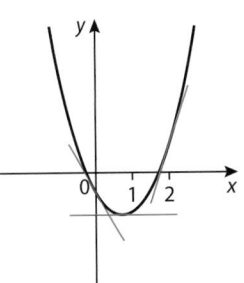

Solution

First we remember that the gradient of the tangent line is equal to the value of the derivative at that point. So we find $f'(x) = 4x - 4$.

a) $f'(0) = 4(0) - 4 = -4$
b) $f'(1) = 4(1) - 4 = 0$
c) $f'(2) = 4(2) - 4 = 4$

We can also determine where a function will have a particular gradient using a different procedure. Instead of substituting for the value of x, we can substitute for $f'(x)$ and solve for the value of x.

Example 13.9

Find the values of x for which the derivative of $f(x) = x^3 - 9x^2 + 13x - 2$ has a gradient of -2.

Solution
First we find the derivative of $f(x)$ which is $f'(x) = 3x^2 - 18x + 13$. We then solve the following:

$$-2 = 3x^2 + 18x + 13$$
$$0 = 3x^2 - 18x + 15$$
$$0 = x^2 - 6x + 5$$
$$0 = (x - 5)(x - 1)$$

Therefore, at $x = 1$ and $x = 5$ the derivative of $f(x)$ is -2.

We can also find the places where the tangent line is horizontal, which has special significance, as we will discover in Section 13.4.

Example 13.10

Find the values of x where the function, $g(x) = x^3 - \frac{3}{2}x^2 - 18x - 1$, has a horizontal tangent line.

Solution
This is the same as asking where the derivative is 0. Hence, we find the derivative of $g(x)$ and substitute 0 for $g'(x)$ and solve for x.

$$g'(x) = 3x^2 - 3x - 18$$
$$0 = 3x^2 - 3x - 18$$
$$0 = 3(x^2 - x - 6)$$
$$0 = x^2 - x - 6$$
$$0 = (x - 3)(x + 2)$$
$$x - 3 = 0 \text{ or } x + 2 = 0$$

Hence at $x = 3$ and $x = -2$ the function has a derivative of 0 and the tangent line will be horizontal.

The gradient of the tangent line to $f(x)$ at a is equal to $f'(a)$.

We can find the equation of the tangent line to a point on a function. We already know how to find the gradient of the function using the derivative. We also know how to find the coordinates of the point given the value of x and given $f(x)$. From this, we can use the point-gradient form of a line, and substitute to find the equation of the line.

Example 13.11

Find the equation of the tangent line to $f(x) = -2x^3 + 2x$ at $x = 2$.

Solution

First we find the coordinates of the point.

$f(2) = -2(2)^3 + 2(2) = -16 + 4 = -12$, so the coordinates are $(2, -12)$.

Next we find the derivative of the function at $x = 2$ to find the gradient of the tangent line.

$$f'(x) = -6x^2 + 2$$

Hence, $f'(2) = -6(2)^2 + 2 = -24 + 2 = -22$.

Finally we substitute both of these into the point-gradient form of a line: $y - y_1 = m(x - x_1)$.

$$y - (-12) = -22(x - 2)$$
$$y + 12 = -22x + 44$$

Hence, the equation is $y = -22x + 32$.

Once we have found the tangent we can also find the equation of the normal to the curve at that point. Recall that the normal is perpendicular to the tangent. Visually this is shown in the diagram below.

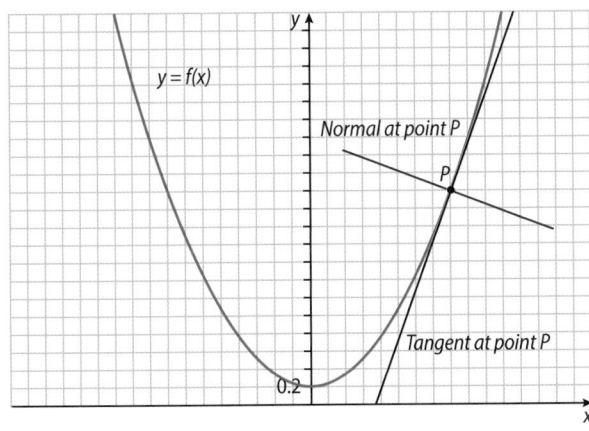

We already know how to find the gradient of the tangent using the derivative and recall from section 3.5 that the gradient of the normal can be found by $m = -\dfrac{1}{m_1}$ where m_1 is the gradient of the tangent.

Example 13.12

Revisit Example 13.11.

This time find the equation of the **normal** to the curve $f(x) = -2x^3 + 2x$ at $x = 2$.

Solution

First we find the coordinates of the point.

$f(2) = -2(2)^3 + 2(2) = -12$, hence the coordinates are $(2, -12)$.

The derivative is given by $f'(x) = -6x^2 + 2$

The gradient of the tangent at $x = 2$ is $f'(2) = -6(2)^2 + 2 = -22$

The gradient of the normal $m = -\dfrac{1}{-22} = \dfrac{1}{22}$

Equation of the normal is found using the point gradient form of the line:

$$y - y_1 = m(x - x_1)$$
$$y - (-12) = \frac{1}{22}(x - 2)$$
$$y + 12 = \frac{1}{22}(x - 2)$$
$$22(y + 12) = (x - 2)$$
$$22y + 264 = x - 2$$
$$y = \frac{1}{22}(x - 266)$$

Hence, the equation of the normal is $y = \dfrac{1}{22}(x - 266)$

Exercise 13.3

1. Find the gradient of the tangent line of $f(x) = \frac{4}{5}x^5 + 2x^2$ at:
   a) $x = 0$ b) $x = 1$ c) $x = -1$ d) $x = 10$

2. Find the gradient of the tangent line at $x = 2$ for:
   a) $f(x) = 100 - x^2$ b) $y = 20x + 50x^2$
   c) $g(x) = (100 - x)(100 - x)$ d) $f(t) = 10t - 5t^3 + 2$

3. Find the equation of the normal at $x = 2$ for each part in question 2.

4. Find the values of x for which the functions below have horizontal tangent lines.
   a) $P(x) = 10\,000x - x^2$ b) $g(x) = 3x^2 - 5x + 6$
   c) $f(x) = x^3 - \frac{3}{2}x^2 - 18x + 20$ d) $h(x) = x + \frac{9}{x}$

5. Find the values of x for which the functions below have a derivative of 0.
   a) $f(x) = x^2 - 6x$ b) $g(x) = x^3 - 27x$
   c) $y = \frac{1}{3}x^3 - 2x^2 - 5x$ d) $y = 3$

6. Find the equations of the tangent and normal lines at $x = 3$ for each of the following functions.
   a) $y = 3x + 5$ b) $y = x^2$
   c) $f(x) = 20 - x^2$ d) $f(x) = x^3 - x^2 + 2$

7. $f(x) = 3x^2 - 4x + 2$
   a) Graph $f(x)$ for $0 \leqslant x \leqslant 4$.
   b) (i) Find $f'(x)$.
        (ii) Find the equation of the tangent line at $x = 2$.
        (iii) Hence, graph the tangent line at $x = 2$ on the same set of axes as part a).

8. $g(x) = 2x^3 - 15x^2 + 24x$
   a) Find $g'(x)$.
   b) (i) Find the equation of the tangent line to $g(x)$ at $x = 2$.
        (ii) Find the equation of a line perpendicular to your answer to part b)(i) passing through the point $(2, 4)$.

9. $H(x) = 10x^2 - \frac{2}{3}x^3$
   a) Find $H'(x)$.

b) Solve $H'(x) = 0$ for x.

c) Write down the equations of the tangent lines from your answers to part b).

10. Consider the function $f(x) = x^3 + 7x^2 - 5x + 4$.

   a) Differentiate $f(x)$ with respect to x.

   b) Calculate $f'(x)$ when $x = 1$.

   c) Calculate the values of x when $f'(x) = 0$.

11. Let $y = x^3 + x^2 - 3x + 4$.

   a) Find $\dfrac{dy}{dx}$.

   b) Find the gradient of the curve when $x = 2$.

   c) Is the function increasing or decreasing when $x = 2$?

   d) Find the equation of the tangent line when $x = 2$.

   e) Find the value(s) of x when the gradient of the curve is 3.

   f) Find the equation of the normal when $x = 2$.

12. $f(x) = x^2 - 8x$

   a) (i) Find $f'(x)$.

   (ii) Find the value of x that makes $f'(x) = 0$.

   b) (i) Find the value of $f(2)$.

   (ii) Find the slope of the secant line through $(0, 0)$ and $(4, -16)$.

   (iii) Find the value of x for which $f'(x)$ is equal to your answer to part b)(ii).

   (iv) Find the equation of the tangent line to $f(x)$ through your answer to part b)(iii).

13. A function has the form $f(x) = ax^2 - 4x + 3$.

   a) (i) If $a \in \mathbb{Z}$, find $f'(x)$.

   (ii) It is known that $f'(2) = 0$. Find the value of a.

   (iii) Graph $f(x)$ for $0 \leqslant x \leqslant 5$.

   b) (i) Find the equation of the secant line through $(2, -1)$ and $(4, 3)$.

   (ii) Find the equation of the tangent line at $x = 3$.

   (iii) Graph the equation of the tangent line on the same set axis as your graph of $f(x)$.

13.4 Increasing and decreasing functions

7.4	Increasing and decreasing functions. Graphical interpretation of $f'(x) > 0$, $f'(x) = 0$, $f'(x) < 0$.
7.5	Values of x where the gradient of a curve is zero. Solution of $f'(x) = 0$. Stationary points. Local maximum and minimum points.

One use of calculus is to determine the intervals on which a function is increasing or decreasing. We learnt in Chapter 5 how to recognize whether a function is increasing or decreasing, using our GDC. When our function has a positive gradient, then it is increasing, and when it has a negative gradient then it is decreasing. At points of transition, where the gradient of the function is 0, the function could be both increasing and decreasing, depending on the direction from which you approach the transition point. These points of transistion are known as **stationary points**.

 Why would a point with a horizontal derivative be considered a stationary point?

For the mathematical definitions of increasing and decreasing functions, visit www.pearsonhotlinks.co.uk, enter the ISBN for this book and click on weblink 13.1.

Example 13.13

Identify the interval(s) where the function $f'(x) = 3x^2 - 12x$ is increasing.

Solution

First we take the derivative of the function and find that $f'(x) = 6x - 12$. We set $f'(x) = 0$ and solve for the value of x that is our transition point.

$$0 = 6x - 12$$
$$6x = 12$$
$$x = 2$$

We know that at $x = 2$ we have a sign change in the derivative (since we did not have a double root) and we can confirm this from the graph of the function.

Figure 13.3 From the graph, it is clear that $f(x)$ is increasing when $x \geqslant 2$.

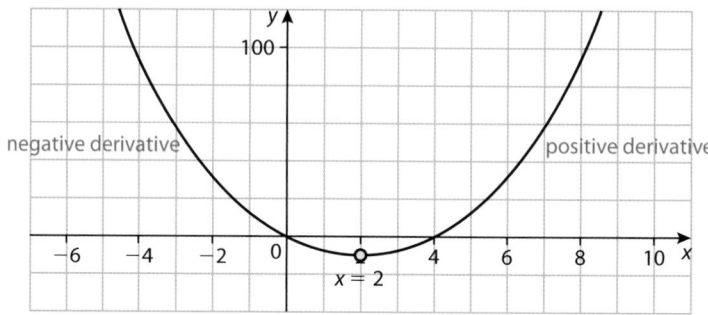

Hence, the interval on which $f(x)$ is increasing is $[2, \infty)$.

Example 13.14

$f(x) = -\frac{2}{3}x^3 + 2x^2 + 6x + 3$

a)   (i)   Find $f'(x)$.
     (ii)   Solve $f'(x) = 0$.

b)   Identify the intervals where the function is increasing or decreasing.

Solution

a)   (i) $f'(x) = -2x^2 + 4x + 6$
     (ii) $0 = -2x^2 + 4x + 6$
          $0 = x^2 - 2x - 3$
          $0 = (x - 3)(x + 1)$
          Hence, $x = 3$ and $x = -1$.

b)   Since our transition points are at $x = 3$ and $x = -1$, our possible intervals where we could be either increasing or decreasing are:

$$(-\infty, -1], [-1, 3] \text{ and } [3, \infty).$$

If we find the value of the derivative at a point in each of these intervals, we can determine if they are intervals on which the function increases or decreases.

We check $x = -2$ (since $-2 \in (-\infty, -1]$) and find that:

$$f'(-2) = -2(-2)^2 + 4(-2) + 6 = -10$$

Hence, over the interval $(-\infty, -1]$, $f(x)$ is decreasing.

We check $x = 0$ (since $0 \in [-1, 3]$) and find that:

$$f'(0) = -2(0)^2 + 4(0) + 6 = 6$$

Hence, over the interval $[-1, 3]$, $f(x)$ is increasing.

Finally we check $x = 4$ (since $4 \in [3, \infty)$) and find that:

$$f'(4) = -2(4)^2 + 4(4) + 6 = -10$$

Hence, over the interval, $[3, \infty)$, $f(x)$ is decreasing.

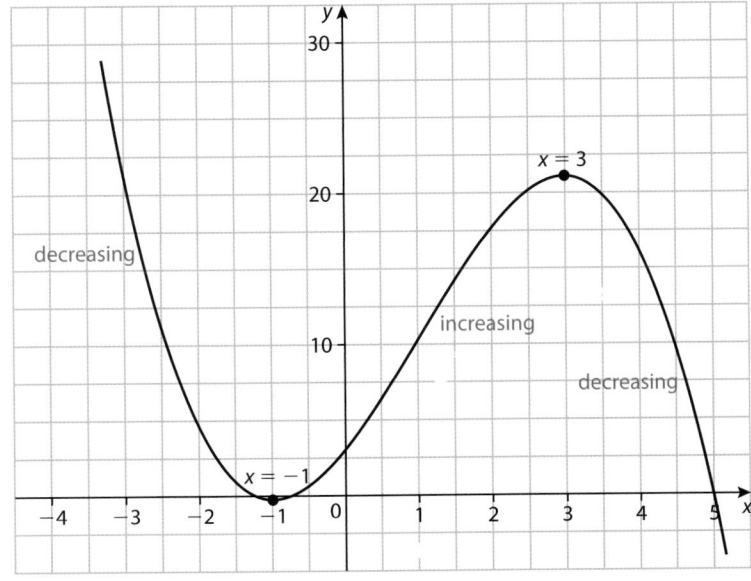

Figure 13.4 The graph of $f(x)$ confirms our calculations.

We can use calculus, and some elementary rules about motion, to determine much more complicated formulae about motion. For example, $D = v_i t + \frac{1}{2}at^2$ represents the distance D of an object with an initial velocity of v_i, at constant acceleration of a at time t. The derivative of D (which is the velocity of the object) is $D'(t) = v_i + at$, which is another familiar formula in physics.

Given the graph of a function, we can use our understanding of the derivative to find the intervals where the function increases or decreases, and identify the values of x for which the function has a derivative of 0.

Example 13.15

Using the graph of $f(x)$, given below, answer the following questions.

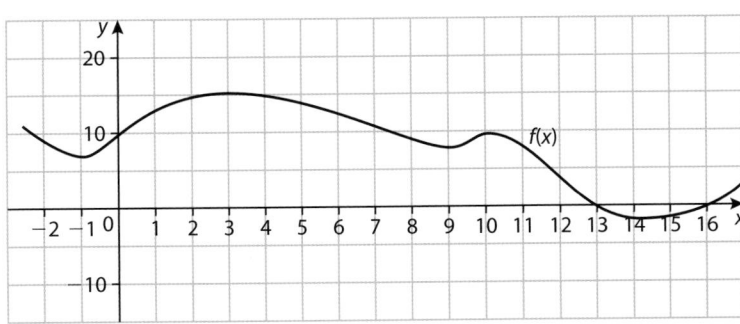

a) Identify the values of x for which $f'(x)$ is 0.

b) (i) Write down the intervals for which $f'(x) > 0$.
    (ii) Hence, identify the intervals for which $f(x)$ is increasing.

c) (i) Write down the intervals for which $f'(x) < 0$.
    (ii) Hence, identify the intervals for which $f(x)$ is decreasing.

Solution

a) We look at the graph of the function to find the places where its gradient is 0.

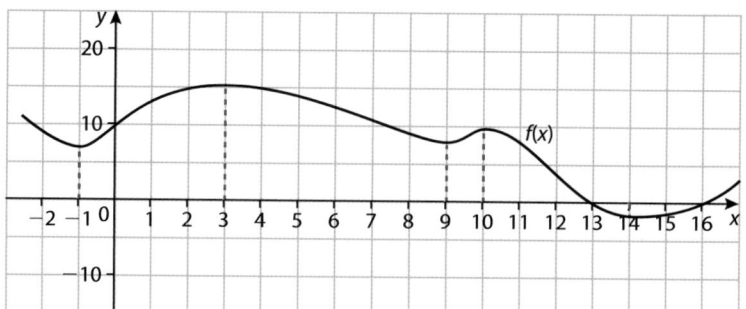

We can see the derivative is zero when $x = -1$, $x = 3$, $x = 9$, $x = 10$ and $x = 14$.

b) (i) From the graph, we see that the gradient of $f(x)$ is positive for the intervals $(-1, 3)$, $(9, 10)$ and $(14, \infty)$. These intervals are also therefore where $f'(x) > 0$.

    (ii) Hence, the intervals where $f(x)$ is increasing are: $[-1, 3]$, $[9, 10]$ and $[14, \infty)$.

c) (i) From the graph, we can see the gradient of $f(x)$ is negative for the intervals $(-\infty, -1)$, $(3, 9)$ and $(10, 14)$. These intervals are also therefore where $f'(x) < 0$.

    (ii) Hence, the intervals where $f(x)$ is decreasing are: $(-\infty, -1]$, $(3, 9)$, and $[10, 14]$.

● **Examiner's hint:** It is important to note that a particular point can be considered to be part of an interval where a function is increasing from one side, and part of an interval where the same function is decreasing on the other side.

Exercise 13.4

1. Identify the interval(s) over which the following functions are increasing.
    a) $g(x) = x^2 + 3x + 2$
    b) $h(x) = \frac{1}{3}x^2 - 2x$
    c) $f(x) = x^3 + 9x^2 + 24x + 30$
    d) $P(x) = 100x - \frac{25}{2}x^2$

2. Identify the interval(s) over which the following functions are decreasing.
    a) $f(x) = -\frac{1}{3}x + \frac{2}{5}$
    b) $D(t) = 100t - 10t^2$
    c) $A(x) = 20x - 5x^2$
    d) $g(x) = x^3 - 12x^2 + 45x - 50$

3. Use the graphs below to identify the coordinates of the points where the derivative of the functions is 0.

a)

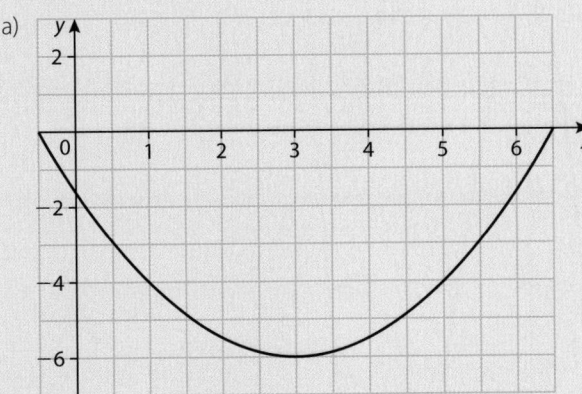

b)

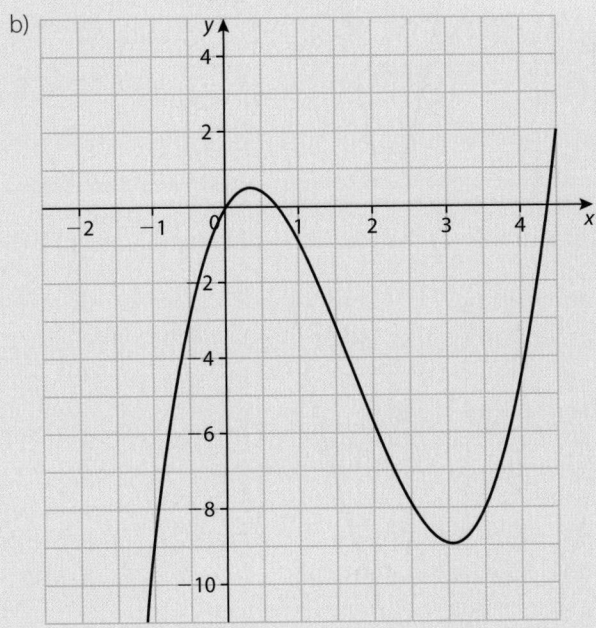

c)

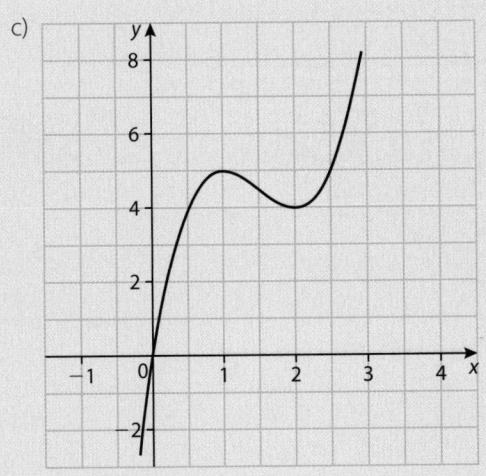

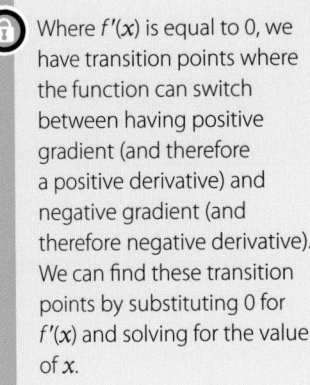

Where $f'(x)$ is equal to 0, we have transition points where the function can switch between having positive gradient (and therefore a positive derivative) and negative gradient (and therefore negative derivative). We can find these transition points by substituting 0 for $f'(x)$ and solving for the value of x.

d)

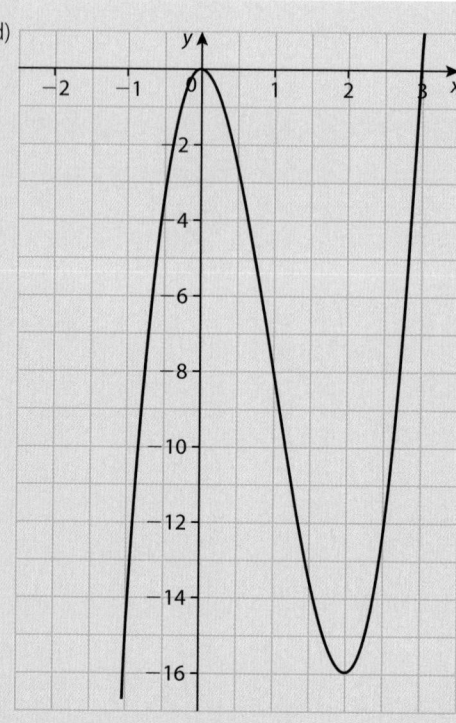

4. Use the graphs below to identify the intervals over which the functions are increasing.

a)

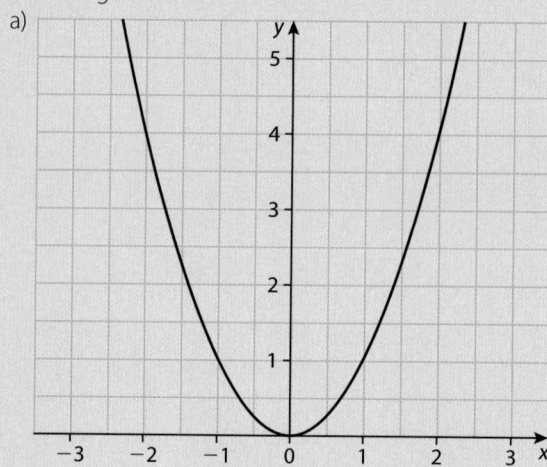

b)

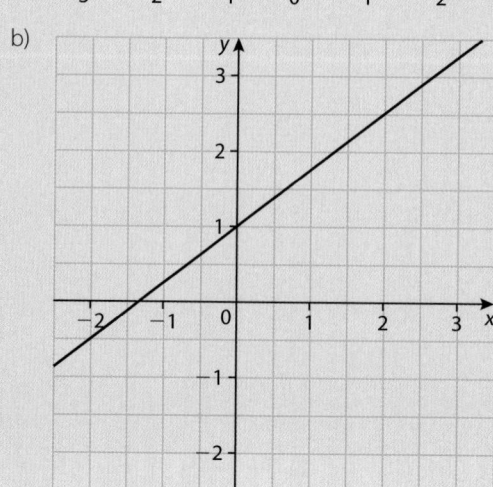

c)

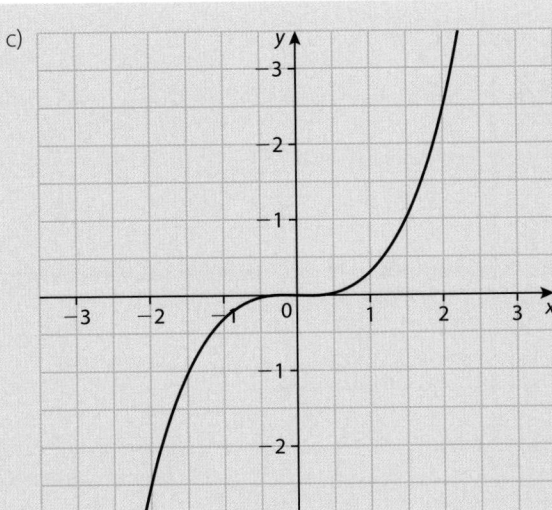

d)

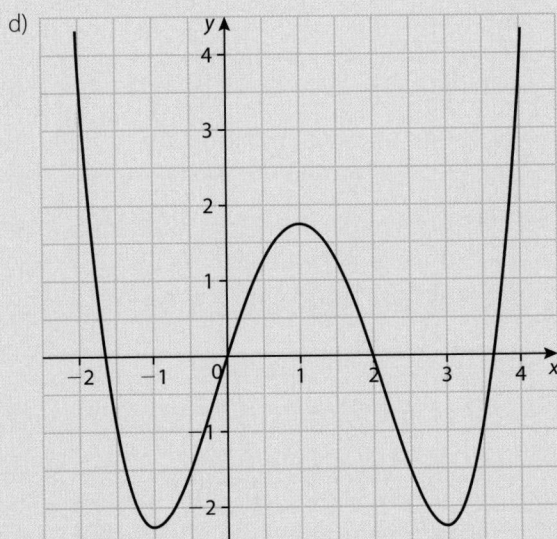

5. Use the graphs below to identify the intervals over which the functions are decreasing.

a)

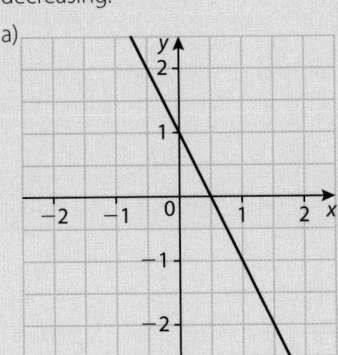

b)

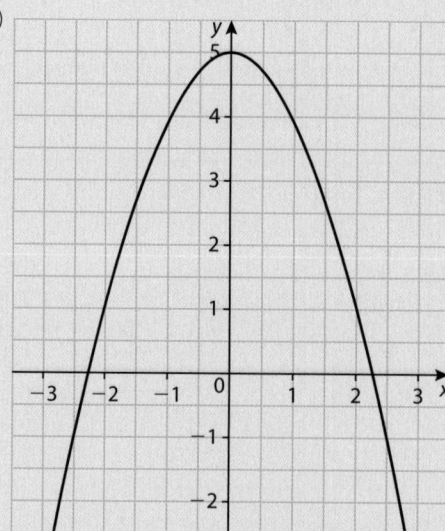

c)

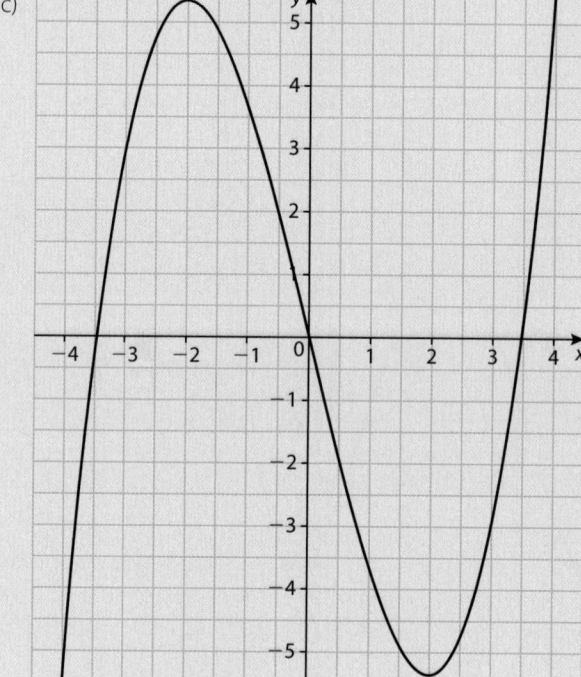

d)

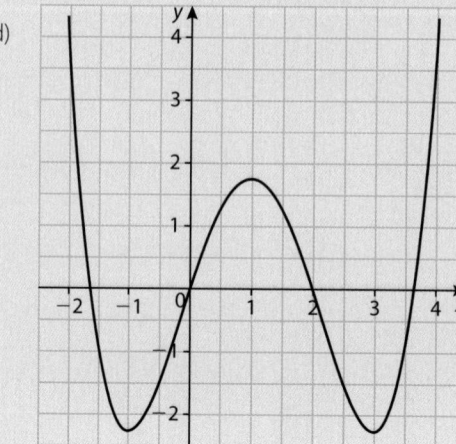

6. The total money in trillions of US dollars held by the Gaussian World Bank from 2000 to 2007 has been graphed below.

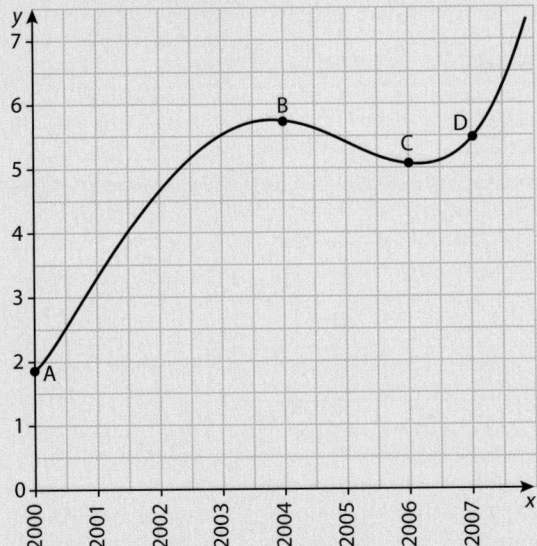

a) Identify the years during which the bank increased its holdings of US dollars.
b) For which points is the gradient of the function zero?
c) In which year between 2000 and 2007 did the bank have the greatest amount of US dollars?

7. It is known that a function is increasing on the interval [0, 3], decreasing on the interval [3, 5], and finally increasing on the interval [5, 8]. The function is only defined on the interval [0, 8] and $f(0) = -5$.
a) Sketch a *possible* curve for $f(x)$.
b) Determine the value(s) of x for which your function has a horizontal derivative.

8. When a golf ball is hit in the 2085 Lunar Olympics, the ball increases in height over the first 600 metres of its flight, then decreases in height thereafter, until it lands on the ground. The maximum height the ball reaches is 200 metres.
a) Sketch a parabola that matches the information given.
b) (i) Write down $f(0)$ and $f(1200)$.
   (ii) Write down the value of x which makes $f'(x) = 0$.
This function has the form $y = ax(1200 - x)$.
c) (i) Find $f'(x)$.
   (ii) Find the value of a.

 Alan Shepard hit two golf balls while on the moon, one of which travelled about 300 metres. This feat was made quite a bit more difficult as the inflexibility of his suit forced him to hit the shot with one hand.

9. The graph of $g(x)$ is shown below.

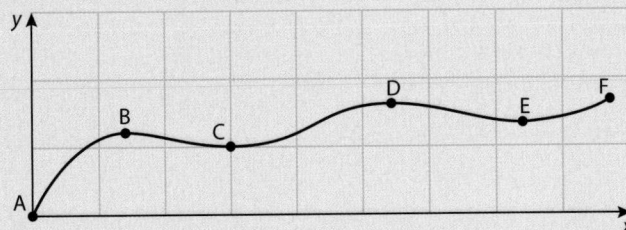

a) Identify the points where a tangent line to $g(x)$ would be horizontal.
b) List the intervals over which $g(x)$ is increasing.
c) Decide whether the following statements are true or false.
   (i) $g'(x)$ at B is equal to $g'(x)$ at C.

(ii) The slope of the secant line through A and B is less than the slope of the secant line through C and D.

(iii) The average rate of change from D to E is negative.

10. Use the following information to *sketch* the graph of $f(x)$.

$f(3) = 1$

$f(x)$ is increasing for all $x \in \mathbb{R}$.

The only value of x for which $f'(x)$ is zero is $x = 3$.

11. The function $f(x)$ is given by $f(x) = x^3 - 3x^2 + 3x$ for $-1 \leqslant x \leqslant 3$.

a) Differentiate $f(x)$ with respect to x.

b) Copy and complete the table below.

x	-1	0	1	2	3
$f(x)$		0	1	2	9
$f'(x)$	12		0		12

c) Use the information from b) to sketch the graph of $f(x)$.

d) Write down the gradient of the tangent to the curve at the point $(3, 9)$.

e) Write down the intervals of x for which $f(x)$ is

(i) increasing

(ii) decreasing.

12. The graph of $f(x)$ is given below.

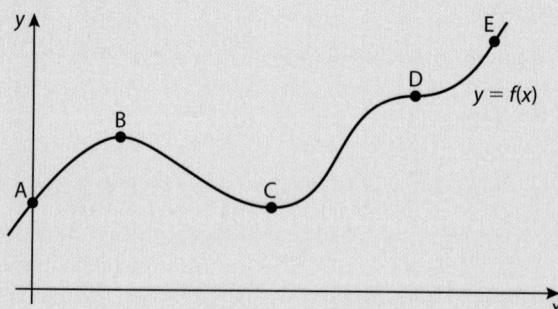

a) For each of the following points, state whether it is a maximum point, minimum point, or just a stationary point.

(i) B

(ii) C

b) Describe the gradient of the curve in passing from point B, through point C to point D.

c) State the intervals over which $f(x)$ is increasing.

d) D has coordinates $(a, f(a))$ and the x-coordinate at E is $a + 4$. Write an expression for the gradient of the line segment DE.

7.5	Values of x where the gradient of a curve is 0 (zero): solution of $f'(x) = 0$. Local maximum and minimum points.
7.6	Optimization problems.

When the gradient of a function is 0, there are three possibilities:

1. The function increases before the function has a gradient of 0, then decreases. In this case, we say that this transition point of $f(x)$ is a local maximum of the function.
2. The function decreases before the function has a gradient of 0, then increases. In this case, we say that this transition point of $f(x)$ is a local minimum of the function.
3. The function increases (or decreases) before the function has a gradient of 0, then continues increasing (or decreasing). In this case, we say that $f(x)$ only has a stationary point at the transition point.

Where the function has a maximum or minimum value, we call this an **extremum** (plural **extrema**) of the function.

 One of the languages from which mathematics 'borrows' a lot of its vocabulary is Latin. For example, the word 'extrema' has its roots in Latin.

Example 13.16

Identify the stationary points of the function $f(x) = \frac{1}{4}x^4 - x^3 - 5x^2 + 1$ and label each as a local maximum, local minimum, or neither.

Solution

First we find the stationary points of $f(x)$. To do this, we find the derivative of $f(x)$, set $f'(x) = 0$ and solve for x.

$$f'(x) = x^3 - 3x^2 - 10x$$
$$0 = x^3 - 3x^2 - 10x$$
$$0 = x(x^2 - 3x - 10)$$
$$0 = x(x - 5)(x + 2)$$

Hence, at $x = -2$, $x = 0$, and $x = 5$ the function $f(x)$ has stationary points. For each of these stationary points, we need to establish whether it is a local maximum, local minimum or neither.

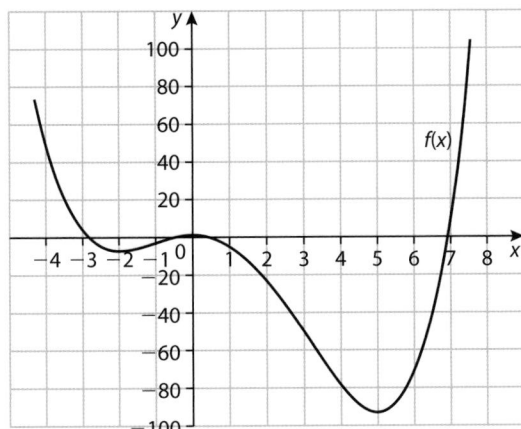

From the graph of $f(x)$, we can see that at $x = -2$ we have a local minimum, at $x = 0$ we have a local maximum, and at $x = 5$ we have a local minimum.

● **Examiner's hint:** The use of the word **absolute** here is *not* referring to the sign of the answer. The use of the word **global** instead is also common.

The largest value of a function over an interval is called the **absolute maximum**, and the smallest value is called the **absolute minimum**. To find these values, we find the local maxima and minima of the function, and then check the end points of the function.

Example 13.17

$f(x) = x^3 - 45x^2 + 600x + 20$

a) Find the local extrema and identify them as either a local maximum or a local minimum.

b) Find the coordinates of the absolute maximum and absolute minimum of the function in the interval $[0, 30]$.

Solution

a) First we find the derivative of $f(x)$.

$$f'(x) = 3x^2 - 90x + 600$$

Next we set $f'(x) = 0$ and solve for x.

$$0 = 3x^2 - 90x + 600$$
$$0 = x^2 - 30x + 200$$
$$0 = (x - 10)(x - 20)$$

Hence, we have local extrema at $x = 10$ and $x = 20$. To identify them as either maxima or minima, we can use the derivative.

$$f'(9) = 3(9)^2 - 90(9) + 600 = 33$$
$$f'(11) = 3(11)^2 - 90(11) + 600 = -27$$

Hence, at $x = 10$ we must have a local maximum, since the function increases (its derivative is positive) up to the value at $x = 10$, then decreases (its derivative is negative).

$$f'(19) = 3(19)^2 - 90(19) + 600 = -27$$
$$f'(21) = 3(21)^2 - 90(21) + 600 = 33$$

Hence, at $x = 20$ we must have a local minimum, since the function decreases down to the value at $x = 20$, then increases.

b) We also need to check the endpoints, $x = 0$ and $x = 30$.

$$f(0) = (0)^3 - 45(0)^2 + 600(0) + 20 = 20$$
$$f(10) = (10)^3 - 45(10)^2 + 600(10) + 20 = 2520$$
$$f(20) = (20)^3 - 45(20)^2 + 600(20) + 20 = 2020$$
$$f(30) = (30)^3 - 45(30)^2 + 600(30) + 20 = 4520$$

So the absolute maximum of the function (in the interval $[0, 30]$) occurs at the point $(30, 4520)$ and the absolute minimum occurs at $(0, 20)$.

One application of using calculus to find extrema is in the maximization and minimization of functions from real-life problems. In all of these problems we will have to identify the function that needs to be maximized or minimized, then use our procedure from before to find its maximum.

Such applications are known as optimization problems where the calculus is intended to find the ideal solution, whether it be a maximum or minimum.

Example 13.18

Ben has 200 metres of electric fence. With this fence he creates a rectangular enclosure.

a) Write an expression for the length, l, of the enclosure, in terms of the width, w.

b) Use your expression from part a) to write an expression for the area of the enclosure in terms of the width of the enclosure.

c) (i) Using your answer from part b), find the dimensions of the enclosure Ben can create with the maximum possible area.

(ii) What is the maximum area Ben can enclose with his electric fence?

Solution

a) Since the perimeter of the fence must be 200 metres, we know that $200 = 2l + 2w$ and so $200 - 2w = 2l$ and $l = 100 - w$.

b) The area of a rectangle is length × width, so $A = w(100 - w)$.

c) (i) We know $A(w) = 100w - w^2$. We need to find the derivative of $A(w)$, and use it to find the extrema of the function. We then find the absolute maximum on the interval $[0, 100]$. We know the minimum width of the rectangle is 0 and the maximum width is 100.

$$A'(w) = 100 - 2w$$
$$0 = 100 - 2w$$
$$2w = 100$$
$$w = 50$$

Hence, we need to check $w = 0$, $w = 50$ and $w = 100$.

$$A(0) = 0$$
$$A(50) = 2500$$
$$A(100) = 0$$

We can see very easily that the dimensions of the enclosure should be 50 metres by 50 metres.

(ii) Using our work from part c)(i) we see that the maximum area is 2500 m^2.

Up until now many of the functions we have used have been functions of x such as $f(x) = ax^n + bx + c$. In these cases the gradient function or derivative is written in the form $f'(x) = anx^{n-1} + b$.

However, when working with applied problems it is often better to write the equations in terms of the variables being used.

e.g. If you are looking at Area (A) and length (l) then the original function may be better written as $A(l) = cl^n + dl$ where c and $d \in \mathbb{Z}$, and the derivative would be expressed as $A'(l) = cnl^{n-1} + d$.

Other examples relate to time, $f(t)$ when the derivative is written as $f'(t)$.

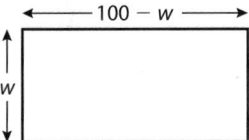

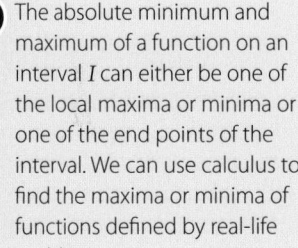

The absolute minimum and maximum of a function on an interval I can either be one of the local maxima or minima or one of the end points of the interval. We can use calculus to find the maxima or minima of functions defined by real-life problems.

Calculus is not the only way to find the maximum or minimum of a function, but it does seem to be one of the fastest ways. Is the greater flexibility of calculus to solve problems worth the extra difficulty it presents in learning it?

Example 13.19

A ball is thrown upward from the ground so that its distance, s metres after t seconds in motion is given by $s(t) = 30t - 4.9t^2$.

    a) Sketch the graph of $s(t)$.

    b) What is the object's initial position, that is when $t = 0$?

    c) Determine $\dfrac{ds}{dt}$.

    d) What is the maximum height of the ball?

    e) How much time will elapse before the ball returns to the ground?

Solution

a)

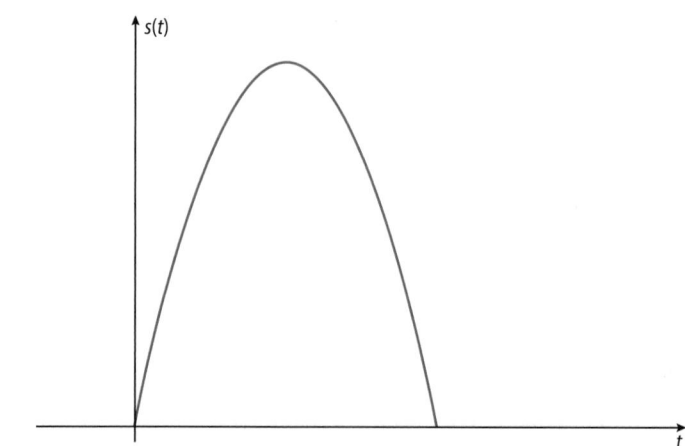

b) Initial position occurs when $t = 0$.

$$s(0) = 30(0) - 4.9(0)^2 = 0$$

c) $s(t) = 30t - 4.9t^2$

$$\frac{ds}{dt} = s'(t) = 30 - 9.8t$$

d) Maximum height occurs when

$$\frac{ds}{dt} = 0$$

$30 - 9.8t = 0$

$9.8t = 30$

$t = \dfrac{30}{9.8} = 3.06$ seconds

Maximum height occurs at 3.06 seconds.

We can substitute into the original equation.

$s(t) = 30t - 4.9t^2$

$s(3.06) = 30(3.06) - 4.9(3.06)^2 = 0 = 45.9$ metres to 3 s.f.

Alternatively the GDC can be used to find both the time to reach the maximum height and the value of the maximum height.

Enter the equation $Y1 = 30x - 4.9x^2$. Set an appropriate window such as the one given below, then Press 2nd, TRACE (CALC), ∨ …, 4: maximum, ENTER,

Move cursor to left side of the maximum point, ENTER,

Move cursor to right side of turning point, as shown, ENTER,

Move cursor to Maximum Point and press ENTER

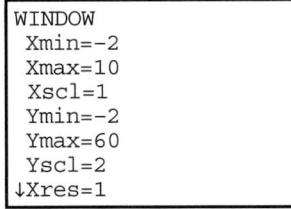

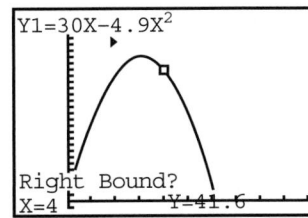

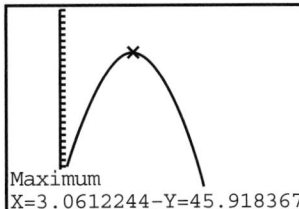

Maximum point is $(3.06, 45.9)$

That is, the time to reach the highest point is 3.06 seconds.

Maximum height reached by the ball is 45.9 metres.

e) The time taken for the ball to land again is given by
$s(t) = 30t - 4.9t^2 = 0$

Algebraically we can solve this equation by firstly factorizing the equation.

$s(t) = t(30 - 4.9t) = 0$

$t(30 - 4.9t) = 0$

Then $t = 0$ or $30 - 4.9t = 0$

$t = 0$ is our initial position.

Solving $30 - 4.9t = 0$ will give the time taken for the ball to return to
$t = \dfrac{30}{4.9} = 6.12$ seconds.

Alternatively we can use the symmetry of the parabola to determine the time taken. The maximum height is reached at 3.06 seconds, so the total time taken for the ball to return to the ground is $2 \times 3.06 = 6.12$ seconds.

Exercise 13.5

1. Find the coordinates of the local maxima and minima of the functions below.

a) $f(x) = 4x^2 - 8x - 3$

b) $f(x) = 80x - x^2$

c) $f(x) = x^3 - \dfrac{15}{2}x^2 - 18x + 12$

d) $f(x) = 3x^4 - 6x^2$

2. Estimate the coordinates of the local maxima and minima of the graphs of the functions below.

a)

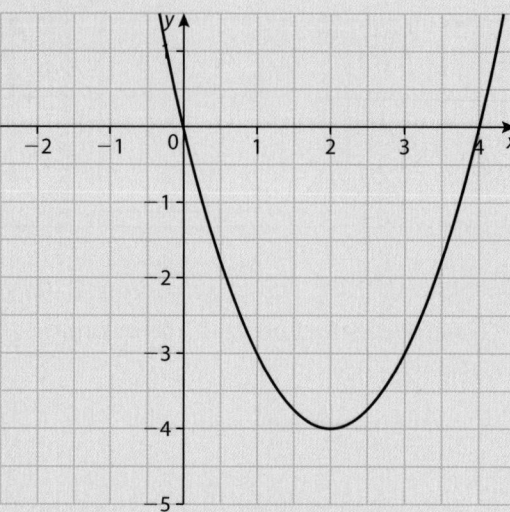

b)

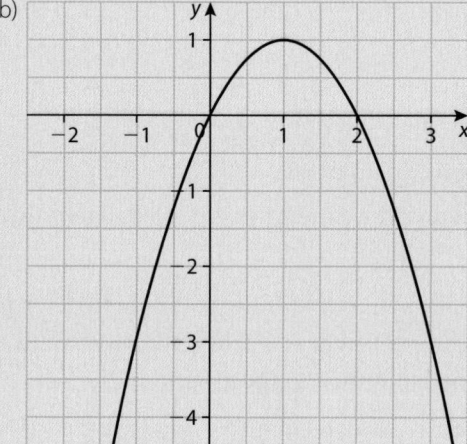

c)

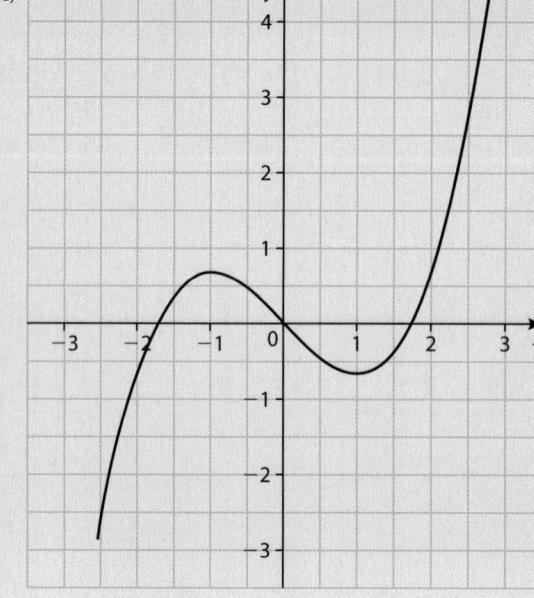

d)

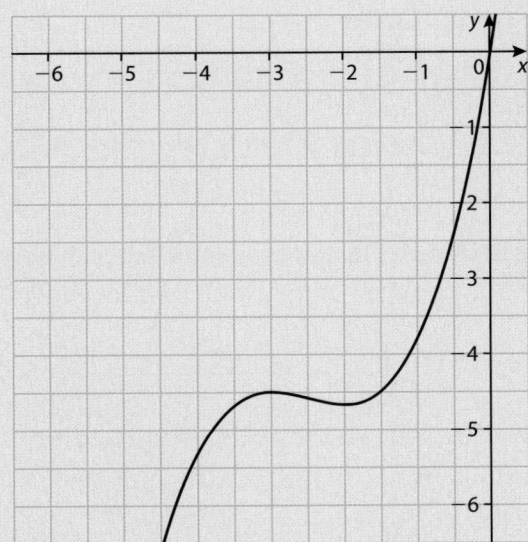

3. Find the coordinates of the absolute maximum functions over the indicated intervals.

a) $f(x) = \frac{3}{4}x + 3$ over $[-2, 5]$

b) $f(x) = \frac{1}{2}x(6 - x)$ over $[1, 6]$

c) $f(x) = x^2 - 4$ over $[-3, 4]$

d) $f(x) = 16x^3 + 84x^2 + 72x$ over $[-4, 0]$

4. Find the absolute maximum and minimum from the graphs of the functions given below, over the indicated intervals.

a) $I = [0, 4]$

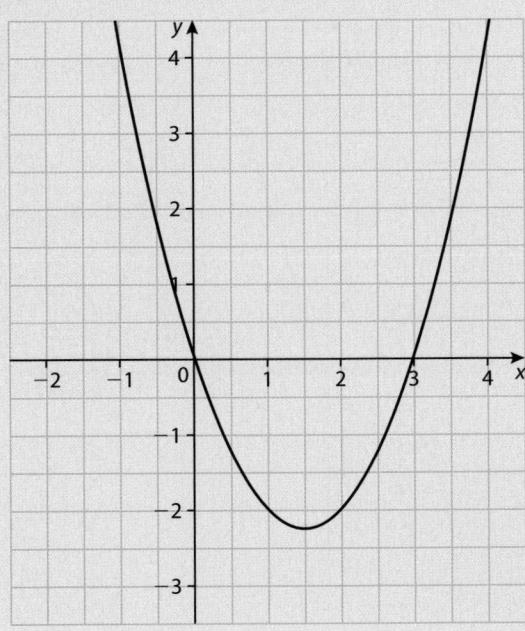

b) $I = [0, 4]$

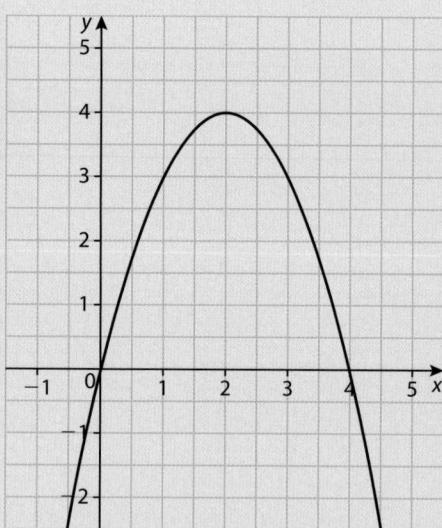

c) $I = [0, 4]$

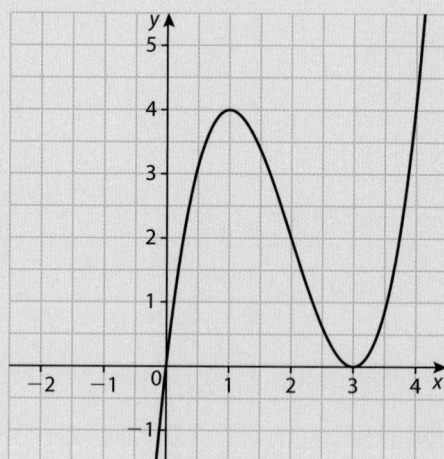

d) $I = [-2, 4]$

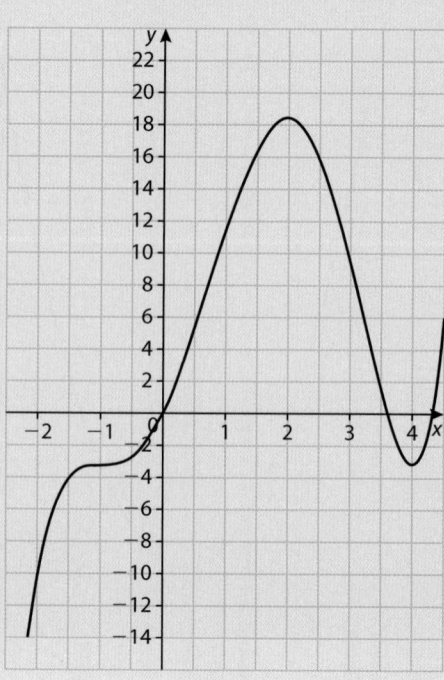

5. Let $f(x) = x^2 - 4x + 4$.

   a) Find the values of x such that $f(x) = 0$.

   b) (i) Find $f'(x)$.

      (ii) Solve $f'(x) = 0$ for x.

   c) Find the coordinates of the absolute maximum of $f(x)$ on the interval $[0, 5]$.

6. The Euclidean Widget Company creates x widgets per month. The profit, in thousands of dollars, the company makes each month can be found using the formula $P(x) = x(200 - x)$.

   a) Find $P'(x)$.

   b) Find the number of widgets for which the profit is a maximum.

   c) Find the maximum profit (in millions of dollars) earned by the company in a month.

7. A square piece of cardboard that measures 1.2 m by 1.2 m has four identical squares cut out of its corners, as shown in the diagram. The edges of the cardboard are then folded up to make a box with its top open.

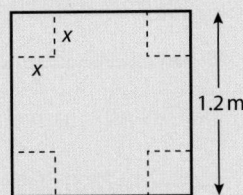

   a) Write an expression for the length of the base of the box in terms of x.

   b) Write an expression for the volume of the resulting box.

   c) Find the maximum volume of the box that can be constructed in this fashion.

8. The curve $y = f(x)$ has its only local minimum value at $x = a$ and its only local maximum value at $x = b$.

   a) If $a < 0$ and $b > 0$, *sketch* a possible curve of $y = f(x)$ indicating clearly the points $(a, f(a))$ and $(b, f(b))$.

   b) Given that $0 < h < 1$ and $b - a > 1$, are the following statements about the curve $y = f(x)$ true or false?

      (i) $f(a + h) < f(a)$

      (ii) $f'(b - h)$ is positive.

      (iii) The tangent to the curve at the point $(a, f(a))$ is parallel to the vertical axis.

      (iv) The gradient of the tangent to the curve at the point $(a, f(a))$ is equal to zero.

      (v) $f(a - h) < f(a) < f(a + h)$

9. The function g is defined as follows

$$g : x \mapsto px^2 + qx + c, \qquad \text{where } p, q, c \in \mathbb{R}.$$

   a) Find $g'(x)$.

   b) If $g'(x) = 2x + 6$, find the values of p and q.

   c) $g(x)$ has a minimum value of -12 at the point A. Find

      (i) the x-coordinate of A

      (ii) the value of c.

10. The perimeter of a rectangle is 24 metres.

a) The table shows some of the possible dimensions of the rectangle. Find the values of a, b, c, d, and e.

Length (m)	Width (m)	Area (m²)
1	11	11
a	10	b
3	c	27
4	d	e

b) If the length of the rectangle is x m and the area is A m², express A in terms of x only.

c) What are length and width of the rectangle if the area is to be a maximum?

11. A farmer wishes to enclose a rectangular field using an existing fence for one of the four sides.

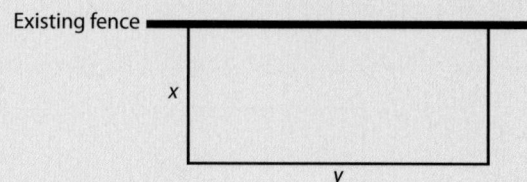

Existing fence

x

y

a) Write an expression, in terms of x and y, that shows the total length of the new fence.

b) The farmer has enough material for 2500 metres of fence. Show that $y = 2500 - 2x$.

c) $A(x)$ represents the area of the field in terms of x.

   (i) Show that $A(x) = 2500x - 2x^2$.

   (ii) Find $A'(x)$.

   (iii) Hence or otherwise, find the value of x that produces the maximum area of the field.

   (iv) Find the maximum area of the field.

12. The velocity v, in ms⁻¹, of a kite, after t seconds, is given by
$$v = t^3 - 4t^2 + 4t$$

a) What is the velocity of the kite after

   (i) one second?

   (ii) half a second?

b) Calculate the values of a and b in the table below.

t	0	0.5	1	1.5	2	2.5	3	3.5	4
v	0			a	0	0.625	b	7.88	16

c) (i) Find $\dfrac{dv}{dt}$ in terms of t. Find the value of t at the local maximum and minimum values of the function.

   (ii) Explain what is happening to the function at its local maximum point. Write down the gradient of the tangent to its curve at this point.

d) On graph paper, draw the graph of the function $v = t^3 - 4t^2 + 4t$, for $0 \leqslant t \leqslant 4$.

   Use a scale of 2 cm to represent 1 second on the horizontal axis and 2 cm to represent 2 ms⁻¹ on the vertical axis.

e) Describe the motion of the kite at different times during the first 4 seconds. Write down the intervals corresponding to changes in motion.

The questions below are meant as a review of the entire Mathematical Studies syllabus content. For the most part, there will be one question for each content item. After starting with Prior Learning Topics (PLT), the questions will be ordered in accordance with the Mathematical Studies SL Guide and the syllabus content therein as much as possible. The questions will be designated with the Assessment Statement numbering (e.g. AS2.1). They are not (necessarily) meant to be challenging, or new and intriguing. They are simply review questions to help you remember the concepts of the required content and to help prepare you for the upcoming IB Mathematical Studies exam.

1. Simplify: (PLT)
   a) $-(2 - 3 \times 7 + 5)$
   b) -2^2
   c) $\frac{2}{3} + \frac{6}{7}$
   d) $3.7 + 2\frac{4}{5}$
   e) $0.25\overline{25}$
   f) $\sqrt{3^2 + 4^2}$

2. List the first ten prime numbers. (PLT)

3. Write the prime factorization for: (PLT)
   a) 72
   b) 244

4. Write the positive integer factors for: (PLT)
   a) 80
   b) 68

5. Write the first five multiples for: (PLT)
   a) 7
   b) 26

6. Find the GCF and the LCM for: (PLT)
   a) 72, 244
   b) 24, 80, 144

7. 8% of a number is 20. Find the number. (PLT)

8. Greg wants to make a profit of $300\,000$ from the sale of his house, with an agent receiving a 7% commission on the sale. What should he sell his house for to the nearest dollar? (PLT)

9. Diane, who is 1.8 metres tall, wanted to determine the height of the tree in her front yard. She measured her shadow to be 3 metres and the shadow of the tree to be 8 metres. How tall is the tree? (PLT)

10. Expand and simplify: (PLT)
    a) $-(2x - 7)$
    b) $(x - 3)(x + 5)$
    c) $(a + 5)^2$
    d) $(2x - 7)^2(x + 1)$
    e) $3(x + 4)(x - 4)$
    f) $(y + 2)^3$

11. Factorize: (PLT)
    a) $3x - 6y + 15$
    b) $x^2 - x - 56$
    c) $x^2 + 4x + 21$
    d) $x^2 - 36$
    e) $3x^2 - 27$
    f) $2x^2 - 9x - 5$

12. Solve for the indicated variable in each equation. (PLT)

    a) For y: $ax + by + d = 0$ b) For b: $P = \dfrac{360}{b}$

    c) For r: $A = 4\pi r^2$, $r > 0$ d) For a: $A = \frac{1}{2}ab\sin C$

13. Evaluate each formula with the given information. (PLT)

    a) $V = \frac{1}{3}BH$; $B = 4\pi, H = 6$

    b) $A = \pi r l$; $r = 4.5$, $l = 6.8$

    c) $y = -x^2 + 3x - 5$; $x = -2$

14. Solve each equation. (PLT)

    a) $5x - 1 = 8$

    b) $-2(x - 5) = -12$

    c) $3x + 5 = 7x - 19$

15. Solve each system of equations. (PLT)

    a) $3x + 4y = 9$ b) $x - y = 7$

       $5x - 2y = 15$ $y = 2x + 1$

16. Evaluate each of the following. Leave the answer in fractional form. (PLT)

    a) 2^{-1} b) 3^{-3} c) 5^0

    d) 4^5 e) 2^{10} f) 7^{-3}

17. Solve each inequality and graph the solution on a number line. (PLT)

    a) $2x - 1 > 7$ b) $-3x + 2 \leqslant 11$

18. Graph each interval on a number line. (PLT)

    a) $x > 5$ b) $-3 < x \leqslant 5$ c) $6 > x$

19. a) Draw an angle whose measure is:

       (i) $30°$ (ii) $120°$.

    b) Construct a line (AB) that is perpendicular to line (CD) at the point E.

    c) Sketch a diagram of line (EF) that intersects plane G at point H.
      (PLT)

20. Use a sheet of 2-mm IB graph paper. (PLT)

    a) Write the equation of the x-axis.

    b) Write the equation of the y-axis.

    c) Name each quadrant.

    d) Plot each ordered pair:

       (i) $(2, 3)$ (ii) $(-4, 5)$

       (iii) $(-2, -4)$ (iv) $(3, -1)$

21. Find the midpoint of the line segment whose end points are:

    a) the ordered pairs $(-2, 5)$ and $(4, -3)$

    b) the ordered triples $(5, 6, 8)$ and $(-2, -3, 4)$. (PLT)

22. Find the length of the line segment whose end points are:

    a) the ordered pairs $(-2, 5)$ and $(4, -3)$

    b) the ordered triples $(5, 6, 8)$ and $(-2, -3, 4)$. (PLT)

23. The following data indicates the number of television shows 100 people watch in a randomly selected community. (PLT)

Television show	Number
Sitcom	15
Reality	30
Drama	20
Action	10
News	25

a) Construct a bar chart that depicts the data collected.
b) Construct a pie chart that describes the data collected.

24. Colleen took a survey of home sales during 2007 and reported the following data. (PLT)

City	Home sales
City A	10 000
City B	8000
City C	5000
City D	2500

Let one house picture represent 1000 homes and construct a pictogram that describes the data.

25. Solve each problem. (PLT)
a) If 1 US$ = 0.6794 €, how many euros is 1500 US$?
b) If 1 ¥ = 0.010 146 AUS$, how many ¥ is 500 AUS$?

26. List in roster form:
a) the set of $\mathbb{N}$ (Hint: Don't forget the '…')
b) the set of non-negative integers
c) the set of positive rational numbers, $\mathbb{Q}^+$
d) the set of integers, $\mathbb{Z}$ (AS1.1)

27. Let $x = 2.31 \times 10^3$ and $y = -4.67 \times 10^{-1}$, find:
a) $x \cdot y$ as an integer to 3 significant figures.
b) $x + y$ in the form $a \times 10^k$, where $1 \leqslant a < 10$ and $k \in \mathbb{Z}$.
c) $x - y$ to the nearest integer.
d) $\frac{x}{y}$ as a real number correct to 2 decimal places. (AS1.2, 1.3)

28. In finding the area of a triangle by using the formula $A = \frac{1}{2}ab \sin C$, a was found to be 8.0243, b was found to be 10.548 and $\sin C$ was found to be 0.838 47, all correct to 5 s.f. In the computation of the area, the student calculated the area to be 35.3 square units to 3 s.f.
a) What numbers, for a, b, and $\sin C$, were (probably) used to arrive at the answer 35.3?
b) What should the correct answer be to 3 s.f.?
c) What did the student do wrong when he arrived at his answer in part a)? (AS1.2)

29. The number of spectators at a football game was counted as 64 258. Over the loudspeaker it was announced that the attendance was 64 000.

a) What was the error?

b) What was the absolute percentage error? (AS1.2)

30. Is $\dfrac{5}{2}$ a good estimate for the gradient of the line in the diagram? Explain your answer.

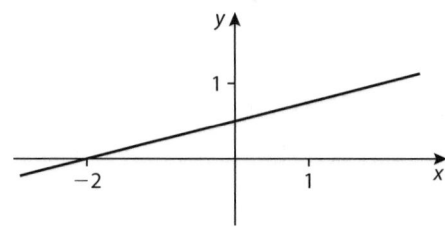

(AS1.2)

31. Perform the following conversions.
    a) 30 litres to cubic feet, where $1 \text{ ft}^3 = 28.32 \text{ l}$
    b) 4 quarts to litres, where $1 \text{ l} = 1.057$ qt
    c) 150 pounds to kilograms, where $1 \text{ kg} = 2.205 \text{ lb}$
    d) 200 kilometres to miles, where $1 \text{ mile} = 1.609 \text{ km}$
    e) 350 milligrams to grams
    f) 55 km/hr to ms^{-1} (metres per second) (AS1.4)
    g) Given $F° = \frac{9}{5}C° + 32$, find the Celsius temperature if the Fahrenheit temperature is 75°.

32. Consider the currency conversion: 1 Swiss franc = 0.6094 euro.
    a) Find how many euro you would receive for 2500 Swiss francs.
    b) Find how many Swiss francs you would receive for 2000 euro.
    (AS1.5)

33. A currency exchange company will exchange Canadian dollars (CAN$) for British pounds (GB£). The company charges a 1.5% commission of the Canadian dollar amount to be exchanged. 2000 CAN$ are to be exchanged for GB£.
    a) What is the amount of commission in Canadian dollars?
    b) How many Canadian dollars are to be exchanged?
    c) Find the number of pounds that will be received when the exchange rate is 1 CAN$ = 0.5113 GB£. (AS1.5)

34. Using your GDC, find the solution for the following pair of linear equations, correct to 3 s.f.
    $$2x + 3y = 11$$
    $$3x - 5y = 3$$
    (AS1.6)

35. Solve for x by factorizing: $2x^2 - 5x - 3 = 0$. (AS1.6)

36. Solve for x by using the quadratic formula: $2x^2 - 7x + 1 = 0$. Give your answer
    a) exactly
    b) to 3 significant figures. (AS1.6)

37. Find the 50th term of the arithmetic sequence: 5, 8, 11, 14, ... (AS1.7)

38. Pam deposits £5000 in an account on January 1. On the first of each month thereafter, she deposits £200. How much money will be in the account at the *end* of the 20th month?
(Hint: Draw a diagram.) (AS1.7)

39. Find the sum of the first 50 terms of $3 + 7 + 11 + 15 + ...$ (AS1.7)

40. Jane starts a job on a salary of €26 000. She is given an annual salary increase of €500 at the end of each calendar year. Determine the total amount she earns over the first 10 years of employment. (AS1.7)

41. Find the 8th term of the geometric sequence: 1, 3, 9, ... (AS1.8)

42. A boat depreciates in value by 12% per year. If the purchase price was $175 000,
a) what would its value be at the end of 10 years?
b) when, to the nearest year, would its value be less than $10 000? (AS1.8)

43. Expand and then find the sum of the geometric series: $\sum_{i=1}^{10} 3^i$. (AS1.8)

44. Keaton drops a ball from a height of 20 metres. It always rebounds to a height of 90% of the height from which it dropped. (Hint: Draw and label an accurate diagram!)
a) How high will the fifth bounce be?
b) How many (total) metres did the ball travel when it hit the ground for the seventh time? (AS1.8)

45. Alexa invested $2300 in an account that pays 8% per year compounded monthly for 10 years.
a) What will be the total amount in the account after 10 years, assuming no additional money is invested and no money is withdrawn?
b) How much interest is earned on the initial investment? (AS1.9)

46. Sharlene bought a new car for $25 000. If the car loses 15% of its value per year, then
a) how much of its original value did the car lose after one year?
b) what was the value of the car after one year?
c) what was the value of the car after six years? (AS1.9)

47. When will $400 double if the annual interest rate is 5% compounded yearly? (AS1.9)

48. Classify each as discrete or continuous data:
a) the weight of a child from age 5 to 6
b) the time it takes to watch a TV programme
c) the number of students in a room

d) the height of a girl from age 3 to 6
e) the number of airline flights per day
f) incomes of first-year teachers
g) lifetime of an AA battery
h) capacity of water used monthly by residents of London   (AS2.1)

49. Stella collected data for the number of meals per week 50 newly-wed couples ate outside the home, during the first month of marriage in the city of Knotlonsum. The results are tabulated below.

Meals per week	Frequency	Cumulative frequency
1	4	
2	7	
3	9	
4	12	
5	10	
6	8	

a) Draw a bar graph (a frequency polygon) for the data. Use the scale 1 cm = 1 division on the vertical axis and 1 cm = 1 bar width on the horizontal axis. Space the bars 1 cm apart.
b) Find the mean number of meals per week.
   (Hint: L_1 = Meals per week and L_2 = Frequency, and then 1-Var Stats.)
c) What is the modal number of meals eaten?
d) What is the median number of meals eaten?
e) Find the standard deviation of the number of meals per week.
f) Fill in the 'Cumulative frequency' column.
g) In the Cumulative frequency column, what does the number '20' represent?   (AS2.2, 2.5, 2.6)

50. The following data represents the number of pages read on the day just before term finals for 100 randomly selected high school students.

Pages read	Frequency f	Mid-interval MI	$f \cdot MI$	Cumulative frequency
0–10	15			
11–21	20			
22–32	30			
33–43	22			
44–54	13			
Totals				

a) Fill in the table above.
b) Calculate the mean number of pages read, to the nearest page.
c) Find the modal class.

d) Find the median class.

e) In the Cumulative frequency column, what does the number '87' describe?

(AS2.3, 2.4, 2.5)

51.

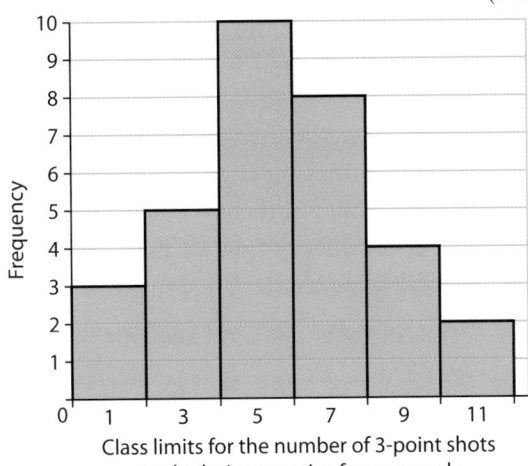

Class limits for the number of 3-point shots made during practice for one week.

(AS2.2, 2.5, 2.6)

The numbers on the horizontal axis represent the class mid-intervals.

a) Find the mean of the data. (Hint: Show work.)

b) Find the median *class* of the data.

c) Find the modal *class* of the data.

d) Find the standard deviation of the data.

(Hint: L_1 = mid-interval, L_2 = Frequency, and then 1-Var Stats.)

52. Find the range, median, first and third quartiles, the interquartile range, the standard deviation, and the mode of each data set.

a) 5, 8, 10, 11, 14, 16, 17, 18, 19, 20, 29

b) 6, 7, 3, 9, 10, 3, 8, 1, 6, 2, 0, 18

c) 253, 278, 254, 237, 293, 200, 283, 275, 255, 274, 203, 212, 294, 264, 101

(AS2.4)

53. Draw a box-and-whisker diagram (a box plot) for the data in Questions 52 a), b), and c) above.

(AS2.4)

54. Warren collected the following data, which represents the keying speed (in number of words per minute) recorded during a final test for a data entry job.

12	69	21	58	42	39	29	44	54	33	12	35	56
50	43	32	14	36	46	36	67	28	22	27	37	57
22	48	65	62	25	36	41	51	38	44	45	39	45

a) Find the median.

b) Find the mode.

c) Find the range.

d) Find the interquartile range.

(AS2.5, 2.6)

The following problem is very long as it involves many graphs and poses many questions. It reviews all of the concepts in AS 2.3, 2.4 and 2.5 for continuous data. Carefully fill in the chart below. All of the questions and graphs rely on successful completion of the chart. The data values are the same as in Question 54 to help you fill in the chart accurately. The data is now continuous instead of discrete.

55. The following data were obtained from a survey of 39 randomly selected IB students from King High School. Each value represents the time, in minutes, that a student travels in order to get to school. Construct a frequency distribution for the data, using seven classes, by completing the following chart.

12	69	21	58	42	39	29	44	54	33	12	35	56
50	43	32	14	36	46	36	67	28	22	27	37	57
22	48	65	62	25	36	41	51	38	44	45	39	45

Class limits	Class boundaries	Tally marks	Frequency	Class mid-interval	Cumulative frequency (cf)	Relative cf
12–20					3	
		𝓣𝓗𝓛 //				
	29.5–38.5					
			10			0.718
					33	
				61		
66–74			2			

Using the above information, construct the following graphs.
(Hint: Be very careful! Wrong chart means wrong graphs!)

a) Histogram. Use the scale, 2 cm = 9 minutes on the x-axis and 1 cm = 1 student on the y-axis. (Hint: It might be helpful to turn your graph paper horizontally. Label the x-axis starting with 11.5 at the end of the second cm from 0, and then 20.5 at the end of the fourth, etc.)

b) Cumulative frequency graph (cf). Use the scale 2 cm = 9 minutes on the x-axis and 1 cm = 5 students on the y-axis. (Hint: Make the division marks on the x-axis the same as in a) above and don't forget to use the upper boundaries.)

c) Percentile graph. Use the scale, 2 cm = 9 minutes on the x-axis and 1 cm = 0.10 marks (10%) on the y-axis. (Hint: Make the division marks on the x-axis the same as in a) above and don't forget to use the upper boundaries.)

Answer the following questions, using, when necessary, the above chart and/or graphs.

    d) What is the interval width?

    e) What is the upper boundary for the first class?

    f) What is the lower boundary for the seventh class?

    g) What does a 'cf' graph help you find?

    h) Find the number of minutes that represents the 80th percentile. (Hint: Draw lines on your percentile graph. Note that each small division mark on the x-axis $= \frac{9}{10} = 0.9$.)

    i) Comment on the meaning of the number you found in part h) above.

    j) Which number represents the:

      (i) 50th percentile? (Hint: Draw lines on your percentile graph.)

      (ii) first quartile? (Hint: Draw lines on your percentile graph.)

      (iii) third quartile? (Hint: Draw lines on your percentile graph.)

    k) Use the mid-interval values and find the mean of the data to 4 s.f. (Hint: 1-Var Stats, L_1, L_2.)

    l) Find the mean of the data by adding the data points together and dividing by 39. (Hint: 1-Var Stats.)

    m) Comment on your answers from k) and l) above.

    n) Find the median by drawing lines on the cumulative frequency graph. (Hint: Each small division mark on the x-axis $= \frac{9}{10} = 0.9$.)

    o) Find the median by using your GDC.

    p) Comment on your answers from n) and o) above. (AS2.3, 2.4, 2.5)

56. Given the following numbers: 20, 25, 40, 70, 82, without using your calculator, find:

    a) $\bar{x}$

    b) σ_x (AS2.6)

57. Construct the truth table property for each symbol: $\neg, \wedge, \vee, \Rightarrow, \underline{\vee}, \Leftrightarrow$. (AS3.1)

58. Use one or more words to describe each symbol: $\neg, \wedge, \vee, \Rightarrow, \underline{\vee}, \Leftrightarrow$. (AS3.1)

59. Construct a truth table for the following:

    a) $\neg P \Rightarrow Q$

    b) $(P \Rightarrow Q) \Leftrightarrow (\neg Q \Rightarrow P)$

    c) $\neg (P \vee Q) \Rightarrow (\neg P \wedge \neg Q)$

    d) $(P \underline{\vee} Q) \wedge P$ (AS3.1, 3.2)

60. Draw the corresponding Venn diagram for each logic statement.

    a) $\neg P$

    b) $P \wedge Q$

    c) $P \vee Q$

    d) $P \underline{\vee} Q$

    e) $P \Rightarrow Q$ (Hint: Show a point x in the inside circle.)

    f) $P \Leftrightarrow Q$ (AS3.2)

61. Translate the following argument to symbolic form.
    'If it is raining outside, I will get wet. I got wet. Therefore, it is raining outside.'
    (Use R: 'It is raining outside' and W: 'I got wet'.)   (AS3.2)

62. Is the argument, in Question 52 above, valid? Explain your answer.
    (Hint: Make a truth table.)   (AS3.3)

63. a) Complete the truth table shown.
    b) Under what conditions is the compound statement true? (Hint: Look for 'T' in the right column and then look at the two left columns. Find the corresponding values of P and Q.)   (AS3.3)

P	Q	¬ P ∧ Q
T	T	
T	F	
F	T	
F	F	

64. Consider the argument: If this is a maths test, then I will pass it.
    a) Using M for maths and P for pass, symbolize the statement.
    b) Write the inverse in both symbolic form and word form.
    c) Write the converse in both symbolic form and word form.
    d) Write the contrapositive in both symbolic form and word form.
    e) Which of b), c), d) will always be equivalent to the original statement?   (AS3.4)

65. Fill in each blank using your knowledge of symbolic logic.
    a) (P ⇒ Q) and (Q ⇒ R), therefore _____.
    b) (A ∨ B) and ¬B, therefore _____.
    c) (M ⇒ N) and M, therefore _____.
    d) ¬(¬P) ⇔ _____.
    e) (A ⇒ B) and ¬B, therefore _____. (AS3.4)

66. Let A = {a, b, c, d}. List all the subsets of A.   (AS3.5)

67. a) Draw a Venn diagram of the set of real numbers. Include the sets of natural, integer, rational, and irrational numbers in the diagram.
    b) Put examples of each type of number in each part of the diagram.
                                                                           (AS1.1, 3.5)

Hint: ⟶ ☐☐

68. Let the sets A and B intersect so that neither of them is a subset of the other and the two are not disjoint. Shade a Venn diagram for:
    a) A' ∩ B
    b) (A ∪ B)'
    c) (A ∩ B)'   (AS3.5)

69. Let U = {0, 1, 2, …, 11}, A = {0, 1, 2, 3, 4, 5}, B = {3, 4, 5, 6, 7, 9}
    and C = {1, 2, 3, 6, 8}.

    Let A, B, and C intersect (inside U) so that there are eight distinct
    regions formed.
    a) Draw a Venn diagram for the sets listed above.
    b) Find:
        (i) $A \cap B$
        (ii) $A \cup C$
        (iii) A′ (the complement of A)
        (iv) $(B \cup C)′$
        (v) $B \cup C′$ (AS3.5, 3.6)

70. Four coins are tossed at the same time. Draw a sample space for all of
    the outcomes. Find each probability.
    (Hint: T_1 T_2 T_3 T_4, $2^4 = 16$ rows.)
    a) No heads appear.   b) Exactly one tail appears.
    c) At least one tail appears.   d) One head appears.
    e) Two or more heads appear.   (AS3.6)

71. If P(passing the Mathematical Studies exam) = 0.95, find P(not
    passing the Mathematical Studies exam).   (AS3.6)

72. An urn contains 10 red, 6 green, and 3 blue marbles. Ron chooses three
    marbles from the urn.
    a) Assume that Ron does not replace the marbles and find each
       probability: (Hint: Don't forget to use the correct notation.)
        (i)   Ron chooses all three red marbles.
        (ii) Ron chooses three red or three blue marbles.
        (iii) Ron chooses at least one green marble.
        (iv) Ron chooses one of each colour.
        (v)   Ron chooses exactly two green.
    b) Assume that Ron chooses a marble, notes its colour, replaces it, and
       then chooses another. Find each of the above probabilities again.
                                                         (AS3.7)

73. The town of Alright has a population of 200 adults. 60% are women
    and 10% of them are more than six feet tall. Of the men, 30% are more
    than six feet tall. Draw a tree diagram for the town's population.

    (Hint: ⟨diagram⟩ and label with correct numbers.)   (AS3.7)

74. From the town of Alright, a basketball coach chooses 3 people at random
    to play in a 3-on-3 basketball tournament. Find each probability.

    (Hint: Remember to use correct notation, e.g. P(all three are men)
    $= P(\text{man}) \cdot P(\text{man}) \cdot P(\text{man}) = M \cdot M \cdot M$ or simply MMM.)
    a) All three players will be men.
    b) All three players will be women.
    c) A player will be more than six feet tall.
    d) A player will be less than six feet tall given that the player is a man.

e) One of the players will be a woman.

f) At least one of the players will be a woman.

g) Exactly two of the players will be women.   (AS3.7)

75. Given $P(A \cup B) = 0.82$, $P(A) = 0.45$ and $P(B) = 0.57$,

a) find $P(A \cap B)$

b) draw a Venn diagram that depicts the above information.   (AS3.7)

76. One card is drawn from a standard 52-card deck. Find the probability that the card is:

a) red   b) black or an ace

c) a ten or a seven   d) a diamond or a face card.   (AS3.7)

77. The probability that you will solve any given maths problem correctly is 0.80. Find the probability that you will solve five randomly selected maths problems correctly.   (AS3.7)

78. A study of IB students' after-school study time and Pre-IB or IB status showed the following results.

Student status	After school study time per week (in hours)		
	1–4	5–8	9–12
Pre-IB	7	9	8
IB	6	8	12

If a student is selected at random, find these probabilities. (Hint: When solving d) and e), make sure to use the formula provided in the formula packet for conditional probability and then check that answer with the 'cover and reduce the sample space' method.)

a) The student has Pre-IB status.

b) The student studies 5–8 hours per week.

c) The student studies at least 5–8 hours per week.

d) The student has Pre-IB status given that he or she studies 5–8 hours per week.

e) Given that the student has IB status, the student studies 9–12 hours per week .   (AS3.7)

79. The weight of tins of beans is normally distributed with a mean of 453 g and a standard deviation of 4 g.

What percentage of cans is under the labelled weight of 450 g? (AS4.1)

80. Daniel collected data to try to determine if there was a linear correlation between height and hand size for seniors in high school. His raw data is listed below.

Height x (inches)	72	70	66	61	65	61	75	61
Hand size y (inches)	9.25	8.25	7	6.75	7	7	8.75	7.5

Height x (inches)	70	68	69	70	66	68	66
Hand size y (inches)	8.5	8	7.75	9.25	7.5	8.5	8

a) Draw a scatter plot. Use the IB 2-mm graph paper in a vertical position. Draw the y-axis on the leftmost edge of the graph paper. Start the x-scale at the end of the first cm. Label that division mark as 58. Use the scale 1 cm = 1 inch on the x-axis and 2 cm = 1 inch on the y-axis.

b) Determine the r-value using your GDC:

c) Interpret the r-value as it relates to the relationship between the variables.

d) Determine the equation of the regression line using your GDC.

e) Construct a regression line if the r-value suggests a correlation between the variables. (Hint: The r-value does suggest a correlation!)

f) By drawing lines on your graph, predict:

   (i) the hand size, to the nearest 0.5 inches, for a senior who is 64 inches tall.

   (ii) the height, to the nearest 0.5 inches, of a senior whose hand size is 6.5 inches. Comment on this answer.

g) Use the equation of the regression line and predict:

   (i) the hand size, to the nearest 0.5 inches, for a senior who is 80 inches tall. Comment on this answer.

   (ii) the height of a senior whose hand size is 9 inches. (AS4.2, 4.3)

81. Nick conducted a study to determine if recycling was independent of gender. He recorded his findings 100 randomly selected people in the contingency table below.

	Does recycle	Does not recycle
Female	23	34
Male	25	18

Solve the problem with the use of a GDC.

a) State the null and alternative hypotheses.

b) Find the p-value to 3 s.f.

c) At each significance level below, decide whether to reject or not reject

   (accept) H_0 and explain your reasoning.

   (i) 1%

   (ii) 5%

   (iii) 10%

d) For each decision in c) above, interpret the results.

   (Hint: 'There is (or is not) enough evidence to …'.)   (AS4.4)

82. Write an equation for a line that passes through $(3, 5)$ and $(-2, -3)$. Leave your answer in the form $ax + by + d = 0$, where $a, b, d \in \mathbb{Z}$, and

$a > 0.$ (AS5.1)

83. Find the y-intercept for the line that passes through $(2, 4)$ and $(3, 0)$.
(Hint: Find the gradient.) (AS5.1)

84. Find an equation for the line that is perpendicular to $3x + 2y - 18 = 0$
and which passes through $(1, 5)$. Leave your answer as $ax + by + d = 0$,
where $a, b, d \in \mathbb{Z}$, and $a > 0$. (Hint: Find $\perp m$, and then write as
$y = (\perp m)x + c$.) (AS5.1)

85. By drawing the graph, find to the nearest tenth, the intersection of:
$2x + 3y = 12$ and $3x - 4y = 6$. Use 1 division $= 1$ cm on the x- and
y-axes. (Hint: Draw lines on your graph to show your work.) (AS5.1)

86. Find the intersection, without your calculator, of $3x + y = 7$ and
$2x - 3y = 11$,
a) exactly (Hint: Fraction form.)
b) correct to 3 significant figures. (AS5.1)

87. Write an equation for a line that is parallel to $3x - 5y = 17$ and which
passes through $(2, -1)$. Leave the answer as:
a) $y = mx + c$ (Hint: Find m.)
b) $ax + by + d = 0$, where $a, b, d \in \mathbb{Z}$, and $a > 0$. (AS5.1)

88. Using your GDC, find each of the following correct to 4 s.f. (AS 5.2)
a) $\sin 23°$ b) $\cos 85°$ c) $\tan 70°$

89. Given right triangle ABC, find each of the following. (AS5.2)
a) $\sin A$ b) $\cos A$
c) $\tan A$ d) $\sin B$
e) $\cos B$ f) $\tan B$

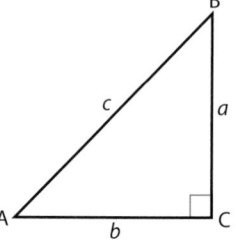

90. Using the triangle in Question 89 and the given information, find each
unknown. (AS 5.2)
a) $a = 3, b = 4, c = ?$ b) $a = 8, c = 17, b = ?$
c) $\sin A = 0.6, a = 12, b = ?$ d) $b = 7, c = 17, a = ?$
e) $\tan B = \frac{3}{4}, b = 12, c = ?$

91. In $\triangle ABC$, $\angle A = 25°$, $\angle C = 90°$ and $BC = 4$. Find,
a) $\angle B$ b) AC to 3 s.f. c) AB to 3 s.f. (AS5.2)

92. Solve $\triangle DEF$ if $DE = 5$, $DF = 8$ and $\angle D = 39°$. Find EF, $\angle F$, and $\angle E$.
(AS5.2)

93. Solve for x to 3 s.f. (Hint: Make sure your GDC is in degree mode.)
a) $\tan x = \frac{2}{3}$ b) $\sin x = 0.234\,5$ c) $\cos x = \frac{\sqrt{2}}{2}$ (AS5.2)

94. The angle of depression from the top of a cliff to a boat in the ocean below is found to be 27°. Using a yardage scope, the boat is measured to be 320 yards from the top of the cliff.
    a) Draw and label a diagram to represent the information.
       (Hint: Draw a right triangle, using ¬, ∡, h, x.)
    b) How far is the boat from the cliff?
    c) What is the height of the cliff?
    d) What is the angle of elevation from the water line on the boat to the top of the cliff? (AS5.2)

95. Find the area of △XYZ to 3 s.f. (AS5.3)

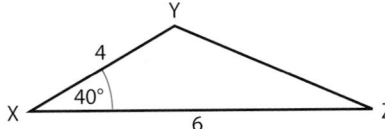

96.

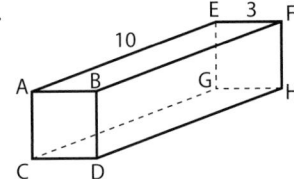

Consider the diagram, where $AE = 10$, $EF = 3$, and $FH = 4$. This figure is a rectangular prism. (Hint: Form right triangles when answering the questions.)

    a) $DH =$
    b) Let P be a point in the centre of plane $ABCD$.
       (i) $AP =$
       (ii) The size of $B\hat{A}P =$
    c) $BC =$
    d) $BH =$
    e) Let Q be the midpoint of $[BF]$.
       (i) $BQ =$
       (ii) The size of $B\hat{Q}D =$
       (iii) The size of $F\hat{Q}H =$
    f) What is the measure of the angle between planes $ECDF$ and $GCDH$?
    g) What is the measure of the angle between $[DE]$ and plane $EFHG$?
    h) Let R be the midpoint of $[GH]$ and S be the midpoint $[CD]$.
       (i) $PR =$
       (ii) The size of $P\hat{R}S =$ (AS5.4)

97.

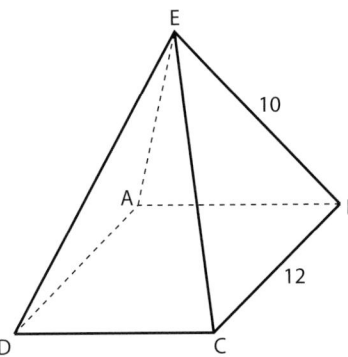

Let $ABCDE$ be a squared-based pyramid with $BC = 12$ and $EB = 10$. (Hint: Carefully mark and label your diagram and form right triangles in your redraws.)

a) Let M be the midpoint of $[BC]$.
     (i) $EM =$ (ii) The size of angle $E\hat{M}B =$
     (iii) The size of $E\hat{B}M =$
b) Let P be the centre of plane $ABCD$.
     (i) $EP =$ (ii) The size of $E\hat{P}M =$
     (iii) The size of $M\hat{E}P =$ (iv) $PM =$
     (v) $PC =$ (vi) The size of $E\hat{C}P =$
c) Find the total surface area of pyramid $ABCDE$.   (AS5.4, 5.5)

98. Evaluate each function.
    a) If $f(x) = x^2 + 3x + 2$, find $f(-1)$.
    b) If $h(x) = \sqrt{x^2 + 16}$, find $h(3)$.
    c) If $f(x) = 2^x + 1$ and $g(x) = x^2 + 2$, find $f(g(-1))$.
    d) If $h(t) = t^3 - 1$ and $k(t) = 2t^2 - 10$, find $(h \circ k)(2)$.   (AS6.1)

99. Find the domain and range of each function.
    (Hint: Draw the graph with your GDC.)
    a) $y = x^2 + 1$
    b) $f(x) = \sqrt{x}$
    c) $y = \dfrac{1}{x}$ (AS6.1)

100. Draw a mapping diagram for each of the functions listed below.
    (Hint: Draw and label ovals and connect with ⟶.)
    a) $f(x) = x^2 + 1$, where $-2 \leqslant x \leqslant 2, x \in \mathbb{Z}$
    b) $g(t) = 2^t$, where $-2 \leqslant t \leqslant 2, t \in \mathbb{Z}$ (AS6.1)

101. Draw the graph of each of the following on the same coordinate plane.
    Use the scale 1 division $= 1$ cm on both axes.
    a) $45x + 60y = 180$
    b) $f(x) = \frac{2}{3}x + 2$ (AS6.1)

102. By drawing lines on the graph in Question 101, find the point of
    intersection.   (AS5.1)

103. Let $y = x^2 - 2x - 8$.
    a) Find the x-intercepts.
    b) Find the y-intercept.
    c) Find the equation of the axis of symmetry.
    d) Express in 'turning point form'.
    e) Find the turning point (the vertex).
    f) Draw the graph using 2-mm graph paper and the scale
       1 cm $= 1$ division on both the x- and y-axes.
    g) By drawing lines on your graph, find the x-value(s) when $y = 5$.
    h) Find the domain and range.   (AS6.3, 6.6)

104. The value of a surround sound system can be modelled by the formula
    $P = 150 + 200 \times 2.5^{\frac{-t}{2}}$, where P is the purchase price of the system in
    US$ and t is the age of the surround sound system in years after being
    purchased.
    a) What was the initial cost of the system?
    b) What was the value of the system after 3 years?

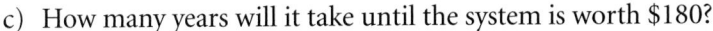

c) How many years will it take until the system is worth $180?

d) According to the model, what is the least amount you should be able to sell it for on the internet 5 years later?

e) How accurate or useful do you think the model is? (AS6.4)

105. Use 2-mm graph paper and a scale of 5 degrees = 1 cm on both axes.

a) Draw the graph of $C = \frac{5}{9}(F - 32)$. In this formula, Celsius temperature is a linear function of Fahrenheit temperature. (Label the vertical axis as $C°$.)

b) By drawing lines on your graph, find:

   (i) the Celsius temperature when the Fahrenheit temperature is 85°

   (ii) the Fahrenheit temperature when the Celsius temperature is 10°.

(AS6.2, 6.6)

106. $y = 3 \cdot 2^x$

a) Draw the graph of the function. Use 2-mm graph paper and the scale 1 division = 1 cm on both the x- and y-axes. Let $-\infty < x \leqslant 2.5$.

b) Write the equation for the horizontal asymptote.

c) Find the domain and range.

d) Solve $10 = 3 \cdot 2^x$, by drawing lines on your graph. (Hint: Graph the line $y = 10$ on your graph.) (AS6.4, 6.6)

107. Find the equations of all lines of asymptote for each of the following.

a) $y = 3 \cdot 2^x - 1$

b) $y = \dfrac{2}{3 + x}$

c) $y = \dfrac{2x - 1}{x}$

d) $y = \dfrac{3x + 1000}{x - 5}$

e) $y = 2 \cdot \left(\frac{3}{4}\right)^x + 2$

(AS6.5)

108. Let $y = \dfrac{2}{x - 1}$, for $x \in \mathbb{R}$, $x \neq 1$.

a) Draw the graph using 1 division = 1 cm on both the x- and y-axes.

b) What value of x is not in the domain of the function?

c) Write the equation of the vertical asymptote. (Hint: $x =$)

d) Write the equation of the horizontal asymptote.

e) Find the domain and range. (AS6.5, 6.6)

109. a) Draw the graph of $f(x) = \begin{cases} 2x - 3, x \geqslant 2 \\ x - 1, x < 2 \end{cases}$.

b) Is f continuous? Explain your answer. (AS6.7)

110. Use your calculator to solve each problem exactly or correct to 3 s.f.

a) Find the point(s) of intersection of the graphs $y = \dfrac{2}{x - 1}$ and $y = 2x - 1$.

b) Solve $0.5^x = 5$ by finding the intersection of $y = 0.5^x$ and $y = 5$.

c) Find the three points of intersection for the functions $y = 2^x$ and $y = x^2$.

d) Solve for x: $5x = 3^x$. (AS6.7)

111. Let $f(x) = x^2 + 3$. (Hint: Draw an accurate diagram to help you visualize this problem.)

a) Find $f(1)$ and let this ordered pair be point P.
   (Hint: Write P as an ordered pair.)
b) Find $f(1.1)$ and let this ordered pair be point Q_1.
   (Hint: Write Q_1 as an ordered pair.)
c) Find $m(PQ_1)$, the gradient of the secant (PQ_1).
d) Find $f(1.01)$ and let this ordered pair be point Q_2.
   (Hint: Write Q_2 as an ordered pair.)
e) Find $m(PQ_2)$, the gradient of the secant (PQ_2).
f) Hence or otherwise, find $\lim\limits_{x \to 1} \dfrac{f(x) - f(1)}{x - 1}$.
   (Hint: As Q moves closer to P, then the gradients of the secants
   seem to be approaching what number?) (AS7.1)

112. Define the derivative as the gradient function. (AS7.1)

113. Name three different ways to say what the concept of $f'(x)$ represents. (AS7.1)

114. List three of the most common notations used to describe the derivative. (AS7.1)

115. Find the derivative of each of the functions below. (Hint: Each answer should be of the form y' or $f'(x)$. Also, be careful of the negative exponents.)
a) $y = 2x^3$
b) $f(x) = -4x^3 + 2x^2 - x + 1$
c) $y = \dfrac{1}{x}$
d) $y = \dfrac{3}{x^2}$
e) $f(x) = \dfrac{2x - 1}{x}$. (Hint: First, make two fractions.)
f) $f(x) = \dfrac{x^3}{5} - \dfrac{4x^2}{3} + \dfrac{x}{2} - 1$
g) $f(x) = \dfrac{3}{x^3} + \dfrac{2x^2}{5} - 3x - 2$
h) $f(x) = x$
i) $f(x) = 3 - 3x^2$ (AS7.2)

116. a) If $y' = 2x^2 + 7x - 2$, find y''.
b) If $f'(x) = -2x^{-3}$, find $f''(x)$. (Hint: Be careful of the negative exponents.) (AS7.2)

117. Given $f(x) = x^2 - 5x - 1$, find the x-value where $f'(x) = 2$. (AS7.3)

118. If $f(x) = 3x^2 - 7x + 2$, write the equation of the line tangent to the curve at $x = 2$. (AS7.3)

119. If $y = \frac{2}{3}x^3 - 2x^2$ has a local maximum at $x = 0$, then consider the interval, $(-\infty, 2)$:
a) when $x < 0$, then $f'(x)$ _____ 0.
b) when $x = 0$, then $f'(x)$ _____ 0.

c) when $x > 0$, then $f'(x)$ _____ 0.
(Hint: Use $<$, $>$ or $=$ signs.) $\hspace{2cm}$ (AS7.4)

120. For $f(x) = x^3 + 2x^2 - x + 1$, and using your GDC, find the interval(s), correct to 3 s.f., where f is:
   a) increasing (Hint: There are two intervals.)
   b) decreasing (Hint: There is only one interval.)
   c) Find the local maximum point, correct to 3 s.f.
   d) Find the local minimum point, correct to 3 s.f.
   (Hint: Draw an accurate diagram to help you visualize the concept.)
   $\hspace{10cm}$ (AS7.4, 7.5)

121. If $f(x) = 2x^2 - x + 1$, then:
   a) find the x-value where $f'(x) = 0$.
   b) find the minimum value of the function.
   c) find the turning point (the vertex). $\hspace{3cm}$ (AS7.5)

122. If $f(x) = 4x^2 + 6x + 1$, find x if $f'(x) = 10$. $\hspace{2cm}$ (AS7.5)

123. A farmer wishes to fence off a rectangular field, using an existing stone wall on one side. The total length of fencing materials available is 270 m.

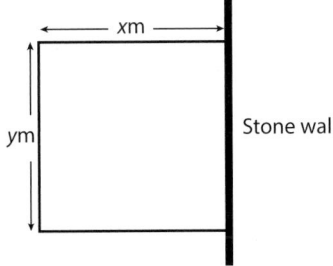

Let x m and y m be the width and length of the field respectively (as shown).

   a) Write down an equation for the perimeter, in terms of x and y.
   b) Write down an equation for the area, in terms of x and y.
   c) Determine an equation for area in terms of x. What are the restrictions on x?
   d) Determine the gradient function.
   e) **Hence**, determine the length and width for the maximum area.
   f) Use your GDC to determine the area. $\hspace{3cm}$ (AS7.6)

Practice Papers

Overview

This chapter is intended to give you some tips about how to take IB examinations, and what to do to prepare for an IB exam. The chapter includes two complete sample papers with complete detailed solutions.

You should attempt these practice papers under actual exam conditions if possible. Each exam is divided into two papers, each of which should take 90 minutes to complete.

Preparing for your IB exams

Some useful exam preparation tips:

1. Get enough sleep in the days leading up to your exams. It is tempting to overstudy during those last few days, but this will only hurt your performance, not help it. Any studying that you do should be spread out over the months leading up to your exams.

2. Be well-fed and hydrated on the day of your exam. If allowed, bring a bottle of water to your exam. Your brain will need lots of energy and water to function properly.

3. Practise under exam conditions as much as possible, then verify the accuracy of your work afterward. If you cannot do a question while studying, make every effort to complete the test in the time allotted. Once you have finished the entire test, review with your teacher the questions you found difficult.

4. Practise using the formula sheet and learn where the information on the pages is stored. There is no need to memorize any of it, since these formulae will be provided during the exam. However, knowing where to look for the information quickly will save you time during the exam.

5. Your calculator is going to be extremely helpful during the exam. Become comfortable with using it. Learn how to use your calculator to solve equations, graph functions, and create tables of values, etc.

6. Use a checklist, such as the list of assessment statements provided in each chapter of this textbook, to keep track of the topics you have studied. Focus on the topics with which you have had the most difficulty. In the weeks before the exam, if you find a topic that you just cannot grasp, no matter how much you try, *stop wasting time on that topic!* You are much better off focusing on the other topics of the test.

First practice exam

Paper 1

1. a) Given s = $\dfrac{v^2 - u^2}{2a}$, $v = 15.5$, $u = 6$, and $a = 9.8$ calculate the value of
    s, and

    (i)   write your answer to two decimal places (2 d.p.).   [*2 marks*]

    (ii) write your answer to 3 significant figures.   [*2 marks*]

    Another student used the values of $v = 16$, $u = 6$, and $a = 10$ and
    calculated the value of s as 11.

    b) Calculate the percentage error in the second student's result,
        compared to your answer to 2 d.p.   [*2 marks*]

2. The following table shows the distribution of the numbers of pens and
    pencils in all grade 10 students' pencil cases.

Number of pens and pencils	Number of students
$0 \leqslant x < 5$	5
$5 \leqslant x < 10$	7
$10 \leqslant x < 15$	15
$15 \leqslant x < 20$	11
$20 \leqslant x < 25$	9
$25 \leqslant x < 30$	1

   a) Construct a histogram to represent the data.   [*3 marks*]

   b) Write down the modal group.   [*1 mark*]

   c) Calculate an estimate of the mean number of pens and pencils in a
       student's pencil case.   [*2 marks*]

3. a) Complete the truth table below.

p	q	$p \wedge q$	$\neg(p \wedge q)$	$\neg p$	$\neg q$	$(\neg p \vee \neg q)$
T	T					
T	F					
F	T					
F	F					

[*4 marks*]

   b) State whether $\neg(p \wedge q)$ and $(\neg p \vee \neg q)$ are logically equivalent.
       [*1 mark*]

   c) Explain your answer to part b).   [*1 mark*]

4. The lines L_1 and L_2 are given on the diagram below where L_2 is normal
    to L_1

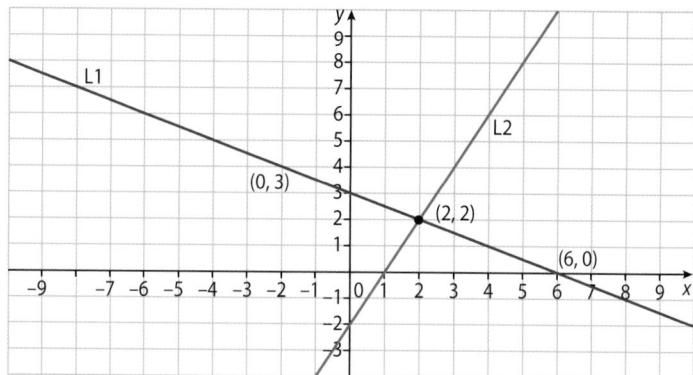

a) What is the gradient of L_1? [2 marks]

b) Write down the equation for L_1. [2 marks]

c) Write down the equation for L_2 in the form $ax + by + c = 0$. [2 marks]

5. Consider the arithmetic sequence 272, 265, 258, 251, …

a) Find the value of the common difference of this sequence. [2 marks]

b) Calculate the sum of the first 10 terms of this sequence. [2 marks]

c) The last number in the sequence is 6. How many terms are in the sequence? [2 marks]

6. The height of a building is 345 m. From the top of the building the angle of depression to the top of a building located horizontally 276 m away is 31°.

a) Draw a diagram to show this information. [2 marks]

b) Calculate the height of the second building. [4 marks]

7. A box contains 11 green balls and 5 red balls. Bjarne chooses a ball at random from the box and does not replace the ball.

a) (i) What is the probability that the ball is red? [1 mark]

(ii) Ali then takes a ball from the box. What is the probability that the ball is green given that Bjarne took a red? [1 mark]

(iii) What is the probability that Bjarne chose a red ball and Ali chose a green ball? [1 mark]

b) Find the probability that Bjarne and Ali chose different coloured balls. [3 marks]

8. A curve of the form $y = ax^2 + bx + c$ is drawn and is shown below.

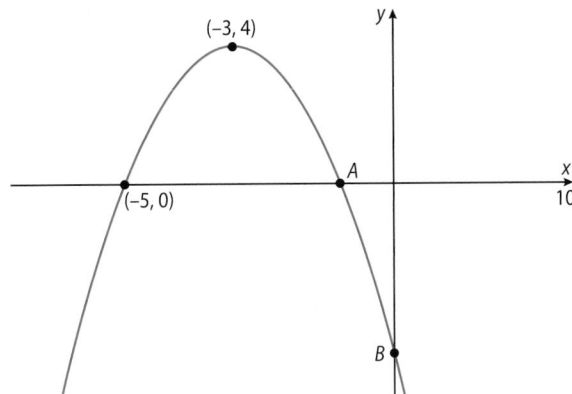

a) Find the coordinates of the point A. [2 marks]

b) Find the coordinates of the point B. [3 marks]

c) Write down the equation of the parabola. [1 mark]

9. The volume of a can of a soft drink is approximately normally distributed with a mean of 358 ml and a standard deviation of 7 ml.

a) It is known that 80% of all cans have a volume less than v ml. Determine the value of v. [2 marks]

b) Sketch a diagram of the distribution of the volume of the cans of drink, indicating the location of v on your diagram. [2 marks]

c) Given that the company produces 10 000 cans of soft drink a day what is the expected number of cans that are below the required volume of 350 ml? [2 marks]

10. The box-and-whisker diagram below shows the statistics for a set of data.

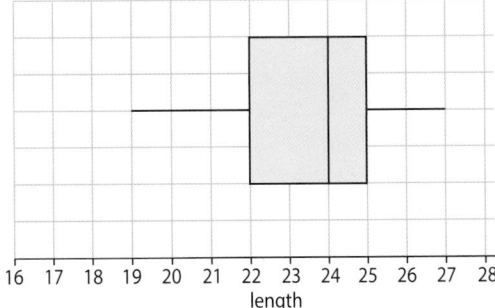

length

a) Write down the value of each of the following:

(i) the median

(ii) the upper quartile

(iii) the minimum value

(iv) the interquartile range. [4 marks]

b) A second box-and-whisker diagram is to be drawn on the same grid. The following information is known about the data for this box-and-whisker diagram: the range is 10, the minimum value is 18, the interquartile range is 6, the lower quartile is 19.5, and the median is 23.

(i) Draw the box-and-whisker diagram on the same grid as the first box-and-whisker diagram.

(ii) Describe the differences in the diagrams. [2 *marks*]

11. A square-based pyramid sits on top of a cube as shown.

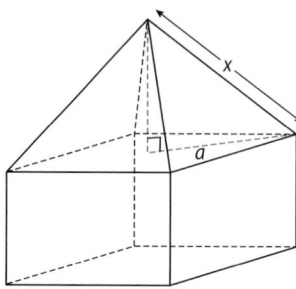

The total height of the object is 25 cm and the cube has a side length of 8 cm.

a) Write down the height of the pyramid. [2 *marks*]

b) Determine the length of a. [2 *marks*]

c) Determine the length of x. [2 *marks*]

12. Data comparing the heights and weights of 190 students were graphed. A researcher was trying to decide if there was any correlation between the height and weight of the students. Indicate which graph(s) show:

No correlation.
Strong negative correlation.
Moderate positive correlation. [6 *marks*]

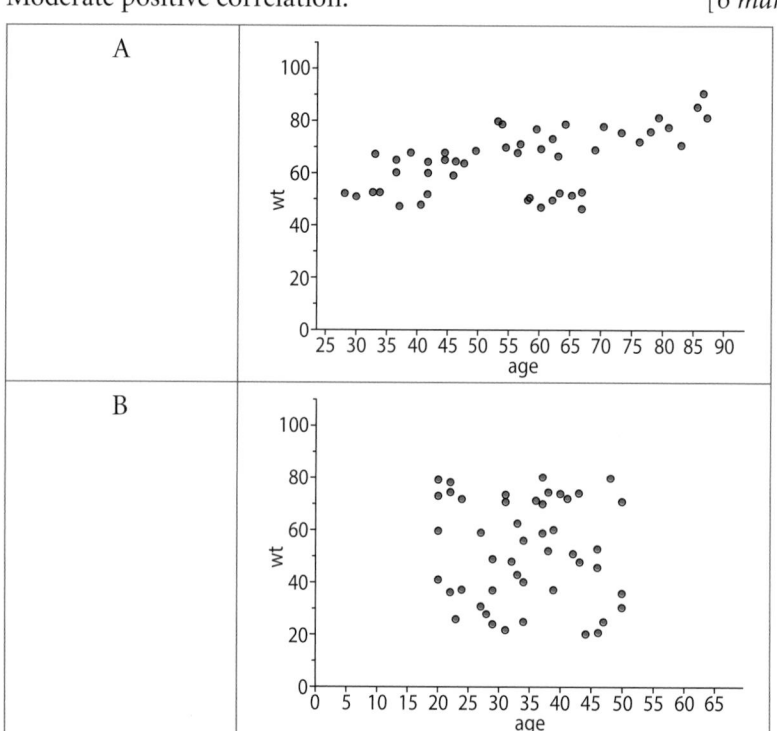

C	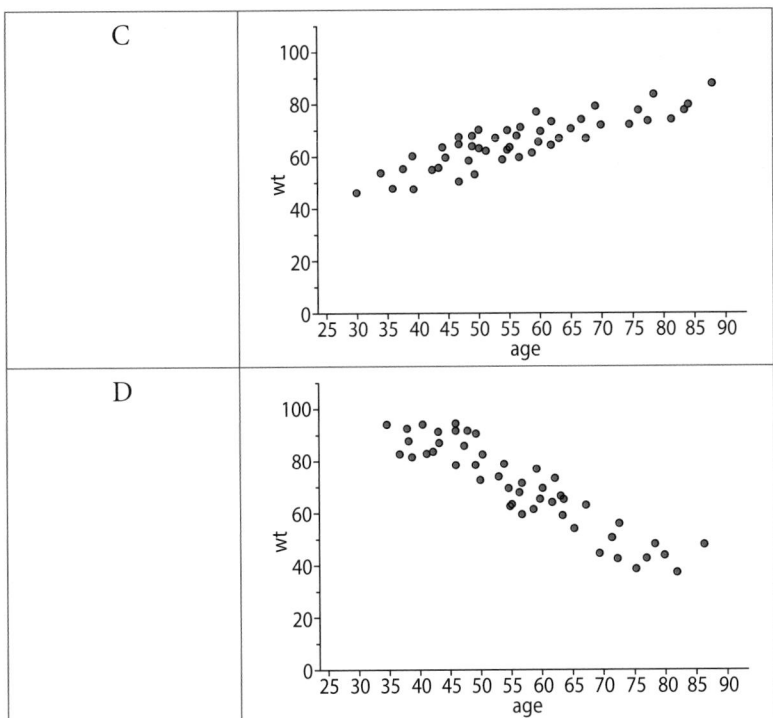
D	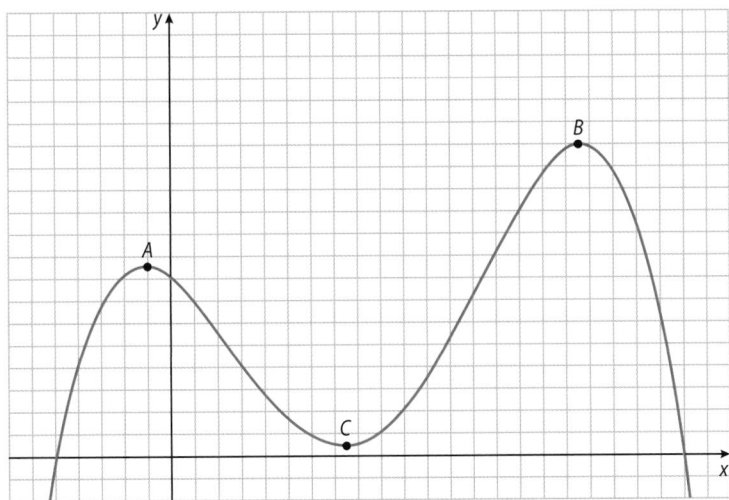

13. A curve is drawn and is given below.

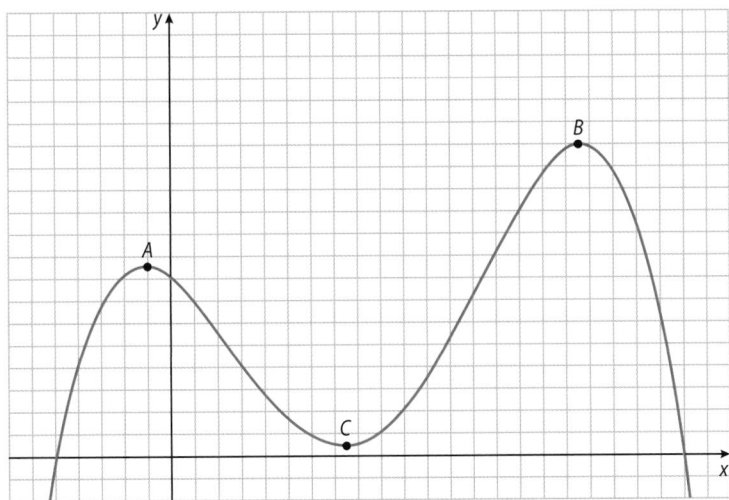

A series of statements is given below. Choose all that are correct.

a) The function has a negative *y*-intercept.

b) The graph has three stationary points.

c) The graph has two *x*-intercepts.

d) The point *C* has zero gradient.

e) The gradient between *A* and *C* is positive.

f) *A* and *B* have the same gradient. [*6 marks*]

14. A tower (*CD*) sits on top of a building, as shown. $A\hat{B}C$ is a right angle.

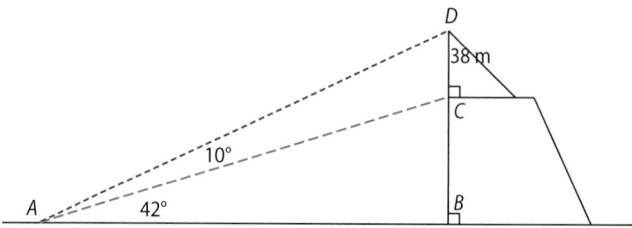

From A, the angles of elevation to *C* and *D* are 42° and 52° respectively, and it is known that the height of the tower is 38 m.

a) Find $A\hat{D}C$. [*3 marks*]

b) Hence, determine the height (*CB*) of the building. [*3 marks*]

15. $f(x) = \frac{1}{3}x^3 + 2x^2 + \frac{2}{x} - \frac{1}{4}$

a) Find $f'(x)$. [*4 marks*]

Carlie claims that $f'(0) = 0$.

b) (i) Is she correct? [*1 mark*]

(ii) Explain your answer to b)(i). [*1 mark*]

Paper 2

1. 113 randomly selected persons were asked what their preferred summer holiday destination was. All of the respondents travelled to the beach, to the mountains, or to another country.

57 respondents travelled to the beach.

21 respondents visited the mountains.

76 respondents visited another country.

Only 7 respondents visited the mountains and visited the beach, but did not visit another country.

10 respondents visited the beach and another country, but did not visit the mountains.

12 respondents visited the beach, the mountains, and another country.

a) Represent this information on a Venn Diagram. [*4 marks*]

b) Calculate the number of respondents:

(i) who travelled to the mountains and visited another country, but did not go to the beach;

(ii) who only visited the mountains. [*4 marks*]

Sana indicated in her response that she travelled to another country for her summer vacation.

c) Find the probability that she also visited the beach. [*2 marks*]

Two other respondents were chosen at random from the 113 students.

d) Find the probability that

(i) both respondents visited the beach;

(ii) only one of the respondents visited the beach. [*7 marks*]

2. A school offers 4 subjects from the IB Diploma Group 3. These are business management, economics, history, and psychology. The number of students choosing each subject by gender is given below.

	Business Management	Economics	History	Psychology
Male	27	15	7	14
Female	13	7	15	11

A χ^2 (chi-squared) test at the 5% significance level is used to determine whether the choice of subject is independent of gender.

a) What is the total number of persons in the sample and how many males and females were there in the sample? [*3 marks*]

b) Show that the expected frequency for females choosing economics is 9.2844. [*3 marks*]

c) Write down the contingency table for the expected frequencies. [*4 marks*]

d) Write down the p-value for the test. [*2 marks*]

e) State whether the null hypothesis is accepted or rejected. Give reasons for your answer. [*3 marks*]

3. Consider the function $f(x) = \frac{3}{2}x^2 - 5x - 2$.

a) Calculate $f(4)$. [*2 marks*]

b) Write down the y-intercept. [*1 mark*]

c) Determine the x-intercepts. [*3 marks*]

d) Sketch the graph of the function $y = f(x)$ for $-5 \leqslant x \leqslant 5$ and $-10 \leqslant y \leqslant 10$. [*4 marks*]

e) Find $f'(x)$. [*2 marks*]

f) Find the coordinates of the minimum point. [*2 marks*]

g) Find the gradient of the tangent at $x = 4$. [*2 marks*]

h) Determine the equation of the tangent at $x = 4$. [*2 marks*]

i) Determine the equation of the normal at $x = 4$. [*4 marks*]

4. Bob is trying to decide whether to invest $40 000 or to purchase a new car which will cost $80 000. Bob can invest the money, $40 000, at 9% p.a. compounding monthly with the interest added at the end of each month.

a) Write down the investment function $I(t)$ which would be used to calculate the value of the investment at any time. [*2 marks*]

b) Hence, show that he will earn $3752.30 in interest after the first year. [*2 marks*]

c) Write down how much money Bob will have in the bank at the end of year 2. [*1 mark*]

d) How many years will it take for his money to double?   [*2 marks*]

However, Bob has realised that if he buys a new car it will gradually depreciate in value. He has been advised that the rate of depreciation for the car will be 7% per annum.

e) Write down the depreciation function $D(t)$ which would be used to calculate the value of the car at any time.   [*2 marks*]

f) Hence, calculate the value of the car after one year, to the nearest dollar.   [*1 mark*]

g) Sketch a graph showing both the investment function $I(t)$ and the depreciation function $D(t)$ on the same set of axes, with $0 \leqslant t \leqslant 10$. Indicate on the graph the initial values for each function.   [*4 marks*]

h) Determine the value of the car when the investment function $I(t)$ and the depreciation function $D(t)$ are equal.   [*4 marks*]

i) When will the value of the car equal $40 000?   [*4 marks*]

5. A function is given by the equation $f(x) = 35 \times 1.3^{-0.5x} + 20$.

a) Show that at $x = 0$, $f(x) = 55$.   [*2 marks*]

The following table shows values for x and $f(x)$.

x	1	2	3	4	5	6	8	10
$f(x)$	50.7	46.9	43.6	40.7	p	35.9	32.3	q

b)

   (i) Write down the values of p and q.   [*2 marks*]

   (ii) Draw the graph of $f(x)$ for $0 \leqslant x \leqslant 10$. Use a scale of 1 cm to represent 1 on the horizontal axis and a scale of 1 cm to represent 10 on the vertical axis.   [*4 marks*]

   (iii) **Use your graph** to find how long it takes for $f(x)$ to decrease to 39. Show your method clearly.   [*2 marks*]

   (iv) Write down the horizontal asymptote. Justify your answer.   [*2 marks*]

Consider the function $g(x) = 4x - 9$ for $0 \leqslant x \leqslant 10$.

c) Draw the graph of $g(x)$ on the same set of axes used for part (b).   [*2 marks*]

d) **Use your graph** to solve the equation $g(x) = f(x)$. Show your method clearly.   [*2 marks*]

Second Practice Exam

Paper 1

1. A mathematical studies textbook weighs 1.469 kg.

    A school has ordered 50 textbooks for its new Mathematical Studies class.

    a) What is the weight of the parcel? [3 marks]

    The delivery company, however, charges to the nearest 5 kg, so they will charge the package at 75 kg.

    b) What is the percentage error in the company's weight calculation? [3 marks]

2. Frederick had to change British pounds (GB£) into Swiss francs (CHF) in a bank. The exchange rate is 1 GB£ = 1.5 CHF. There is also a bank charge of 3 GB£ for each transaction.

    a) How many Swiss francs would Frederick buy with 133 GB£? [2 marks]

    b) Let s be the number of Swiss francs received in exchange for b GB£. Express s in terms of b. [2 marks]

    c) Frederick received 430 CHF. How many British pounds did he exchange? [2 marks]

3. In an arithmetic sequence $u_9 = -23$ and $u_{25} = 25$.

    a) Find

       (i) the common difference;

       (ii) the first term. [4 marks]

    b) Find S_{25}. [2 marks]

4. Weights of 100 apples, in grams, from an orchard were recorded

Weight of apples (g)	Number of apples	Cumulative frequency
$110 \leqslant x < 115$	13	13
$115 \leqslant x < 120$	14	27
$120 \leqslant x < 125$	18	45
$125 \leqslant x < 130$	22	p
$130 \leqslant x < 135$	17	84
$135 \leqslant x < 140$	10	94
$140 \leqslant x < 145$	5	99
$145 \leqslant x < 150$	1	100

    a) Write down the missing value p in the cumulative frequency column. [1 mark]

    The cumulative frequency graph showing the weights in grams of the apples is given below.

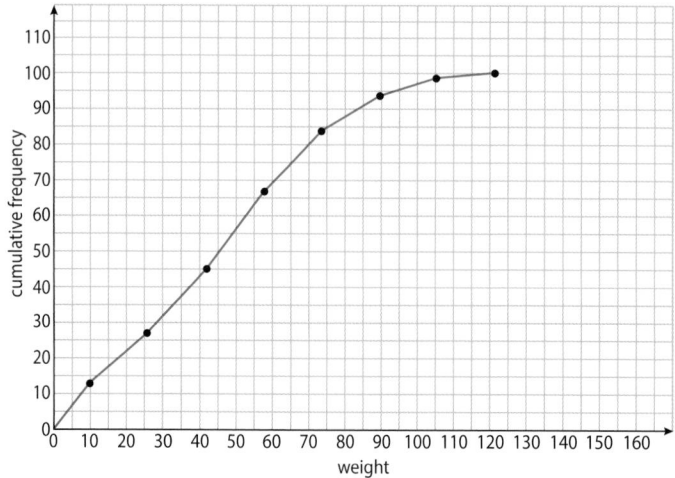

b) Write down

   (i) the median

   (ii) the lower quartile

   (iii) the upper quartile. [*3 marks*]

c) Draw a box-and-whisker diagram to represent the information.

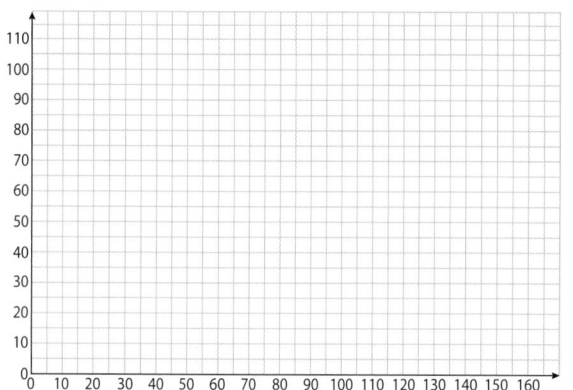

[*2 marks*]

5. An unbiased, red, five-sided die has the numbers 1, 2, 3, 4, and 5 written on its faces. An unbiased, blue, four-sided die has the numbers 1, 2, 3, and 4 written on its faces. The two dice are rolled.

   a) Complete the sample space diagram. [*3 marks*]

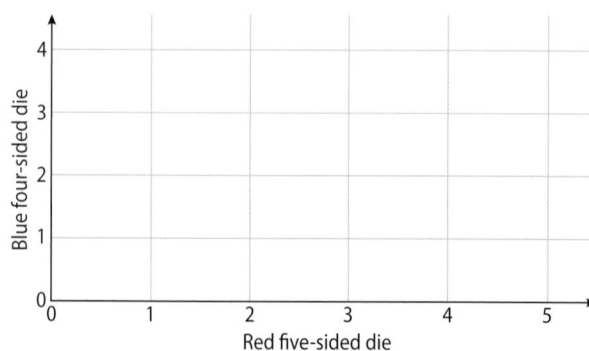

   b) Find the probability that the number on the upper face of the blue die is greater than the number on the upper face of the red die.

[*1 mark*]

c) Find the probability that the red die shows a prime number and the blue die shows a 3. [*2 marks*]

6. A drinking cup has been designed in the shape of a slanted cylinder as shown below.

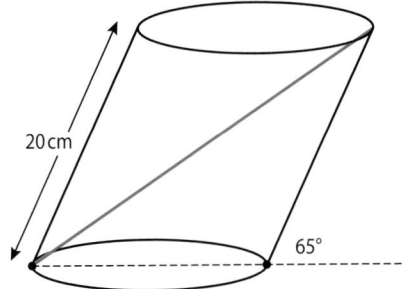

20 cm

65°

The diameter of the base is 12 cm and the slant height is 20 cm. What is the longest straw that will be required to go from top to bottom as shown in the diagram? [*6 marks*]

7. a) Complete the truth table shown below. [*3 marks*]

p	q	$\neg q$	$p \Rightarrow \neg q$	$\neg p$	$(p \Rightarrow \neg q) \wedge \neg p$	$[(p \Rightarrow \neg q) \wedge \neg p] \Rightarrow q$
T	T	F				
T	F	T	T	F		
F	T		T	T		
F	F	T				

b) State whether the compound proposition $[(p \Rightarrow \neg q) \wedge \neg p] \Rightarrow q$ is a contradiction, a tautology, or neither. [*1 mark*]

Consider the following propositions

p: If the trains are running
q: Sasha will go to school.

c) Write in symbolic form the following proposition. [*2 marks*]

If Sasha does not go to school then the trains are not running.

8. The number of hours that a professional footballer trains each day in the month of June is represented in the histogram.

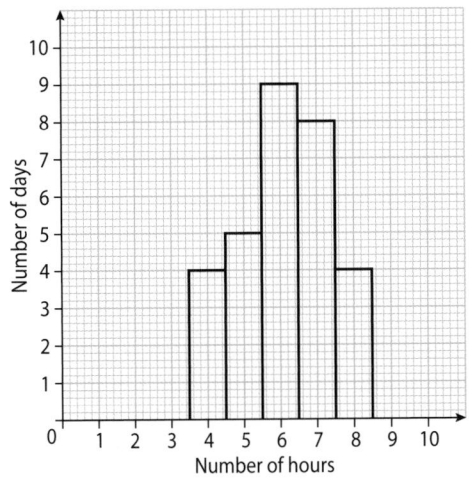

a) Write down the modal number of hours trained each day. [*2 marks*]

b) Calculate the mean number of hours he trains each day. [*4 marks*]

9. The heights and weights of 10 football players are given below.

Height cm	193	173	178	184	189	185	179	177	174	182
Weight kg	84	61	71	77	79	72	67	66	68	73

a) Draw the data points on the following grid. [*3 marks*]

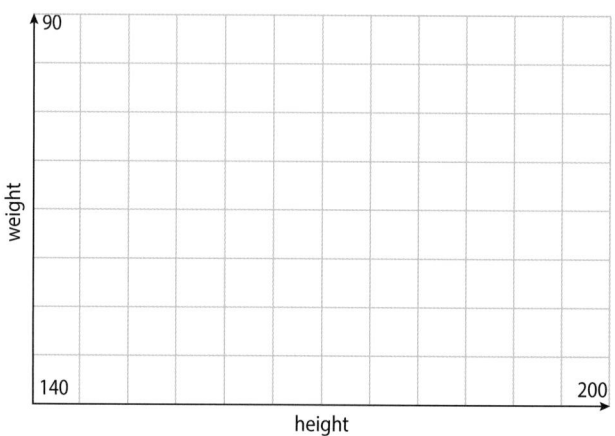

b) Determine $(\bar{x}, \bar{y})$ and mark the point on the grid. [*2 marks*]

c) Draw a line of best fit. [*1 mark*]

10. The length, width, and height of a rectangular prism are given as $2x$, $9 - x$, and $9 + x$ respectively.

a) Write an expression for the volume of the box, $V(x)$. [*2 marks*]

b) Given that $V'(x) = 162 - 6x^2$, find the value of x which corresponds to the maximum volume of the rectangular prism. [*4 marks*]

11. The straight line M passes through the point $P(3, 4)$ and is parallel to the line $3y + 2x - 6 = 0$.

a) Calculate the gradient of M. [*2 marks*]

b) Find the equation of M. [*2 marks*]

The line N is perpendicular to the line M, and passes through the point $B(-4, 0)$.

c) Determine the equation of the line N. [*2 marks*]

12. The height, in centimetres, of a type of fast-growing wheat is given by the function $h(t) = 1.3 \times 2^{0.5t}$, where t is the time in weeks.

a) Determine the height of the wheat at the end of 6 weeks. [*3 marks*]

b) When the wheat reaches a height of 1.5 metres it can be harvested. After how many weeks will the wheat be ready to harvest? [*3 marks*]

13. An investor invests $20 000 in an account which pays an annual interest rate of 2% compounded half yearly.

a) Find the value after 3 years. [*3 marks*]

b) Find the value after 3 years, if the interest was compounded quarterly.

[*3marks*]

14. The probability that the sun shines today is 0.3. If the sun shines today, then the probability that the sun shines tomorrow is 0.6. If the sun does not shine today, then the probability that it will shine tomorrow is 0.4.

a) Complete the tree diagram.   [*3 marks*]

Today   **Tomorrow**

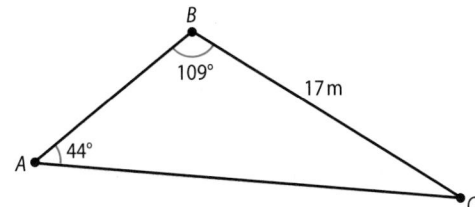

Sunshine 0.3

No sunshine

Sunshine

No sunshine

Sunshine

No sunshine

b) Calculate the probability that the sun does not shine tomorrow.

[*3 marks*]

15. A baseball is hit by the batter and it follows a parabolic trajectory. After t seconds, the vertical height above the ground is given by $H(t) = 14t - 0.25t^2$.

a) Find the height of the ball after 10 seconds.   [*2 marks*]

b) Determine the maximum height of the ball.   [*2 marks*]

c) How much time elapses before the ball lands?   [*2 marks*]

Paper 2

1. A triangular garden, *ABC*, has been constructed with $BC = 17$ m, $A\hat{B}C = 109°$, and $C\hat{A}B = 44°$.

Note: diagram not to scale.

B

109°

17 m

44°

A

C

a) Calculate the length of *AC*, correct to 2 d.p.   [*3 marks*]

b) Write down the angle $A\hat{C}B$.   [*1 mark*]

The gardener wants to build a pathway from *B* to the point *D* which is halfway between *A* and *C*.

c) Calculate the length of the pathway from B to the midpoint of AC to 2 d.p. [*5 marks*]

d) Calculate the area of the triangle BDC to 2 d.p. [*3 marks*]

e) The fence around the triangle BDC is 20 cm high and the gardener intends to fill the area with soil. Find the volume of soil to be used, in cubic metres. [*3 marks*]

2. Two hundred travellers were surveyed to determine their waiting time at check-in at an international airport. These times are given in the table below.

Waiting time in minutes	Number of travellers
$0 \leqslant x < 5$	3
$5 \leqslant x < 10$	11
$10 \leqslant x < 15$	21
$15 \leqslant x < 20$	36
$20 \leqslant x < 25$	43
$25 \leqslant x < 30$	37
$30 \leqslant x < 35$	22
$35 \leqslant x < 40$	18
$40 \leqslant x < 45$	7
$45 \leqslant x < 50$	2

a) Construct a histogram to represent the data. [*3 marks*]

b) Calculate an estimate of the mean waiting time at the airport check-in. [*3 marks*]

c) Construct a cumulative frequency table [*3 marks*]

d) Draw a cumulative frequency curve on the grid below. [*3 marks*]

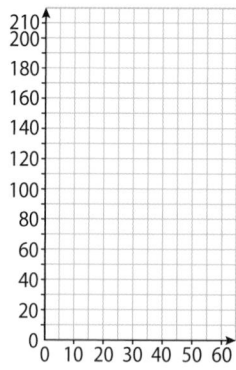

e) Using the cumulative frequency curve, or otherwise, find

(i) the median waiting time

(ii) the inter quartile range of the wait time. [*3 marks*]

A mathematician has determined that the waiting times are normally distributed with a mean of 24.25 and a standard deviation of 9.38.

f) Find the probability that a traveller selected at random has a wait time of less than 25 minutes. [*2 marks*]

g) The probability that a person has a wait time of less than W minutes is 0.10. Find the value of W. *[2 marks]*

h) One hundred travellers are randomly selected. Find the expected number of travellers whose wait time is between 30 and 50 minutes. *[2 marks]*

3. The following graph shows the function $f(x) = x^3 - 6x^2 + 8x + c$.

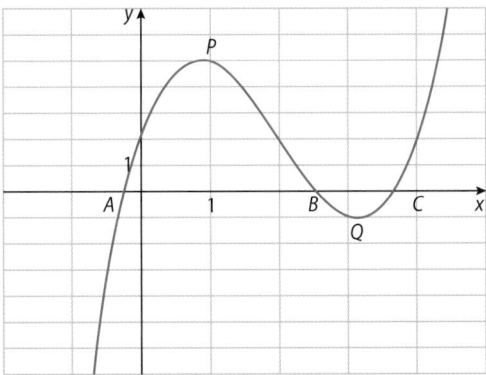

a) Given that $f(1) = 5$, show that $c = 2$. *[3 marks]*

b) The function has 3 solutions for $f(x) = 0$. Write the values of x at these points. *[3 marks]*

c) Write down the gradients at the points P and Q. *[2 marks]*

d) Determine the coordinates of the points P and Q. *[4 marks]*

e) Write down the domain where $f(x)$ is decreasing. *[1 mark]*

A function $g(x) = mx + c$ passes through the points $(0, -2)$ and $(4, 2)$.

f) Determine the equation of the line $g(x)$. *[3 marks]*

g) Using your GDC, graph the functions $f(x)$ and $g(x)$.

   (i) How many solutions are there where $f(x) = g(x)$?

   (ii) Write down the solutions. *[3 marks]*

h) Write down the interval(s) where $f(x) \geqslant g(x)$. *[3 marks]*

4. A circular based cone sits on top of a cylinder.

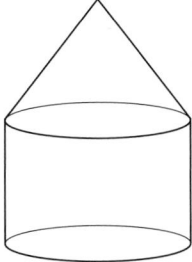

a) (i) Show that the volume of the cone and the cylinder is given by
$$V = \pi r^2 \left(h_{cyl} + \frac{1}{3}h_{cone}\right)$$
[3 marks]

Isaac has designed a new container which consists of a cone placed on top of a cylinder, where the height of the cylinder = the height of the cone.

● **Examiner's hint:** Do not forget to include the appropriate units for the question.

(ii) Show that the volume is now given by $V = \frac{4}{3}\pi r^2 h$.   [*2 marks*]

(iii) Given that the height of the cylinder is 10 cm and the radius is 5 cm, determine the volume of the combined shape. [*3 marks*]

b) A rectangle has dimensions $(5 + 3x)$ metres and $(7 - 6x)$ metres.

   (i) Show that the area, A, of the rectangle can be written as
$$A(x) = 35 - 9x - 18x^2.$$ [*3 marks*]

   (ii) Write down the equation of the axis of symmetry of the curve. [*3 marks*]

   (iii) Hence, or otherwise, determine the lengths of the sides for the maximum area.   [*4 marks*]

5. Data were collected to determine the ages and genders of managers of businesses registered on the local stock exchange. The results are tabulated below.

	< 40 years old	$40 \leqslant$ age < 50	$\geqslant 50$ years old	Total
Male	64	117	67	248
Female	87	112	53	252
Total	151	229	120	500

The table below shows the **expected number** of males and females at each age group.

	< 40 years old	$40 \leqslant$ age < 50	$\geqslant 50$ years old	Total
Male	a	b	60	248
Female	c	d	60	252
Total	151	229	120	500

a) (i) Show that the expected number of males under 40 years old is 75 (to 0 d.p.)   [*2 marks*]

   (ii) Hence find the values of b, c, and d. Write your answers to 0 d.p.   [*3 marks*]

b) (i) Write a suitable null hypothesis for this data.

   (ii) Write a suitable alternate hypothesis for this data.   [*2 marks*]

c) (i) Write down the degrees of freedom.

   (ii) Determine the p-value.

   (iii) What conclusion can be drawn regarding gender and age in management positions at the 5% level of significance?   [*6 marks*]

Solutions to First Practice Exam
Paper 1

1. a) Using our calculators we find that s $= \dfrac{15.5^2 - 6^2}{2 \times 9.8} = 10.4209$.

   (i) To 2 decimal places, this is 10.42.   [*2 marks*]

   (ii) To 3 significant figures, this is 10.4.   [*2 marks*]

b) $V_A = 10.42$ and $V_E = 11$: $s = \left|\dfrac{11 - 10.42}{10.42}\right| \times 100\% = 5.57\%$ to 3 s.f.

[2 marks]

2. a)

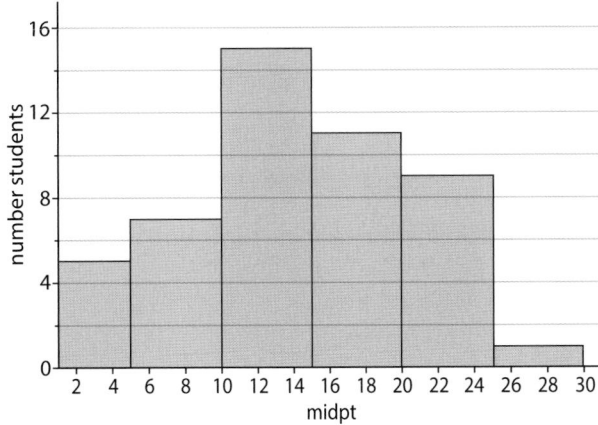

[3 marks]

b) The modal group is $10 \leqslant x < 15$. [1 mark]

c) Mean $= \bar{x} = \dfrac{3 \times 5 + 8 \times 7 + 13 \times 15 + 18 \times 11 + 23 \times 9 + 28 \times 1}{48}$

   $= 14.6$ to 3 s.f. [2 marks]

3. a)

p	q	$p \wedge q$	$\neg(p \wedge q)$	$\neg p$	$\neg q$	$(\neg p \vee \neg q)$
T	T	T	F	F	F	F
T	F	F	T	F	T	T
F	T	F	T	T	F	T
F	F	F	T	T	T	T

[4 marks]

b) Logically equivalent that is $\neg(p \wedge q) \Leftrightarrow (\neg p \vee \neg q)$. [1 mark]

c) Logically equivalent as 4th column and 7th column are the same.

[1 mark]

4. a) Gradient of $L_1 m_{L1} = \dfrac{0 - 3}{6 - 0} = \dfrac{-1}{2}$ [2 marks]

b) Equation L_1. $y = \dfrac{-1}{2}x + 3$ [2 marks]

c) Gradient of $L_2 = \dfrac{-1}{m_{L1}} = 2, -2x + y + 2 = 0$ [2 marks]

5. a) Common difference $d = 265 - 272 = -7$ [2 marks]

b) $n = 10, d = -7$ then $s_n = \dfrac{10}{2}(2 \times 272 + (10 - 1)(-7)) = 2405$

[2 marks]

c) $n = 10, d = -7$ then $u_n = 6 = 272 + (n - 1)(-7) \therefore n = 39$ [2 marks]

6. a)

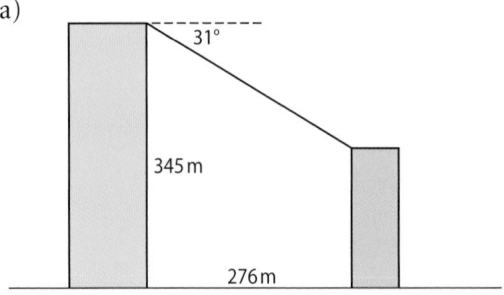

[*2 marks*]

b) Need to solve following triangle for length *l*.

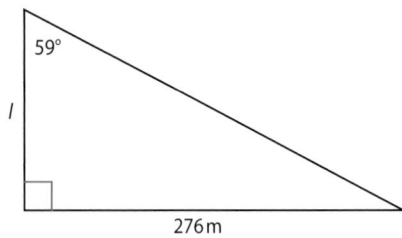

$$l = \frac{276}{\tan(59°)} = 166 \text{ m to 3 s.f.}$$

Height of second building is $345 - 166 = 179$ m. [*4 marks*]

● **Examiner's hint:** The instructions for general solutions for IB Mathematical Studies do not specify that fractions need to be expressed in lowest terms, so do not waste time reducing your fractions, unless a question specifically asks you to do so, or an obvious simplification can be undertaken.

7. a) (i) Probability a ball is red is $\frac{5}{16}$. [*1 mark*]

   (ii) Probability a ball is green given that Bjarne chose a red is $\frac{11}{15}$. [*1 mark*]

   (iii) Probability is $\frac{55}{240} = \frac{11}{48}$. [*1 mark*]

   b) $\frac{55}{240} + \frac{55}{240} = \frac{110}{240} = \frac{11}{24}$. [*3 marks*]

8. a) By symmetry, the coordinates of *A* are $(-1, 0)$. [*2 marks*]

   b) The factorized form of the quadratic is written as

   $y = a(x - p)(x - q) = a(x + 1)(x + 5)$

   Expanding $y = a(x^2 + 6x + 5)$

   Substitute a point in for *x* and *y*, e.g. $(-3, 4)$, to solve for *a*.

   $4 = a(9 - 18 + 5) \therefore a = -1$

   $y = -x^2 - 6x - 5$

   Coordinates of *B* are $(0, -5)$. [*3 marks*]

   c) $y = -x^2 - 6x - 5$ [*1 mark*]

9. a) Using a GDC, $v = 367$ ml. [*2 marks*]

b)

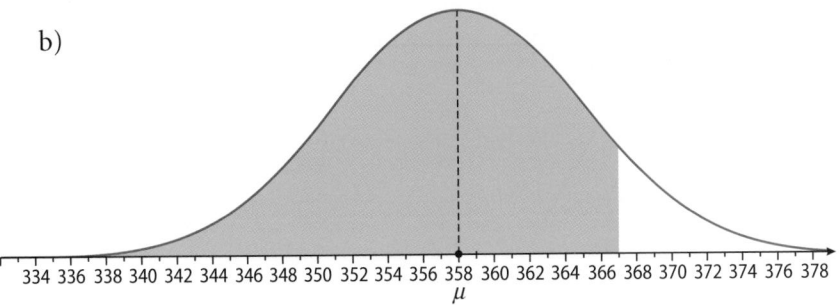

334 336 338 340 342 344 346 348 350 352 354 356 358 360 362 364 366 368 370 372 374 376 378
μ

[*2 marks*]

c) Using a GDC, the probability is 0.126549.

Number of cans = $10\,000 \times 0.126\,549 = 1260$ to 3 s.f.   [*2 marks*]

10. a) (i) the median = 24

(ii) the upper quartile = 25

(iii) the minimum value = 19

(iv) the interquartile range = 3   [*4 marks*]

b) (i)

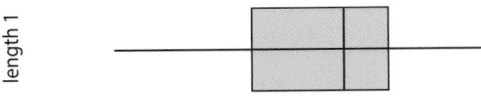

17 18 19 20 21 22 23 24 25 26 27 28

○ length 1   ○ length 2   [*1 mark*]

(ii) There is a greater range, the median is lower, the interquartile range is greater, the maximum value is also larger, the data has a greater spread.   [*1 mark*]

11. a) The height of the pyramid is $25 - 8 = 17$ cm.   [*2 marks*]

b) The length a is found by using a triangle on the top of the cube as shown.

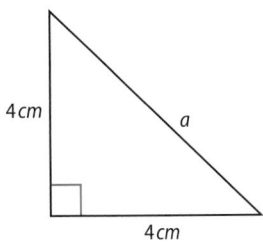

4 cm

a

4 cm

$a^2 = 4^2 + 4^2 = 32 \therefore a = \sqrt{32} = 5.657$ cm   [*2 marks*]

c) The length of the edge of the pyramid is found using the following triangle.

● **Examiner's hint:** Often on IB exams, one will be expected to use the result from one calculation for a further calculation. In this case it is advisable to use the answer from the previous question before it is expressed in 3 s.f.

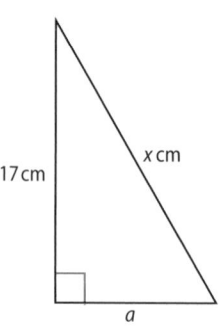

$$x^2 = 32 + 17^2 = 321 \therefore x = \sqrt{321} = 17.9 \text{ cm to 3 s.f.} \qquad [2 \text{ marks}]$$

12.   B: Has no correlation.   [2 marks]

     D: Strong negative correlation.   [2 marks]

     A and C: Moderate positive correlation.   [2 marks]

13.   Correct statements are:

     b)   The graph has three stationary points.

     c)   The graph has two x-intercepts.

     d)   The point C has zero gradient.

     f) A and B have the same gradient.   [6 marks]

14.   A tower (CD) sits on top of a building, as shown. $A\hat{B}C$ is a right angle.

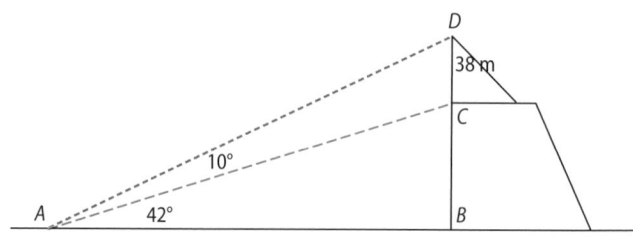

● **Examiner's hint:** Do not forget to include the appropriate units for the question.

     a) $A\hat{C}B = 90° - 42° = 48°$ and $A\hat{C}D = 180° - 48° = 132°$

         $A\hat{D}C = 180° - 132° - 10° = 38°$

         or:

         $B\hat{A}D = 52°$ [3 marks]

         $A\hat{D}C = 90° - 52° = 38°$ [3 marks]

     b)   Need to calculate length AC. Using the sine rule

         $\dfrac{38}{\sin(10)} = \dfrac{AC}{\sin(38)}$, $AC = 134.727$

         $CB = 134.727 \cos(42°) = 100$ m to 3 s.f   [3 marks]

● **Examiner's hint:** For Paper 1 calculus questions, each term you need to differentiate is typically worth 1 mark. Therefore, you should use partial answers for derivatives, even if you are not able to differentiate each term.

15.   a)   Using the power rule: $f'(x) = x^2 + 4x - \dfrac{2}{x^2}$ [4 marks]

     b)   (i)   No, see below.   [1 mark]

         (ii) $f(0) = (0)^2 + 4(0) - \dfrac{2}{(0)^2}$

         Since the last term is undefined, the whole expression is undefined.   [1 mark]

Paper 2

1. a) Let B be respondents who visited the beach, M those respondents who visited the mountains and C those who visited another country.

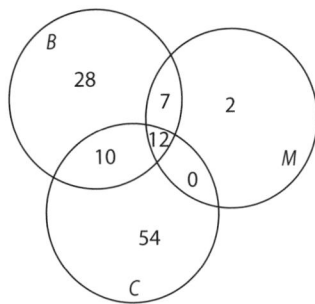

[*4 marks*]

   b) (i) Number of respondents $M \cap C \cap B' = 0$ [*4 marks*]

   (ii) Only visited the mountains $= 2$

   c) $P(B|C) = \dfrac{22}{76} = \dfrac{11}{38}$ [*2 marks*]

   d) (i) $\dfrac{57}{113} \times \dfrac{56}{112} = \dfrac{57}{226}$

   (ii) $\dfrac{57}{113} \times \dfrac{56}{112} + \dfrac{56}{113} \times \dfrac{57}{112} = \dfrac{57}{113}$ [*7 marks*]

2. a) Total surveyed is 109 students, 63 males, and 46 females. [*3 marks*]

   b) Expected frequency $= \dfrac{46}{109} \times \dfrac{22}{109} \times 109 = 9.2844$ [*3 marks*]

   • **Examiner's hint:** Use your calculator to find the expected values, and χ^2 values as the examiners are expecting you to do this.

   c)

23.1	12.7	12.7	14.4
16.9	9.28	9.28	10.6

[*4 marks*]

   d) p-value $= 0.0345$ [*2 marks*]

   e) As $0.0345 < 0.05$ then we reject the null hypothesis. [*3 marks*]

3. Consider the function $f(x) = \dfrac{3}{2}x^2 - 5x - 2$

   a) $f(4) = \dfrac{3(4)^2}{2} - 5(4) - 2 = 2$ [*2 marks*]

   b) y-intercept $= -2$ [*1 mark*]

   c) Using GDC, the roots are $x = -0.360\,921$ and $x = 3.694\,25$ [*3 marks*]

   d)

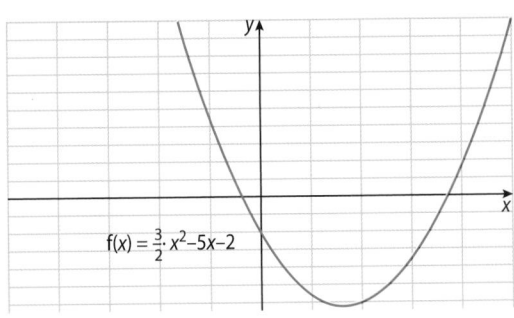

$f(x) = \dfrac{3}{2} \cdot x^2 - 5x - 2$

[*4 marks*]

   e) $f'(x) = 3x - 5$ [*2 marks*]

f) At the minimum point $f'(x) = 0 = 3x - 5$.

$x = \dfrac{5}{3} = 1.67$ to 3 s.f.

By substitution or GDC, $y = -6.17$ to 3 s.f. Coordinates of minimum are $(1.67, -6.17)$.

g) Gradient of the tangent at $x = 4$: $f'(4) = 3(4) - 5 = 7$ [2 marks]

h) Equation of tangent: $y - y_1 = m(x - x_1)$ using $(4, 2)$

$y - (2) = 7(x - 4)$

$y = 7x - 26$ [2 marks]

i) Gradient of normal: at $x = 4$ is $\dfrac{-1}{7}$

Equation of normal: $y - y_1 = m(x - x_1)$ using $(4, 2)$

$y - (2) = \dfrac{-1}{7} \times (x - 4)$

$\therefore y = \dfrac{-1}{7}x + \dfrac{18}{7} = -0.143x + 2.57$ [4 marks]

4. a) $I(t) = 40\,000\left(1 + \dfrac{9}{100 \times 12}\right)^{12t}$ [2 marks]

b) $I(1) = 40\,000\left(1 + \dfrac{9}{100 \times 12}\right)^{12(1)} = 43\,752.30$

Interest earned is $43\,752.30 - 40\,000 = 3752.30$ [2 marks]

c) $I(2) = 40\,000\left(1 + \dfrac{9}{100 \times 12}\right)^{12(2)} = \$47\,856.50$ [1 mark]

d) Solve $80\,000 = 40\,000\left(1 + \dfrac{9}{100 \times 12}\right)^{12t}$ for t. Using GDC:

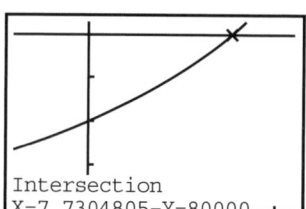

```
Intersection
X=7.7304805 Y=80000
```

Bob will have $80\,000 after 7 years 9 months, approximately.

    [2 marks]

e) $D(t) = 80\,000\left(1 - \dfrac{7}{100}\right)^{t}$ [2 marks]

f) $D(1) = 80\,000\left(1 - \dfrac{7}{100}\right)^{1} = \$74\,400$ [1 mark]

g)

(4.27, 58 700)

$D(t)$

$I(t)$

    [4 marks]

h) As shown in graph, the values will be equal after 4.27 years.

[*4 marks*]

i) Solve $40\,000 = 80\,000\left(1 - \dfrac{7}{100}\right)^t$ for t.

Graphically

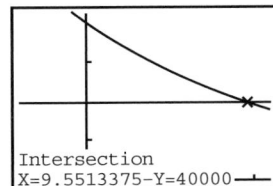

Intersection
X=9.5513375—Y=40000

The car will have a value of $40\,000$ after 9.55 years. [*4 marks*]

5. A function is given by the equation $f(x) = 35 \times 1.3^{-0.5x} + 20$.

   a) $f(0) = 35 \times 1.3^{-0.5x} + 20 = 35(1.3)^{-0.5(0)} + 20 = 35 + 20 = 55$

   [*2 marks*]

   b) (i) $p = 38.2$, $q = 29.43$ [*2 marks*]

   (ii)

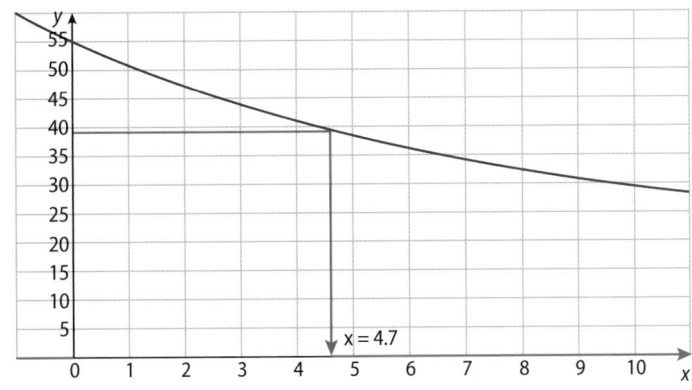

[*4 marks*]

   (iii) $x = 4.7$ [*2 marks*]

   (iv) The horizontal asymptote is $y = 20$. As x becomes larger, the
       graph will approach $y = 20$. [*2 marks*]

   c) Consider the function $g(x) = 4x - 9$ for $0 \leqslant x \leqslant 10$.

[*2 marks*]

   d) $g(x) = f(x)$ when the graphs intersect. Thus, the solution is $(9.7, 30)$
      by reading from the graph. [*2 marks*]

Solutions to Second Practice Exam
Paper 1

1. a) The weight of the parcel $= 50 \times 1.469 = 73.45$ kg [3 marks]

   b) Percentage Error $= \left| \dfrac{75 - 73.45}{73.45} \right| \times 100 = 2.11\%$ [3 marks]

2. a) Since there is a charge of 3 GB£, our £133 becomes £130 before any conversion occurs. We then convert to CHF:

   $130 \times 1.5 = 195$ CHF [2 marks]

   b) First we subtracted 3 GB£, then we multiplied by 1.5. In a formula, this is equivalent to $s = 1.5(b - 3)$. [2 marks]

   c) We substitute our value of 430 in our formula, and solve for b.
   Hence, $450 = 1.5(b - 3)$

   $\dfrac{450}{1.5} = b - 3$

   $300 + 3 = b$

   $b = 303$ GB£ [2 marks]

● **Examiner's hint:** Be careful when asked to create an expression that you do not simplify it unnecessarily. If they do not ask for it simplified, do not waste your valuable exam time simplifying it.

3. a) $u_9 = -23 = u_1 + (9 - 1)d;\ u_{25} = 25 = u_1 + (25 - 1)d$
   $u_1 + 8d = -23$
   $u_1 + 24d = 25$

   Solving equations simultaneously with GDC.

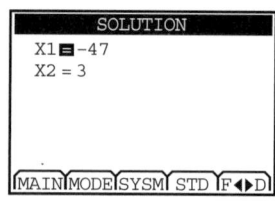

   (i) $u_1 = -47$ and $d = 3$ [4 marks]
   (ii) $S_{25} = \dfrac{25}{2}[2 \times -47 + (25 - 1)3] = -275$ [2 marks]

4. a) $p = 67$ [1 mark]

   b) (i) median $= 128$
   (ii) the lower quartile $= 118$
   (iii) the upper quartile $= 133$ [3 marks]

   c)

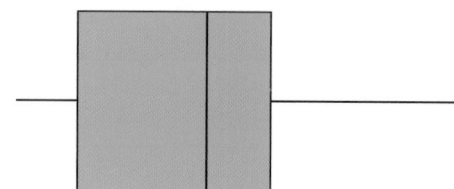

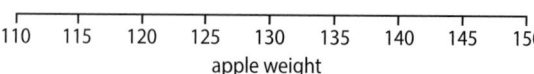

apple weight

[2 marks]

5. a) Sample space diagram is:

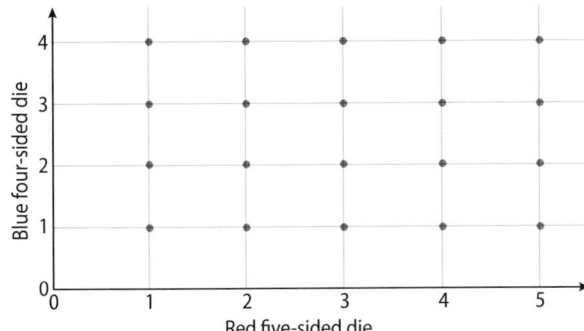

There are 20 possible outcomes.   [*3 marks*]

b) There are 6 relevant outcomes. The probability is $\frac{6}{20} = \frac{3}{10}$.

[*1 marks*]

c) Prime numbers are 2, 3, 5. The probability is $\frac{3}{20}$.   [*2 marks*]

6. Angle between the side and the base is $180° - 65° = 115°$.

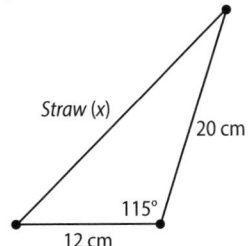

Using the cosine rule
$x^2 = 20^2 + 12^2 - 2 \times 20 \times 12 \times \cos(115°) = 746.857$

$x = \sqrt{746.857} = 27.3$ cm.   [*6 marks*]

7. a)

p	q	$\neg q$	$p \Rightarrow \neg q$	$\neg p$	$(p \Rightarrow \neg q) \wedge \neg p$	$[(p \Rightarrow \neg q) \wedge \neg p] \Rightarrow q$
T	T	F	F	F	F	T
T	F	T	T	F	F	T
F	T	F	T	T	T	T
F	F	T	T	T	T	F

[*3 marks*]

b) As the final column does not contain all T's or all F's then compound proposition is neither a tautology nor a contradiction.

[*1 mark*]

c) $\neg q \Rightarrow \neg p$ [*2 marks*]

8. a) The modal number of hours is the most frequent number, which is 6 hours.   [*2 marks*]

b) The total number of hours is $4 \times 4 + 5 \times 5 + 6 \times 9 + 7 \times 8 + 8 \times 4$
$= 183$.

The total number of days is $4 + 5 + 9 + 8 + 4 = 30$.

The mean is therefore $\frac{183}{30} = 6.10$ hours. [4 marks]

9. a)

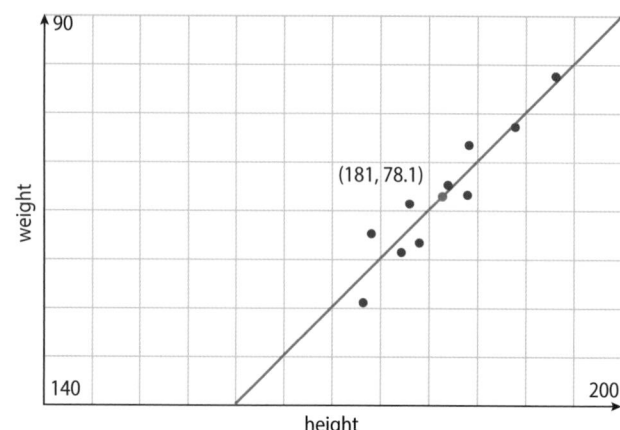

[3 marks]

b) $\bar{x} = 181, \bar{y} = 78.1$ [2 marks]

c) Shown on graph. [1 mark]

● **Examiner's hint:** Often a sketch of a graph can be the fastest way to identify a local extreme as a maximum or a minimum. Be sure to include a reasonably neat sketch with your solution, should you use this method.

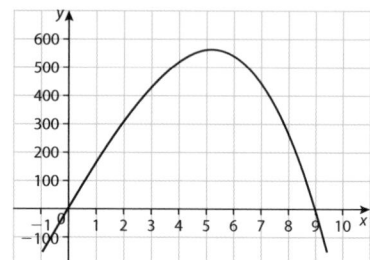

10. a) The volume of a rectangular prism is $V = length \times width \times height$ which becomes
$$V = 2x(9 - x)(9 + x)$$ [2 marks]

b) We set $V'(x) = 0$ and solve for x first. Hence,
$$0 = 162 - 6x^2$$
$$6x^2 = 162$$
$$x^2 = 27$$
$$x = \pm\sqrt{27} \approx \pm 5.20$$

Since we require x to be positive $x = \sqrt{27} \approx 5.20$ is our solution. [4 marks]

11. a) Gradient of $3y + 2x - 8 = 0$, rearrange $y = -\frac{2}{3}x - 8$

gradient of $M = \frac{-2}{3}$ [2 marks]

b) $y - 4 = -\frac{2}{3}(x - 3)$

Equation of line M: $y = -\frac{2}{3}x + 6$ [2 marks]

c) Gradient of line $N = \frac{3}{2}$

$y - 0 = \frac{3}{2}(x + 4)$

Equation of line N: $y = \frac{3}{2}(x + 4)$ [2 marks]

12. a) $h(t) = 1.3 \times 2^{0.5 \times 6} = 10.4$ cm [3 marks]

b) Need to solve for 1.5 metres (150 cm.)
$$1.3 \times 2^{0.5t} = 150$$

Using GDC, draw the graph for $Y1 = 1.3 \times 2^{0.5t}$ and $Y2 = 150$ and find the point of intersection.

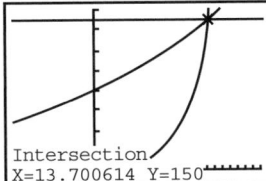

Intersection
X=13.700614 Y=150

The wheat will be ready to harvest at 13.7 weeks.   [*3 marks*]

13. a) Using the compound interest formula,

$$FV = 20\,000 \left(1 + \frac{2}{100 \times 2}\right)^{2 \times 3} = \$21\,230.40 \text{ to the nearest 10 cents.}$$
[*3 marks*]

b) If the interest is compounded quarterly then the equation is given by $FV = 20\,000 \left(1 + \frac{2}{100 \times 4}\right)^{4 \times 3} = \$21\,233.60$

The difference in interest is $\$21\,233.60 - \$21\,230.40 = \$3.20$ [*3 marks*]

14. a) Tree diagram

Today	Tomorrow

Sunshine 0.3

Sunshine 0.6
No sunshine 0.4

No sunshine 0.7

Sunshine 0.4
No sunshine 0.6
[*3 marks*]

b) Probability that the sun does not shine tomorrow is
$0.3 \times 0.4 + 0.7 \times 0.6 = 0.54$ [*3 marks*]

15. a) $H(10) = 14(10) - 0.25(10)^2 = 115$ m   [*2 marks*]

b) Maximum height occurs at axis of symmetry
$$t = \frac{-b}{2a} = \frac{-14}{2 \times (-0.25)} = 28$$

$H(28) = 14(28) - 0.25(28)^2 = 196$ m

Alternatively graphically.

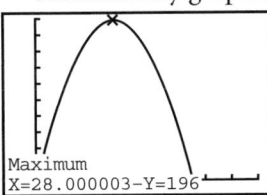

Maximum
X=28.000003 Y=196

[*2 marks*]

c) Need to solve for the x-axis intercepts. i.e. $14t - 0.25t^2 = 0$

$14t - 0.25t^2 = t(14 - 0.25t) = 0, t = 0, 56$

This can also be done by symmetry. As there is an axis of symmetry at $t = 28$, then the ball lands at $t = 2 \times 28 = 56$ seconds. *[2 marks]*

Paper 2

1. a) Using the sine rule, $\dfrac{AC}{\sin(109)} = \dfrac{17}{\sin(44)}$.

$AC = \dfrac{17}{\sin(44)} \sin(109)$

$AC = 23.1$ m *[3 marks]*

b) $A\hat{C}B = 180° - (109° + 44°) = 27°$. *[1 marks]*

c)

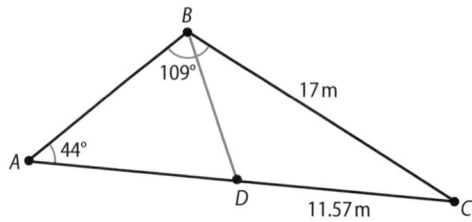

Distance between C and $D = \dfrac{1}{2}AC = \dfrac{1}{2} \times 23.1392 = 11.57$ m

Using the cosine rule

$BD^2 = 17^2 + 11.57^2 - 2 \times 17 \times 11.57 \times \cos(27°) = 72.36$

$BD = 8.51$ m *[5 marks]*

d) Area of triangle $BDC = \dfrac{1}{2} \times 17 \times 11.57 \times \sin(27°) = 44.65$ m². *[3 marks]*

e) Convert 20 cm to metres, i.e. 0.2 m. Volume $= 0.2 \times 44.65 = 8.93$ m³ *[3 marks]*

● **Examiner's hint:** Remember to always express the answer in the form requested, in this case 2 d.p. or to 3 significant figures. Either way, do not forget to include the appropriate units for the question.

● **Examiner's hint:** As this question involves a lot of data, it would be helpful to enter the data into your GDC before doing any calculations.

2. a)

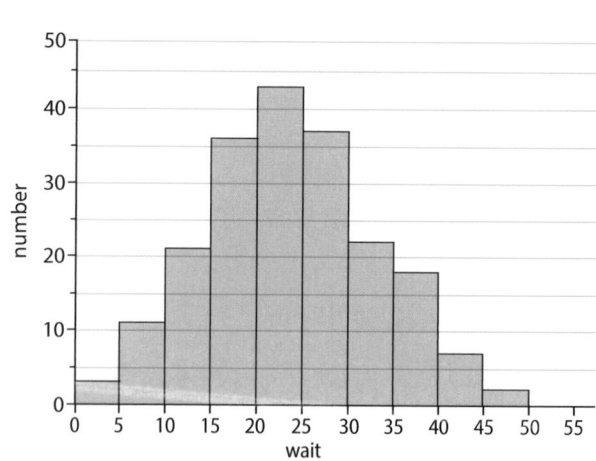

[3 marks]

b) $$\frac{3 \times 3 + 11 \times 8 + 21 \times 13 + 36 \times 18 + 43 \times 23 + 37 \times 28 + 22 \times 33 + 18 \times 38 + 7 \times 43 + 2 \times 48}{200}$$

= 24.25 to 3 s.f. = mean

Alternatively using GDC.

```
1-Var Stats
x̄=24.25
Σx=4850
Σx²=135200
Sx=9.401031698
σx=9.377499667
↓n=200
```

[*3 marks*]

c)

Waiting time in minutes	Number of travellers	Cumulative frequency
$0 \leqslant x < 5$	3	3
$5 \leqslant x < 10$	11	14
$10 \leqslant x < 15$	21	35
$15 \leqslant x < 20$	36	71
$20 \leqslant x < 25$	43	114
$25 \leqslant x < 30$	37	151
$30 \leqslant x < 35$	22	173
$35 \leqslant x < 40$	18	191
$40 \leqslant x < 45$	7	198
$45 \leqslant x < 50$	2	200

```
L1      L2      L3       3
3       3       -------
8       11
13      21
18      36
23      43
28      37
33      22
L3=cumSum(L2)
```

• **Examiner's hint:** cumSum on GDC can be used.

[*3 marks*]

d)

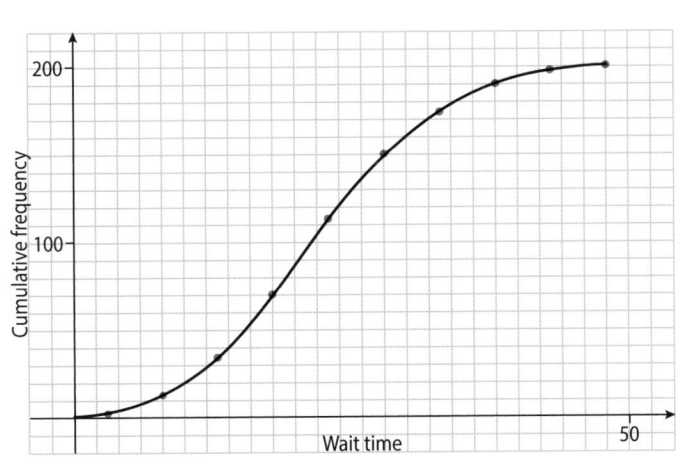

• **Examiner's hint:** When 'otherwise' is used in a question it is left to the candidate to choose their preferred method.

[*3 marks*]

e) (i) Median waiting time is approximately 23 minutes. [*1 mark*]

(ii) Interquartile range is = $Q_3 - Q_1 = 28 - 18 = 10$. [*2 marks*]

f) Normal distribution with mean $\mu = 24.25$ and $\sigma = 9.38$.
Using GDC $P(X < 25) = 0.532$. [*2 marks*]

g) Inverse normal calculation $W = 12.2$ minutes. [*2 marks*]

```
normalcdf(-1E99▶
          .5318644822
invNorm(.1,24.2!▶
          12.22904631
```

h) Using GDC, the lower bound is 30, the upper bound is 50 with $\mu = 24.25$ and $\sigma = 9.38$. Probability $= 0.2669$. Expected number of passengers is $0.2669 \times 100 = 26.7 \approx 27$ passengers.   [*2 marks*]

3. a) $f(x) = x^3 - 6x^2 + 8x + c$

   $f(1) = 1^3 - 6(1)^2 + 8(1) + c = 5$

   $3 + c = 5$

   $c = 5 - 3 = 2$ [*3 marks*]

- **Examiner's hint:** As the question says 'Write down' it is intended that the GDC will be used to assist in the calculation. However, you should show what you are doing and one way to do this is to write the equation being solved.

b) Using GDC and plysmlt2. Solve $x^3 - 6x^2 + 8x + 2 = 0$ for the x-intercepts.

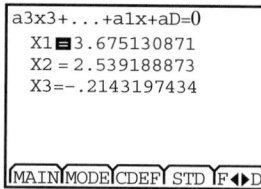 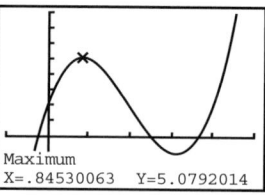 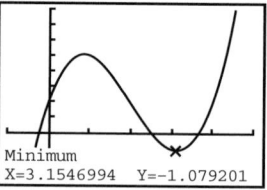

   $A(-0.214, 0)$, $B(2.54, 0)$ and $C(3.68, 0)$ [*3 marks*]

c) P and Q are turning points and the gradient in each case is zero.   [*2 marks*]

d) Using GDC to find maximum and minimum:

   $P(0.845, 5.08)$ and $Q(3.15, -1.08)$ [*4 marks*]

e) $f(x)$ is decreasing when $0.845 < x < 3.15$ [*1 mark*]

f) Gradient $m = \dfrac{2 - (-2)}{4 - 0} = 1$

   $y - y_1 = m(x - x_1)$

   Using $(0, -2)$ $y - (-2) = 1(x - 0)$

   $g(x) = x - 2$ [*3 marks*]

g) (i) There are 3 solutions, points of intersection.

   (ii) Using GDC:

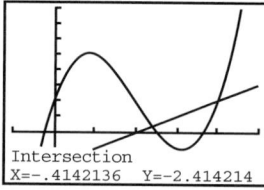 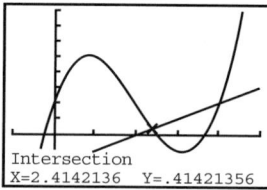 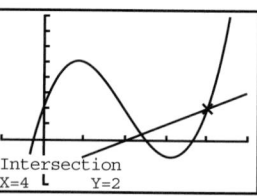

   Solutions are at $x = -0.414, 2.41, 4$ [*3 marks*]

h) $f(x) \geqslant g(x)$ when $-0.414 \leqslant x \leqslant 2.41$ and $x \geqslant 4$ [*3 marks*]

4. a) (i) Volume of the cone $V_{\text{cone}} = \dfrac{1}{3}\pi r^2 h_{\text{cone}}$ and

   Volume of cylinder $V_{\text{cyl}} = \pi r^2 h_{\text{cyl}}$

   Combined volume

   $$V = \pi r^2 h_{\text{cyl}} + \frac{1}{3}\pi r^2 h_{\text{cone}} = \pi r^2\left(h_{\text{cyl}} + \frac{1}{3}h_{\text{cone}}\right)$$ [*3 marks*]

   (ii) New volume $V = \pi r^2\left(h + \dfrac{1}{3}h\right)$

   $V = \dfrac{4}{3}\pi r^2 h$ [*2 marks*]

(iii) $V = \dfrac{4}{3}\pi \times 5^2 \times 10 = 1047.2 \approx 1050$ cubic cm to 3 s.f.   [*3 marks*]

b) A rectangle has dimensions $(5 + 3x)$ metres and $(7 - 6x)$ metres.

(i) $A(x) = (5 + 3x) \times (7 - 6x) = 35 + 21x - 30x - 18x^2$

$\therefore A(x) = 35 - 9x - 18x^2$ [*3 marks*]

(ii) Axis of symmetry $= x = \dfrac{-b}{2a} = \dfrac{9}{2 \times -18} = \dfrac{-1}{4}.$ [*3 marks*]

(iii) Hence, or otherwise, determine the length of the sides for the maximum area.

Using GDC:

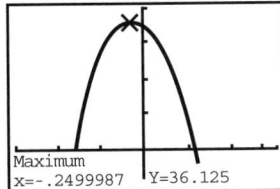

```
Maximum
x=-.2499987   Y=36.125
```

Maxmimum area occurs at $x = -\dfrac{1}{4}$

$(5 + 3x) = 5 + 3 \times \dfrac{-1}{4} = 4\dfrac{1}{4} = 4.25$ m and $(7 - 6x) = 7 - 6 \times \dfrac{-1}{4}$

$= 8\dfrac{1}{2} = 8.5$ m length $= 8.5$ m and width $= 4.25$ m

[*4 marks*]

5. a) (i) Expected number of males under 40 $= \dfrac{248}{500} \times \dfrac{151}{500} \times 500$

$= 74.9 \approx 75$ [*2 marks*]

(ii) b, c, and d are shown in the table.   [*3 marks*]

	< 40 years old	40 ⩽ age < 50	⩾ 50 years old	Total
Male	75	114	60	248
Female	76	115	60	252
Total	151	229	120	500

b) (i) Null hypothesis; The age group of managers is independent of gender.   [*1 mark*]

(ii) Alternate hypothesis: The age group of managers is not independent of gender.   [*1 mark*]

c) (i) degrees of freedom $= 2$

(ii) p-value from GDC $= 0.0591$

(iii) As $0.0591 > 0.05$ then we do not reject the null hypothesis.

[*6 marks*]

The Project

Overview

This chapter provides support and advice for your Mathematical Studies project. This project is worth 20% of your overall IB grade in this course, so it is very important. The chapter includes some helpful tips on time management and a detailed discussion of the assessment criteria for the project — this explains how to maximize your score in each area. The chapter concludes with a sample project.

Time management during the project

Your project can be broken down into smaller tasks, which you must complete.

1. You have to decide what task you are going to do. Your teacher will probably have some suggestions, and at the end of this section there is a list of project ideas (with a brief description of each).

2. Once you have decided which task you are going to do, you will have to create a plan of action for your project. Decide (and write down) what data or information you are going to collect and the mathematics that will be appropriate for your project.

3. Read the assessment criteria for the project carefully.

4. Collect your data and information.

5. Do whatever calculations are necessary and keep track of all of your calculations. Make sure any mathematics you do has a purpose.

6. Write your introduction, validity section, and conclusion for your project.

7. Organize your project into a reasonable order. (Our suggestion is to use exactly the same order as the assessment criteria if you are unsure.)

8. Assess yourself with the criteria to verify you have completed the project.

9. Review your project with your teacher before handing it in.

Your project is supposed to take you about 25 hours to complete, from start to finish. Some students spend too much time working on minute details that will not affect their overall score, while missing important sections of the project.

The assessment criteria

The assessment criteria can be confusing to someone who is not familiar with them, so here is a brief outline of what is expected. Each of the following sections includes the exact wording of the IB project assessment criteria.

A: Introduction (3 marks)

Achievement level	Descriptor
0	The project does not contain a clear statement of the task. *There is no evidence in the project of any statement of what the student is going to do or has done.*
1	The project contains a clear statement of the task. *For this level to be achieved, the task should be stated explicitly.*
2	The project contains a title, a clear statement of the task, and a description of the plan. *The plan need not be highly detailed, but must describe how the task will be performed. If the project does not have a title, this achievement level cannot be awarded.*
3	The project contains a title, a clear statement of the task, and a detailed plan that is followed. *The plan should specify what techniques are to be used at each stage and the purpose behind them, thus lending focus to the task.*

Your introduction should include:

1. A title page with the title of your project, your name, your teacher's name, your IB candidate number, and the date you finished your project.
2. A clear and concise description of your project.
3. A clear description of the steps you will go through to finish your project.
4. Evidence that you have a detailed plan.
5. Explanation of which mathematical operations you are using and importantly why you are using them.

Note that your introduction does not have to be written before you start your project. But to achieve full points you will need to indicate not only the plan but how it is to be followed and justify why you are using the methods (techniques) to undertake the investigation.

● **Examiner's hint:** Make sure you include both **what** you are going to do, and **how** you will do it.

B: Information/measurement (3 marks)

Achievement level	Descriptor
0	The student does not collect relevant information or generate relevant measurements. *No attempt has been made to collect any relevant information or to generate any relevant measurements.*
1	The student collects relevant information or generates relevant measurements. *This achievement level can be awarded even if a fundamental flaw exists in the instrument used to collect the information, for example, a faulty questionnaire or an interview conducted in an invalid way.*

2	The relevant information collected, or set of measurements generated by the student, is organized in a form appropriate for analysis or is sufficient in both quality and quantity. *A satisfactory attempt has been made to structure the information/ measurements ready for the process of analysis, **or** the information/ measurement collection process has been thoroughly described and the quantity of information justified. The raw data must be included for this achievement level to be awarded.*
3	The relevant information collected, or set of measurements generated by the student, is organized in a form appropriate for analysis **and** is sufficient in both quality and quantity. *The information/measurements have been properly structured ready for analysis **and** the information/measurement collection process has been thoroughly described and the quantity of information justified. If the information/ measurements are too sparse or too simple, this achievement level cannot be awarded. If the information/ measurements are from a **secondary** source, then there must be evidence of sampling if appropriate. All sampling processes should be completely described.*

You will very likely need to collect data for your project and/or organize information. You need to:

1. Collect sufficient data in order to be able to draw conclusions from it.
2. Organize your data in a logical fashion.
3. Verify the accuracy of your data.
4. Confirm that the data you are including is relevant.
5. Provide a copy of the raw data in an appendix to your project.
6. Provide a copy of any questionnaires used in the appendix
7. If using data from a secondary source e.g. United Nations then explain how you selected your data, that is did you take a random sample, or did you use another sampling method?

Note that a very popular way for students to collect data is to create a survey and ask their peers a variety of questions. ***We advise strongly against doing this!*** When you collect data from your peers, it is impossible to verify its accuracy, it is difficult and time consuming to collect enough data, and errors are introduced through improper sampling methods.

Better ways of collecting statistical data are:
• creating an experiment and collecting raw numerical data.
• finding a reliable source of data, such as a government statistics database or an encyclopaedia, and using data that a professional has collected. If you do this, you will need to cite the source of your data, and do your own analysis of that data.

C: Mathematical processes (5 marks)

Achievement level	Descriptor
0	The project does not contain any mathematical processes. *For example, the processes have been copied from a book, with no attempt being made to use any collected/generated information. Projects consisting of only historical accounts will achieve this level.*
1	At least two simple mathematical processes have been carried out. *Simple processes are considered to be those that a mathematical studies SL student could carry out easily, for example, percentages, areas of plane shapes, graphs, trigonometry, bar charts, pie charts, mean and standard deviation, substitution into formulae, **and any** calculations and/or graphs using technology only.*
2	At least two simple mathematical processes have been carried out correctly. *A small number of isolated mistakes should not disqualify a student from achieving this level. If there is incorrect use of formulae, or consistent mistakes in using data, this level cannot be awarded.*
3	At least two simple mathematical processes have been carried out correctly. All processes used are relevant. *The simple mathematical processes must be relevant to the stated aim of the project.*
4	The simple relevant mathematical processes have been carried out correctly. In addition, at least one relevant further process has been carried out. *Examples of further processes are differential calculus, mathematical modelling, optimization, analysis of exponential functions, statistical tests and distributions, compound probability. For this level to be achieved, it is not required that the calculations of the further process be without error. At least one further process must be calculated, showing full working.*
5	The simple relevant mathematical processes have been carried out correctly. In addition, at least one relevant further process has been carried out. All processes, both simple and further, that have been carried out are without error. *If the measurements, information, or data are limited in scope, then this achievement level cannot be awarded.*

In order to achieve maximum marks for this criterion you will need to:

1. Avoid just writing about the history of a particular topic. For example, writing about the development and history of the Golden Ratio will not in itself allow you to gain any points in this criterion.

2. Choose mathematical techniques that are appropriate to furthering the aim of your project.

3. Ensure that you provide accurate drawings and graphs, not hand drawn sketches.

4. Verify the accuracy of your results and confirm you have no errors in your calculations.

5. Use at least two simple mathematical processes, such as graphs, trigonometry, descriptive statistics, and ensure their correctness.

● **Examiner's hint:** Make sure you know the difference between 'valid' and 'reliable'. For validity, does the project do what it set out to do? For reliability, if another student performed the same statement of task, would they get the same results?

6. Use at least one further mathematical process such as calculus or a statistical test.

7. Provide reasons for the relevance of your mathematical processes.

Most students lose two or three marks in this section, largely because they do not check their calculations carefully enough, and they do not use sophisticated enough mathematics. Always endeavour to include at least a further mathematical process in your project. This may require you to consider the initial design of the project to take account of the requirement for the use of further mathematical processes.

D: Interpretation of results (3 marks)

Achievement level	Descriptor
0	The project does not contain any interpretations or conclusions. *For the student to be awarded this level, there must be no evidence of interpretation or conclusions anywhere in the project, or a completely false interpretation is given without reference to any of the results obtained.*
1	The project contains at least one interpretation or conclusion. *Only minimal evidence of interpretations or conclusions is required for this level. This level can be achieved by recognizing the need to interpret the results and attempting to do so, but reaching only false or contradictory conclusions.*
2	The project contains interpretations and/or conclusions that are consistent with the mathematical processes used. *A "follow through" procedure should be used and, consequently, it is irrelevant here whether the processes are either correct or appropriate; the only requirement is consistency.*
3	The project contains a meaningful discussion of interpretations and conclusions that are consistent with the mathematical processes used. *To achieve this level, the student would be expected to produce a discussion of the results obtained and the conclusions drawn based on the level of understanding reasonably to be expected from a student of mathematical studies SL. This may lead to a discussion of underlying reasons for results obtained. If the project is a very simple one, with few opportunities for substantial interpretation, this achievement level cannot be awarded.*

This section is about looking at your mathematical results and deciding what they mean. In order to do well in this section, you must:

1. Provide your own reasonable explanations of what your mathematical calculations show.

2. Verify that these interpretations are consistent with your calculations.

3. Provide enough detail in your interpretations so that a reasonable observer would agree with your conclusions.

E: Validity (1 mark)

Achievement level	Descriptor
0	There is no awareness shown that validity plays a part in the project.
1	There is an indication, with reasons, if and where validity plays a part in the project. *There is discussion of the validity of the techniques used **or** recognition of any limitations that might apply. A simple statement such as "I should have used more information/measurements" is not sufficient to achieve this level. If the student considers that validity is not an issue, this must be fully justified.*

● **Examiner's hint:** Make sure you know the difference between 'valid' and 'reliable'. For validity, does the project do what it set out to do? For reliability, if another student performed the same statement of task, would they get the same results?

All students who produce mathematical studies projects are expected to evaluate their work and confirm that it has been successful and without error. In order to get maximum marks on validity for your project, you must:

1. Comment on how you know that the mathematics you have used is correct. Evidence of having checked your calculations is useful here.

2. Comment on whether the mathematical processes used were appropriate. Could/should other techniques have been used? If so, explain why.

3. Comment on how you know that the conclusions you have drawn about your project are accurate. It is useful to quote (and cite in your bibliography) similar research that matches your conclusions here.

4. Provide an explanation of how you can improve your project.

F: Structure and communication (3 marks)

Achievement level	Descriptor
0	No attempt has been made to structure the project. *It is not expected that many students will be awarded this level.*
1	Some attempt has been made to structure the project. *Partially complete and very simple projects would only achieve this level.*
2	The project has been structured in a logical manner so that it is easily followed. *There must be a logical development to the project. The project must reflect the appropriate commitment for this achievement level to be awarded.*
3	The project has been well structured in accordance with the stated plan **and** is communicated in a coherent manner. *To achieve this level, the project would be expected to read well, and contain footnotes and a bibliography, as appropriate. The project must be focused and contain only relevant discussions.*

This criterion focuses on the organization and structure of the project. In order to achieve the maximum for this criterion, you must:

1. Have a logical order to your project. A suggested order is:
   - (i) Title page
   - (ii) Introduction
   - (iii) Organized data
   - (iv) Mathematical processes
   - (v) Interpretation of results
   - (vi) Discussion of validity
   - (vii) Appendices (as required)
   - (viii) Bibliography

2. Include footnotes in your work as appropriate and cite your sources of information.

3. Use appropriate mathematical vocabulary and symbols correctly.

4. Avoid placing repeated calculations in the main text. Include sample calculations, place the remaining calculations of the same type in the appendix, and refer to them where appropriate.

5. Have someone who has not taken this course read your work and comment on its readability. You should consider your audience for this project to be someone who is not familiar with the content of this course.

G: Notation and terminology (2 marks)

Achievement level	Descriptor
0	The project does not contain correct mathematical notation or terminology. *It is not expected that many students will be awarded this level.*
1	The project contains some correct mathematical notation **or** terminology.
2	The project contains correct mathematical notation **and** terminology throughout. *Variables should be explicitly defined. An isolated slip in notation need not preclude a student from achieving this level. If it is a simple project requiring little or no notation and/or terminology, this achievement level cannot be awarded.*

This criterion deals with the use of mathematical notation and terminology. For this criterion you must:

1. Use correct mathematical notation, not GDC notation. For example, $x{\wedge}2$, is inappropriate. Instead, use x^2. Similarly, if using a spreadsheet formula do not use "*" to indicate multiplication.

2. Take the time to look in the textbook to find the correct mathematical words to describe your work.

3. Remember to define what x represents. Do not just write $x = $ a number. Examples include, 'Let x be the length of the hypotenuse of the triangle,' or 'Let p be the number of people travelling by bus to school every day'.

Overall total

When you are working on your project, you should often verify that you are meeting the criteria established by the IBO. Confirm that you have all of the required sections for your project, and that you have done each of them well. It is important to note that even students who are weak in mathematics can achieve reasonably high levels of success in this project if they focus on meeting the criteria.

The maximum score you can receive on a project is 20 marks. A score of 11 or less indicates that you did not seek the help you required and did not put sufficient effort into your project. A score between 12 and 15 probably indicates that you had a major error in your project that you were unable to fix, and that more effort was required. It is also possible that you chose a project topic that was too difficult. Most students should be able to score 16 or above on their project if they follow the guidelines given in this chapter.

Sample project ideas

Many students find it difficult to come up with a project idea. This is often because they feel that their idea should be *'the best idea ever'* and completely new and original. However, the most important criteria for the selection of your idea are:

1. You find it interesting.

2. You will be able to complete the project before the deadline assigned to you.

3. There is sufficient sophisticated mathematics associated with the project idea.

One way to get started in choosing your project idea is to start asking questions about things you are interested in, and see if you want to answer any of these questions. For example, one such project title might be 'Are red M&M's more common than other colours?'.

Here are some questions related to successful projects:
- What is the function for an Ollie (a special skateboard jump)?
- Does location affect the local rate of suicides?
- Is there a relationship between the number of key signature changes in a piece of music and the time period of the music?
- Does getting more sleep the night before an exam improve your score?

- Does the amount of water plants get affect their rate of growth?
- What is the relationship between GNP and AIDS infection rates?
- What is the pattern in the rise and fall of the tides?
- How quickly does water drain from a bucket?
- How long does it take for a cup of hot coffee to cool down to room temperature?
- What is the stopping distance of a Porsche?
- How many bounces can a basketball make before it stops moving?
- What is the equation of a bungee jump?

Project checklist

As you are working on your project, use the checklist below to make sure you have completed all of the minimum requirements for your project.

1. Front cover sheet with name, IB candidate number, teacher's name, name of school, date and title of project.

2. Table of contents.

3. Numbered pages.

4. Criterion A: Introduction
   a) Project task has been clearly identified.
   b) Project plan is clear and easy to follow and contains sufficient detail.

5. Criterion B: Information/measurement
   a) The raw data has been included in an appendix.
   b) Data has been verified.
   c) Data has been organized.
   d) There is at least one paragraph describing the data collection.
   e) Questionnaire or surveys used are provided in the appendix.
   f) Where appropriate, the method of sampling used is described.

6. Criterion C: Mathematical processes
   a) Well organized and clear to follow.
   b) All mathematical processes used are relevant.
   c) At least 2 simple and one sophisticated mathematical process have been used.
   d) Appropriate terminology and symbols have been used.
   e) Calculations have been checked for accuracy.
   f) The use of the mathematical processes has been justified.

7. Criterion D: Interpretation of results
   a) Results have been interpreted.
   b) A comprehensive discussion of results has been included.

8. Criterion E: Validity
   a) All positive and negative aspects of the project's validity have been discussed.
   b) At least one realistic suggestion for improvement has been made.
   c) Comments on the mathematical processes used, have been made.
   d) Comments on the conclusions drawn, have been made.

9. Criterion F: Structure and communication
   a) Project has a logical order.
   b) The project has been checked carefully for grammatical errors.
   c) Footnotes have been used where appropriate.
   d) An appendix has been included.
   e) A bibliography has been included.

10. Criterion G: Notation and terminology
    a) All mathematical notation is in the correct form with no use of GDC or spreadsheet notation.
    b) Correct mathematical terms are used.
    c) All variables have been defined.

Sample IB project – Relationship between temperature and precipitation

Introduction

It is my intention in this project to find out if there is a relationship between the average temperature in the United States and the average precipitation. In order to do this, I will do the following:

1. Collect data from an online resource.

2. Organize this data into a table.

3. Plot this data in a scatter plot to visually analyze the relationship.

4. Find the line of best fit of the data.

5. Examine the Pearson's correlation coefficient of the data.

6. Group my data into four groups using their relationship with the mean.

7. Analyze my data using the χ^2 test.

8. Draw conclusions from my analysis.

9. Confirm that my conclusions and the mathematics used are valid.

10. Discuss any flaws or shortcomings with my project.

● **Examiner's hint:** Include explicit section headings in your project. It will make it easier to mark and may help you organize your project.

 To access a useful statistics website, visit www.pearsonhotlinks.co.uk, enter the ISBN for this book and click on weblink 16.1.

● **Examiner's hint:** The steps given on the left are a good sequence to follow for almost any statistics project. Remember also that it is perfectly acceptable to write your introduction *after* you have done the rest of your project.

Data collection

Year	Temperature (Fahrenheit)	Precipitation (inches)
1970	74.04	2.38
1971	72.77	2.76
1972	72.47	2.59
1973	73.29	2.04
1974	70.99	3.44
1975	72.26	2.81
1976	71.46	2.03
1977	72.48	3.55
1978	72.5	2.73
1979	71.72	2.96
1980	73.54	2.57
1981	72.77	2.79
1982	72.78	2.55
1983	75.84	2.38
1984	73.89	2.41
1985	71.7	2.92
1986	72.38	2.83
1987	72.55	2.75
1988	74.42	2.59
1989	72.07	2.71
1990	73.02	2.64
1991	73.73	2.61
1992	70.47	2.92
1993	72.68	2.74
1994	72.88	2.54
1995	74.98	2.67
1996	73.03	2.67
1997	72.48	2.68
1998	74.7	2.42
1999	73.75	2.29
2000	74.75	1.92
2001	74.93	2.46
2002	73.54	2.44
2003	75.44	2.59
2004	70.95	3.06
2005	73.98	3.08
2006	74.19	2.86
2007	75.45	2.63

● **Examiner's hint:** Remember that you need to include your raw data with your project. In this case, since this data was collected using an online resource, it is included in the body of the project. This is because the data was already organized when it was collected.

This data shows the mean temperature in the US (in degrees Fahrenheit) and the mean precipitation in the US (in inches) from 1970 to 2007. The data was collected using an online form and represents official US government records, according to the website.

• **Examiner's hint:** A short discussion of why the data set 1970 to 2007 was chosen, would be helpful as would comments on the sampling process.

Calculations

One obvious thing to do to first examine the data is to create a scatter plot, using temperature vs precipitation, since presumably the amount of precipitation depends on the temperature.

Using an Excel spreadsheet, I created the graph that you can see below.

Ⓦ For an alternative to the spreadsheet program Excel, download the free Calc program from www. pearsonhotlinks.co.uk, enter the ISBN for this book and click on weblink 16.2.

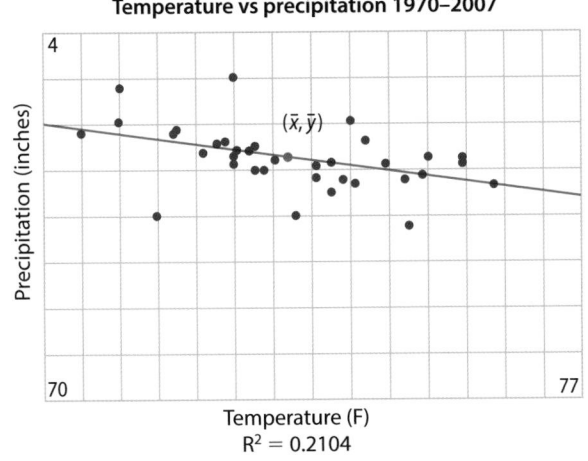

I then calculated the means for temperature and precipitation on my GDC. These were:

mean of temperature $\bar{x} = 73.1808$ and mean of rainfall $\bar{y} = 2.6582$.

I plotted these points on my graph and drew a line of best through the mean point, which is shown in red on the graph. From this graph, we can see any relationship between temperature and precipitation is at best a weak negative correlation.

I then calculated the gradient for my line of best fit by choosing another point that my line passed through. This was (75.8, 2.38).

Gradient $m = \dfrac{2.38 - 2.6582}{75.8 - 73.1808} = -0.106$ 3 s.f.

Equation is then $y - y_1 = m(x - x_1)$

$y - 2.6582 = -0.106(x - 73.1808)$

$y = -0.106x + 10.4$

My line of best fit is $y = -0.106x + 10.4$, where y is precipitation and x is temperature.

I will now calculate Pearson's product–moment correlation coefficient using the following formula:

• **Examiner's hint:** Determining a line of best fit manually by finding the means, drawing a line of best fit on a graph, and calculating the equation of the line of best fit is to be encouraged.

$$r = \frac{s_{xy}}{s_x s_y} \text{ where}$$

$$s_x = \sqrt{\frac{\sum_{i=1}^{n}(x_i - \bar{x})^2}{n}}$$

$$s_y = \sqrt{\frac{\sum_{i=1}^{n}(y_i - \bar{y})^2}{n}}$$

and $\quad s_{xy} = \frac{1}{n}\sum_{i=1}^{n}(x_i - \bar{x})(x_i - \bar{y})$

Using these formulae and our spreadsheet program, we find the following:

Year	Temperature (Fahrenheit)	Precipitation (inches)	$(x - \bar{x})$	$(y - \bar{y})$	$(x - \bar{x})^2$	$(y - \bar{y})^2$	$(x - \bar{x})(y - \bar{y})$
1970	74.04	2.38	−0.8592	0.2782	0.7382	0.0774	−0.2390
1971	72.77	2.76	0.4108	−0.1018	0.1687	0.0104	−0.0418
1972	72.47	2.59	0.7108	0.0682	0.5052	0.0046	0.0484
1973	73.29	2.04	−0.1092	0.6182	0.0119	0.3821	−0.0675
1974	70.99	3.44	2.1908	−0.7818	4.7996	0.6113	−1.7129
1975	72.26	2.81	0.9208	−0.1518	0.8479	0.0231	−0.1398
1976	71.46	2.03	1.7208	0.6282	2.9611	0.3946	1.0809
1977	72.48	3.55	0.7008	−0.8918	0.4911	0.7954	−0.6250
1978	72.5	2.73	0.6808	−0.0718	0.4635	0.0052	−0.0489
1979	71.72	2.96	1.4608	−0.3018	2.1339	0.0911	−0.4409
1980	73.54	2.57	−0.3592	0.0882	0.1290	0.0078	−0.0317
1981	72.77	2.79	0.4108	−0.1318	0.1687	0.0174	−0.0542
1982	72.78	2.55	0.4008	0.1082	0.1606	0.0117	0.0433
1983	75.84	2.38	−2.6592	0.2782	7.0714	0.0774	−0.7397
1984	73.89	2.41	−0.7092	0.2482	0.5030	0.0616	−0.1760
1985	71.7	2.92	1.4808	−0.2618	2.1927	0.0686	−0.3877
1986	72.38	2.83	0.8008	−0.1718	0.6413	0.0295	−0.1376
1987	72.55	2.75	0.6308	−0.0918	0.3979	0.0084	−0.0579
1988	74.42	2.59	−1.2392	0.0682	1.5356	0.0046	−0.0845
1989	72.07	2.71	1.1108	−0.0518	1.2339	0.0027	−0.0576
1990	73.02	2.64	0.1608	0.0182	0.0259	0.0003	0.0029
1991	73.73	2.61	−0.5492	0.0482	0.3016	0.0023	−0.0264
1992	70.47	2.92	2.7108	−0.2618	7.3484	0.0686	−0.7098
1993	72.68	2.74	0.5008	−0.0818	0.2508	0.0067	−0.0410
1994	72.88	2.54	0.3008	0.1182	0.0905	0.0140	0.0355
1995	74.98	2.67	−1.7992	−0.0118	3.2372	0.0001	0.0213
1996	73.03	2.67	0.1508	−0.0118	0.0227	0.0001	−0.0018
1997	72.48	2.68	0.7008	−0.0218	0.4911	0.0005	−0.0153
1998	74.7	2.42	−1.5192	0.2382	2.3080	0.0567	−0.3618
1999	73.75	2.29	−0.5692	0.3682	0.3240	0.1355	−0.2096
2000	74.75	1.92	−1.5692	0.7382	2.4624	0.5449	−1.1583
2001	74.93	2.46	−1.7492	0.1982	3.0597	0.0393	−0.3466
2002	73.54	2.44	−0.3592	0.2182	0.1290	0.0476	−0.0784
2003	75.44	2.59	−2.2592	0.0682	5.1040	0.0046	−0.1540
2004	70.95	3.06	2.2308	−0.4018	4.9764	0.1615	−0.8964
2005	73.98	3.08	−0.7992	−0.4218	0.6387	0.1780	0.3371
2006	74.19	2.86	−1.0092	−0.2018	1.0185	0.0407	0.2037
2007	75.45	2.63	−2.2692	0.0282	5.1493	0.0008	−0.0639

where $\bar{x} = 73.1808$ and $\bar{y} = 2.6582$.

From the table, we can calculate $s_x = 8.0059$, $s_y = 1.9967$, and $s_{xy} = -7.3326$ so $r = -0.4587$ and $r^2 = 0.2104$, which is the same as calculated using the spreadsheet program.

This value of r^2 corresponds to a weak correlation, and, from $r = -0.4587$, we say that we have a weak negative correlation between the temperature and amount of precipitation between 1970 and 2007. Therefore, any relationship between temperature and precipitation is weak at best.

Since our results seem somewhat inconclusive, it makes sense to attempt a different test on the data. We can use the χ^2 test if we can group the data appropriately. A reasonable grouping seems to be based on $\bar{x}$ and $\bar{y}$.

	Below $\bar{y}$	Above $\bar{y}$
Below $\bar{x}$	16	5
Above $\bar{x}$	14	3

Unfortunately, with this grouping, we have a group with fewer than 5 years, so we should use a different grouping if we choose to use the χ^2 test. Another possibility is that we can use the **Yates' correction for continuity**, which is a modification of the χ^2 test when we are dealing with a 2 by 2 contingency table and one cell that is less than 5. The formula for the Yates' correction is:

$$\chi^2_{calc} = \sum_{i=n}^{n} \frac{(|O_i - E_i| - 0.5)^2}{E_i}$$

Our null hypothesis is that temperature and precipitation are independent factors within a 5% significance level, and our alternative hypothesis is that temperature and precipitation are dependent factors within a 5% significance level.

First we calculate the expected values from the observed values table and we find:

	Below $\bar{y}$	Above $\bar{y}$
Below $\bar{x}$	16.579	4.4211
Above $\bar{x}$	13.421	3.5789

From this we use the Yates' correction formula (see Yates calculation).

$$\chi^2_{calc} = \sum_{i=n}^{n} \frac{(|O_i - E_i| - 0.5)^2}{E_i} =$$

$$\frac{(|16 - 16.579| - 0.5)^2}{16.579} + \frac{(|14 - 13.421| - 0.5)^2}{13.421}$$

$$\frac{(|5 - 4.4211| - 0.5)^2}{4.4211} + \frac{(|3 - 3.5789| - 0.5)^2}{3.5789}$$

$$\chi^2_{calc} = 0.003\,989$$

● **Examiner's hint:** This table of calculations could be placed in the appendix, and referred to in the test.

● **Examiner's hint:** Beware of making any inaccurate generalizations, especially when using either the χ^2 test or linear regression.

● **Examiner's hint:** This way of grouping the data is quite useful, if no other obvious grouping can be found.

● **Examiner's hint:** The calculation of Yates' continuity showing full working provides access to the higher achievement levels in criterion C. Comments on the relevance of Yates' continuity are also important here.

The degrees of freedom is 1, so our $\chi^2_{crit} = 3.841$.

Since $\chi^2_{calc} < \chi^2_{crit}$ we can accept the null hypothesis, and therefore temperature and precipitation are independent factors for any given year.

Interpretation

So our linear regression was not conclusive but our χ^2 test was definitively negative. This leads us to believe that there is no relationship between the average temperature in the United States and the average precipitation for any given year between 1970 and 2007.

We have double-checked our calculations for the linear regression and confirmed that our calculations are correct. As for the χ^2 test, since our results from it seem to match what we found with the linear regression, we tend to assume that we have calculated it correctly. By using the Yates' correction, we have made our use of the χ^2 test valid.

Validity

Linear regression allows us to determine the linear relationship (if any) between two variables. So our use of this test here was appropriate, since from the scatter plot a linear relationship (and not some other kind of relationship) seemed appropriate. However, since our value of r^2 is so small (less than 0.5) and our χ^2 test was conclusive, we can see that our conclusion must be correct. There is no relationship between temperature and precipitation.

One reason why we may not see this relationship is because of our choice of data. The average temperature and average precipitation in the United States are two variables that vary quite widely locally, as in from state to state, but we are trying to measure a relationship between them. This may not be very appropriate. We would probably do better if we restricted our data to a specific state, or even area of a state.

Besides the wide geographic variation in our data, there is large climatic variation. The United States includes areas of desert, arctic tundra, and wetlands, all of which vary greatly in both average temperature *and* average precipitation. We should again choose one type of climate and focus on our two variables for that one type of climate.

It is also possible that there is error in our data. We did not collect the data ourselves, and it was probably collected using different instruments between 1970 and 2007. There are unknown differences in collection methods between different areas of the United States, and unknown differences in collection techniques between 1970 and 2007. Both of these factors could contribute to our data being questionable.

● **Examiner's hint:** Remember, when discussing validity, you must examine both why you know your conclusions drawn are correct and why your mathematics is correct.

● **Examiner's hint:** When looking for possible sources of error in your project, examine where your data came from, how it was collected and any false assumptions you have made in organizing it.

● **Examiner's hint:** Always include an area of improvement that can actually be done. Do not say something vague like 'I should have worked harder' or 'My time management could have been better', since these are probably true of anyone who does a Mathematical Studies project. Focus on things that will make your *project* better, not things that will make *you* a better person.

Areas of improvement

One area of improvement for this project would be in the choice of data collected. Our sample size was just from too broad a geographic region to be reasonable. We could also have chosen data from a wider selection of years. As well, instead of looking at average temperatures against average precipitation for a given year, we could have collected daily statistics, and examined a few years worth of data (although in this case we would have been forced to use a spreadsheet program for all of our calculations).

Bibliography

The United States Climate Survey, http://lwf.ncdc.noaa.gov/oa/climate/research/cag3/na.html, retrieved on January 26th, 2008 at 10:00 am.

Yates' correction for continuity, http://en.wikipedia.org/wiki/Yates'_correction for continuity, retrieved on February 3rd, 2008 at 7:28 pm.

The chi-square-test,

http://en.wikipedia.org/wiki/Pearson's_chi-squared_test, retrieved on February 3rd, 2008 at 8:00 pm.

● **Examiner's hint:** Always include a bibliography in your project. Any websites that you looked at for inspiration, or for information on how to use a formula, or from which you collected your data are appropriate for inclusion. Also, an easy resource to include is this textbook.

Theory of Knowledge

Introduction

'Without mathematics, one cannot fathom the depths of philosophy; without philosophy, one cannot fathom the depths of mathematics; without the two, one cannot fathom anything.'
Bordas-Demoulins

This chapter will look at mathematics from a more philosophical point of view than the previous chapters. It will ask questions you can think about, discuss with your maths or TOK teacher, or even amongst yourselves. You have been studying mathematics for over a decade (almost two if you include the time when your parents first began teaching you to count) and you have learned many ideas and concepts. Why do think it is so important to your parents, teachers and society that you study mathematics for so many years?

Can you think of any areas of our society that use mathematics in a practical way? How about in a philosophical way?

Are there any aspects of our social structure that make no use of mathematics?

▲ What do you think he is pondering in reference to the quote to the left?

Below are several quotes about maths. Discuss each from a philosophical standpoint.

'There is something sublime in the secrecy in which the really great deeds of the mathematician are done.'
Thomas Hill

'Even today, the true importance of mathematics as an element in the history of thought is not fully appreciated.'
Whitehead

'Mathematics is one of those subjects which obliges one to make up his mind.'
L.G. Des Lauriers

468

Ralph Waldo Emerson

Deciphering ancient stone carvings is a difficult task. Knowledge of early mathematics is necessary.

The beginning of numbers

The time when mankind began to count, when we began to use the concept of numbers, is not perfectly clear. There is some evidence that Neanderthal man (50 000 B.C.E.) used scratches or laid out sticks to keep track of things. There is substantial evidence of Egyptian hieroglyphics being used to write numerals, dating as far back as approximately 3000 B.C.E. Why do you think that early man had reason to count at all?

Do you think that knowing how to count is genetically coded in our DNA?

There have been many number systems used by different cultures since the Egyptians used hieroglyphics. A few are listed below:

Babylonians, Greek, Roman, Chinese-Japanese, Mayan.

The Hindu-Arabic numeral system that is in use today has its origin approximately 300 C.E. It was invented by the Hindus, and adopted by the Arabs.

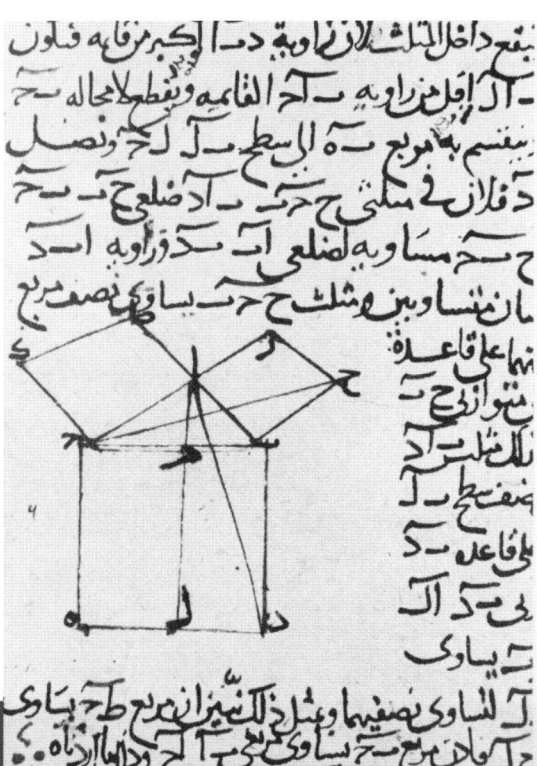

Some say that mathematics is the universal language. The writing is not readable by most, but the geometric diagram has some meaning to all. Can you glimpse the concept the diagram is explaining?

For more information on early number systems, visit www.pearsonhotlinks.co.uk, enter the ISBN for this book and click on weblinks 17.1 or 17.2.

What is the difference between 'number' and 'numeral'?

Think of the Roman numeral numbering system. Visit www.pearsonhotlinks.co.uk, enter the ISBN for this book and click on weblink 17.3.

Will mankind ever adopt a different system than the current Hindu-Arabic system? Why or why not?

The number concept	Base 10 numeral	Base 5 numeral
One	1	1
Two	2	2
Three	3	3
Four	4	4
Five	5	10
Six	6	11
Seven	7	12
Eight	8	13
Nine	9	14
Ten	10	20

Different number systems use different bases to group numbers together. For example, we use a base 10 system. This means that there are 10 and only 10 different symbols used to create any number you can think of. The current Hindu-Arabic symbols are: **0, 1, 2, 3, 4, 5, 6, 7, 8, 9**. A base 5 system (using the Hindu-Arabic symbols) would only have these 5 symbols: **0, 1, 2, 3, 4**. A comparison is shown in the table above.

The number concept column is how you *say* the number concept.

The base 10 and the base 5 columns indicate how you *write* the numeral for the number concept you say.

How ingrained are the numerals you use associated with the numbers you think of?

Are there any other number bases in use today? (Hint: Think of a turning light on and then off again using a wall switch.)

Discuss the future of number systems and base number systems as they apply to the world of technology.

Did mankind discover or invent number systems?

More on numbers

The number zero has not always been part of mathematical language. The ancient Greeks reasoned that if there was nothing there, how could we discuss its properties?

The concept of zero has a long history. A symbol for zero seems to have appeared in Indian mathematics in the middle of the 7th century.

zero   nought   nil   cipher   null   o   ∅

When you think of zero, what are you thinking about? If you divide 1 by 0.1 you get 10, 1 divided by 0.01 equals 100, 1 divided by 0.001 equals 1000, and so on. What do you think is the answer to 1 divided by 0?

What would you get if you 'cut an apple into zero pieces'?

Not everyone agrees on the names for numbers. In the American value system, the number 'billion' is written as 1 000 000 000 (10^9) and in the British system, 'billion' is written as 1 000 000 000 000 (10^{12}). How is it that the same word for a number can have different values? Is this logical? Can you think of any other similar situations in which this apparent dichotomy exists?

For a complete list of names for large numbers, visit www.pearsonhotlinks.co.uk, enter the ISBN for this book and click on weblink 17.4.

However, both the American and the British systems agree on the name for the *very* large number 10^{100}. This number is called a **googol**. It was reportedly named when the American mathematician Edward Kasner asked his 9-year-old nephew to give a name to the number.

It has been estimated that a googol is larger than the number of *all* of the particles in the known universe!

A number larger than a googol was named as a 'googolplex'. This number is written as $10^{10^{100}} = 10^{googol}$. This number is pretty hard to think about. A googol can be written as:

10 000 000 000 000 000 000 000 000 000 000 000 000 000 000 000 000 000 000 000 000 000 000 000 000 000 000 000 000 000 000 000 000 000 000 000 000 000

A googolplex is a '1' followed by a googol number of zeros!

Why is this number so hard to think about?

Try to estimate, and then calculate, how long it would take you to write down:

a) a million number of 1's if you could write one '1' per second
b) a googol number of 1's if you could write one '1' per second.
(Hint: Your calculator may be of no use here.)

'God created the integers, the rest is the work of man.'
Leopold Kronecker

'Numbers constitute the only universal language.'
Nathanael West

To find out more about the googol, visit www.pearsonhotlinks.co.uk, enter the ISBN for this book and click on weblinks 17.5 and 17.6.

Are there numbers larger than a googolplex?

If numbers were originally conceived to count objects and to keep objects organized, would there be any use for a number that is larger than any number of objects (including atoms, electrons, etc.) that exist?

Why does mankind continue to think about such ideas as large numbers, how to find the next prime number, or the next digit in pi's decimal expansion?

Sometimes, in trying to estimate an answer to a problem, it is helpful to use an analogy. Use the following information to describe the size of a molecule.

Avogadro's number $= 6.02 \times 10^{23}$ molecules in a mole.
Cargo ship container size $= 12\,m \times 2.4\,m \times 2.4\,m$
BB size: diameter $= 4.5\,mm$

The question: Suppose there is a pile of BB's, (small, spherical balls used as projectiles in a BB-gun), equal in amount to Avogadro's number, next to a container ship. How many cargo ship container(s) will it take to load all of the BB's?

The plan: Construct a cardboard cube, of side 3 cm, fill the cube with BB's and then calculate the number of container(s).

The answer: Once you calculate the number of container(s), you will be able to get a sense of the size of a molecule.

▲ In the infamous words of Carl Sagan, 'billions and billions'

This is one representation of a molecule. How would you start to describe the size of a molecule, an atom or an electron? ▶

'The infinite! No other question has ever moved so profoundly the spirit of man.'
David Hilbert

Intuition versus logic

The examples about large numbers have demonstrated that our intuition may not always be reliable. In other words, what you feel is correct may, in fact, not be. Below is one more example to demonstrate this idea.

Consider the two sets of numbers

$\mathbb{Z}^+ = 1, 2, 3, 4, 5, \ldots$ and

$\mathbb{N} = 0, 1, 2, 3, 4, 5, \ldots$

Which set has *more* numbers in it?

If you answered the set of natural numbers, you would have answered as most people do. It does *feel* that since $\mathbb{N}$ has the 'extra' element '0', then $\mathbb{N}$ has more numbers than $\mathbb{Z}^+$. This is what our intuition tries to tell us. The German mathematician George Cantor (1845 – 1918) logically showed otherwise.

▲
George Cantor

Consider the following scenario:

Suppose a school principal wanted to know if there were more boys than girls, or more girls than boys, or an equal number at a school dance.

The problem was that, as they were always moving around, it was difficult to count. Fortunately, a maths teacher offered her help. She made this announcement: 'Will each girl please hold hands with one and only one boy and walk to the north side of the gym'.

After a few minutes, the maths teacher turned to the principal and said, 'There you have it.' To which the principal replied, 'Ah, yes! Thank you so much.'

What did the principal see?

Now, let's apply this idea to our sets above.

```
1   2   3   4   5   6   7   8 ...
↕   ↕   ↕   ↕   ↕   ↕   ↕   ↕
0   1   2   3   4   5   6   7 ...
```

If you think of the positive integers 'holding hands' with the 'natural numbers', can you see that there will never be any numbers of either set left over?

What conclusions can you make concerning the sets of numbers $\mathbb{Z}^+$ and $\mathbb{N}$ based on the above discussion?

George Cantor went on to prove that, even though $\mathbb{Z}^+$ and $\mathbb{N}$ (as well as any set that could be placed in a one-to-one correspondence with the set of $\mathbb{Z}^+$) had the same number of elements, there was a set of numbers that had *more* numbers in it than $\mathbb{Z}^+$! The set he found was $\mathbb{R}$, the set of real numbers.

Thus, Cantor was able to demonstrate that there is more than one level of infinity! (To help your peace of mind, the answer lies with the understanding of irrational numbers and their decimal representations that never end and never terminate.)

Discuss the idea of different levels of infinity and how you 'feel' about it. Do you believe Cantor's assertion? Do you need verification? If your intuition misled you on this account, are there possibly other areas of knowledge in which it has also misled you?

Discuss the quote shown right in relation to the fact that there is a 'small' infinity and a 'larger' infinity.

'The notion of infinity is our greatest friend; it is also the greatest enemy of our peace of mind.'
James Pierpont

The axiomatic system

'There is no royal road to geometry.'
Euclid

The axiomatic system of verifying conjectures (theorems) is largely attributed to the Greeks, and in particular to Euclid (circa 300 B.C.E.). He is widely considered as the 'Father of Geometry' and is best known for his textbook, *Elements*. In almost every geometry classroom, you will hear the phrase 'Euclidean geometry'. He is credited with organizing mathematical thinking into structured reasoning, whereby one starts with a minimum amount of 'self-evident' knowledge (axioms or postulates) and logically deduces (or verifies) conjectures (theorems).

'It is the glory of geometry that from so few principles, fetched from without … it is able to accomplish so much.'
Sir Isaac Newton

Discuss the above two quotes as they pertain to your IB experience as a whole or to your own experience in geometry class.

Language, true to its own structure, is circular in nature. A word is necessarily defined in terms of other words until, eventually, the original word appears as a definition again. Mathematical knowledge is structured in a 'linear' fashion. Each branch of mathematics has a beginning point, known as *undefined terms*; it proceeds to add *definitions* (to tell us how to think about an idea), and finally it states several strongly self-evident statements called *axioms*.

Undefined terms → Definitions → Axioms (postulates) → Theorems (conjectures)

For example, in the branch of algebra, the system would be:

Undefined terms: '+', '−', and the variables *a*, *b* and *c*.

Defined terms: 'Addition' is (partially) defined below:

+	1	2	3	4	5	6	7	8	9
1	2	3	4	5	6	7	8	9	10
2	3	4	5	6	7	8	9	10	11
3	4	5	6	7	8	9	10	11	12
4	5	6	7	8	9	10	11	12	13
5	6	7	8	9	10	11	12	13	14

Axioms: The eleven field axioms (see the list to the right).

Euclid used the term 'self-evident' for those statements called 'axioms'. Consider the commutative axiom for addition:

For all $a, b \in \mathbb{R}$, $a + b = b + a$:

It certainly seems 'self-evident' that regardless of the order in which you add two numbers the result is the same. For example: $3 + 7 = 7 + 3$ or $0.25 + 1.3 = 1.3 + 0.25$.

However, how can you be sure that $\sqrt{2} + \sqrt{3} = \sqrt{3} + \sqrt{2}$ when both the $\sqrt{2}$ and $\sqrt{3}$ have decimal representations that do not repeat and never end?

Consider other areas in your life where you simply have to believe the concept/idea is correct. Do we sometimes so firmly believe in something that we simply make that idea a 'fact' without any type of formal proof other than our feelings or our observations?

'Even though I see the horizon as a straight ▶ line, I *know* the Earth is not flat'.

The eleven field axioms

1 **Closure under addition**: real numbers are *closed* under addition.

2 **Closure under multiplication**: real numbers are *closed* under multiplication.

3 **Additive commutativity**: $x + y = y + x$.

4 **Multiplicative commutativity**: $x \cdot y = y \cdot x$.

5 **Additive associativity**: $(x + y) + z = x + (y + z)$.

6 **Multiplicative associativity**: $(xy)z = x(yz)$.

7 **Distributivity**: Multiplication distributes over addition: $x(y + z) = xy + xz$.

8 **Additive identity element**: The additive identity is a *unique* element, which can be added to any element without altering it The additive identity is zero (0): $x + 0 = x$.

9 **Multiplicative identity element**: The multiplicative identity is unique; it is one (1): $x \cdot 1 = x$.

10 **Additive inverses**: For every real number, there exists a *unique* inverse such that when added together the result is the additive identity (0). The additive inverse is the opposite (negative) of the given real number: $x + (-x) = 0$.

11 **Multiplicative inverses**: For every real number not equal to zero, there exists a *unique* inverse such that when multiplied together the result is the multiplicative identity: $x \cdot x^{-1} = 1$.

Proving theorems in mathematics

The first proofs constructed make use almost entirely of the axioms and definitions. See the examples below:

Prove, if $a = b$, then $a + c = b + c$, where $a, b, c \in \mathbb{R}$.

Statements	Reasons
1. $a, b, c \in \mathbb{R}$	1. Given information
2. $a = b$	2. Given information
3. $a + c \in \mathbb{R}$	3. Closure axiom for addition in the real number system. (This is one of the eleven field axioms for algebra.)
4. $a + c = a + c$	4. The reflexive axiom
5. $\therefore a + c = b + c$	5. Substitution principle

'**Each problem that I solved became a rule which served afterwards to solve other problems**'.
Rene Descartes

The next proof will use the result of the first proof as a 'reason' for statement 3.

Prove, if $a + c = b + c$, then $a = b$, for all $a, b, c \in \mathbb{R}$.

Statements	Reasons
1. $a + c = b + c$ and $a, b, c \in \mathbb{R}$	1. Hypothesis (the given information)
2. $-c \in \mathbb{R}$	2. Axiom of additive inverses
3. $(a + c) + -c = (b + c) + -c$	3. This is the result of the last theorem we just proved
4. $a + (c + -c) = b + (c + -c)$	4. Associative axiom for addition
5. $a + 0 = b + 0$	5. Axiom of additive inverses
6. $\therefore a = b$	6. Identity axiom for addition

'**I have found a very great number of exceedingly beautiful theorems.**'
P. Fermat

'**[Concerning $e^{\pi i} + 1 = 0$] … Gentlemen, that it is surely true, it is absolutely paradoxical; we cannot understand it, and we don't know what it means, but we have proved it, and therefore we know it must be the truth.**'
Benjamin Pierce

Like a poem that gives hope or a picture that speaks to us in a 'thousand words', a proof that is devised using a minimum number of statements and yields a great amount of insight can also be considered beautiful. The 'less is more' idiom is clearly understood when this occurs.

For example: Prove that $\frac{a}{a} = 1, a \in \mathbb{R}, a \neq 0$.

Proof: $\frac{a}{a} = a \cdot \frac{1}{a} = 1$.

This proof is very easy, yet elegant in its simplicity.
(Often the reasons are not given as they are clearly understood.)

Pythagoras' theorem is perhaps the most famous theorem since the time of Euclid. The result of the theorem impacts our lives on a daily basis. Many beautiful proofs have been constructed for the theorem, many by the process of 'dissection' – the taking apart, rearranging and putting back together of the individual geometric components.

Considered as perhaps the most elegant and beautiful proof devised is the one constructed by Euclid concerning whether there are finitely or infinitely many primes.
What does your intuition 'tell' you about the number of prime numbers?

Before you answer, consider the following discussion. A prime number is defined as a natural number greater that one, whose only divisors are one and itself. Clearly, the larger a number becomes, the more 'opportunities' other numbers have to divide into it (evenly) thereby making it composite. What does your intuition tell you now? Is there, or better yet, *should* there be a largest prime number or not?

The following proof has been modified from its original form.

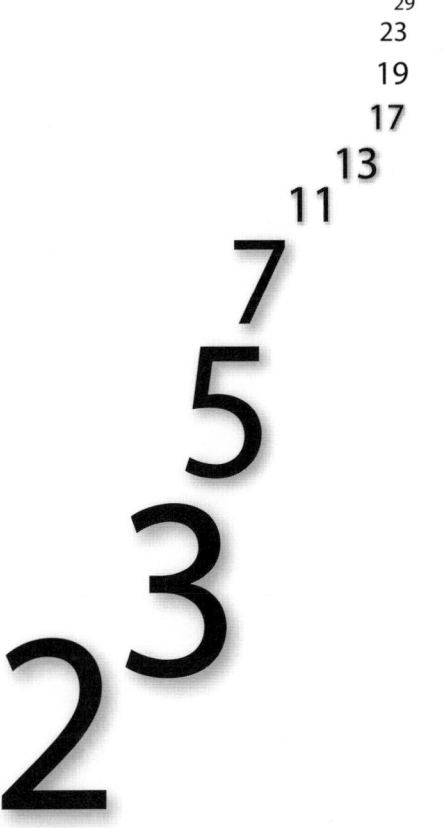

▲
Currently the largest prime number is reported to have close to 10 million digits!

Let's suppose there are a finite number of prime numbers, in this case five, but that list could be as long as you desire. Also suppose that there exists a number, call it P, that is equal to the product of all of those finite prime numbers plus one. In other words,

$$P = a \cdot b \cdot c \cdot d \cdot e + 1$$

At this point, one of two statements must occur. Either

1. P is a prime number or
2. P is not a prime number (it would then be called a composite number).

If P is prime then it cannot be any of a, b, c, d, e, since none of a, b, c, d, e would divide evenly into P and hence there are infinitely many prime numbers.

If P is not a prime, then it must be composite and, by definition, there must be a prime number (other than a, b, c, d or e) that will divide evenly into P. Hence there are infinitely many prime numbers.

◀ Like the petals of a rose, exposing the inner layers of mathematics reveals its true beauty.

Beautiful theorems and equations

There are many elegant theorems in mathematics. Listed here are some of the ones that are considered beautiful in nature, structure and importance.

1　Fermat's last theorem
2　Kepler's conjecture
3　The four-colour map theorem
4　Gödel's incompleteness theorems
5　Pythagoras' theorem
6　Euclid's infinitely many primes theorem
7　Cantor's continuum hypothesis (levels of infinity)
8　$\sqrt{2}$ is irrational
9　π is irrational
10　The sum of the measures of the angles of a triangle is 180 degrees.

Why do you think that these theorems are considered beautiful?
What do you think the criteria should be for a proof to be classified as 'beautiful' or 'elegant'?
Can you think of any ways in which constructing a 'beautiful' proof can improve the quality of life?

'The golden age of mathematics – that was not the age of Euclid, it is ours.'
C.J. Keyser

To see the 20 greatest equations, as proposed by the people who submitted them, visit www.pearsonhotlinks.co.uk, enter the ISBN for this book and click on weblink 17.8.

Included in the 'Top 20 Greatest Equations Ever' are:

$e^{i\pi} + 1 = 0$

$F = ma$

$a^2 + b^2 = c^2$

$E = mc^2$

$1 + 1 = 2$

$\dfrac{a}{b} = \dfrac{c}{d}$

For more about the four-colour map theorem, visit www.pearsonhotlinks.co.uk, enter the ISBN for this book and click on weblink 17.7.

Why do you think the above equations deserve to be included in this list?
Make a list of your own 'Top 10' of the greatest equations and explain why you have chosen them.

Unsolved problems

There are many unsolved problems, often referred to as 'conjectures', some of which are listed below:

Bunyakovsky conjecture
Hilbert's ninth problem
Palindromic prime
Quasiperfect number
Sierpinski number

To view a list of unsolved problems in mathematics, visit www.pearsonhotlinks.co.uk, enter the ISBN for this book and click on weblink 17.9.

Perhaps the most popular unsolved problem discussed is Goldbach's conjecture. It states that every even number greater than 2 is the sum of two prime numbers.

Test the conjecture by experimenting with a few more even numbers than the ones in the box below. What pattern do you see?

What does your intuition tell you about the rest of the even numbers?
What does logic demand that you do?

$$4 = 2 + 2$$
$$6 = 3 + 3$$
$$8 = 3 + 5$$
$$10 = 5 + 5$$
$$12 = 5 + 7$$

The physicist William Coolidge with an early prototype of an X-ray tube. How many of our recent technological inventions have come about because of mankind's impulse to solve problems?

The Goldbach conjecture has been around for about 250 years and it still has not been proved. Discuss the philosophical aspect of each question below:

1 Does the 'proof' for his conjecture simply not exist?
2 Are we simply not knowledgeable enough (yet) to 'see' the solution?

Keep in mind that Fermat's last theorem and the four-colour map theorem have just recently been proven.

These questions lead us to one of the main philosophical ideas asked about mathematical knowledge: Are the concepts, ideas and structure that make up mathematics invented or discovered? In other words, is mathematical knowledge out there waiting to be found, or is there already a structure from within, which we can make our own mathematical ideas happen?

Paradoxes

A paradox is a self-contradictory statement that at first seems true. There are several famous paradoxes in mathematics.

1 Zeno's dichotomy paradox, simply put, is that before you try to walk out of a room you must first go halfway to the doorway, and then halfway again, and so on, thus never leaving the room.

2 Zeno's most famous paradox is about Achilles and the tortoise. The tortoise demands to race the fleet-footed Achilles. Achilles, being very confident, gives the tortoise a generous head start. The paradox is that Achilles will never be able to catch the tortoise since when Achilles has advanced to a certain point on the track, the tortoise has moved forward again. Yet Achilles does, eventually, pass the tortoise. How is that possible?

3 Examine this statement for its truthfulness or falseness: 'This statement is false!'.

4 Consider the 'barber paradox'. A barber in a small town is given the task of shaving all men who do not shave themselves. Will the barber be able to shave himself?

5 $1 = \sqrt{1} = \sqrt{-1 \cdot -1} = \sqrt{-1} \cdot \sqrt{-1} = i \cdot i = i^2 = -1!?$

> **'Perhaps the greatest paradox of all is that there are paradoxes in mathematics.'**
> *J.R. Newman and E. Kasner*

> You might wish to research the imaginary number i.

Where do we go from here?

As evidenced by C.J. Keyser's quote and the number of conjectures still not proved, mathematics is still alive and well. Just recently (clearly this term is relative) the word 'fractal' entered into the mathematical realm.

A fractal refers to geometric shapes that are 'irregular' in nature. Examples include the coastline of Britain, the shapes of clouds, and of mountain ranges.

◀ A fractal is often self-similar; each piece or part when zoomed in on will be an exact replica of the larger piece.

Perhaps the most famous is the Mandelbrot set, named after Benoit Mandelbrot who used fractals to describe the self-similar nature of such graphs.

Try to imagine the mathematics required to make cellphone usage, wireless internet access or satellite TV a reality in your everyday lives.

Exactly how does your voice travel from your cellphone through the walls, through the trees, to a tower and back to your friend almost instantaneously?

What do you think the technological future holds for communication? Do you think mathematics will play a part? If so, how?

> 'If you ask mathematicians what they do, you always get the same answer. They think. They think about difficult and unusual problems. They do not think about ordinary problems: they just write down the answers.'
> *M. Egrafov*

> 'All human knowledge thus begins with intuitions, proceeds thence to concepts, and ends with ideas.'
> *David Hilbert*

Acquisition of knowledge

Knowledge is obtained in many ways and in many forms. In mathematics, you intuit a concept, maybe while under an apple tree, maybe while in the bath, or maybe while sitting in maths class. Once you have a concept, then you guess-and-check, or gather data and experiment, or put pencil to paper to find a solution. The verification process is usually called 'proof' and that may come in many different forms: axiomatic, diagrammatic, computer-generated or logistically.

Whatever the case may be, mathematics is a language, often called the universal language, that science uses to describe our biological and physical world around us, that language arts uses to make sure we 'say what we mean and mean what we say', that history uses to keep track of past events and helps us make the best logical decisions about the future, and that psychology uses to run experiments about why we behave as we do. It is a practical mechanism to help us find real world and abstract answers to problems, but its philosophical side helps us in our quest for understanding why we are here and where we are headed.

Answers

Chapter 1

Exercise 1.1

1. $\{-5, -4, -3, -2, -1, 0, 1, 2, 3\}$
2. Starting from 0, move 16 columns to the right and 27 rows down.
3. $\ldots, -\frac{4}{3}, -\frac{4}{2}, -\frac{4}{1}, \frac{4}{1}, \frac{4}{2}, \frac{4}{3}, \ldots$
4. 1
5. There is no **first** rational number to the right of 0.
6. There is no **first** real number to the right of 0.
7. $\{2, 4, 6, 8, 10, \ldots\}$
8. $\{3, 5, 7, 9, 11, \ldots\}$
9. $\{2, 3, 5, 7, 11, 13, 17, 19, 23, 29, 31, 37, 41, 43, 47, 53, 59, 61, 67, 71\}$
10. a) $\mathbb{N}, \mathbb{Z}, \mathbb{Z}^+, \mathbb{Q}, \mathbb{Q}^+, \mathbb{R}, \mathbb{R}^+$
    b) $\mathbb{Z}, \mathbb{Q}, \mathbb{R}$
    c) $\mathbb{Q}, \mathbb{Q}^+, \mathbb{R}, \mathbb{R}^+$
    d) $\mathbb{Q}', \mathbb{R}, \mathbb{R}^+$
    e) $\mathbb{N}, \mathbb{Z}, \mathbb{Z}^+, \mathbb{Q}, \mathbb{Q}^+, \mathbb{R}, \mathbb{R}^+$
    f) $\mathbb{Q}', \mathbb{R}, \mathbb{R}^+$
    g) $\mathbb{Q}, \mathbb{Q}^+, \mathbb{R}, \mathbb{R}^+$
    h) $\mathbb{N}, \mathbb{Z}, \mathbb{Q}, \mathbb{R}$
    i) $\mathbb{N}, \mathbb{Z}, \mathbb{Z}^+, \mathbb{Q}, \mathbb{Q}^+, \mathbb{R}, \mathbb{R}^+$
    j) $\mathbb{Q}, \mathbb{R}$
11. a) $1, 2, 3, 6, 9, 18$
    b) $1, 3, 5, 9, 15, 45$
    c) $1, 2, 3, 4, 6, 8, 9, 12, 18, 24, 36, 72$
    d) $1, 2, 4, 5, 10, 20, 25, 50, 100$
12. a) $3, 6, 9, 12, 15$
    b) $13, 26, 39, 52, 65$
    c) $19, 38, 57, 76, 95$
13. a) $2^2 \cdot 3^3$
    b) $3 \cdot 5^2$
    c) $3^2 \cdot 7^2 \cdot 13$
14. a) not prime b) prime   c) not prime d) prime
15. a) GCF = 12; LCM = 144
    b) GCF = 2; LCM = 1800
    c) GCF = 3; LCM = 396
16. a) 13.25
    b) 10.5
    c) 6993.57
    d) -4.74
    e) 6
    f) -1.37
    g) $\sqrt{200} = 10\sqrt{2}$
    h) 6
17. a) $\frac{720}{120}$
    b) $\frac{720}{120} = \frac{2160}{360}$
    c) $\frac{90}{100} = \frac{x}{360}$
18. a) $\frac{12}{365.25}$
    b) $\frac{16}{487}$
    c) $\frac{12}{365.25} = \frac{48}{1461}$
    d) $\frac{16}{487}$
    e) $\frac{25}{100} = \frac{x}{48}$

Exercise 1.2

1. a) $\frac{2}{3}$ b) $\frac{3}{5}$ c) 4   d) $\frac{5}{3}$
2. a) $\frac{9}{14}$ b) $\frac{7}{8}$ c) $-\frac{48}{5}$ d) 10   e) -1
   f) $\frac{3}{7}$ g) $\frac{ac}{bd}$ h) $\frac{eh}{fg}$ i) 4   j) $\frac{5}{32}$
3. a) $\frac{11}{10}$ b) $-\frac{5}{44}$ c) $\frac{34}{5}$ d) $-\frac{33}{10}$
   e) $\frac{73}{10}$ f) $-\frac{31}{9}$ g) $\frac{ad + bc}{bd}$ h) $\frac{x^2 - y^2}{xy}$
   i) $\frac{yz + 1}{y}$ j) $\frac{x - 1}{x}$

Exercise 1.3

1. a) $5x + 15$ b) $-3y + 21$ c) $x^2 + xy$
   d) $zw - zt$ e) $x^2 + 11x + 18$ f) $r^2 - 6r - 7$
   g) $6y^2 + 5y - 6$ h) $x^2 - 16$ i) $4a^2 + 12a + 9$
   j) $9z^2 - 1$ k) $x^3 + 5x^2 + 10x + 12$
   l) $g^3 - 12g^2 + 34g + 5$

2. a) $5(x - 1)$ b) $3(y + 2)$
   c) $2(x^2 + 3x + 4)$ d) $5(z^2 - 3z + 9)$
   e) $(x + 2)(x + 3)$ f) $(y + 3)(y + 5)$
   g) $(z - 2)(z + 1)$ h) $(w + 7)(w - 3)$
   i) $(x + 4)(x - 4)$ j) $(r + 5)(r - 5)$
   k) $(2x + 3)(x - 7)$ l) $(3m + 1)(m + 3)$

3. a) $y = x + 5$ b) $y = 7 - 2x$ c) $z = \frac{7 - 4w}{2}$
   d) $r = \frac{7 + 5s}{3}$ e) $t = \frac{d}{r}$ f) $b = \frac{360}{P}$
   g) $n = \frac{u - a + d}{d}$ h) $d = \frac{u - a}{n - 1}$ i) $b = -2ax$
   j) $a = \frac{-b}{2x}$ k) $h = \frac{2A}{a + b}$ l) $S_y = \frac{S_{xy}}{r \cdot S_x}$

4. a) 16π b) 36π c) 15   d) 12
   e) 20   f) 200   g) 2000   h) Undefined
   i) 99   j) $\frac{1023}{256}$ k) 4743.49 to 2 dp.
   l) 5022.58 to 2 d.p.   m) $\frac{1}{343}$ n) $\frac{1}{32}$
   o) 2   p) $\frac{16}{9}$

5. a) $x = 2$ b) $y = 5$ c) $z = 2$ d) $x = 3$
   e) $r = 4$ f) $t = -1$ g) $x = -1$ h) $w = \frac{-6}{7}$
   i) $y = \frac{27}{2}$ j) $t = \frac{-7}{8}$ k) $x = \frac{15}{8}$ l) $y = \frac{-12}{7}$
   m) $r = \frac{44}{9}$ n) $x = \frac{4}{21}$ o) $z = \frac{23}{8}$ p) $x = \frac{41}{9}$

Exercise 1.4

1. a) $y = -2x + 9$
   b) $y = \frac{3}{2}x + \frac{-7}{2}$ or $y = \frac{3}{2}x - \frac{7}{2}$
   c) $y = \frac{-3}{5}x + \frac{8}{5}$
   d) $y = \frac{21}{2}x - \frac{7}{4}$
   e) $y = \frac{-1}{14}x + \frac{5}{28}$

2. a) $4x + -5y - 6 = 0$ or $4x - 5y - 6 = 0$
b) $-4x + y + 5 = 0$ c) $5x + 12y - 7 = 0$
d) $3x + 7y - 2 = 0$

3. a) $2x - 4y - 3 = 0$ b) $3x - 10y + 5 = 0$

4. a) $x = 3, y = 2$ b) $y = -5, x = -8$
c) $x = 1, y = \frac{1}{2}$ d) $y = \frac{23}{8}, x = \frac{-25}{8}$

5. a) $x = 2, y = 5$ b) $x = -3, y = 5$
c) $x = \frac{8}{11}, y = \frac{29}{11}$ d) $x = -\frac{105}{2}, y = -41$

6. a) $x > 5$;

b) $z \leqslant -7$;

c) $t < 6$;

d) $r \leqslant -6$;

e) $m \geqslant \frac{-14}{15}$;

f) $w < 9$;

7. a) $(7, \infty)$ b) $(-\infty, 4]$
c) $[-5, 6]$ d) $(13, 25]$
e) $[-3, \infty)$ f) $(8, 12)$

8. a) $2 \leqslant x < \infty$ or $x \geqslant 2$ b) $-\infty < x < 9$ or $x < 9$
c) $-2 \leqslant x < 8$ d) $3 < x < 10$
e) $-\infty < x < \infty$ or $x \in \mathbb{R}$ f) $-6 < x \leqslant -1$
g) $3 \leqslant x \leqslant 4$ h) $-5 < x < 0$

Exercise 1.5A

1. Answers will vary: the corner of a table, computer pixel, the tip of an arrowhead.
2. Answers will vary: the horizon, a railroad track, centerline of a highway.
3. Answers will vary: the flat screen of a computer monitor, a table top, a wall of a room.
4. Circular reasoning is the type of reasoning in which a concept (or an argument) is used to explain that concept (or defend the argument) that is being used. Logical reasoning is the type of reasoning in which only statements set forth previously are allowed to be used to support or defend new statements or conjectures.
5. Subtraction can be **defined** in terms of addition as follows: $a - b = a + -b$, for all $a, b \in \mathbb{R}$
6. Answers will vary: a) The **segment** from A to B, $[AB]$, is defined as the set of all points that are between A and B inclusive. b) The **length** $[AB]$ is defined as AB. c) The **bisector** of $[AB]$ is a point C such that $AC = CB$.
7. A Postulate is a statement that is accepted as true without proof. An undefined term is a concept we know to be true, but only have psychological ideas to provide as evidence.
8. Answers will vary: a) A unique straight line may be drawn between two points. b) A plane contains at least three non-collinear points. c) The intersection of two planes is a line.

9. Postulates are statements that are accepted as true without proof. Theorems are statements that can be proved true based on previous undefined terms, definitions, postulates, and previous theorems.
10. Answers will vary: a) The sum of the measures of the angles in a triangle is 180. b) In a parallelogram, the opposite sides are equal in length. c) The area of a triangle is equal to one-half the product of its base and height.

Exercise 1.5B

1. a) 70 m   b) 159 cm   c) 85 mm
d) 37π ft   e) 28 m
2. a) 5000 m²   b) 117 cm²   c) 908 ft²
d) 166.2 m²   e) 298 cm²   f) 272.5 m²

Exercise 1.5C

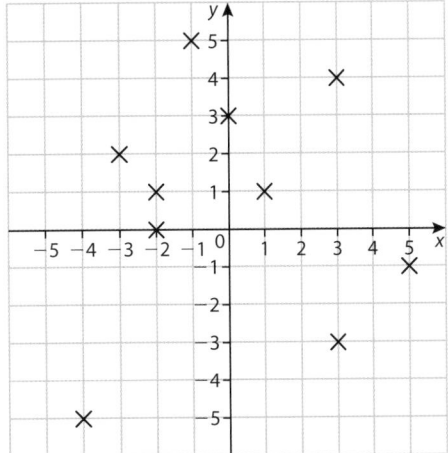

Exercise 1.5D

1. a) $M\left(\dfrac{2+5}{2}, \dfrac{3+6}{2}\right) = M(3.5, 4.5)$
b) $M\left(\dfrac{-4+4}{2}, \dfrac{-4+4}{2}\right) = M(0, 0)$
2. $M_{AB}(0, 0.5)$, $M_{AC}(1.5, -2)$ and $M_{BC}(4.5, 0.5)$
3.

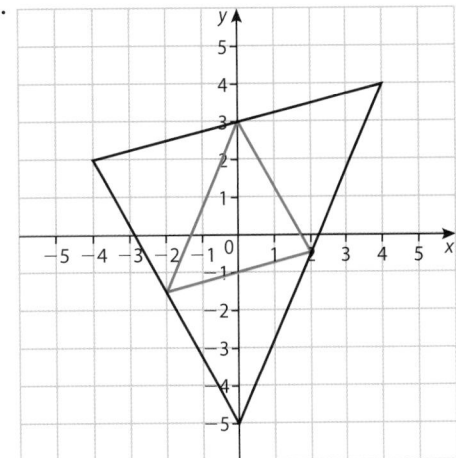

4. $(-2, 8)$

5. a) $(1, -1)$ b) $(-1.5, 0.5)$

6. $M_{AC} = \left(\dfrac{-3 + 3}{2}, \dfrac{4 - 1}{2}\right) = (0, 1.5)$

     $M_{DB} = \left(\dfrac{-5 + 5}{2}, \dfrac{-1 + 4}{2}\right) = (0, 1.5)$

7. 80 metres north of Redha and 60 metres west.

Exercise 1.5E

1. a) $[RS]$ b) $\widehat{T}, R\widehat{T}S, S\widehat{T}R$ c) $[RT], [ST]$
    d) 90   e) RS, t f) ST, r
    g) RT, s

2. a), c), e), f), g)

3. a), b), d), f), g)

4. a) 10   b) $\sqrt{170}$ c) 14.4
    d) 41.9   e) 24   f) $\sqrt{611}$
    g) 8.9

Exercise 1.5F

1. a) 5   b) $3\sqrt{2}$ c) 5

2. $AB = \sqrt{(3 - 3)^2 + (6 - 2)^2} = 4$
    $CD = \sqrt{(-2 - 1)^2 + (3 - 2)^2} = \sqrt{10}$
    $EF = \sqrt{(-1 - 1)^2 + (-1 - 1)^2} = \sqrt{8} = 2\sqrt{2}$
    $GH = \sqrt{(-1 - 3)^2 + (-2 + 2)^2} = \sqrt{16} = 4$

3. A parallelogram has opposite sides of equal length.
    $d = \sqrt{(-3 + 5)^2 + (-1 - 4)^2} = \sqrt{29}$
    $d = \sqrt{(-3 - 5)^2 + (4 - 4)^2} = 8$
    $d = \sqrt{(-5 - 3)^2 + (-1 + 1)^2} = 8$
    $d = \sqrt{(3 - 5)^2 + (-1 - 4)^2} = \sqrt{29}$
    Since $AD \cong BC$ and $AB \cong DC$, $ABCD$ is a parallelogram.

4. a) $2\sqrt{13} \approx 7.21$ blocks
    b) 10 blocks.
    c) 2.79 blocks.

5. a) $d_{AB} = \sqrt{65}$
      $d_{AC} = 8$
      $d_{BC} = \sqrt{65}$
    b) 8, 8, 8

Exercise 1.6

1. a) ATS   b) DKK   c) DEM   d) ILS
    e) MXN   f) SAR   g) TWD

2. Answers will vary. Given rates are as of July 7, 2007.
    a) Franc; 29.607 23   b) Markka; 4.363 833
    c) Dollar; 7.810 527   d) Lira; 1421.114
    e) Guilder; 1.617 400   f) Ruble; 25.712 25
    g) Franc; 1.218 064

3. a) 2432.40   b) 32.38

4. a) 747.81   b) 1537.82

5. a) 6246   b) 790

6. a) 28   b) 35 985

Exercise 1.7

1. a) N   b) O
    c) I   d) I
    e) N   f) N

    g) O   h) I
    i) N   j) I

2. a) Car Companies

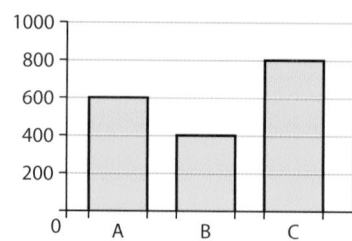

    b) Read-a-Lot

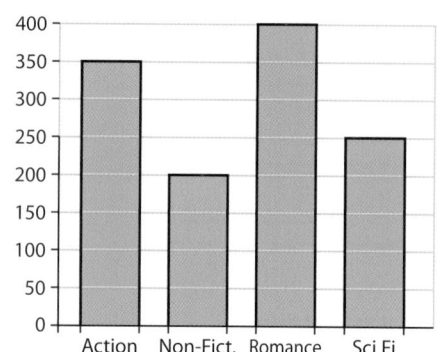

3. a) Hair Colour

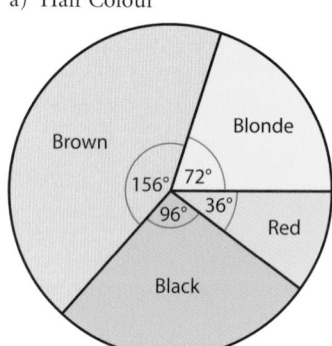

    b) Golf Balls

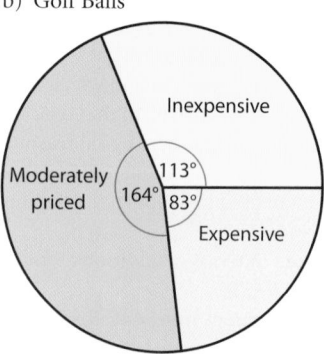

4. a) School Buses

b) Basketballs

March
April
May
June

Chapter 2

Exercise 2.1

1. Answers may vary, but some examples of possible answers are:
   a) $1, 2, 3, 4$
   b) $-1, 5, 6, 10$
   c) $\frac{2}{5}, \frac{7}{4}, 4, \frac{-1}{7}$
   d) $\pi, \sqrt{2}, \sqrt{7}, e$

2. Since 1.9 has a digit in the tenths position, we should use $\frac{19}{10}$

3. $5.17 = \frac{517}{100}$ hence it is rational by definition.

4. $\{\ldots, -2, -1, 0, 1, 2, 3, 4, 5, 6, 7, 8, 9\}$ Note that this includes all of the negative counting numbers.

5. $3.15 = \frac{315}{100} = \frac{63}{20}$ is one example. There are infinitely more examples.

6.
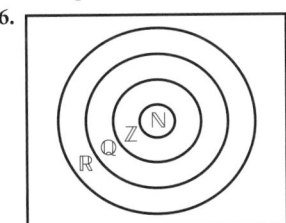

7. a) $-2.7, 35, 300\,000, 0$
   b) $35, 300\,000, 0$
   c) $\frac{\pi}{3}, \sqrt{17}$

8. $2^{10} = 1024$, therefore it belongs to all of the sets except $\mathbb{Q}'$.

9. $\sqrt{19}, \frac{\pi}{3}, 0.12112111211112\ldots, \frac{\sqrt{3}}{2}$ are irrational.

10. a) $\sqrt{3}, 2\pi$
    b) $6.3, 5.\dot{5}, \frac{3}{2}, \sqrt{3}, 2\pi$
    c) $-5, 6.3, 5.\dot{5}, \frac{3}{2}, \sqrt{3}, 2\pi$
    d) $\sqrt{3}, 2\pi$
    e) Answers may vary, for example: 1, 2, 3

11. a) any one of 2, 3, 5, 7
    b) 2, 3, 5, 7, 11, 13, 17, 19
    c)
    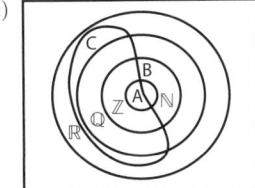

Exercise 2.2

1. a) 412.5
   b) 412.46
   c) 412.456
   d) 412
   e) 412
   f) 400

2. a) 346
   b) 34 600
   c) 0.000 346
   d) 1010
   e) 1200
   f) 19.0

3. a) $\frac{199.5 - 200}{199.5} \times 100 = 0.251\%$ (to 3 s.f.)

   b) $\frac{2.5 - 2.3}{2.5} \times 100 = 8.00\%$

c) $\frac{99.995 - 100}{99.995} \times 100 = 0.00500\%$ (to 3 s.f.)

d) $\dfrac{3.141\,592\,654 - \frac{22}{7}}{3.141\,592\,654} \times 100 = 0.0402\%$ (to 3 s.f.)

4. $\frac{13\,600\,000\,000 - 15\,000\,000\,000}{13\,600\,000\,000} \times 100 = 10.3\%$ (to 3 s.f.)

5. Estimated volume $= 2 \times 6 \times 6 = 72\,\text{m}^3$, actual volume $2.1 \times 5.9 \times 6.4 = 79.296\,\text{m}^3$,

   percent error $= \frac{79.296 - 72}{79.296} \times 100 = 9.20\%$ (to 3 s.f.)

6. a) $V = (44)(32)(25) = 35\,200\,\text{cm}^3$
   b) $V = (44.3)(32.4)(24.6) = 35\,308.872\,\text{cm}^3$
   c) percent error $= \frac{35\,308.872 - 35\,200}{35\,308.872} \times 100 = 0.308\%$ (to 3 s.f.)

7. a) $2\,\text{m}^2$
   b) $2.4\,\text{m}^2$
   c) $\frac{2.35 - 2}{2.35} \times 100 = 14.9\%$, $\frac{2.35 - 2.4}{2.35} \times 100 = 2.13\%$ (both to 3 s.f.)

8. a) 4.515 625
   b) 2.1
   c) x^2 to 2 s.f. is 4.5, so hence
   $\frac{4.515\,625 - 4.5}{4.515\,625} \times 100 = 0.346\%$ (to 3 s.f.)

9. a) 748.8
   b) 750
   c) $\frac{748.8 - 750}{748.8} \times 100 = 0.160\%$ (to 3 s.f.)

10. a) 0.030 642 m
    b) 0.0306 kg
    c) -0.137%

11. a) $\frac{2(2.03)^3 + 3(2.03)}{(4.51) + (3.92)} \approx 2.71$ (to 3 s.f.)
    b) $\frac{2(2.0)^3 + 3(2.0)}{(4.5) + (3.9)} \approx 2.62$ (to 3 s.f.)
    c) $\frac{2(2)^3 + 3(2)}{(5) + (4)} \approx 2.44$ (to 3 s.f.)
    d) $\frac{2.7071 - 2.62}{2.7071} \times 100 = 3.22\%$
    e) $\frac{2.44 - 2.7071}{2.7071} \times 100 = -9.96$. The percent error is 9.96% (to 3 s.f.)

12. a) $\sqrt{(2.78)^2 + (3.06)^2} \approx 4.13$
    b) (i) 4.1   (ii) 4.13
    c) (i) 4.2   (ii) 4.18
    d) $\frac{4.134\,247 - 4.18}{4.134\,247} \times 100 = 1.11\%$

Exercise 2.3

1. a) 4.35×10^5
   b) 1.9×10^3
   c) 8.7×10^{-3}
   d) -1.2005×10^4
   e) 2×10^{-1}
   f) 3.23×10^2
   g) -4.56×10^{-3}
   h) -1.96×10^1
   i) 1.9×10^4
   j) 1.57×10^{-1}

2. a) 20 060 000
   b) 0.00088
   c) -2000
   d) 506 000
   e) 0.000 0339
   f) -75
   g) 0.09
   h) $-0.000\,033$
   i) 4.3
   j) $-5\,400\,000$

3. a) 8×10^8
   b) -1.5×10^{11}
   c) 1.02×10^9
   d) 3×10^3
   e) 2×10^{-4}
   f) 5×10^{-8}

4. a) 5.4034×10^6
   b) 3.31×10^5
   c) 9.817×10^{-3}
   d) 5.01×10^7
   e) -1.26×10^{-2}
   f) 1.698×10^2

5. a) 8×10^1
   b) 4×10^2
   c) 1×10^{-7}

6. a) $A = l \times w = (3.4 \times 10^4 \times 2.7 \times 10^4) = 9.18 \times 10^8\,\text{m}^2$

b) $A = l \times w = (3.4 \times 10^1 \times 2.7 \times 10^1) = 918\,\text{km}^2$

c) $900\,\text{m}^2$

d) $\dfrac{918 - 900}{918} \times 100 = 2\%$

7. a) $x = \dfrac{2a+b}{c} = \dfrac{2(2.5 \times 10^5) + (5 \times 10^5)}{(2 \times 10^4)} = \dfrac{(1 \times 10^6)}{(2 \times 10^4)} = 50$

b) 5×10^1

8. a) $V = \frac{4}{3}\pi(6.373 \times 10^3)^3 = 1.08 \times 10^{12}$ (to 3 s.f.)

b) $V = \frac{4}{3}\pi(6.3 \times 10^3)^3 = 1.05 \times 10^{12}$ (to 3 s.f.)

c) $\dfrac{1.08 \times 10^{12} - 1.05 \times 10^{12}}{1.08 \times 10^{12}} = 2.78\%$ (to 3 s.f.)

9.

$6.6 \times 10^9 \times 102\% = 6.6 \times 10^9 \times 1.02$	6.732×10^9
$6.732 \times 10^9 \times 1.02$	6.86664×10^9
$6.86664 \times 10^9 \times 1.02$	7.0039728×10^9
$7.0039728 \times 10^9 \times 1.02$	7.144052256×10^9
$7.144052256 \times 10^9 \times 1.02$	7.286933301×10^9

To 3 s.f., this is 7.29×10^9 or 7.29 billion people.

10. a) $\dfrac{(2.3 \times 10^7)}{(1.2 \times 10^3)} \approx 19\,200$ (to 3 s.f.)

b) 120% of $2.3 \times 10^7 = 2.76 \times 10^7$, 70% of

$1.2 \times 10^3 = 8.4 \times 10^2$, therefore $\dfrac{(2.76 \times 10^7)}{(8.4 \times 10^2)} \approx 33\,000$.

11. a) $1.2 \times 10^{57} \times 400\,000\,000\,000 \times 8.0 \times 10^{10} = 3.84 \times 10^{79}$

b) $\dfrac{3.84 \times 10^{79}}{6.02} \times 10^{23} = 6.38 \times 10^{55}$ grams.

12. a) 333 000 times (to 3 significant digits)

b) 9.19×10^{22}

c) 2.99×10^{21} launches.

d) 7.48×10^{19} years.

Exercise 2.4

1. a) $26\,\text{m}$ b) $16\,000\,\text{m}$ c) 8.5×10^{-5}

d) $25\,\text{m}$ e) $1.2 \times 10^9\,\text{m}$

2. a) $20\,000\,\text{cm}^2$ b) $1.9\,\text{cm}^2$ c) $3.6 \times 10^{11}\,\text{cm}^2$

d) $9.2 \times 10^9\,\text{cm}^2$ e) $5.6 \times 10^8\,\text{cm}^2$

3. a) $3000\,\text{l}$ b) $0.25\,\text{l}$ c) $0.345\,\text{l}$

d) $25\,000$ e) $0.67\,\text{l}$

4. a) $2 \times 10^{-4}\,\text{kg}$ b) $0.034\,\text{kg}$

c) $59\,000\,\text{kg}$ d) $3.93 \times 10^{-7}\,\text{kg}$

e) $0.0625\,\text{kg}$

5. a) 10 b) 10^8 c) 10^{-2} d) 10^{-4} e) 10^5

6. a) $1.35 \times 1000 \times 1000 = 1.35 \times 10^6\,\text{m}^2$

b) $0.5\,\text{km}^2$

c) $\dfrac{1.345 - 1.350}{1.345} \times 100 = 0.372\%$

d) $\dfrac{1\,349\,999.5 - 1\,350\,000}{1\,349\,999.5} \times 100 = 3.70 \times 10^{-5}\%$

7. $20\text{Mg} \times 10^6 = 2.0 \times 10^7\,\text{kg}$ hence

$E = \frac{1}{2}(2.0 \times 10^7\,\text{kg})(13\,\text{ms}^{-1})^2$

$E = 1.69 \times 10^6\,\dfrac{\text{kgm}^2}{\text{s}^2}$

8. a) $0.5\,\text{m}$

b) $\dfrac{99.5 - 100}{99.5} \times 100 = 5.03\%$

c) $\dfrac{10\,022.5 - 10\,023}{10\,022.5} \times 100 = 0.00499\%$

9. $140\,\text{dam}^2 \times 10^2 = 14\,000\,\text{m}^2$ hence $\dfrac{14\,000}{250} = 56$, and $56 \times 2 = 1121$.

10. a) $150\,000\,000 \times 3 \times 10^9 \times 60 \times 60 \times 24 \times 365 \div 1000 = 1.42 \times 10^{21}\,\text{km}$.

b) 1.80×10^{10} years.

11. a) $1\,\text{hm} = 100\,\text{m}$ hence $1\,\text{hm}^2 = (100\,\text{m})^2 = 10\,000\,\text{m}^2$

b) Since $10\,\text{hm} = 1\,\text{km}$, then $1\,\text{km}^2 = (10\,\text{hm})^2 = 100\,\text{hm}^2$

c) Using answer to b), $17\,000 \times 100 = 1\,700\,000\,\text{ha}$

d) $S_A = 4\pi(6378.1\,\text{km})^2 = 5.11 \times 10^8\,\text{km}^2$

e) $\dfrac{(5.11 \times 10^8)}{17\,000} \approx 30\,100$ years

12. a) $1\,\text{newton} = \dfrac{\text{kgm}}{\text{s}^2}$ b) $10\,\text{kg} \times 9.8\,\dfrac{\text{m}}{\text{s}^2} = 98\,\text{N}$

c) $\dfrac{1000}{65} \approx 15.4\,\dfrac{\text{m}}{\text{s}^2}$

Exercise 2.5

1. a) $2000   b) $20\,000

c) $1860   d) $6.90

2. a) 1225 baht   b) 28000 baht

c) 8750 baht   d) 4200 baht

3. a) $3   b) $0.20   c) $191.67   d) $8.33

4. a) 10 yoldas   b) 316.8 anthmas

c) 0.35 yoldas   d) 9768 anthmas

5. a) €333.3   b) £240 UK

c) $480 US   d) $240 US

6. a) 8232 kyat   b) 7925.88 kyat

7. a) 5190 Turkish lira   b) 1843 Swiss Francs

8. a) £16 UK   b) 1678.80 DM

c) £299.51 UK

9. a) 862.50   b) 1 US = $ 1.85 AUD

10. a) $ 1413.16   b) 1288.34 MD

11. a) €1212.96   b) $ 780.24

c) $19.76   d) $20.45 CAD

12. a) $143\,843.50

b) 1500 RUB

c) 50\,345.23 RUB

d) 73\,004.77 RUB

Exercise 2.6

1. a) $x = 2$ b) $x = 15$ c) $x = 0.65$

   $y = 2$ $y = 12$ $y = 0.75$

d) $x = 2.3$ e) $x = 2.5$ f) $x = 15$

   $y = 2.1$ $y = 3.5$ $y = 8$

2. a) $x = 5$ b) $x = 9$ c) $x = 6$

   $y = 5$ $y = 33$ $y = 5$

d) $x = -8$ e) $x = -5$ f) $x = 8$

   $y = -4.7$ $y = 9$ $y = 10$

3. Using GDC (to 3 s.f. where appropriate)

a) $x = 3.6$ b) $x \approx 1.40$ c) $x \approx -115$

   $y = 0.8$ $y \approx 0.709$ $y \approx 218$

d) $x \approx 3.78$ e) $x \approx 0.0470$ f) $x \approx 0.803$

   $y \approx 81.3$ $y \approx 0.0197$ $y = 3.01$

4. Using GDC (to 3 s.f. where appropriate)

a) $x = 3$ b) $x = 0.3$ c) $x \approx -0.529$

   $y = 14$ $y = 3.1$ $y \approx 3.65$

d) $x = 6.875$ e) $x \approx -5.67$ f) $x \approx -20.5$

   $y = -2.5$ $y \approx -12.7$ $y \approx -1.43$

5. Using GDC (to 3 s.f. where appropriate)

a) $x \approx 2.17$ b) $x = -2.5$ c) $x \approx 2.09$
 $y \approx -3.33$ $y \approx -5.5$ $y \approx 5.18$
d) $x = 0.155$ e) $x \approx -1.07$ f) $x \approx -20.5$
 $y = 8$ $y \approx 0.1$ $y \approx -3.43$

6. a) $3x + 5y = 2.55$
  $5x + 4y = 2.69$
 b) $x = £0.25$ therefore the cost of an apple is £0.25 and
  $y = £0.36$ the cost of a banana is £0.36.

7. a) $47 = a(3)^2 + b(3) + 5$ $42 = 9a + 3b$
  $69 = a(4)^2 + b(4) + 5$ hence $64 = 16a + 4b$
 b) Using GDC, $a = 2$ $b = 8$

8. a) $6c + 3v = 163.17$ b) Using GDC, $c = \$18.35$
  $9c + 2v = 200.53$ $v = \$17.69$
 c) $9 \times 18.35 = 165.15$
  $180 - 165.15 = \$14.85$

9. a) $L = 2W$
  $2L + 2W = P$
 b) $L = 2W$ and $2L + 2W = 60$ therefore
  $2L + L = 60$ and $L = 20$ cm. Hence $W = 10$ cm

10. a) $n = 2(m + 7)$ b) $2(m + 7) = 3m + 6$
  $n = 3m + 6$ $2m + 14 = 3m + 6$
         $m = 8$ hence $n = 30$

11. a) $2m + 3k = 5.1$ b) Using GDC, $m = 0.87$
  $5m + 7k = 12.19$ $k = 1.12$
 c) Using b) the actual cost of 6 mangoes and 10 kiwi is
  $16.42. Therefore:
  $\dfrac{16.42 - 16}{16.42} \times 100 = 2.56\%$ (to 3 s.f.)

12. a) Using GDC: $p = 1.75$ b) $\dfrac{2(1.75) + (3.6)}{(1.75)^2} \approx 2.32$
     $q = 3.6$
 c) (i) 2   (ii) 2.3

Exercise 2.7

1. Using GDC.
 a) $x = -2$ b) $x = 2$ c) $x = -5$
  $x = -4$ $x = 5$
 d) $x = -2$ e) $x = 1$ f) $x = 4$
       $x = -6$ $x = 5$

2. a) $(2x - 1)(x - 3) = 0$ b) $(2x + 5)(2x + 7) = 0$
  $x = \frac{1}{2}$ or $x = 3$ $x = -\frac{5}{2}$ or $x = -\frac{7}{2}$
 c) $(3x + 1)(x + 5) = 0$ d) $(5x - 1)(x - 4) = 0$
  $x = -\frac{1}{3}$ or $x = -5$ $x = \frac{1}{5}$ or $x = 4$
 e) $(3x + 2)(2x + 8) = 0$ f) $(4x - 9)(2x - 5) = 0$
  $x = -\frac{2}{3}$ or $x = -4$ $x = \frac{9}{4}$ or $x = \frac{5}{2}$

3. a) $(x + 4)(x + 4) = 0$ b) $(x - 4)(x - 5) = 0$
  $x = -4$ $x = 4$ or $x = 5$
 c) $(x + 4)(x + 1) = 0$ d) $(x - 10)(x + 10) = 0$
  $x = -4$ or $x = -1$ $x = 10$ or $x = -10$
 e) $(x - 6)(x + 6) = 0$ f) $x(x - 9) = 0$
  $x = 6$ or $x = -6$ $x = 0$ or $x = 9$

4. Using GDC.
 a) $x = -2$ or $x = -5$ b) $x = -2$ or $x = -3$
 c) $x = -1$ or $x = -6$ d) $x = 2$
 e) $x = 6$ or $x = -8$ f) $x = -5$ or $x = 8$

5. Using GDC.
 a) $x = -1$ b) $x = -\frac{1}{3}$ c) $x = -10$
  $x = -6$ $x = -3$ $x = 10$
 d) $x = -\frac{4}{3}$ e) $x = -\frac{1}{2}$ f) $x = 3$
  $x = -2$ $x = -1$ $x = 5$

6. a) Using GDC, $t = 0$ seconds or $t = 10$ seconds.
 b) This is just before the ball is fired into the air.

7. $x = 2$ or $x = 5$ hence
 $(x - 2)(x - 5) = 0$
 $x^2 - 7x + 10 = 0$
 $a = -7$ and $b = 10$

8. a) Using GDC, $x = 170$ or $x = 0$
 b) The factory is not producing any widgets at all.

9. a) $(x - 5)(x + 5)$ b) $(x - 4)(x + 1)$
 c) $x - 4 = 0$ or $x + 1 = 0$ hence $x = 4$ or $x = -1$

10. $0 = x^2 - 11x + 18$
 $0 = (x - 9)(x - 2)$
 $x = 9$ or $x = 2$
 Since $x = 2$ is too small to fit the diagram, $x = 9$ m.

11. a) Area is length times width hence:
  $A = (5 + 2x)(7 - 2x)$
  $A = 35 + 4x - 4x^2$
 b) (i) $p = 11$
   $q = 35$
   $r = 27$
   $s = -13$
  (ii)

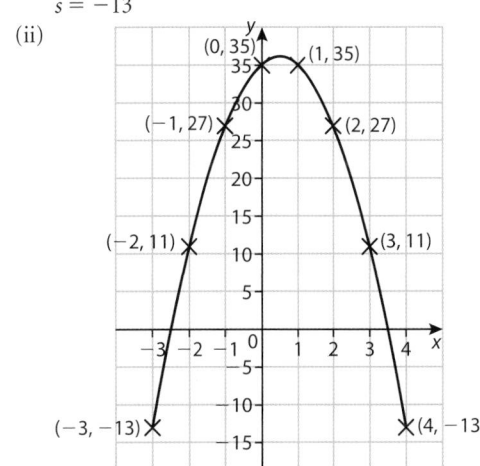

 c) (i) From the graph our axis of symmetry is half way
   between $x = 0$ and $x = 1$, hence $x = 0.5$
  (ii) From the graph either $x = -1$ or $x = 2$.
  (iii) $(5 + 2(-1)) = 3$ and $(7 - 2(-1)) = 9$ or
   $(5 + 2(2)) = 9$ and $(7 - 2(2)) = 3$
 d) (i) Graph shown below:

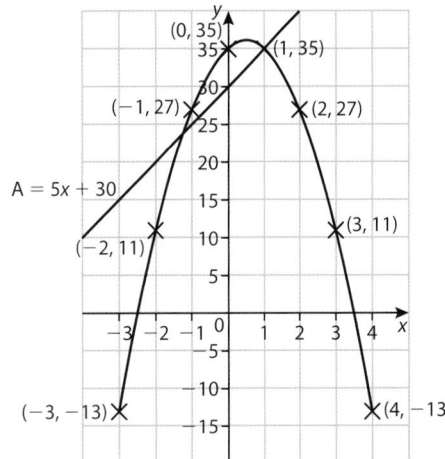

(ii) From the graph it is clear that $x = 1$ is a solution. Hence we know $(x - 1)$ is a factor of
$$4x^2 + x - 5 = 0.$$
$$(x - 1)(4x + 5) = 0$$
$$x - 1 = 0 \text{ or } 4x + 5 = 0$$
$$x = 1 \text{ or } x = -\tfrac{5}{4}$$

12. a) $A = 600 - x(x + 1)$

b) $0 = 600 - x^2 - x$
$$x^2 + x - 600 = 0$$
$$(x + 25)(x - 24) = 0$$
$$x = -25 \text{ or } x = 24$$

c) Neither of these solutions makes sense. $x = -25$ is impossible because a length cannot be negative. $x = 24$ does not make sense because it would mean the dashed box had a dimension larger than the original large box, which is impossible.

Chapter 3

Exercise 3.1

1. a) $y = 3x - 2$ b) $y = \tfrac{1}{3}x + 5$ c) $y = 2x$

   d) $y = x - 5$ e) $y = \tfrac{4}{3}x + 2$ f) $y = -x - 1$

2. a) $-3x + y - 1 = 0$ b) $y + 2x + 2 = 0$

   c) $x + 2y + 2 = 0$ d) $3x + y - 2 = 0$

   e) $-5x + y - 2 = 0$ f) $3x + y - 7 = 0$

3. a) $y = 2x + 4$ b) $y = \tfrac{3}{2}x$ c) $y = -\tfrac{4}{3}x + 1$

   d) $y = -\tfrac{1}{2}x + \tfrac{1}{2}$ e) $y = x + 6$ f) $y = -\tfrac{2}{3}x + 2$

4. a) $-2x + 3y + 12 = 0$ b) $3x + 4y + 2 = 0$

   c) $-5x + 15y - 3 = 0$ d) $15x + 10y - 2 = 0$

   e) $70x - 7y - 4 = 0$ f) $-4x + 6y - 1 = 0$

5. a) (i) $y = -2x + 3$ b) (i) $y = \tfrac{2}{3}x + \tfrac{1}{3}$

    (ii) $2x + y - 3 = 0$ (ii) $2x - 3y + 1 = 0$

   c) (i) $y = \tfrac{1}{12}x - \tfrac{2}{3}$ d) (i) $y = \tfrac{3}{2}x + 3$

    (ii) $x - 12y - 8 = 0$ (ii) $-3x + 2y - 6 = 0$

   e) (i) $y = -x + 4$ f) (i) $y = 9x + 12$

    (ii) $x + y - 4 = 0$ (ii) $-9x + y - 12 = 0$

6. a) $y = x - d$

   b) $22 = 10 - d$
$$d = -12$$

7. a) $y = -\dfrac{2}{3b}x + \dfrac{30}{b}$

   b) $(8) = -\dfrac{2}{3b}(15) + \dfrac{30}{b}$
$$8b = -10 + 30$$
$$b = 2.5$$

8. a) $h = \tfrac{5}{2}a + 10$

   b) $h = \tfrac{5}{2}(50) + 10 = 135\,\text{cm}$

   c) $160 = \tfrac{5}{2}a + 10$
$$a = 60\,\text{cm}$$

9. a) $m = 20y + 10$

   b) $m = 20(7) + 10 = 150\,\text{minutes}.$

   c) $250 = 20y + 10$
$$y = 12$$

10. a) $P = 20\,000x$

   b) $P = 20\,000(36) = 720\,000$

   c) \$20 000

   d) $250\,000 = 20\,000x$
$$x = 12.5$$

So by the end of the 13$^\text{th}$ month, the company has earned at least \$250,000 profit.

11. a) $L = \tfrac{1}{2}P - 20$ b) $L = \tfrac{1}{2}(200) - 20 = 80$

   c) $P = 2L + 40$ d) $P = 2(25) + 40 = 90$

   e) $2L + 2W - 200 = 0$

Exercise 3.2

1. a) $\tfrac{4}{3}$ b) $-\tfrac{2}{3}$ c) 1   d) 3

2. $m_{AB} = 0$ $m_{DC} = 0$ $m_{DA} = \tfrac{5}{2}$ $m_{CB} = \tfrac{5}{2}$

3. a) $m = \tfrac{3}{4}$ b) $m = \tfrac{1}{3}$ c) $m = -\tfrac{2}{5}$

   d) $m = \tfrac{4}{5}$ e) $m = -\tfrac{1}{2}$ f) $m = 3$

   g) $m = 4$ h) $m = 0$ i) $m = 1$

4. a) $-\tfrac{1}{2}$ b) $\tfrac{5}{2}$ c) -1 d) 4   e) $-\tfrac{3}{5}$ f) $\tfrac{2}{3}$

5. $m_{AB} = \tfrac{2}{3}$ $m_{BC} = 0$ $m_{CD} = -\tfrac{2}{3}$

   $m_{DE} = $ undefined $m_{AH} = $ undefined

   $m_{HG} = -\tfrac{2}{3}$ $m_{GF} = 0$ $m_{FE} = \tfrac{2}{3}$

6. a) 400 m   b) 20 m/s.   c) 150 seconds

7. a) $V = 30t + 60$ b) 6 seconds

8. a) $A = 40t$ b) 2000 m^2 c) 10 hours

9. a) $S = 30t$

   b) $550 = 30t$
$$t = 18.\dot{3} \text{ minutes} = 18 \text{ minutes } 20 \text{ seconds.}$$

   c) 15

10. a) $-\tfrac{1}{2}$

   b)

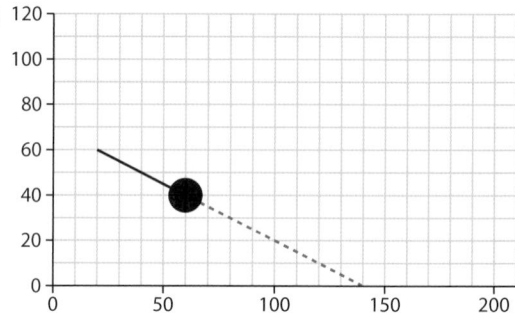

   c) $(140, 0)$ d) $\tfrac{1}{2}$

   e) $(210, 35)$

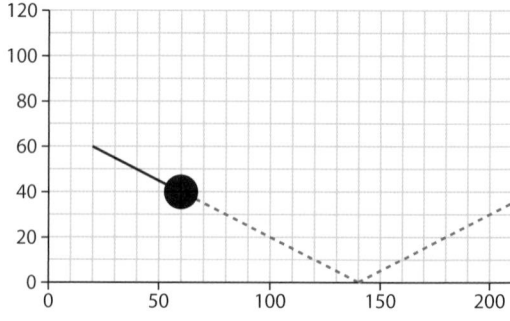

11. a) $-\tfrac{3}{4}$ b) $\tfrac{4}{3}$ c) 5   d) $(-2, \tfrac{3}{2})$

Exercise 3.3

1. a) 1   b) -3 c) 5   d) 0   e) $\tfrac{1}{2}$ f) 0

2. a) -2 b) 10   c) -3 d) 0   e) 3   f) 6

3. a) $(0, 2)$ b) $(0, 6)$ c) $(0, 2.5)$

   d) $(0, 10)$ e) $(0, 5)$ f) $(0, -1.5)$

4. a) $(0, -4)$ b) $(0, -1)$ c) $(0, 2)$ d) $(0, 3)$

5. a) $(-3, 0)$ b) $(-3, 0)$ c) $(1, 0)$ d) $(4, 0)$

6. a) $m = \dfrac{2 - 5}{4 - 1} = -1$ b) $y = -x + 6$
   c) $(0, 6)$ d) $(6, 0)$

7. a) $y = -\frac{4}{5}(0) + 8$ b) $0 = -\frac{4}{5}x + 8$
   $y = 8$ feet $x = 10$ feet

8. a) $y = -\frac{5}{2}(0) + 10$
   $(0, 10)$
   b) $0 = -\frac{5}{2}x + 10$
   $(4, 0)$
   c) $d = \sqrt{(0 - 10)^2 + (4 - 0)^2} = \sqrt{116}$

9. a) $m = \dfrac{7 - 10}{4 - 0} = -\dfrac{3}{4}$ b) $y = -\frac{3}{4}x + 10$
   c) $0 = -\frac{3}{4}x + 10$
   $(13\frac{1}{3}, 0)$

10. a) $r = 200$ AUD b) $s = 525$ AUD

Exercise 3.4

1. a) $(2, 5)$ b) $(1, -1)$ c) $(3, 0)$ d) $(1, 3)$
2. a) $(0, 1)$ b) $(4, 1)$ c) $(2, 1)$ d) $(3, 0)$
3. a) $(0, 0)$ b) $(2, 6)$ c) $(0, -1)$ d) $(5, -2)$
4. a) $(1, 1)$ b) $(0, 0)$ c) $(-5, 3)$ d) $(-31, -4)$
5. $A(1, 4); B(5, 4); C(9, 1); D(5, -2); E(1, -2); F(-3, 1)$
6. a) Video store A: \$10, Video store B: \$4. b) 6
7. a) $y = \frac{2}{3}x - 1$ b) $x = 15, y = 9$
8. a)

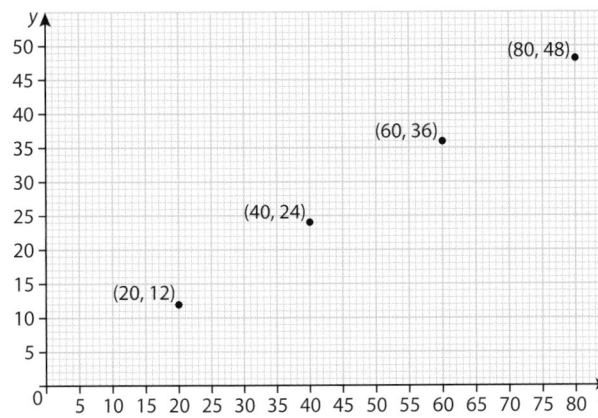

   b) $\frac{3}{5}$ c) $y = \frac{3}{5}x$ d) $(100, 60)$
9. a) smaller: $y = 2x + 6$
   taller: $y = \frac{3}{2}x + 9$
   b) $x = 6$ years
10. a) Akira: $y = 100x + 1000$
   Yuki: $y = 50x + 1500$
   b) After 10 years they have an equal amount, after 11 years Akira will have more.
11. a) $(-1, 0), (4, -1), (0, 5)$ and $(5, 4)$.
   b) $d_{AB} = \sqrt{(5 - 0)^2 + (4 - 5)^2} = \sqrt{26}$
   $d_{BC} = \sqrt{(4 - 5)^2 + (-1 - 4)^2} = \sqrt{26}$
   $d_{CD} = \sqrt{(-1 - 4)^2 + (0 + 1)^2} = \sqrt{26}$
   $d_{DA} = \sqrt{(0 + 1)^2 + (5 - 0)^2} = \sqrt{26}$

Exercise 3.5

1. a) $m_i = 3 = m_{ii}$; therefore line (i) and line (ii) are parallel.
   b) $m_i = -\frac{2}{3}, m_{ii} = \frac{2}{3}$; therefore line (i) and line (ii) are not parallel.
   c) $m_i = -2, m_{ii} = -\frac{1}{2}$; therefore line (i) and line (ii) are not parallel.
   d) $m_i = 2, m_{ii} = \frac{3}{2}$; therefore line (i) and line (ii) are not parallel.
   e) $m_i = \frac{1}{2}, m_{ii} = \frac{1}{2}$; therefore line (i) and line (ii) are parallel.
   f) $m_i = \frac{3}{4}, m_{ii} = -\frac{4}{3}$; therefore line (i) and line (ii) are not parallel.

2. a) $m_i = \frac{1}{2} = -\dfrac{1}{m_{ii}}$; therefore line (i) and line (ii) are perpendicular.
   b) $m_i = 3 \neq -\dfrac{1}{m_{ii}}$; therefore line (i) and line (ii) are not perpendicular.
   c) $m_i = 3 \neq -\dfrac{1}{m_{ii}}$; therefore line (i) and line (ii) are not perpendicular.
   d) $m_i = -5 \neq -\dfrac{1}{m_{ii}}$; therefore line (i) and line (ii) are not perpendicular.
   e) $m_i = -\dfrac{8}{5} = -\dfrac{1}{m_{ii}}$; therefore line (i) and line (ii) are perpendicular.
   f) $m_i = 7 \neq -\dfrac{1}{m_{ii}}$; therefore line (i) and line (ii) are not perpendicular.

3. a) $m_i = 4, m_{ii} = -\frac{1}{4}$, since $m_i = -\dfrac{1}{m_{ii}}$ lines (i) and (ii) are perpendicular.
   b) $m_i = 2, m_{ii} = -\frac{1}{3}$, since $m_i \neq -\dfrac{1}{m_{ii}}$ lines (i) and (ii) are not perpendicular.
   c) $m_i = 1, m_{ii} = -1$, since $m_i = -\dfrac{1}{m_{ii}}$ lines (i) and (ii) are perpendicular.
   d) $m_i = \frac{2}{3}, m_{ii} = -3$, since $m_i \neq -\dfrac{1}{m_{ii}}$ lines (i) and (ii) are not perpendicular.

4. a) a) $m_i = 3, m_{ii} = 3$; therefore line (i) is parallel to line (ii).
   b) $m_i = 3, m_{ii} = \frac{5}{2}$; therefore line (i) is not parallel to line (ii).
   c) $m_i = 0, m_{ii} = 0$; therefore line (i) is parallel to line (ii).
   d) $m_i = -\frac{7}{2}, m_{ii} = -3$; therefore line (i) is not parallel to line (ii).

5. a) $y = 3x - 1$ b) $y = -x + 9$
   c) $y = -\frac{1}{2}x - 3$ d) $y = -\frac{5}{4}x + \frac{21}{4}$

6. a) $m_{AB} = \frac{4}{3}, m_{BC} = -\frac{3}{4}, m_{CD} = \frac{4}{3}, m_{DA} = -\frac{3}{4}$
   b) Since the gradient of AB is equal to the gradient of CD and similarly $m_{BC} = m_{DA}$ so opposite sides are parallel.

7. a) Any adjacent sides of $ABCD$ have either gradient $\frac{4}{3}$ or $-\frac{3}{4}$ and $\frac{4}{3} \times -\frac{3}{4} = 1$ so any sides are perpendicular, which means they form right angles.
   b) $d_{AB} = d_{BC} = d_{CD} = d_{DA} = 5$ so $ABCD$ is a square.

8. a)

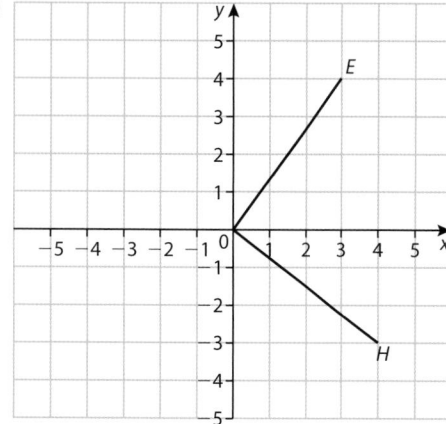

b) $m_{OE} = \frac{4}{3}$, $m_{OH} = -\frac{3}{4}$

c) $\frac{4}{3} \times -\frac{3}{4} = -1$ so OE is perpendicular to OH and therefore $\angle EOH = 90°$.

9. a) $m_A = \dfrac{(400 - 100)}{(6 - 0)} = 50$

$m_B = \dfrac{(350 - 50)}{(6 - 0)} = 50$

b) y-intercept of A is 100

y-intercept of B is 50

c) Since the gradients of the two lines are equal, the lines are parallel, therefore they will never meet.

10. a) $y = -\frac{3}{4}x + 3$

b)

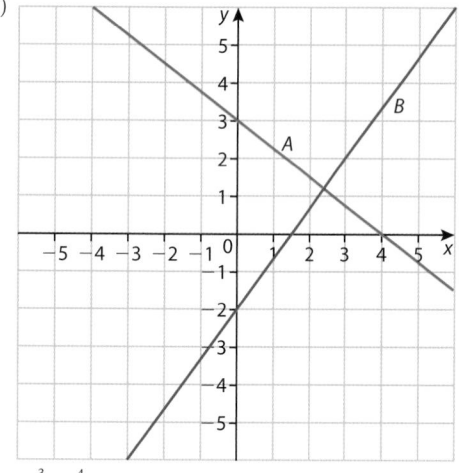

c) $-\frac{3}{4} \times \frac{4}{3} = -1$ so their paths are perpendicular which means they meet at a 90°.

11. a) $-\frac{2}{3}$

b) $(4, 3.5)$

c) $\sqrt{65}$

d) $7x + 4y - 42 = 0$.

e) $m_{BC} = \dfrac{3 - 0}{8 - 6} = \dfrac{3}{2}$. Since $m_{BC} = -\dfrac{1}{m_{AB}}$ the two lines are perpendicular.

12. a) $D(2.5, 6.5)$, $E(3.5, 1)$

b)

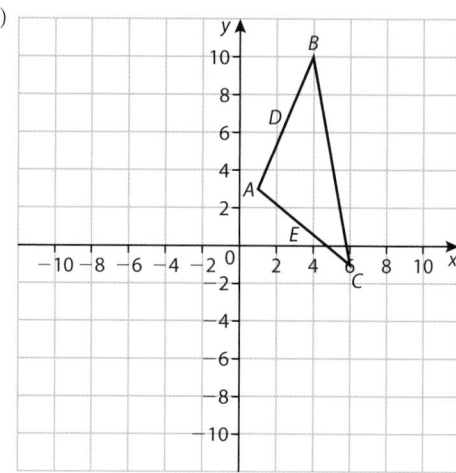

13. a) $y = -ax + d$

b) $y = -(3)(20) + (-5) = -65$

14. a)

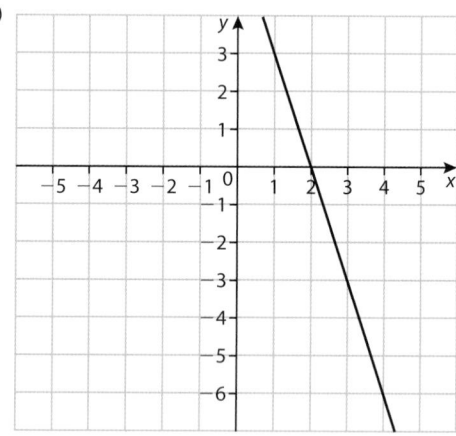

b) $y = -3x + 6$

15. a) -2 　　　 b) $(0, -4)$ 　　　 c) $(-2, 0)$

16. a) (i) 6 units

(ii) 6 units

(iii) 3.11 units

b) Triangle ABC is an equilateral triangle.

c) (i) $m = \sqrt{3}$ 　　　 (ii) $y = \sqrt{3}\, x$

17. a) $-x + 6y = 10$

b) $(2, 2)$.

c) $-\frac{2}{11}$

d) DC is not parallel to AB since they have different gradients.

Chapter 4

Exercise 4.1

1. A relation is any set of ordered pairs.
2. A function is a relation in which no two ordered pairs have the same first element.
3. Weight is a function of calorie intake.
4. Example: Speed of a car is a function of horsepower.
5. Example: $\{(0,1), (0,2), (0,3), (0,4), (0,5)\}$
6. Example: $\{(1,2), (3,4), (5,6), (7,8), (9,10)\}$

7.

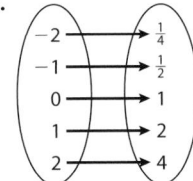

   a) Is a function. Each x-value is paired with one and only one y-value.
   b) $\{-2,-1,0,1,2\}$
   c) $\{\frac{1}{4},\frac{1}{2},1,2,4\}$

8.

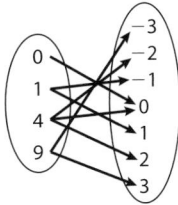

   a) It is not a function. Some of the x-values are paired with more than one y-value.
   b) $\{0,1,4,9\}$
   c) $\{-3,-2,-1,0,1,2,3\}$

9. a) Function. All x-values are the same.
   b) Function
   c) Function
   d) Not a function
   e) Function
   f) Not a function
   g) Not a function
   h) Function

10. Answers vary. Fails the vertical line test.

11. Answers vary. Passes the vertical line test.

12. a) yes
                   b) no; $(4,2)$; $(4,-2)$
   c) yes
                   d) yes
   e) yes
                   f) yes; (principal root)
   g) no; $(5,3)$; $(5,-3)$
       h) no; $(2,2)$; $(2,-2)$

Exercise 4.2

1. a) Domain $= \{0,1,2,3,4\}$
   b) Range $= \{1,2,4,8,16\}$
   c) It is a function. Each x-value is paired with a unique y-value.

2. a) Domain $= \{1,30,45,60,90,120,135,150,180\}$
   b) Range $= \left\{0,\frac{1}{2},\frac{\sqrt{2}}{2},\frac{\sqrt{3}}{2}\right\}$
   c) It is a function. Each x-value is paired with a unique y-value.

3. a) Domain $= \{0,1,4\}$
   b) Range $= \{-2,-1,0,1,2\}$
   c) It is not a function. One x-value is paired different y-values.

4. a) Domain $= \{-2,-1,0,1,2\}$
   b) Range $= \{0,1,4\}$
   c) It is a function. Each x-value is paired with a unique y-value.

5. $D = \{x \,|\, x \in \mathbb{R}\}$

6. $D = \{x \,|\, x \in \mathbb{R}\}$

7. $D = \{x \,|\, x \in \mathbb{R}, x \neq 0\}$

8. $D = \{x \,|\, x \in \mathbb{R}, x \neq 1\}$

9. $D = \{x \,|\, x \geqslant 0\}$

10. $D = \{x \,|\, x \geqslant \frac{3}{2}\}$

11. $D = \{x \,|\, x \in \mathbb{R}\}$
    $R = \{y \,|\, y \geqslant -9\}$

12. $D = \{x \,|\, -4 \leqslant x \leqslant 4\}$
    $R = \{y \,|\, -4 \leqslant y \leqslant 4\}$

13. $D = \{x \,|\, x \leqslant -2 \text{ or } x \geqslant 2\}$
    $R = \{y \,|\, y \in \mathbb{R}\}$

14. $D = \{x \,|\, x \in \mathbb{R}\}$
    $R = \{y \,|\, y \in \mathbb{Z}\}$

15. $D = \{t \,|\, t \geqslant 0\}$, $R = \{y \,|\, y \geqslant 1000\}$

16. $D = \{x \,|\, x \geqslant 0\}$, $R = \{y \,|\, y > 25\}$

Exercise 4.3

1. a) f at one equals five.
   b) r of negative four equals negative two.
   c) g at a equals b.

2. a) $(2,9) \in f$
                      b) $(-3,7) \in g$
   c) $(c,d) \in v$

3. a) -1
                         b) -3
   c) $2a^2 - 3$
                   d) $2h^2 + 8h + 5$
   e) $2\pi^2 - 3$

4. a) $\sqrt{5}$
                        b) 0
   c) $\sqrt{-3} \notin \mathbb{R}$
               d) $\sqrt{r^2 - 4}$
   e) $\sqrt{h^2 + 2h - 3}$

5. $x^2 + 2hx + h^2 + 1$

6. $2hx + h^2 - h$

7. $21\,871.46$ or $21\,900$ to 3 s.f.

8. 1260 to 3 s.f.

9. 9

10. 5

11. 25

12. $\sqrt{15}$

13. 19

14. $\sqrt{2}$

15. 1

16. 2

17. 11

18. 1

19. 0

20. 3

Exercise 4.4

1. Infinitely many

2. Answers vary. $y = 2x + 1$, $f(x) = 3x - 2$

3. Table of values, Functional values, x and y-intercepts, one point and the gradient

4.

x	y
-2	-8
-1	-5
0	-2
1	1
2	4

5.

x	y
0	1
0.5	−1.5
1	−4
1.5	−6.5
2	−9
2.5	−11.5
3	−14

6.

x	y
0	−3.47
0.2	−2.958
0.4	−2.446
0.6	−1.934
0.8	−1.422
1	−0.91
1.2	−0.398
1.4	0.114

7.

x	y
−2	3
−1.6	1.8
−1.2	0.6
−0.8	−0.6
−0.4	−1.8
0	−3
0.4	−4.2
0.8	−5.4
1.2	−6.6
1.6	−7.8
2	−9

8. −2; 2; 6

9. −2; 0; 2

10. Answers vary. $(-2,0)$; $(6,6)$

11. Answers vary. $(-9, 8)$; $(1, 4)$

12. a) $-\frac{1}{2}$; −1

   b) $\frac{1}{2}$; −2

   c) $-\frac{h}{d}$; h

13. a) $(5,0)$; x-axis;   **b)** $(0,4)$; y-axis

14. a) Answers vary.   **b)** $y = 0$

   c) 0 is paired with many values: $(0,1)$, $(0,2)$, $(0,3)$, etc.

15. a)

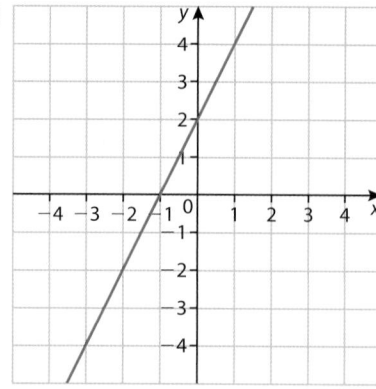

b)

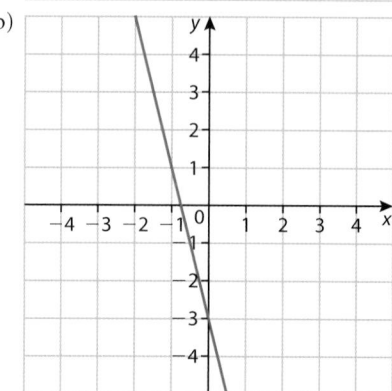

c)

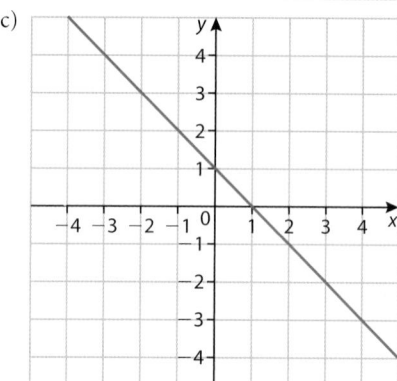

d)

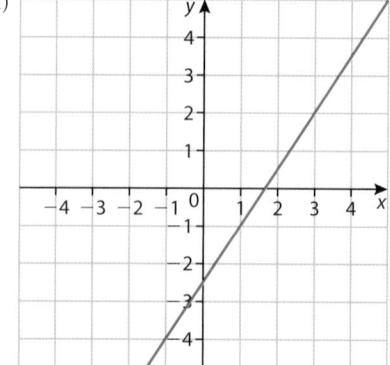

16. a) $f(x) = (\frac{2}{3})x + \frac{7}{3}$ **b)** $f(x) = (-\frac{1}{4})x + 3$

   c) $f(x) = -3x + 5$ **d)** $f(x) = (\frac{4}{5})x + \frac{3}{5}$

   e) $f(x) = -2x + 1$ **f)** $f(x) = (-\frac{6}{7})x + \frac{26}{7}$

   g) $f(x) = (-\frac{4}{3})x + 4$ **h)** $f(x) = (\frac{1}{5})x + 1$

   i) $f(x) = x$

Exercise 4.5

1. a) $6250 = 12r + 850$
   b) Rent for 1 week is 450 AUD.
   c) $C = 450n + s$
   d) Sasha pays a 1700 AUD security deposit.
   e) No, Sasha would have spent more money if she had rented from Beachside Apartments.

2. Depth of snow after 24 hours = 23.2 cm.

3. a) $150
   b) Profit(x) = Sales(x) − C(x) = $0.75x - (0.45x + 150)$ = $0.3x - 150$
   c) When 100 cups are sold there is a loss of $120.
   d) Solve Profit(x) = $0.3x - 150 = 0$. Therefore, the minimum number of cups to make a profit is 501.

4. a) Answer given: $C(t) = 50 + 36t$
   b) $C(4) = 50 + 36(4) = £194$
   c) $C(t) = 50 + 36t = £365$, $t = 8.75$ hrs. Jane collected the car at 7.30 a.m. + 8.57 = 16:45 (or 4.45 p.m.).
   d) Garage $B(t) = 25 + 41t$
   e)

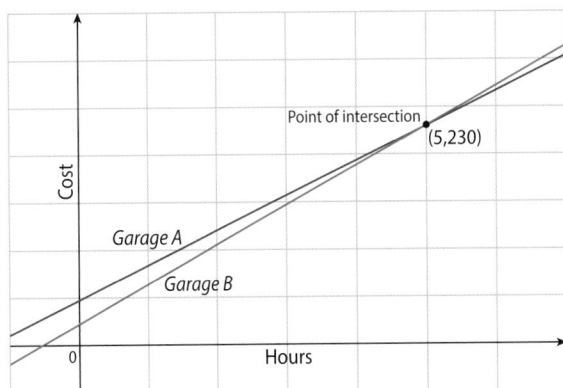

   The cost of repair at either garage is equal after 5 hours, with a cost of £230.

5. a) The gradient or slope is −12. That means the builder moves 12 bricks every hour.
   b) The linear model is thus: number of bricks $B(t) = -12t + 1000$.
   c) Need to solve $B(t) = -12t + 1000$ for $B(t) = 0$. That is $0 = -12t + 1000$, $t = 83.3$ hours.
   d) 144 bricks will be moved in 12 hours.
   e) New model $y(t) = -90t + 856$, where t is the time taken on the second day.
   f) Solve $y(t) = -90t + 856$ for $y(t) = 0$. That is $0 = -90t + 856$, $t = 9.5$ hours.

Exercise 4.6

1. a)   b)

2. Three, preferably five.
3. It will turn down.

---

4. It does not pass the vertical line test.

5.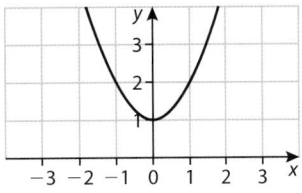

6.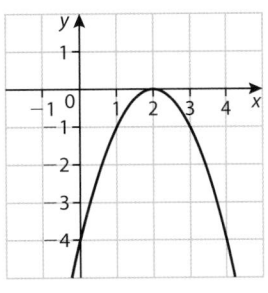

7.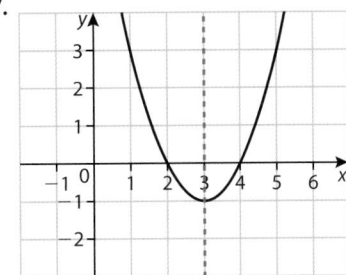

8.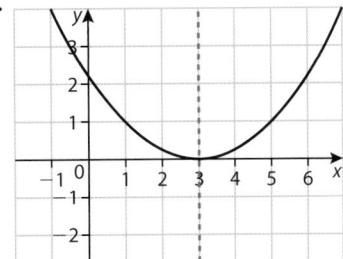

9.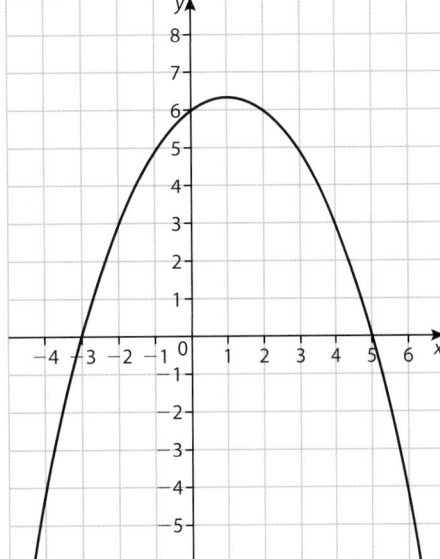

10.

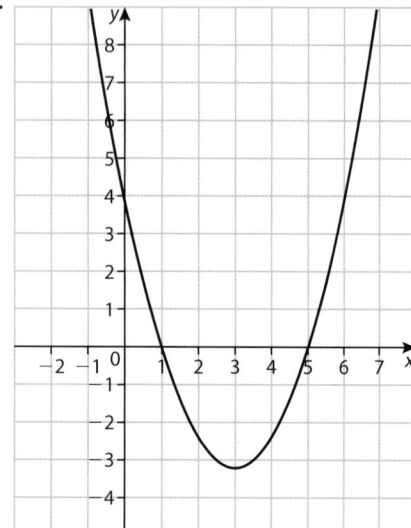

11.

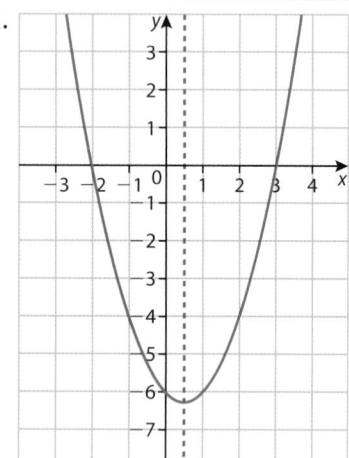

12.

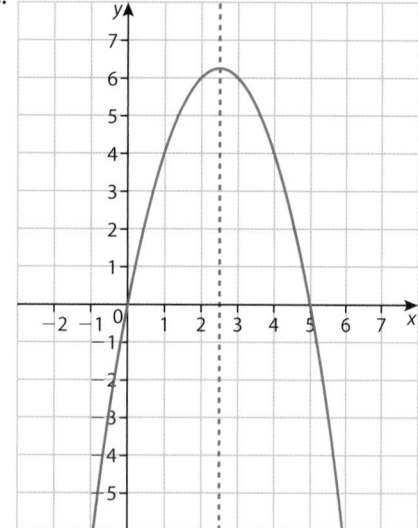

13. a) (i) $x = -7$ or $x = 1$
(ii) $y = -7$
(iii) $x = -3$
(iv) $(-3, -16)$
(v) Answers vary. $(-4, -15); (-2, -15)$
b) (i) None (ii) $y = 15$

(iii) $x = \frac{1}{2}$ (iv) $\left(\frac{1}{2}, \frac{29}{2}\right)$
(v) Answers vary. $(-1, 19); (2, 19)$

c) (i) $x = \frac{3}{2}$ or $x = 6$ (ii) $y = 18$
(iii) $x = \frac{15}{4}$ (iv) $\left(\frac{15}{4}, -\frac{81}{8}\right) = (3.75, -10.13)$
(v) Answers vary. $(3, -9); (5, -7)$

d) (i) $x = \frac{4}{3}$ or $x = -2$ (ii) $y = 8$
(iii) $x = -\frac{1}{3}$ (iv) $\left(-\frac{1}{3}, \frac{25}{3}\right) = (-0.333, 8.33)$
(v) Answers vary. $(-1, 7); (1, 3)$

14. a) (i) $x = 1 \pm \sqrt{3}$ (ii) $y = -2$
(iii) $x = 1$ (iv) $(1, -3)$
(v) Answers vary. $(-1, 1); (3, 1)$

b) (i) $x = 3$ or $x = 1$ (ii) $y = -6$
(iii) $x = 2$ (iv) $(2, 2)$
(v) Answers vary. $(-1, -16); (5, -16)$

c) (i) $x = -3 \pm \sqrt{8}$ (ii) $y = 0.5$
(iii) $x = -3$ (iv) $(-3, -4)$
(v) Answers vary. $(-4, -3.5); (-2, -3.5)$

d) (i) $x = \frac{(-4 \pm \sqrt{3})}{2} = -2 \pm \frac{\sqrt{3}}{2}$. ($x = -1.13$ or
$x = -2.87$ to 3 s.f.)
(ii) $y = -\frac{13}{3}$
(iii) $x = -2$
(iv) $(-2, 1)$
(v) Answers vary. $\left(-3, -\frac{1}{3}\right); \left(-1, -\frac{1}{3}\right)$

15. a)

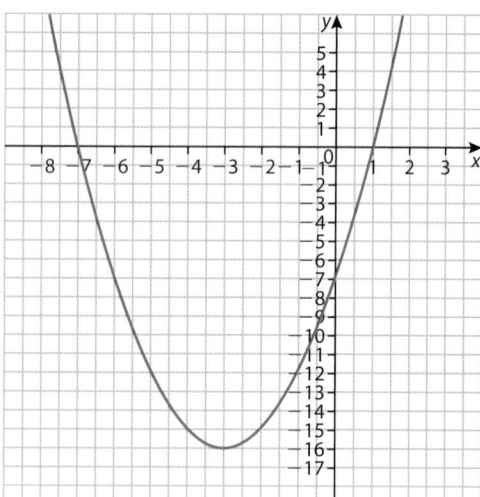

b)

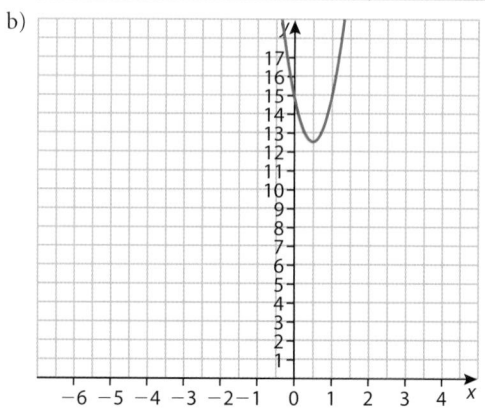

c)

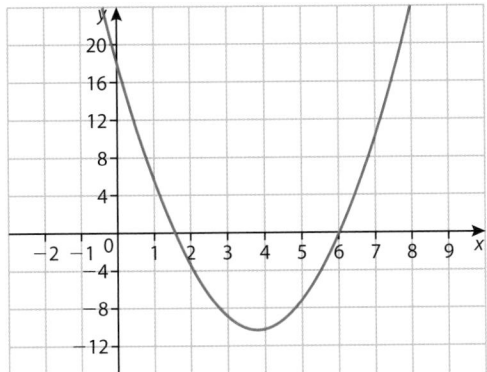

d)

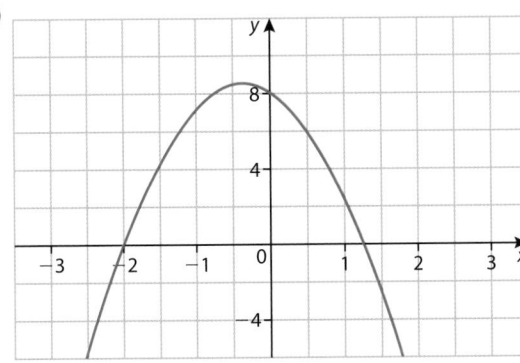

c)

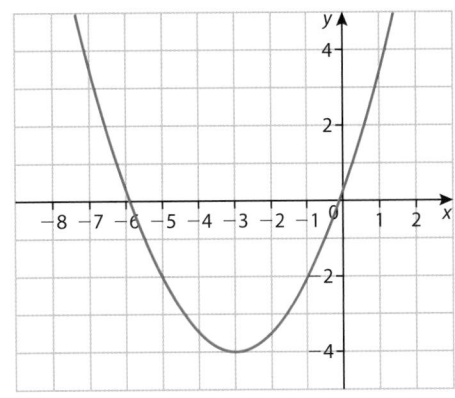

d)

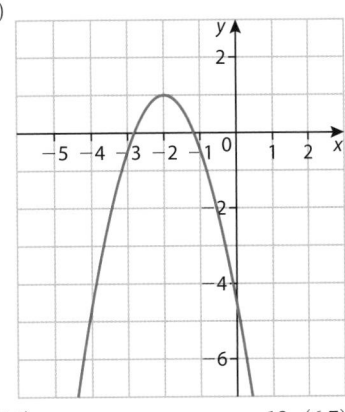

16. a)

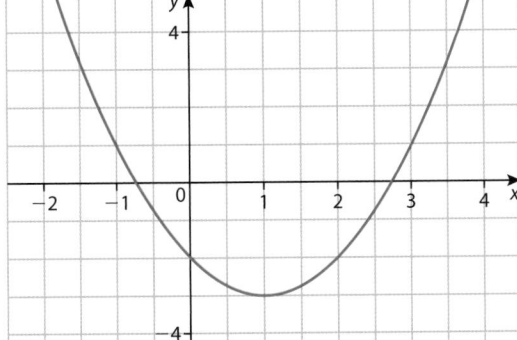

b)

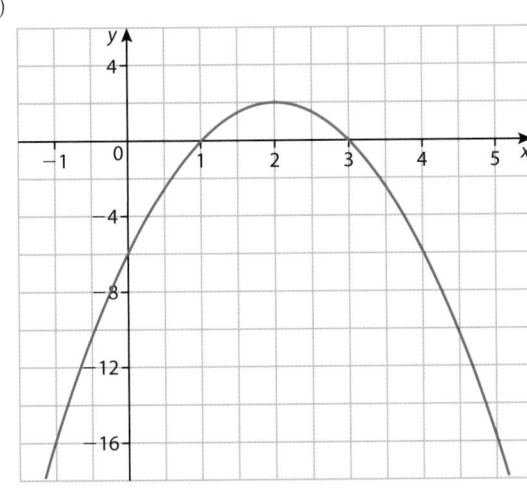

17. (5,3)

18. (6,7)

19. a)

$y = x^2 + 1$

$y = x^2$

b)

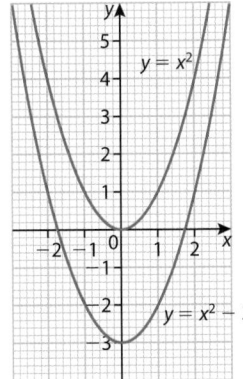

$y = x^2$

$y = x^2 - 3$

c)

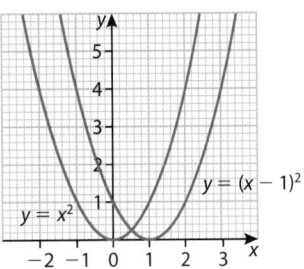

$y = (x - 1)^2$

$y = x^2$

d)

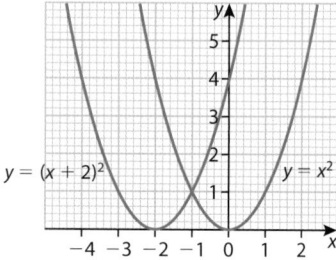

e)

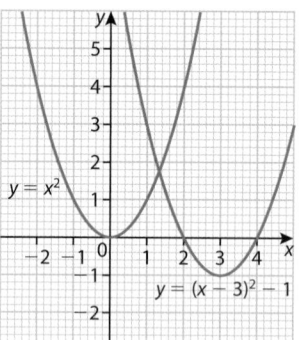

f)

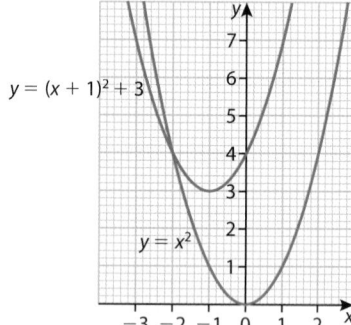

g)

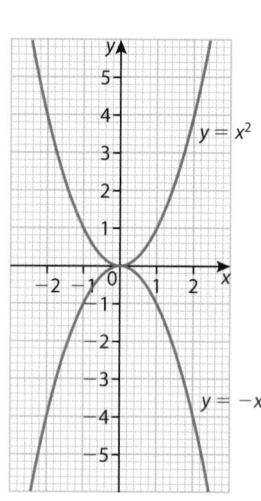

h)

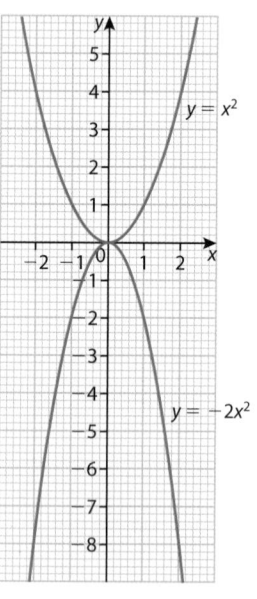

i)

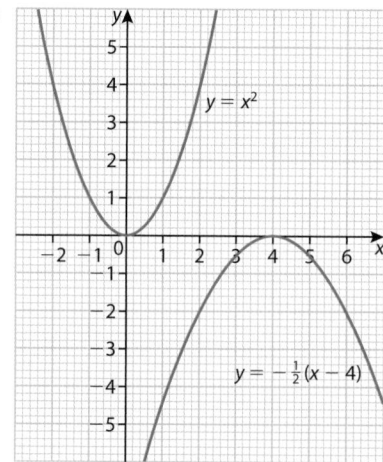

j)

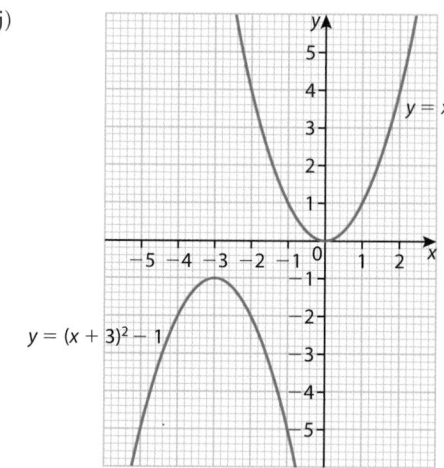

20. a) $y = \frac{1}{2}x^2 - x - \frac{5}{2}$
x-intercepts: $-1.45, 3.45$ to 3 s.f.
y-intercept: -2.5
axis of symmetry: $x = 1$
vertex: $(1, -3)$

x	y
-1.45	0
3.45	0
1	-3
0	-2.5
2	-2.5
4	1.5
-2	1.5

b) $y = -x^2 - 3x + 4$
x-intercepts: $-4, 1$
y-intercept: 4
axis of symmetry: $x = -1.5$
vertex: $(-1.5, 6.25)$

x	y
-4	0
1	0
-1.5	6.25
0	4
-3	4
1.5	-2.75
-4.5	-2.75
-1	6
-2	6

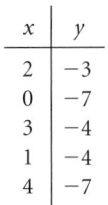
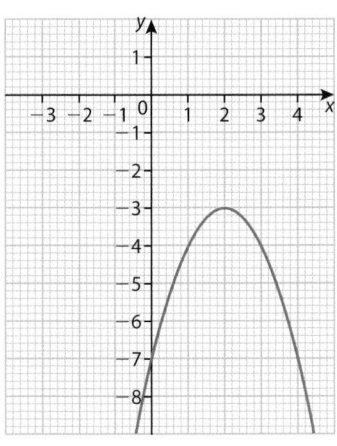

$y = -x^2 - 3x + 4$

x	y
2	-3
0	-7
3	-4
1	-4
4	-7

c) $y = x^2 + 2x - 5$
  x-intercepts: -3.45, 1.45 to 3 s.f.
  y-intercept: -5
  axis of symmetry: $x = -1$
  vertex: $(-1, -6)$

x	y
-3.45	0
1.45	0
0	-5
-2	-5
-1	-6
1	-2
-3	-2

$y = x^2 + 2x - 5$

d) $y = x^2 + 3x + 7$
  x-intercepts: (none)
  y-intercept: 7
  axis of symmetry: $x = -1.5$
  vertex: $(-1.5, 4.75)$

x	y
-1.5	4.75
0	7
-3	7
-1	5
-2	5
$.5$	8.75
-3.5	8.75

$y = x^2 + 3x + 7$

e) $y = -x^2 + 4x - 7$
  x-intercepts: none
  y-intercept: -7
  axis of symmetry: $x = 2$
  vertex: $(2, -3)$

f) $y = x^2 - 4x + 4$
  x-intercept: 2
  y-intercept: 4
  axis of symmetry: $x = 2$
  vertex: $(2, 0)$

x	y
0	4
1	1
2	0
3	1
4	4

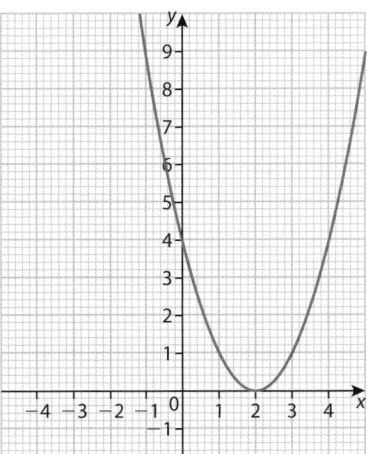

21. See lines on each graph of 20a–f.
  a) $x = -3$ or $x = 5$

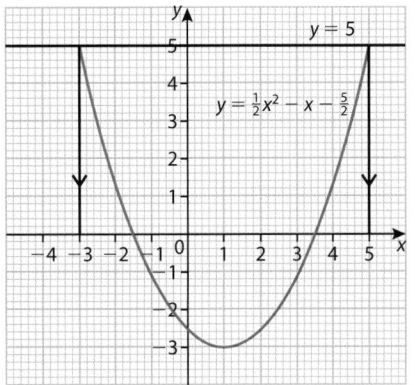

$y = 5$

$y = \frac{1}{2}x^2 - x - \frac{5}{2}$

b) $x = -2$ or $x = -1$

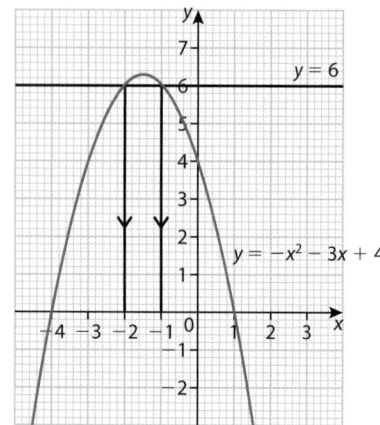

c) $x = -3$ or $x = 1$

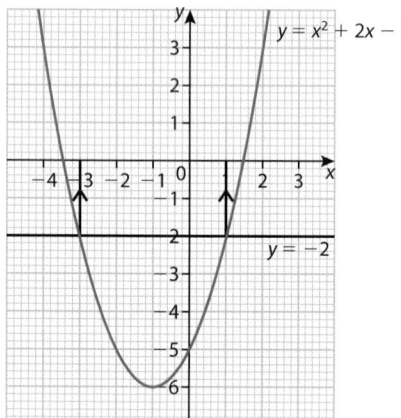

d) $x = -2.6$ or $x = -0.38$ to 2 s.f.

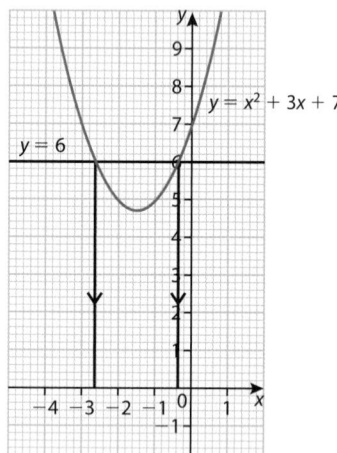

e) $x = 1.3$ or $x = 2.7$ to 2 s.f.

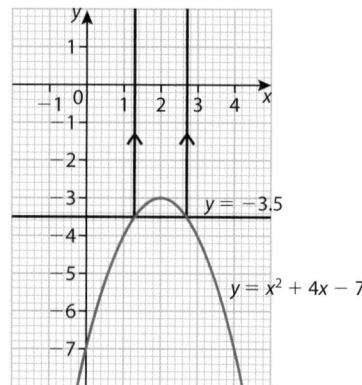

f) $x = -0.32$ or $x = 4.3$ to 2 s.f.

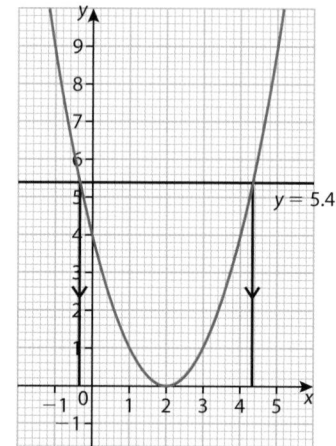

22. a) $x = -3$ or $x = 5$
    b) $x = -2$ or $x = -1$
    c) $x = -3$ or $x = 1$
    d) $x = -2.62$ or $x = -0.382$ to 3 s.f.
    e) $x = 1.29$ or $x = 2.71$ to 3 s.f.
    f) $x = -0.324$ or $x = 4.32$ to 3 s.f.
23. a) $a = 2$
    b) $c = 10$, $y = 2x^2 + 12x + 10$
    c) $A(-3, 0)$ $B(-2, 0)$
24. a) $A(1, 0)$
    b) $B(0, -7)$
    c) $y = -x^2 + 8x - 7$

Exercise 4.7

1. a) 2 p.m. (14:00).
   b) The minimum number of passengers is 86.
   c) The most number of passengers is carried at 6 a.m.
   d) At 8 a.m. 4 buses and at 5 p.m. 2 buses.
2. a) At $t = 0$ there were 50 marsupials.
   b) After 12 months there are 127 marsupials.
   c) After 24 months there are 261 marsupials.
   d) There will be 500 marsupials after 38.5 months (39 months).
3. a) $L + 2W = 200$
      $\therefore L = 200 - 2W$
   b) $A = LW = (200 - 2W)W = 200W - 2W^2$
      $\therefore f_A(W) = 200W - 2W^2$ or $A = 200W - 2W^2$

c) $W = \dfrac{-b}{2a} = \dfrac{-200}{2 \cdot -2} = 50$

d) $\therefore\ L = 200 - 2 \times 50 = 100$

$\therefore\ A = LW = 100 \times 50 = 5000$ sq. ft.

4. $H = -4.9T^2 + VT + B$

$\therefore\ H = -4.9T^2 + 40T + 10$

a)

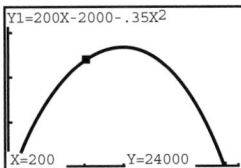

b) 91.6 m to 3 s.f.

5. a) $P(x) = S(x) - C(x)$

$P(x) = 200x - 0.25x^2 - 2000 - 0.1x^2$

$= 200x - 2000 - 0.35x^2$

b)

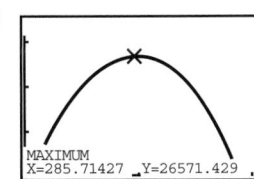

c)

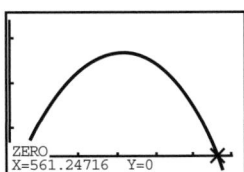

Maximum number of vases to be sold is 286 to 3 s.f.

Profit is $26 571.43

d) For the final question we require the x-intercept, that is the point where the profit becomes negative.

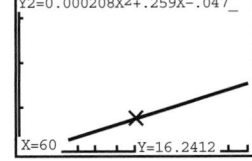

From the graph we can see that the x-intercept occurs at $x = 561.25$

Thus when 562 vases are manufactured the company will be losing money.

6. a)

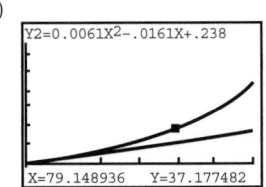

b) The stopping distance is 16.2412 m.

c) 106.8 km/hr

d)

Y2=0.0061X²-.0161X+.238

X=79.148936   Y=37.177482

The driver's reaction time graph climbs more steeply than the stopping time graph, indicating that the driver's reaction time is more important at higher speeds than at lower speeds.

e) 21.232 m

f) $T(v) = 0.00631v^2 + 0.243v + 0.190$

g)

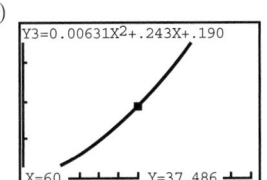

h) The stopping distance is 37.5 m (3 s.f.).

Exercise 4.8

1. b, d, e, g, h, j, k

2. a) $\sqrt{3}$ b) $\dfrac{\sqrt[3]{25}}{5}$ c) $\sqrt[3]{49}$

d) $\sqrt[5]{25}$ e) $\sqrt{\dfrac{3}{2}} = \dfrac{\sqrt{6}}{2}$

f) $\dfrac{2\sqrt[3]{6}}{3}$ $\left(\text{accept: } \sqrt[3]{\dfrac{16}{9}}, \dfrac{\sqrt[3]{48}}{3}, \dfrac{2\sqrt[3]{6}}{3}\right)$

3. a) $3^{\frac{1}{2}}$ b) $5^{\frac{2}{3}}$ c) $(-29)^{\frac{1}{3}}$

d) $\left(\dfrac{3}{4}\right)^{\frac{1}{2}}$ e) $\left(\dfrac{1}{2}\right)^{\frac{1}{4}}$ f) $(-13)^{\frac{3}{5}}$

4. a) 8   b) $\dfrac{1}{2}$ c) 125   d) $\dfrac{16}{9}$

e) $\sqrt{-4}$. This is not a real number.   f) 9

5.

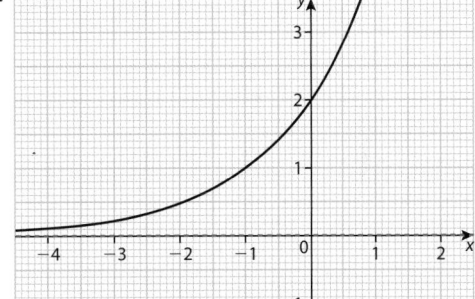

6.

7. Increasing. As x gets larger, y becomes larger.

8. Decreasing. As x gets larger, y becomes smaller.

9. $y = 3$

10. $y = -1$

11. $y = 5$

12. $y = -2.5$

13. Decreasing: $(-\infty, 0]$

Increasing: $[0, -\infty)$

14. Increasing: $(-\infty, -1]$ and $[2, \infty)$
Decreasing: $[-1, 2)$

15. a)

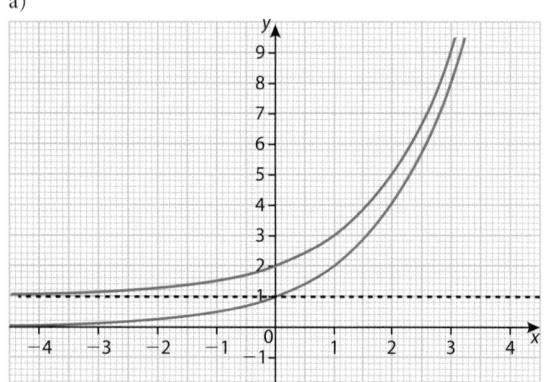

b)

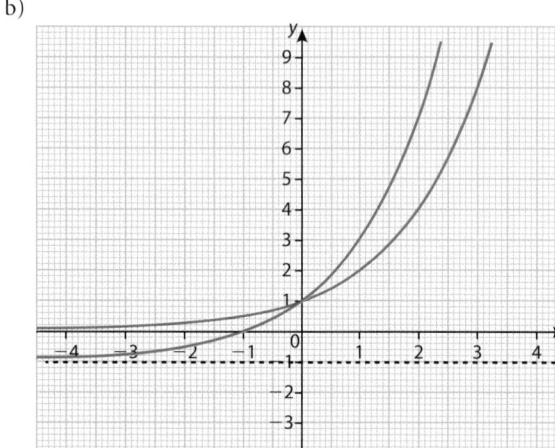

c)

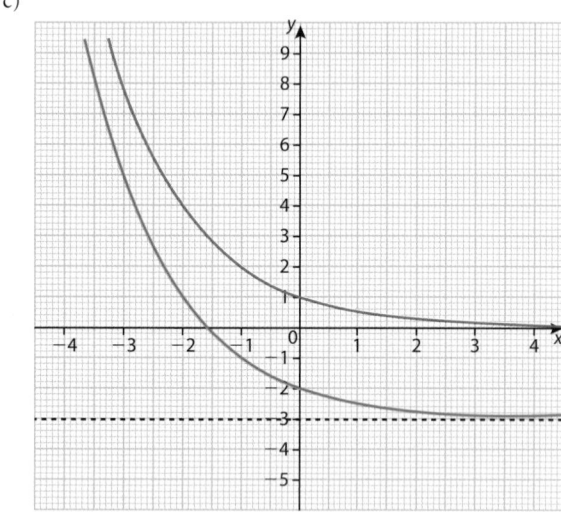

d)

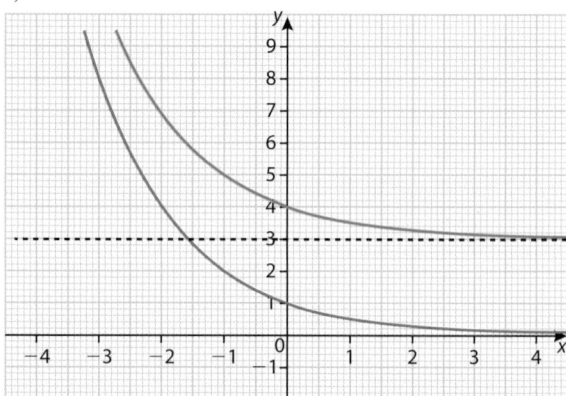

e)

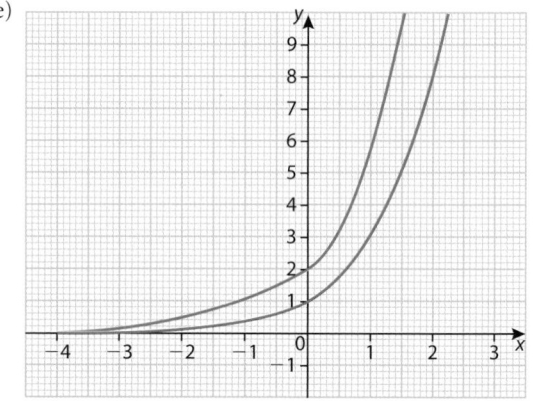

f)

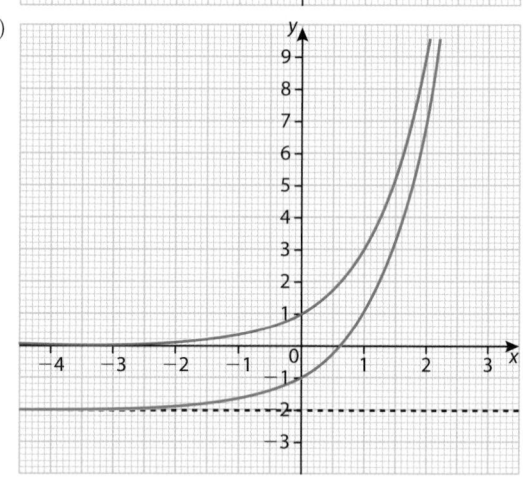

16. a)

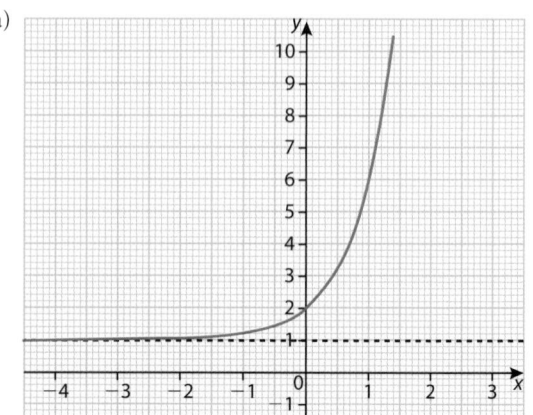

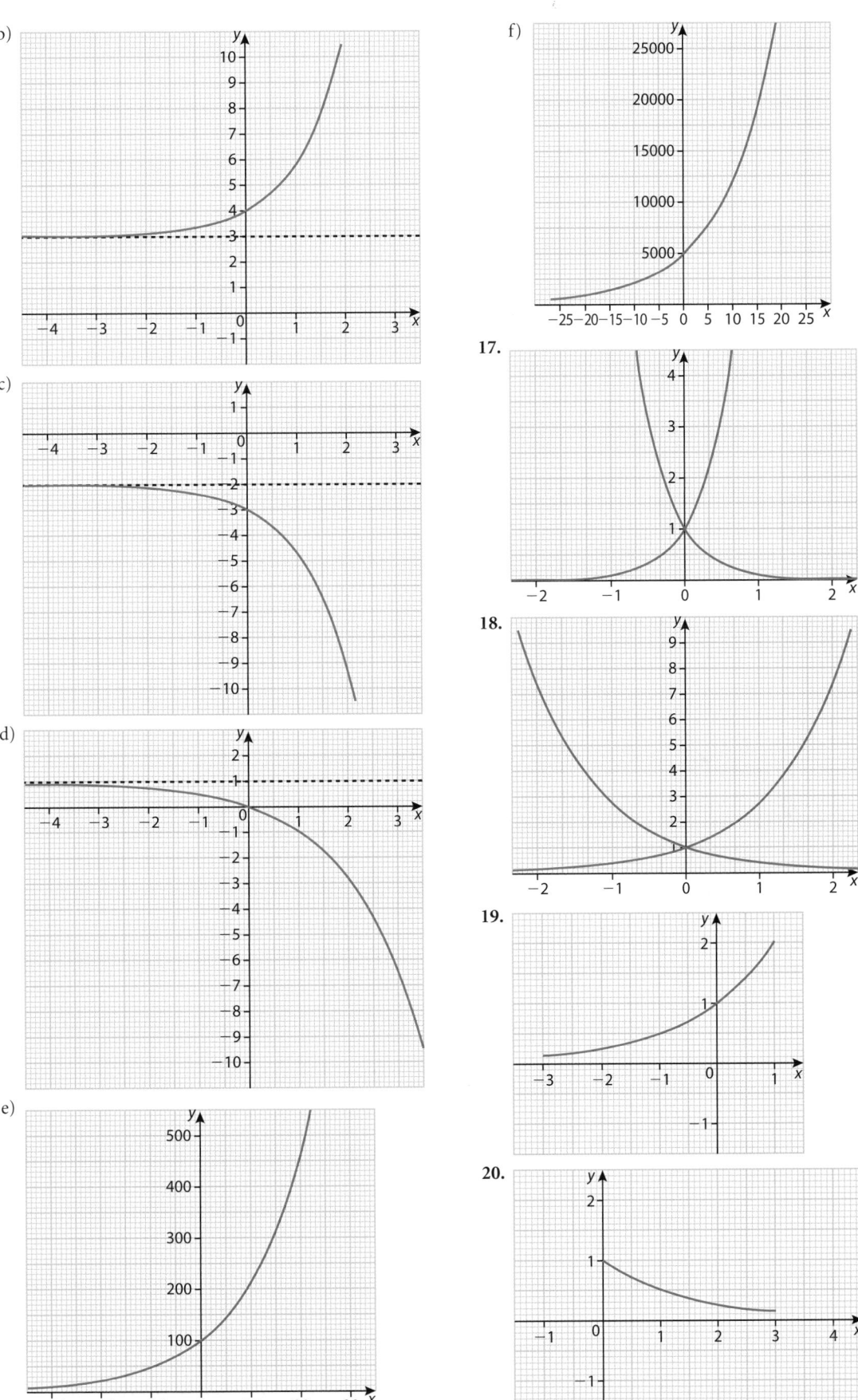

b)

c)

d)

e)

f)

17.

18.

19.

20.

21. $x = -1.6$

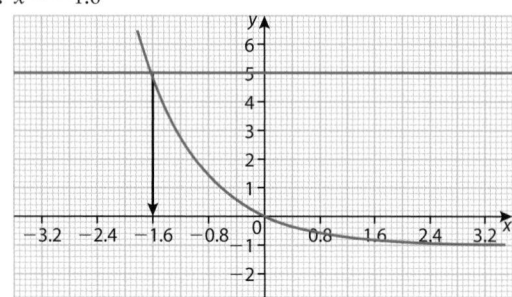

22. $x = 10.2$

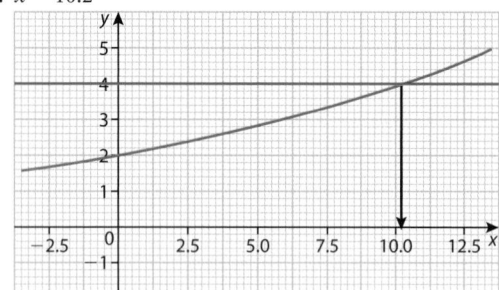

23. $x = 3.5$

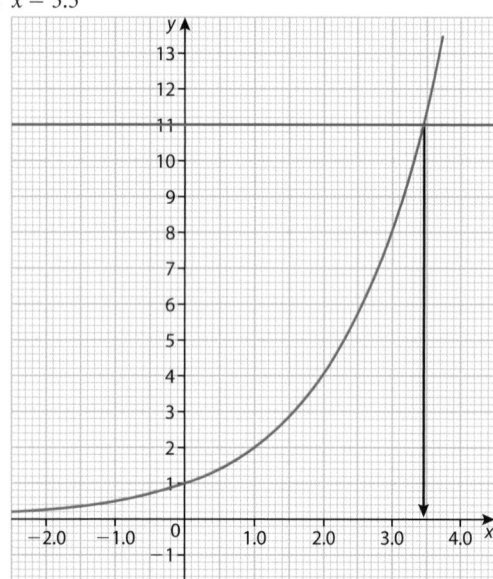

24. $x = -6$

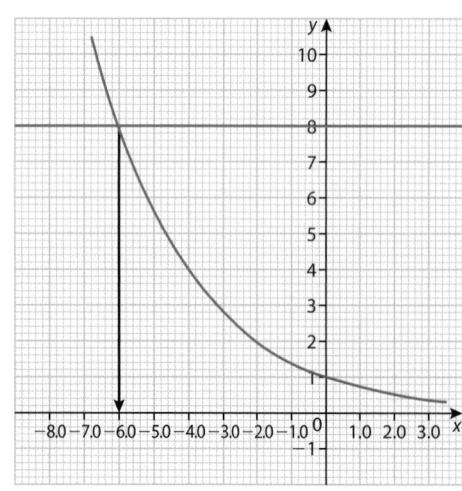

25. $x = -1.63$ to 3 s.f.

26. $x = 10.2$ to 3 s.f.

27. $x = 3.46$ to 3 s.f.

28. $x = -6$

29. $x = 2.58$ to 2 d.p.

30. $x = 14.21$ to 2 d.p.

31. $x = 2.50$ to 2 d.p.

32. $x = 9.69$ to 2 d.p.

33. $\dfrac{\log 37}{\log 5}$

34. $\dfrac{\log 107}{\log 17}$

35. $\dfrac{\log 2}{\log 1.06}$

36. $\dfrac{\log 6}{\log 3}$

37. $\dfrac{-\log 91}{\log 4}$

38. $\dfrac{\log 20}{\log 2}$

39. $\dfrac{\log 7}{\log 2}$

40. $\dfrac{-\log 19/4}{\log 3}$

41. $\dfrac{\log 5}{12 \log(151/150)}$

42. $\dfrac{\log(2/9)}{\log(17/20)}$

Exercise 4.9

1. a)

t	0	1	2	3	4	5
$p(t)$	200	214	223	245	262	281

b)

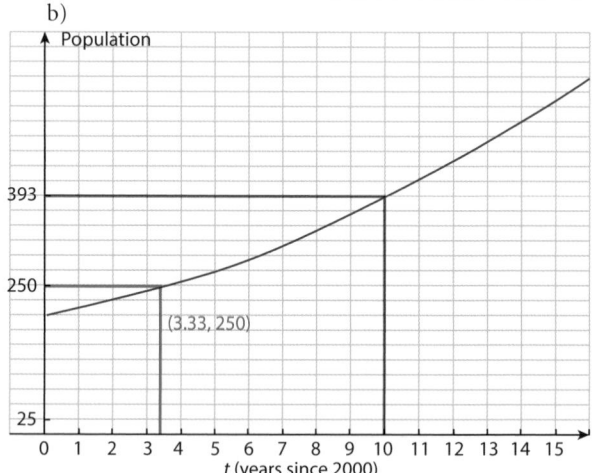

c) (i) Approximately 3.33 years (the green line)

  (ii) Approximately 393 (3 s.f.) (the red line)

d) The population will exceed 1000 in the year 2023.

2. a) $A = 2000(1.08)^n$.

b) 4317.85

  9321.91

  20 125.31

3. a) \$22 000.

b) \$9540 to 3 s.f.

c) $s(t) = 25\,000\,(0.85)^t$

d) y-intercepts when $t = 0$

e)

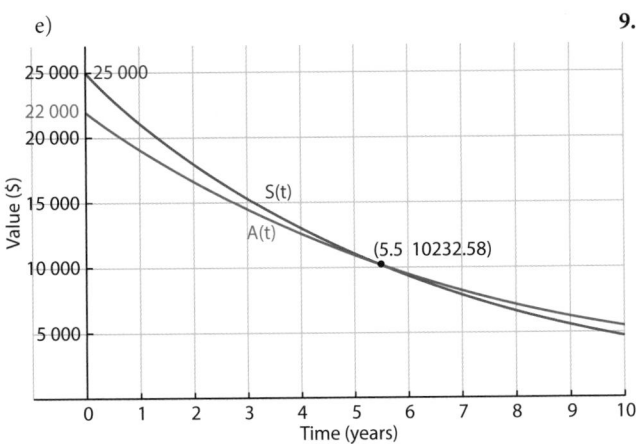

d) 5.5 years

f) Shana's car 9.90 years (3 s.f.).

Chapter 5

Exercise 5.1

1. a) F   b) F   c) F   d) F   e) T
   f) T   g) T   h) T   i) T   j) F

2. a) T   b) T   c) F   d) T   e) F
   f) T   g) T   h) F   i) T   j) F

3. a) F   b) F   c) F   d) T   e) F
   f) T   g) T   h) F   i) F   j) F

4. a) T   b) T   c) F   d) T   e) T
   f) T   g) F   h) F   i) T   j) T

5. a) $-1, 0, 2$ b) 0
   c) $(-\infty, -0.5)$ and $(1.2, \infty)$ d) $(-0.5, 1.2)$
   e) $0.6; (-0.5, 0.6)$ f) $-2.1; (1.2, -2.1)$
   g) $(0.3, -0.7)$

6. a) $-2.5, -1.3, 2$ b) 6.5
   c) $(-1.9, 0.7)$ d) $(-\infty, -1.9)$ and $(0.7, \infty)$
   e) $8.3; (0.7, 8.3)$ f) $-1.4; (-1.9, -1.4)$
   g) $(-0.6, 3.5)$

7. a) $-3, -2, 0, 2$
   b) 0
   c) $(-2.6, -0.9)$ and $(1.3, \infty)$ d) $(-\infty, -2.6)$ and $(-0.9, 1.3)$
   e) $6.0; (-0.9, 6.0)$
   f) $-2.9; (-2.6, -2.9)$ and $-12.9; (1.3, -12.9)$
   g) $(-1.9, 0.8)$ and $(0.4, -5.2)$

8. a) $-7.3, -3.7, 2.$
   b) -11
   c) $(-\infty, -5.9)$ and $(-1.4, 2)$
   d) $(-5.9, -1.4)$ and $(2, \infty)$
   e) $(-5.9, 18.5)$ and $(2, 0)$
   f) $(-1.4, -16)$
   g) $(-3.6, -1.5)$ and $(0.4, -8)$

9.

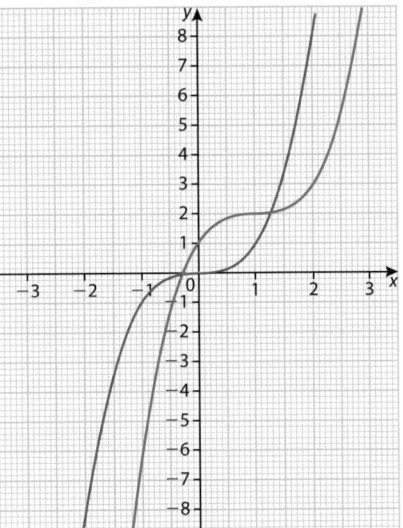

10.

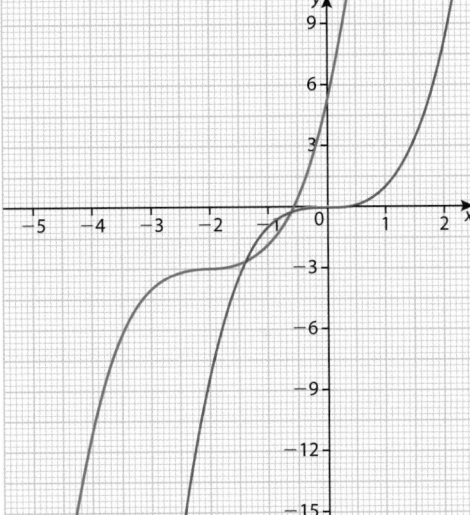

11.

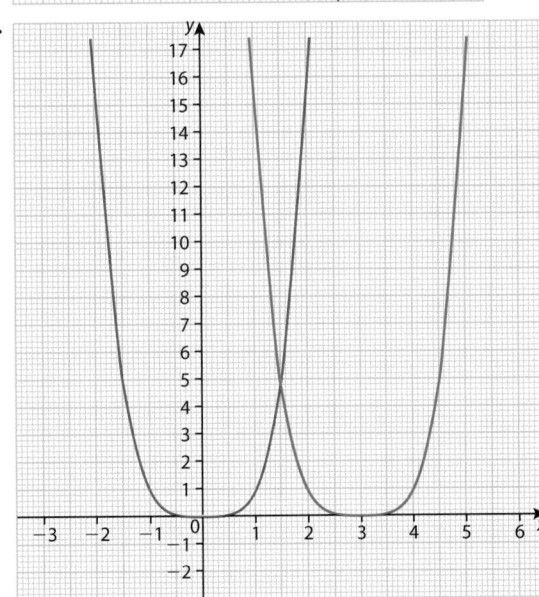

12.

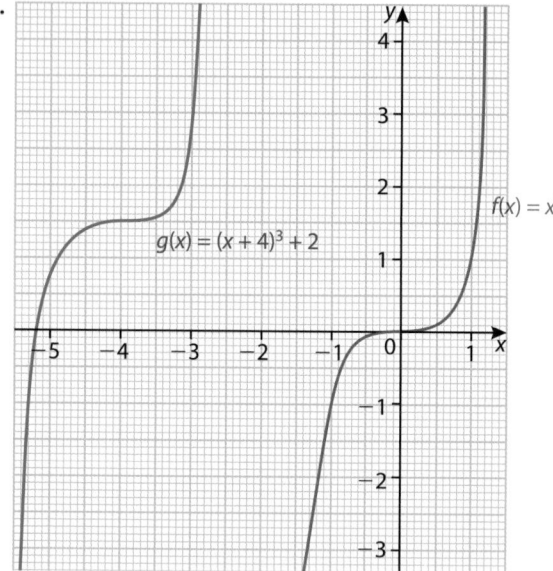

$g(x) = (x + 4)^3 + 2$

$f(x) = x^5$

14.

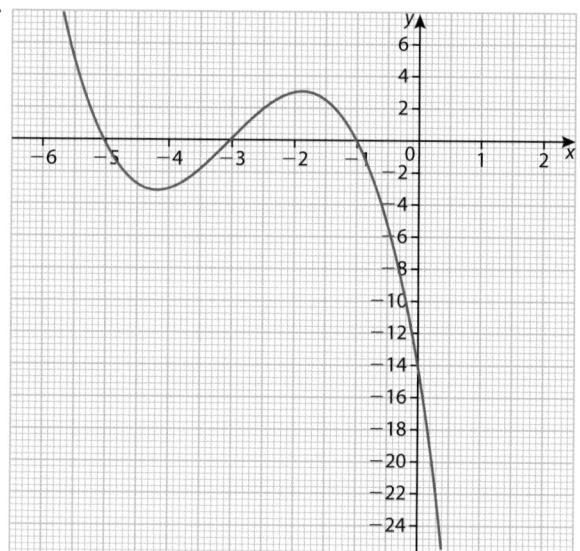

13.

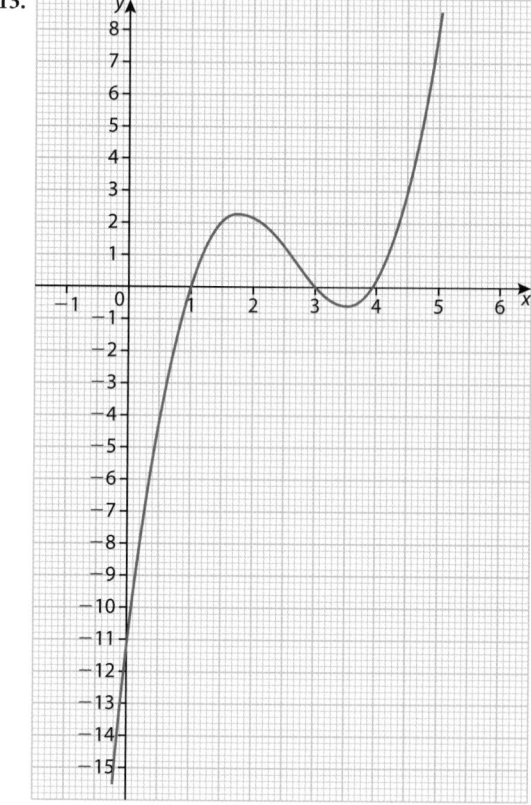

15.

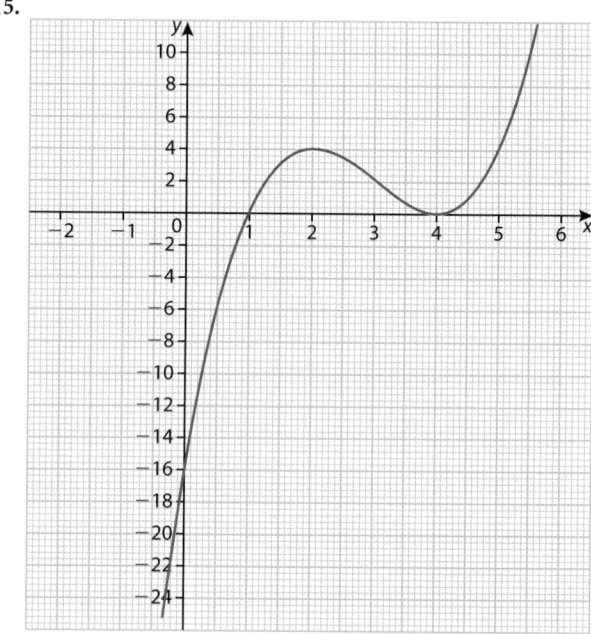

16.

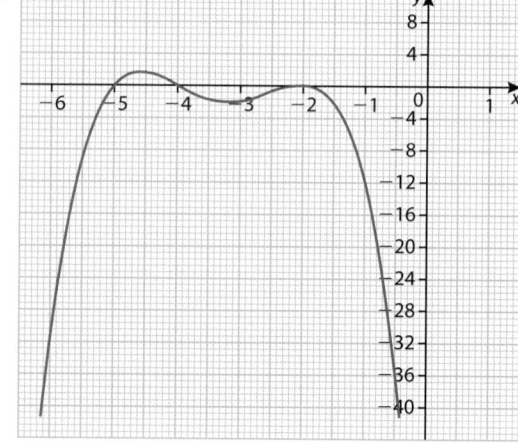

17.

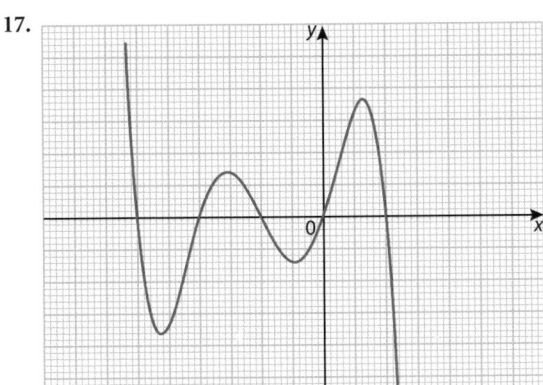

18.

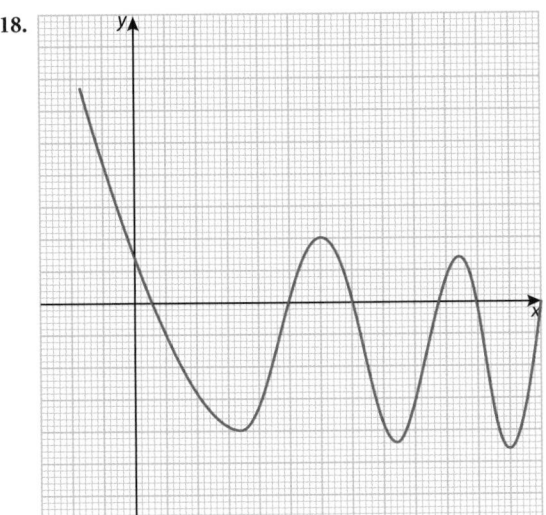

19.

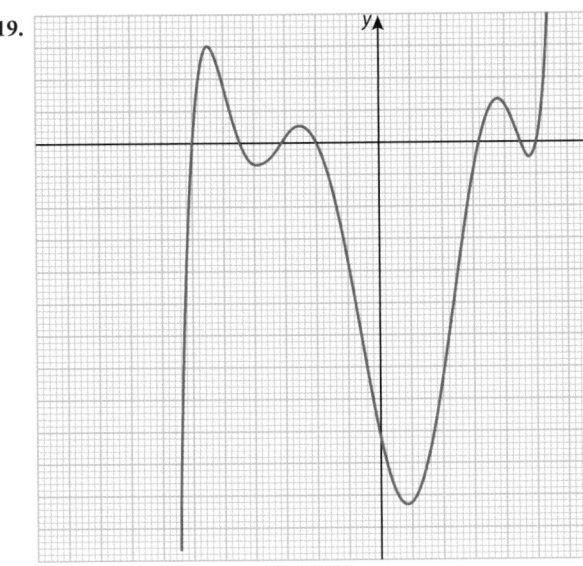

20.

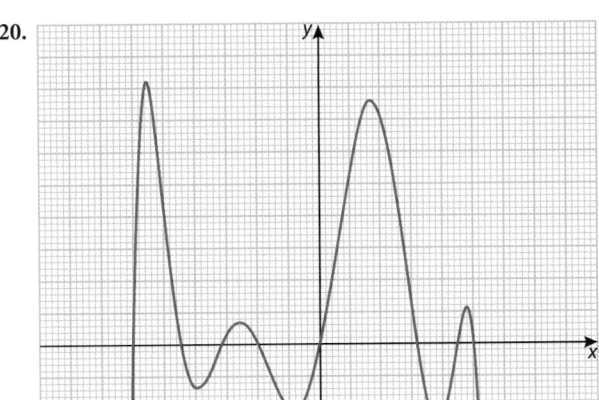

Exercise 5.2

1. An asymptote is a boundary (line) that a curve approaches, but never touches, at the extremes of the number line.
2. A horizontal asymptote is a boundary (line) that a curve approaches and that is parallel to the x-axis.
3. A vertical asymptote is a boundary (line) that a curve approaches and that is parallel to the y-axis.
4. Asymptotes will help you see "end-behaviour", (how the graph behaves near infinity), and thus helps you draw the graph of the function.
5. Large
6. Denominator; zero
7. Close   **8.** F
9. F   **10.** T
11. a) No stretch
    b) No slide
    c) Vertical shift $= +2$ (2 units up)
    d) Vertical asymptote: $x = 0$
    e) Horizontal asymptote: $y = 2$
    f) Domain: $\{x \mid x \in \mathbb{R}, x \neq 0\}$
    g) Range: $\{y \mid y \in \mathbb{R}, y \neq 2\}$
12. a) By a factor of 3
    b) 1 unit to the right
    c) Vertical shift $= -2$ (2 units down)
    d) Vertical asymptote: $x = 1$
    e) Horizontal asymptote: $y = -2$
    f) Domain: $\{x \mid x \in \mathbb{R}, x \neq 1\}$
    g) Range: $\{y \mid y \in \mathbb{R}, y \neq -2\}$
13. a) By a factor of -2
    b) 3 units to the left
    c) No Vertical shift
    d) Vertical asymptote: $x = -3$
    e) Horizontal asymptote: $y = 0$
    f) Domain: $\{x \mid x \in \mathbb{R}, x \neq -3\}$
    g) Range: $\{y \mid y \in \mathbb{R}, y \neq 0\}$
14. a) By a factor of -1
    b) 2 units to the left
    c) Vertical shift $= -3$
    d) Vertical asymptote: $x = -2$
    e) Horizontal asymptote: $y = -3$
    f) Domain: $\{x \mid x \in \mathbb{R}, x \neq -2\}$

g) Range: $\{y | y \in \mathbb{R}, y \neq -3\}$

15.

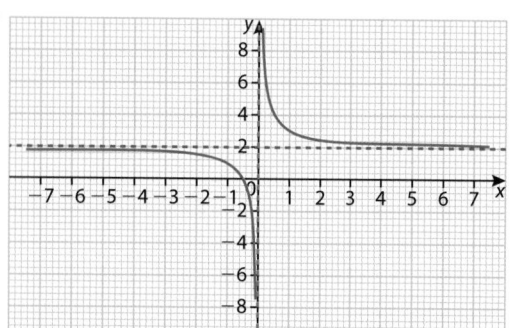

16.

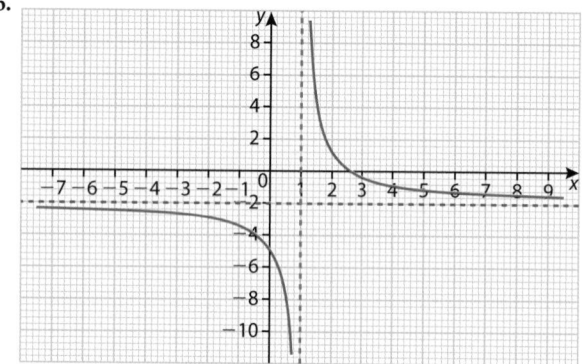

17.

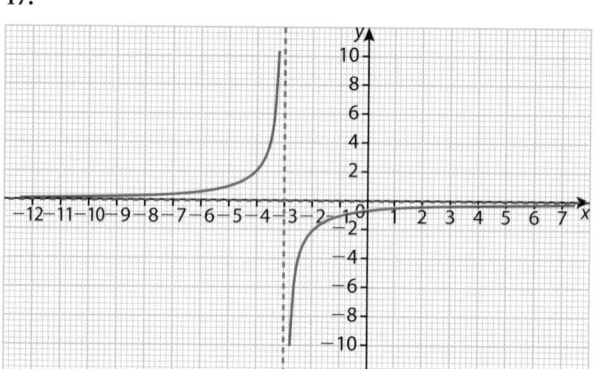

18.

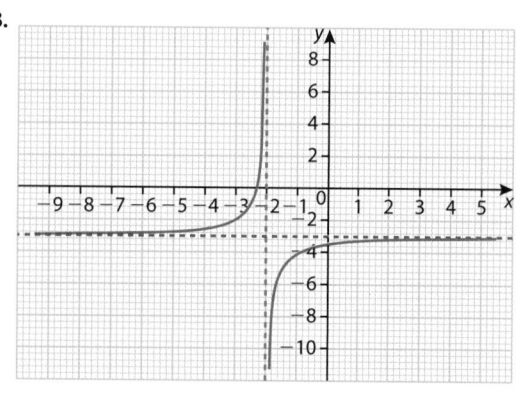

19.

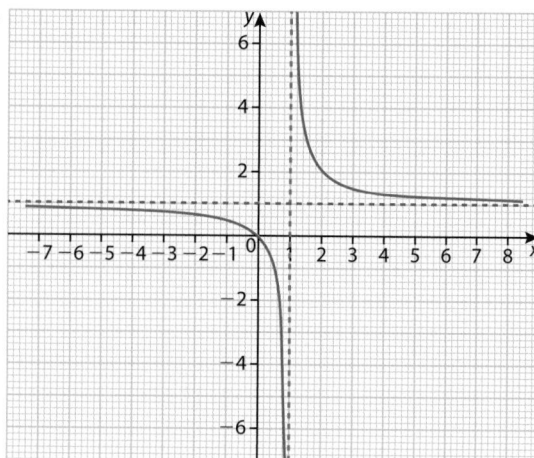

20.

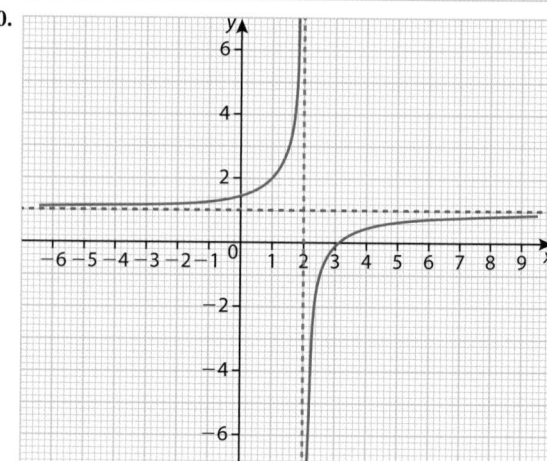

21.

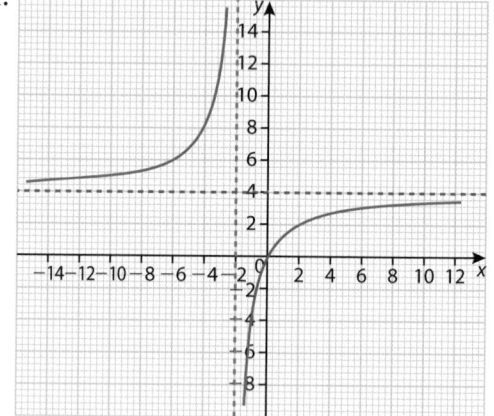

22.

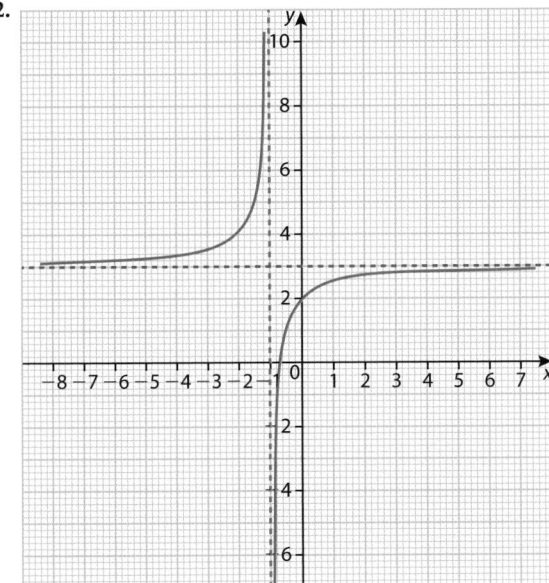

Exercise 5.3

1. a) Logarithmic   b) Not classified   c) Absolute value
    d) Rational   e) Piecewise   f) Polynomial

2. Answers vary: Examples include:
   Linear: $l(x) = 2x + 3$
   Polynomial: $p(x) = x^2 + 1$
   Root: $r(x) = \sqrt{x}$

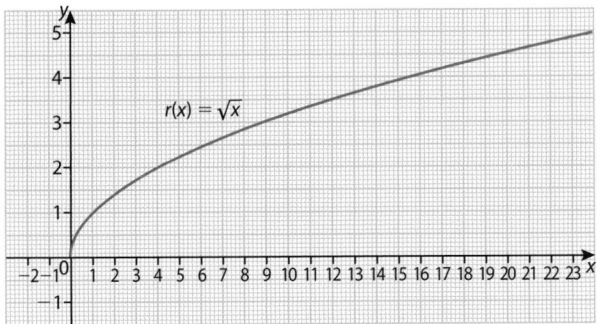

3. Answers vary: $f(x) = x^3 e^x,\ -5 < x < 3$

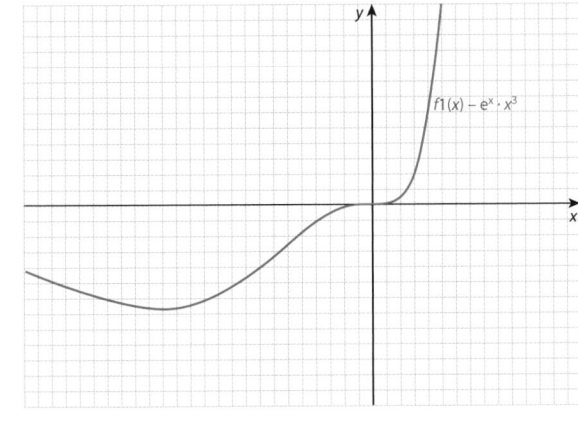

4. a)

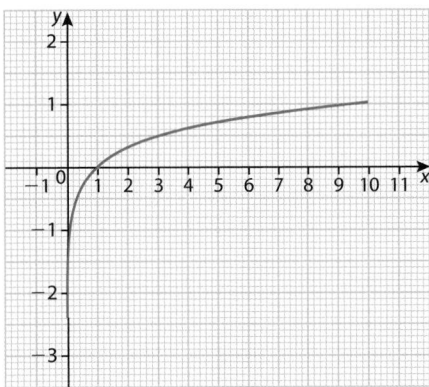

b)

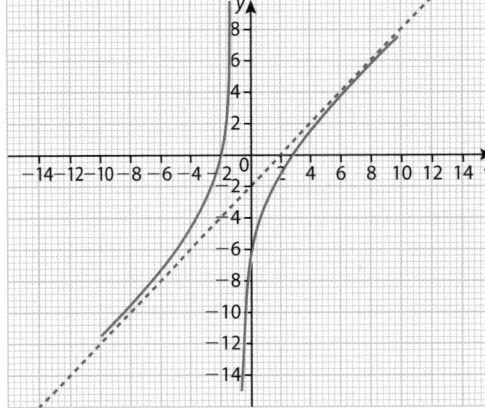

c)

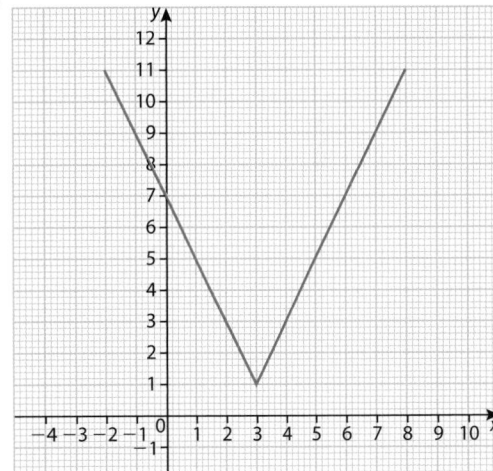

5.

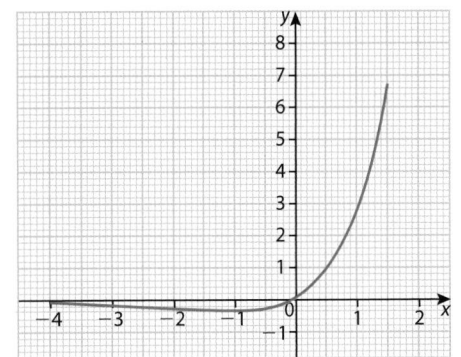

6. a) 2   b) 2   c) 2   d) 2   e) 2   f) 2   g) 3

7. a) Domain: $-3 \leq x \leq 3$

b) Range: $y = -3, -2, -1, 0, 1, 2, 3$ or $\{-3 \leq y \leq 3, y \in \mathbb{Z}\}$

c)

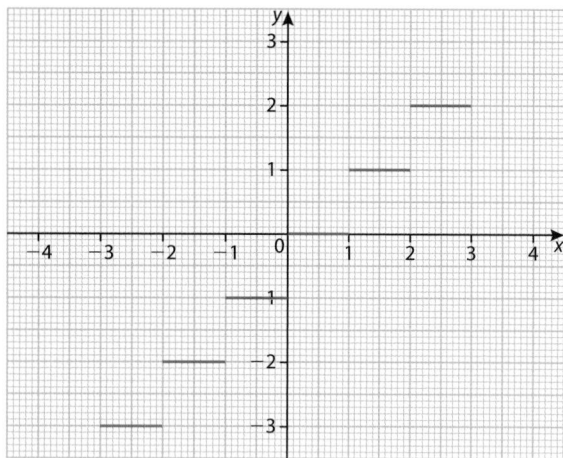

Exercise 5.4

1. a) $x = 2$

b) $x = \pm 3$

c) $x = -3$ or $x = -4$

d) $x = \dfrac{\log 41}{\log 3}$

e) $x = 16$

2. a) On the left hand side the variable, x, is an exponent and on the RHS it is a factor.

b) The LHS is exponential in x and the RHS is rational in x. The equations a and b do not have well-known (or perhaps any known) algebraic techniques that can be used to solve them.

3. For $x^{\frac{2}{3}} - 7x^{\frac{1}{3}} = -12$, let $y = x^{\frac{1}{3}}$.

Therefore, $y^2 - 7y + 12 = 0$

$\qquad\qquad (y - 3)(y - 4) = 0$

Hence, $y = 3$ or $y = 4$.

And therefore, $3 = x^{\frac{1}{3}}$ or $4 = x^{\frac{1}{3}}$.

Hence, $x = 27$ or $x = 64$.

4. $x = \pm 0.816$ to 3 s.f.

5. $x = -1$

6. $x = -0.5, 0$

7. $x = -1, 1$

8. a)

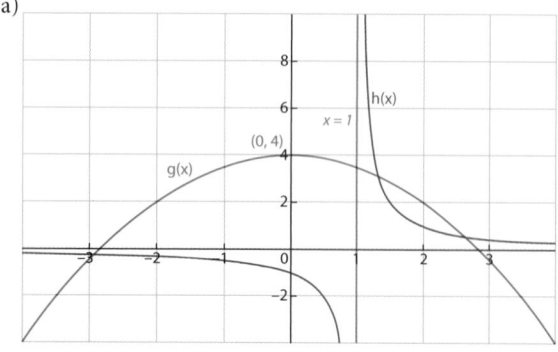

b) see graph $x = 1$

c) $x = -2.92$, $x = 1.32$, $x = 2.60$

9. a) The archer's coordinates are $(0, 2)$.

b) The coordinates of the target are $(30, 23)$.

c)

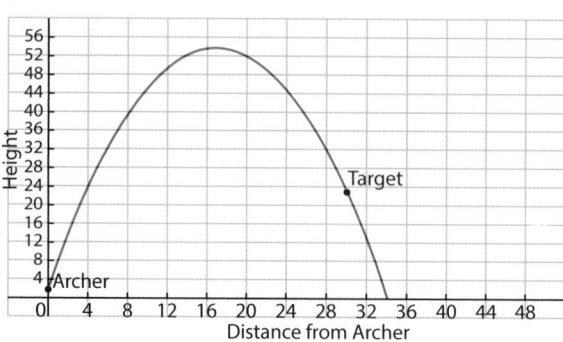

d) The highest point is located 16.9 (3 s.f.) metres horizontally from the archer.

e) The maximum vertical height reached by the arrow is 53.7 metres (3 s.f.).

f) The arrow lands 34.2 metres (3 s.f.) from the archer.

10. a)

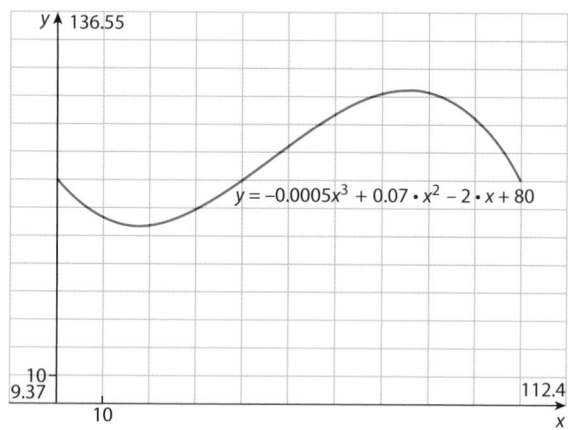

b) 80 metres, which is the y-intercept.

c) 63.8 metres. (3 s.f.)

d) 113 metres. (3 s.f.)

e) $x = 40$ m and $x = 100$ m

11. a)

t (weeks)	0	1	2	3	4	5	6	7	8
h(t)	11	11.11	11.22	11.36	11.5	11.66	11.84	12.03	12.25

b) The height of grass is given by the green curve.

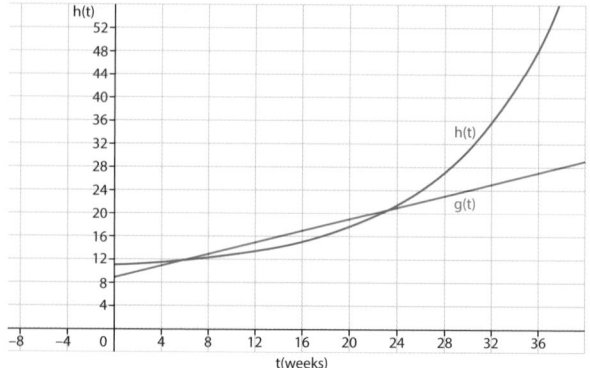

c) (i) 11 cm
   (ii) 12.75 cm.
d) 5.49 weeks and $x = 23.4$ weeks
e) Answer required

Chapter 6

Exercise 6.1

1. a) $\frac{6}{10}$ b) $\frac{8}{10}$ c) $\frac{6}{8}$ d) $\frac{8}{10}$ e) $\frac{6}{10}$ f) $\frac{8}{6}$

2. a) $\sin 40° = \frac{a}{c}$ b) $\cos 40° = \frac{b}{c}$ c) $\tan 40° = \frac{a}{b}$

   d) $\sin 50° = \frac{b}{c}$ e) $\cos 50° = \frac{a}{c}$ f) $\tan 50° = \frac{b}{a}$

3. a) 0.6018   b) 0.0872   c) 0.1228
   d) 0.6782   e) 0.5000   f) 1.0000
   g) 0.5000   h) 0.6018   i) undefined

4. a) 5.95   b) 71.8   c) 21.8
   d) 26.5   e) 10.4   f) 0.427

Exercise 6.2

1. a) $e \approx 31.1$ b) $e \approx 14.5$ c) $a \approx 12.3$ d) $c \approx 3.69$
2. a) $b \approx 7.24$ b) $e \approx 7.83$ c) $d \approx 140$ d) $c \approx 30\,300$
3. a) $j \approx 8.51$ b) $i \approx 13.8$
   c) $\angle ECH \approx 48.2°$ d) $c \approx 679$
4. a) $k \approx 874$ b) $j \approx 79.5$
   c) $\angle EFD \approx 10.3°$ d) $\angle DEF \approx 46.1°$
5. a) $g \approx 119$ b) $d \approx 69.3$
6. a) height is 50 m   b) distance ≈ 167 m
   c) angle $\approx 32.0°$
7. She is ≈ 213 m from her starting position
8. a) $AD \approx 83.9$ m   b) $BC \approx 57.7$ m
   c) (i) Distance is ≈ 173 m   (ii) Distance ≈ 188 m
9. 41.1 m
10. a) ≈ 80 feet   b) ≈ 220 feet   c) ≈ 300 feet
11. a) $\sin(65°) = \frac{32}{a}$ b) $\cos(15°) = \frac{80}{b}$

     $a \approx 35.3$ feet $b \approx 82.8$ feet

   c) $\cos(40) = \frac{120}{c}$ d) $a + 3b + c \approx 440$ feet

     $c \approx 157$ feet

12. a) (i) $\tan(53°) = \frac{BC}{1.2\,\text{km}}$ b) (i) $\cos(63°) = \frac{CD}{1.4\,\text{km}}$

       $BC \approx 1.59$ km $CD \approx 0.636$ km

     (ii) $\cos(53°) = \frac{1.2\,\text{km}}{AC}$ (ii) $\sin(63°) = \frac{DE}{1.4\,\text{km}}$

       $AC \approx 1.99$ km $DE \approx 1.25$ km

   c) $(AB + CD)^2 + (BC + DE)^2 = (AE)^2$
     $AE \approx 3.38$ km

Exercise 6.3

1. a) $AC \approx 14.1$ b) $BC \approx 18.4$
2. a) $\angle BCA \approx 53.5°$ b) $\angle BAC \approx 44.8°$
3. a) $\angle ABC \approx 86.5°$ b) $\angle BCA \approx 65.2°$
4. a) $AC \approx 90.9$ b) $AB \approx 649$
5. a) $\angle DAC \approx 47.9$ b) $AC \approx 38.5$ c) $BC \approx 21.1$
6. a) $\angle ACB \approx 25°$ b) 26 seconds
7. a) $AC \approx 4.24$ cm
   b) Perimeter is approximately 12.4 cm.
8. a) $\angle ACB \approx 49.1°$ b) $BC \approx 301$ m   c) $4.85\,\text{ms}^{-1}$

9. a) $\angle DBA \approx 30.0°$ b) $AD \approx 2.83$ km   c) ≈ 22.2 minutes
10. $\angle BCA \approx 64.0°$, $\angle BAC \approx 57.1°$, $BD \approx 2.56$, $\angle CBD \approx 45.4°$,
   $\angle BDC \approx 94.0°$
11. a) (i) 4.70   (ii) 2.85   (iii) 12.35
   b) (i) $AB \approx 5$, $\angle CAB \approx 71°$, $\angle ACB \approx 74°$
     (ii) 4.92   (iii) 2.98   (iv) 12.9
   c) 4.45%
12. a) $\angle ACB \approx 39.2°$ b) 4.28×10^4 light years
   c) $BC \approx 4.05 \times 10^{20}$ m

Exercise 6.4

1. a) $AC \approx 19.7$ cm   b) $AC \approx 7.97$ m
2. a) $AC \approx 1840$ m   b) $AC \approx 60.3$ feet
3. a) $\angle BAC \approx 45.2°$ b) $\angle ABC \approx 91.8°$
4. a) $BC \approx 20.3$ mm   b) $BC \approx 234$ m
5. a) $BD \approx 17.0$ cm   b) $\angle BDC \approx 71.7°$
6. a) ≈ 13.0 km   b) 64 minutes
7. a) 33°   b) ≈ 13.6 feet
   c) 41.9 feet   d) \$326.25
8. a) 1236 cm   b) 47 years
9. a) 3.1 m   b) 6 seconds
10. a) $c \approx 4.95$ m   b) $v \approx 6.97\,\text{ms}^{-1}$
11. a) 5.66 feet   b) $BC \approx 5.66$ feet   c) 90°   d) 45°
12. a) 70 m   b) $\angle ABC = 44.4°$

Exercise 6.5

1. a) $A \approx 62.4\,\text{m}^2$ b) $A \approx 260\,000\,\text{km}^2$
2. a) $A \approx 673\,\text{cm}^2$ b) $A \approx 5.37\,\text{m}^2$
3. a) $A \approx 0.945\,\text{feet}^2$ b) $A \approx 51.7\,\text{inches}^2$
4. a) $A \approx 9.35\,\text{m}^2$ b) $A \approx 493\,\text{dm}^2$
5. a) $A \approx 3.13\,\text{miles}^2$ b) $A \approx 14.9\,\text{cm}^2$
6. a) $A \approx 55.5\,\text{m}^2$ b) 46 outfits.
7. a) $A \approx 16.6\,\text{m}^2$ b) 6 km per hour.
8. a) $A \approx 693\,\text{cm}^2$ b) 1.9%
9. a) $BD \approx 9.10$ feet   b) $A \approx 39.5\,\text{feet}^2$
   c) $\angle BCD \approx 54.0°$ d) $A \approx 40.1\,\text{feet}^2$
10. a) $124\,000\,\text{feet}^2$ b) Approximately 1880 blocks
11. a)

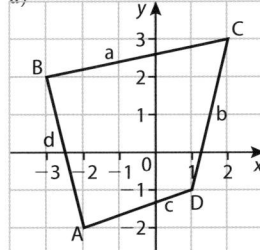

   b) (i) $\sqrt{17}$
     (ii) $\sqrt{10}$
     (iii) 5
     (iv) $\sqrt{17}$
   c) (i) $\angle BAD \approx 85.6°$
       $BC = \sqrt{26}$
     (ii) $\angle BCD \approx 67.2°$
   d) 16 square units
12. a) (i) 6.03   (ii) 6.03
     (iii) $AB = \sqrt{(0 - 3.02)^2 + (-2.5 + 0.32)^2} \approx 3.72$
   b) (i) 36°   (ii) 36°   (iii) 36°
   c) (i) 6.60   (ii) 10.7   (iii) 6.60
   d) 23.9 square units

Exercise 6.6

1. a) $V = 1800 \text{ m}^3$
   $A = 900 \text{ m}^2$
   b) $V = 3.36 \text{ cm}^3$
   $A = 13.6 \text{ cm}^2$
   c) $V = 18\,100\,000 \text{ mm}^3$
   $A = 415\,000 \text{ mm}^2$

2. a) $V = 140 \text{ feet}^3$
   $A = 150 \text{ feet}^2$
   b) $V = 330\,000\,000 \text{ dm}^3$
   $A = 2\,700\,000 \text{ dm}^2$
   c) $V = 50 \text{ inches}^3$
   $A = 75 \text{ inches}^2$

3. a) $V = 3156.55 \text{ m}^3$
   $A = 1040.62 \text{ m}^2$
   b) $V = 7238.23 \text{ inches}^3$
   $A = 1809.56 \text{ inches}^2$
   c) $V = 33.51 \text{ cm}^3$
   $A = 50.27 \text{ cm}^2$

4. a) $V = 61.7 \text{ cm}^3$
   $A = 102 \text{ cm}^2$
   b) $V = 2480 \text{ mm}^3$
   $A = 1210 \text{ mm}^2$

5. a) $V = 16.4 \text{ cm}^3$
   $A = 104 \text{ cm}^2$
   b) $V = 226.4 \text{ m}^3$
   $A = 246.3 \text{ m}^2$

6. a) $V = 1\,090\,000\,000\,000 \text{ km}^3$
   b) $V = 1.09 \times 10^{12} \text{ km}^3$
   c) $A = 510\,000\,000 \text{ km}^2$

7. a) $A \approx 1.73 \text{ cm}^2$
   b) $V \approx 6.93 \text{ cm}^3$
   c) Cost = $60,621.78$

8. a) $V \approx 65.7 \text{ cm}^3$ b) $V = 1370 \text{ cm}^3$ c) $A = 675 \text{ cm}^2$

9. a) $V \approx 15\,700 \text{ feet}^3$
   b) $V \approx 16\,500 \text{ feet}^3$
   c) Approximately 74 500

10. $V \approx 56\,200 \text{ m}^3$

11. a) $V = \frac{4}{3}\pi(1.7)^3 \approx 20.6 \text{ cm}^3$
    b) $V = \pi(6.8)^2(3.4 \times 3.8) \approx 1880 \text{ cm}^3$
    c) $A = 4\pi(1.7)^2 \approx 36.3 \text{ cm}^2$
    d) $\frac{V_{hole}}{V_{ball}} = 91.2$, so approximately 91 balls.
    e) $20\,000 \times 91 = 1\,820\,000$

12. a) (i) $AD^2 + FD^2 = AF^2$
    $12^2 + 5^2 = AF^2$
    $AF = 13 \text{ cm.}$
    (ii) $AF^2 + FM^2 = AM^2$
    $13^2 + 10^2 = AM^2$
    $AM = 16.4 \text{ cm}$
    b) $\sin(\angle MAN) = \frac{MN}{AM}$
    $\sin(\angle MAN) \approx \frac{5}{16.4012}$
    $\angle MAN \approx \sin^{-1}\left(\frac{5}{16.4012}\right)$
    $\angle MAN \approx 17.7°$
    c) Volume = area of base times height
    $V = \frac{1}{2}AD \times FD \times DC$
    $V = \frac{1}{2}(12 \text{ cm})(5 \text{ cm})(20 \text{ cm})$
    $V = 600 \text{ cm}^3$

13. a) 980 cm^3 (3 s.f.)
    b) 641 cm^2 (3 s.f.)

14. a) 4.5 metres
    b) 9.5 metres
    c) 636 m^2 (3 s.f.)

Exercise 6.7

1. a) $AG = 26.2 \text{ cm}$
   b) $AG = 8.69 \text{ m}$
2. a) $36.9°$
   b) $22.6°$
3. a) $\angle FAE \approx 51.3°$
   b) $\angle FAE \approx 54.7°$

4. a) $\approx 102.1°$
   b) $63.5°$
5. a) $\angle AMD \approx 63.4°$
   b) $\angle AMC \approx 54.2°$
6. $60.9°$
7. $14.5°$
8. $23.6°$
9. $69.1°$
10. a) $44.4°$
    b) 21.7 feet
11. a) $\sqrt{164}$
    b) $\sqrt{164}$
    c) M should be drawn be at the centre of the rectangular prism.
    d) $\angle BMA \approx 55.9°$
12. a) (i) $GM = 4 \text{ cm}$ since it is half of SR.
    (ii) $GM^2 + VG^2 + VM^2$
    $4^2 + 12^2 + VM^2$
    $VM \approx 12.6 \text{ cm}$
    b) (i) Surface area = area of base + 4 times area of each triangle face
    $S \approx (8 \text{ cm})^2 + 4\left(\frac{1}{2}(8 \text{ cm})(12.64911 \text{ cm})\right)$
    $S \approx 266 \text{ cm}^2$
    (ii) $\tan(\angle VMG) = \frac{VG}{GM}$
    $\angle VMG = \tan^{-1}\left(\frac{12 \text{ cm}}{4 \text{ cm}}\right)$
    $\angle VMG \approx 71.6°$
    (iii) $V = \frac{1}{3} \times$ area of base $\times$ height
    $V = \frac{1}{3}(8 \text{ cm})^2(12 \text{ cm})$
    $V = 256 \text{ cm}^3$

Chapter 7
Exercise 7.1

1. a, d, e, f
2. a) 2, 10, 18, 26; arithmetic
   b) 5, 10, 20, 40
   c) $-3, 6, -3, 6$
   d) $1, -1, 1, -1$
   e) $-3, 2, 7, 12$; arithmetic
   f) $4, 2, \sqrt{2}, \sqrt[4]{2}$
3. a) $u_1 = 2, u_{n+1} = u_n + 2$
   b) $u_1 = 15, u_{n+1} = u_n - 4$
   c) $u_1 = 2, u_{n+1} = 2(u_n)$
   d) $u_1 = 1, u_{n+1} = \frac{1}{2}u_n$
   e) $u_1 = 1, u_{n+1} = \left(\sqrt[3]{u_n} + 1\right)^3$
   f) $u_1 = 5, u_{n+1} = u_n + 2$
4. a) 61, 8
   b) $-55, -15$
   c) $-73, -12$
   d) 54, 7
   e) 79, 10
   f) $-32, -4$
5. a) 75
   b) -17
   c) 1,888
   d) 896
   e) $-7,924$
   f) 38
6. a) 151
   b) 921
   c) 624
   d) $\frac{251}{42}$
7. a) -7
   b) 45
   c) -251
   d) 223
8. a) 10, 17
   b) $-51, -23, 33, 61, 89$
   c) $\frac{3}{2}, \frac{11}{4}$
   d) $\frac{41}{96}$
9. a) 8
   b) $\frac{-9}{2}$
10. a) 5, 8, 11, 14
    b) $-33, -37, -41, -45, -49$
11. 1, 1, 2, 3, 5, 8, 13, 21, 34, 55, 89, 144, 233, 377, 610
12. a) 1, 2, 1.5, 1.666667, 1.6, 1.625, 1.615385, 1.619048, 1.617647, 1.618182, 1.617978, 1.618056, 1.618026, 1.618037
    b) Increasing, decreasing … c) The golden ratio

13.

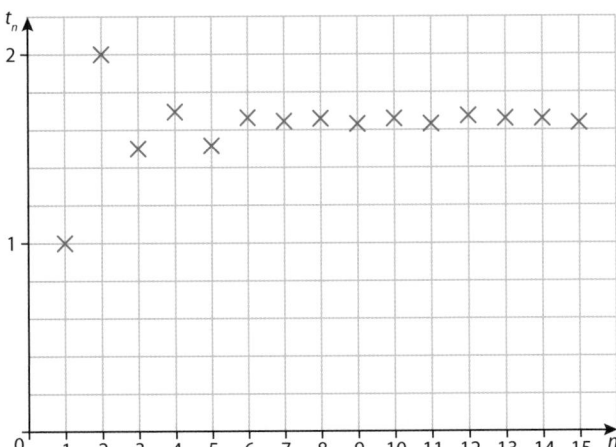

14. The Golden ratio (The Golden Section) $= \dfrac{1 + \sqrt{5}}{2} =$
1.618 034 to 7 s.f. Therefore the sequence generated in Q 12
seems to be approaching the Golden Ratio.

15. a) 5 years: 2700€
b) 10 years: 3400€
c) 30 years: 6200€

16. $u_{24} = 22$

17. $21

18. 15 years

Exercises 7.2

1. a) 500 500
b) 15 555
c) 154 200
d) −111 500
e) 55 909
f) 22 450

2. a) $(1 + 7) + (2 + 7) + (3 + 7) + (4 + 7) = 8 + 9 +$
$10 + 11$
b) $(2 \cdot 1 - 1) + (2 \cdot 2 - 1) + (2 \cdot 3 - 1) + (2 \cdot 4 - 1) +$
$(2 \cdot 5 - 1) = 1 + 3 + 5 + 7 + 9$
c) $(2 - 3 \cdot 1) + (2 - 3 \cdot 2) + (2 - 3 \cdot 3) + (2 - 3 \cdot 4) +$
$(2 - 3 \cdot 5) + (2 - 3 \cdot 6) = -1 - 4 - 7 - 10 - 13 - 16$
d) $5 \cdot 1 + 5 \cdot 2 + 5 \cdot 3 + 5 \cdot 4 + 5 \cdot 5 + 5 \cdot 6 + 5 \cdot 7 = 5 + 10 +$
$15 + 20 + 25 + 30 + 35$

3. a) $\displaystyle\sum_{i=1}^{6}(13i - 11)$ b) $\displaystyle\sum_{i=1}^{5}(17 - 19i)$

c) $\displaystyle\sum_{i=1}^{50}(2i - 1)$ d) $\displaystyle\sum_{i=1}^{75}(5i + 12)$

4. a) $u_1 = 2$ b) $u_1 = -2$ c) $u_1 = 9$
$u_{100} = 299$ $u_{120} = -835$ $u_{50} = 450$
$d = 3$ $d = -7$ $d = 9$
d) $u_1 = -12$ e) $u_1 = 1$ f) $u_1 = 1$
$u_{120} = -1440$ $u_n = n$ $u_n = 2n - 1$
$d = -12$ $d = 1$ $d = 2$

5. a) 15 050   b) −50 220   c) 11 475
d) −87 120   e) $\dfrac{n(n + 1)}{2}$ f) n^2

6. a) 855   b) −98 800   c) 2710   d) 272

7. 1035

8. $300

9. a) (i) $42 = 6a + 15d$
(ii) $210 = 14a + 91d$
b) (i) $a = 2$ (ii) $d = 2$
(iii) sequence of even numbers

10. a) constant addition of $0.50
b) $u_1 = \$0.50\ d = \$0.50$
c) 136 coins ($68)
d) 63 days

Exercises 7.3

1. b, c, d, f

2. a) 3, 6, 12, 24; geometric
b) 100, 50, 25, 12.5; geometric
c) 1, 4, 9, 16
d) 4, −12, 36, −108; geometric
e) 2, 4, 8, 32
f) $1, \frac{1}{3}, \frac{1}{9}, \frac{1}{27}$; geometric

3. a) $u_1 = 2$ b) $u_1 = 1$
$u_{n+1} = 3u_n$ $u_{n+1} = \frac{1}{10}u_n$
c) $u_1 = 64$ d) $u_1 = 100$
$u_{n+1} = \frac{-1}{2}u_n$ $u_{n+1} = 1.08u_n$

4. a) $128; r = -2$ b) $\frac{1}{160}; r = \frac{1}{2}$
c) $\frac{243}{64}; r = \frac{3}{2}$ d) $\frac{-32}{3}; r = -2$

5. a) $u_{10} = 64$
b) $u_{10} = 512$
c) $u_{30} = 711.43 = 711$ to 3 s.f.
d) $u_{12} = 244\,140\,625$

6. a) $u_4 = 104$ b) $u_1 = 3$ c) $n = 10$ d) $r = \frac{3}{4}$

7. a) $n = 13$
b) $n = 5$
c) $n = \dfrac{\log 30}{\log 3} = 3.10$ to 3 s.f.
d) $n = \dfrac{\log 20}{\log 2} = 4.32$ to 3 s.f.

8. a) $n = \dfrac{\log 2}{\log 1.06} = 11.9$ to 3 s.f.
b) $n = 21.9$ to 3 s.f.
c) $r = 0.0760$ to 3 s.f.
d) $r = 0.0444$ to 3 s.f.
e) $n = \dfrac{\log 17}{\log 3} + 1 = 3.58$ to 3 s.f.
f) $n = \dfrac{\log 37}{\log 3} - 1 = 2.29$ to 3 s.f.
g) $r = \pm 2$
h) $r = \sqrt[5]{112} = 2.57$ to 3 s.f.
i) $r = \sqrt[7]{317} = 2.28$ to 3 s.f.
j) $r = 4.54$ to 3 s.f.

9. a) $u_1 = 5$ b) $u_1 = 1$
c) $u_1 = 4$ d) $u_1 = \pm\frac{1}{2}$

10. a) $r = 8$
b) $-0.05, 0.1, -0.2, \mathbf{0.4}, -0.8, 1.6, -3.2, 6.4, \mathbf{-12.8}$
c) $10\,000, 8000, \mathbf{6400}, 5120, 4096, 3276.8, 2621.44, 2097.152,$
$\mathbf{1677.7216}$
$10\,000, -8000, \mathbf{6400}, -5120, 4096, -3276.8, 2621.44,$
$-2097.152, \mathbf{1677.7216}$
d) $13, \mathbf{39}, 117, 351, 1053, \mathbf{3159}$
$-13, \mathbf{39}, -117, 351, -1053, \mathbf{3159}$

11. a) $x = 1$ b) $x = 2$ c) Answers may vary

12. a) 6, 12

b) $2\sqrt[3]{6}, 2\sqrt[3]{36}$

c) $r = 2: 2, 4, 8$

$r = -2: -2, 4, -8$

d) $r = \sqrt[4]{10}: \sqrt[4]{10}, \sqrt[4]{100}, \sqrt[4]{1000}$

$r = -\sqrt[4]{10}: -\sqrt[4]{10}, \sqrt[4]{100}, -\sqrt[4]{1000}$

13. a) \$5.12 b) \$163.84 c) \$5242.88

d) \$10 485.76 e) \$5 368 709.12

14. a) $44\,100 = 40\,000 \cdot r^2$

Hence, the population at the end of 1997 was

$40\,000(1.05) = 42\,000$.

b) $40\,000 = u_1(1.05)^4$

Hence, the population at the end of 1992 was

$\dfrac{40\,000}{1.05^4} = 32\,908$ to the nearest person.

Exercises 7.4

1. a) 65,535

b) $\dfrac{29\,524}{243} = 121.50$ to 2 d.p.

c) 81 380 208

d) $\dfrac{255}{128} = 1.99$ to 3 s.f.

e) 78 432

f) $-1\,627\,604$

2. a) $2 + 4 + 8 + 16 + 32$

b) $1 + 3 + 9 + 27 + 81 + 243$

c) $50 + 250 + 1250 + 6250$

d) $1 + \frac{1}{2} + \frac{1}{4} + \frac{1}{8} + \frac{1}{16} + \frac{1}{32} + \frac{1}{64}$

e) $u_1 + u_2 + u_3 + \ldots + u_n$

f) $u_1 + u_1 r + u_1 r^2 + \ldots + u_1 r^{n-1}$

3. a) $\displaystyle\sum_{i=1}^{7}(9 \cdot 2^{i-1})$ b) $\displaystyle\sum_{i=1}^{6}(-3 \cdot 4^{i-1})$

c) $\displaystyle\sum_{i=1}^{8}(13 \cdot (-6)^{i-1})$ d) $\displaystyle\sum_{i=1}^{7}\left(\frac{8}{3} \cdot \left(\frac{3}{4}\right)^{i-1}\right)$

4. a) $u_1 = 3$ b) $u_1 = 1$

$u_4 = 81$ $u_5 = \frac{1}{625}$

$r = 3$ $r = 0.2 = \frac{1}{5}$

c) $u_1 = 8$ d) $u_1 = \frac{-3}{2}$

$u_9 = 2048$ $u_6 = \frac{-3}{64}$

$r = 2$ $r = \frac{1}{2}$

e) $u_1 = u_1$

$u_n = u_1 r^{n-1}$

$r = r$

5. a) 120

b) $\frac{781}{625} = 1.2496$

c) 4088

d) $\frac{-189}{64} = 2.953\,125$

e) $S_n = u_1 \cdot \dfrac{r^n + 1}{r - 1}$

6. a) $S_1 = 1, S_2 = 4, S_3 = 13, S_4 = 40, S_5 = 121, S_6 = 364$

b) They are getting larger.

c) Yes

d) By larger amounts

7. a) $S_1 = 1, S_2 = 1.5, S_3 = 1.75, S_4 = 1.875, S_5 = 1.9375,$

$S_6 = 1.968\,75, S_7 = 1.984\,375, S_8 = 1.992\,1875$

b) They are getting larger. c) Yes

d) By smaller amounts e) 2

8. a) $S_1 = 1, S_2 = 1.333\,3333, S_3 = 1.444\,4444, S_4 = 1.481\,481,$

$S_5 = 1.493\,827, S_6 = 1.497\,942, S_7 = 1.499\,314,$

$S_8 = 1.499\,771$

b) They are getting larger.

c) Yes

d) By smaller amounts.

e) 1.5

f) No

g) 1.5 is a limiting value.

9. a) $0.1 = \frac{1}{10}$

b) $S_4 = 0.3333$

c) Percentage error $= \dfrac{0.3333 - \frac{3}{9}}{\frac{3}{9}} \times 100 = 0.01\%$

10. a) $n = 0.7 + 0.77 + 0.777 + \ldots$

b) $S_5 = 0.777\,77$

c) Percentage error $= \dfrac{0.777\,77 - \frac{7}{9}}{\frac{7}{9}} \times 100 = 0.001\%$

11. You can rewrite a decimal number that is repeating in "blocks" of 1 by dividing one of the blocks by 9.

12. a) $\frac{5}{9}$ b) $\frac{9}{9} = 1.$ (Yes, **equal to** one!)

c) $\frac{25}{99}$ d) $\frac{36}{99} = \frac{4}{11}$

13. a) After you go halfway to the doorway, you must then continue going half the distance from each successive halfway point, thus never (mathematically) reaching the doorway.

b) The threshold is the limiting value of the nth term of the sequence: 4, 2, 1, 0.5, 0.25, 0.125, 0.625, …, which is 0. (Using calculus notation,

$u_n = \displaystyle\lim_{n \to 8} 4\left(\frac{1}{2}\right)^{n-1} = 4 \lim_{n \to 8} \frac{1}{2^{n-1}} = 4 \cdot 0 = 0$)

14. \$10,737,418.23

15. a) Answers vary

b) $1.125\,899\,91 \times 10^{13}$ cm

c) 112 589 991 km

d) This is about $\frac{3}{4}$ of the distance from the Earth to the sun.

16. a) (i) 153 (ii) 900 (iii) 3600

b) No. She cannot walk 153 miles on the 21st day, and probably cannot walk 900 miles in 21 days.

c) Answers vary. Decrease the total days walked, or decrease the % increase for the next day, or decrease the number of miles walked the first day, or a combination of all three.

17. a) 0.95.

b) 1.1376 m

c) After 59 bounces

Exercise 7.5

1. a) \$2762.82 b) \$1025

c) \$6288.95 d) \$1862.13

2. a) £40 317.49 b) £77 812.27

c) £33 800.75 d) £185 752.09

3. a) ¥11 982.79 b) ¥12216.10

c) ¥12 270.47

4. a) \$9450.79 b) \$6605.40

c) \$11 081.80 d) \$7901.03

5. Option 1 is £2000, option 2 is £1814.49, so option 1

6. Approximately 5.32%.

7. £8615.24

8. £9033.86

9. a) €3440 b) €3819.72

10. a) $P = X\left(1 + \dfrac{0.6}{1200}\right)^{12}$ or $P = X(1.005)^{12}$

   b) 6.17%

11.

Month	Balance at beginning of month	Interest earned during month	Balance at end of month
January	600	4.50	604.50
February	1904.5	14.28	1918.78
March	2148.78	16.12	2164.90
April	2874.90	21.56	2896.46

a) AUD 1918.78

b) AUD 2896.46

c) $C = 3074.88$

Hence, there will be AUD 3074.88 in her account at the end of December 1999.

d)

Year	Balance at beginning of year	Interest earned during year	Balance at end of year
2000	3074.88	107.62	3182.50
2001	3182.50	111.38	3293.88
2002	3293.88	115.29	3409.17

So from our table we can see that at the end of the year 2002, her balance will finally exceed AUD 3300. Hence it will take 3 full years.

Exercise 7.6

1. a) FV = 14 344.63 b) I% = 6.94
   c) PV = −74 062.76 d) PMT = −624.11

2. a) $1508.49 b) $1532.33
   c) $1537.76

3. a) £1521.31 b) £1602.94
   c) £1718.47 d) £1842.11

4. a) $ 306.75 b) $ 490.80
   c) $ 920.25 d) $ 1840.51

5. a) 417 months b) 168 months
   c) 84 months d) 43 months

6. a) 4.52% b) 8 years 5 months

7. a) $94 629.66 b) $1 124 000

8. 5.5%

9. Option 1: $ 3083.55, option 2: $ 3091.02, option 3: $ 3097.50, so option 3 is best.

10. 8.63%

11. a) $1276.28
   b) 15 years.
   c) 7.0 %

12. a) For each account, we will calculate 1 year's worth of interest on $1000.
   Account A: $ 1040.74
   Account B: $ 1047.50

Account C: $ 1045.77

So as long as we deposit our money for at least a year, account B is the best deal.

   b) $I = $ $ 6137.26

   c) $ 93 934.49

13. a) (i) $6.7 \times 10^9 \times 2^1 = 1.34 \times 10^{10}$ people
      (ii) $6.7 \times 10^9 \times 2^5 \approx 2.14 \times 10^{11}$ people
      (iii) $6.7 \times 10^9 \times 2^{\frac{1}{6}} \approx 7.52 \times 10^9$ people

   b) (i) $ $6.7 \times 10^9 \times 2^1 = $ 1.34×10^{10} US
      (ii) Using TVM solver with n = 30, PV = -6.7×10^9, PMT = 0, FV = 1.34×10^{10}, P/Y = 1 and C/Y = 1, solving for I%. Therefore I% ≈ 2.34%

   c) (i) The two answers are the same.
      (ii) Since Ms. Rich's money increases by 2.34% each year, so must the human population.

Exercise 7.7

1. a) $ 10 404 b) $ 11 040.81
   c) $ 12189.94 d) $ 14859.47

2. a) $ 975.20 b) $ 1075.13
   c) $ 1303.12 d) $ 2074.99

3. a) $ 335.04 b) £485.80
   c) $ 12 563.87 d) €8.21

4. a) $ 2.12 b) $ 1.87 c) $ 1.29 d) $ 0.70

5. a) 1.84% b) 3.05% c) 5.75% d) 7.18%

6. a) $ 1037.50 b) $ 549.90

7. $312.48

8. $14 171.76

9. 325 778.93 SAR

10. −20.6% per year.

11. a) $ 392 211.13 b) $48 381.66 less money

12. a) 9.6% b) 4 times.

13. a) $ 20 214.62 b) $ 10 628.82
   c) Difference = $ 11 894.43 d) $ 142

14. a) (i) $C = 47 009.78$ Euros
      (ii) 4.594 %.
   b) 294 months.
   c) 4.14 %.
   d) $\dfrac{50\,000}{0.667} = \$74\,962.52$
   e) $C = \$44\,882.83$
   f) (i) $C = \$91\,378.89$
      (ii) $46 496.06

Chapter 8

Exercise 8.1

1. a) {2, 3, 5, 7, 11, 13, 17, 19, 23, 29, 31, 37}
   b) {3, 6, 9, 12, 15, 18, 21, 24, 27, 30, 33, 36}
   c) {Africa, Antarctica, Asia, Australia, Europe, North America, South America}
   d) {a, e, i, o, u}

2. a) 12 b) 12 c) 24 d) 10

3. a) 90 b) 20 c) 25 d) 99

4. a) {0, 1, 2, 3, 4, 5, 6, 7}
   b) {−3, −2, −1, 0, 1, 2, 3}
   c) {2, 3, 4, 5, 6, 7, 8}
   d) {2, 3, 5, 7, 11, 13, 17, 19, 23, 29, 31}

5. a) $\{x : x \in \mathbb{Z}, -10 < x < 10\}$
   b) $\{x : x \in \mathbb{Q}, -1 < x < 2\}$

c) $\{x: x \in \mathbb{R}, x \geqslant 7\}$
d) $\{x: x \in \mathbb{Q}', 0 < x < \pi\}$

6. a) $\{5, 7, 9, 11, 13, 15\}$
   b) $\{1, 2, 4, 5, 10, 20, 25, 50, 100\}$
   c) $\{ACE, AEC, EAC, ECA, CAE, CEA\}$
   d) $\{21, 22, 24, 25, 26, 27, 28, 30, 32, 33, 34\}$

7. a) True   b) True   c) False   d) False

Exercise 8.2

1. a) Not a subset   b) Subset
   c) Subset   d) Subset

2. a) $\{\}, \{1\}, \{2\}, \{3\}, \{1, 2\}, \{2, 3\}, \{1, 3\}$
   b) $\{\}, \{A\}, \{B\}, \{C\}, \{A, B\}, \{A, C\}, \{B, C\}$
   c) $\{\}, \{10\}, \{20\}$
   d) $\{\}, \{blue\}, \{red\}, \{yellow\}, \{green\}, \{blue, red\},$
      $\{blue, yellow\}, \{blue, green\}, \{red, yellow\}, \{red, green\},$
      $\{yellow, green\}, \{blue, red, yellow\}, \{blue, red, green\},$
      $\{blue, yellow, green\}, \{red, yellow, green\}$

3. a) $A' = \{1, 4, 6, 8, 9, 10, 12, 14, 15, 16, 18, 20\}$
   b) $B' = \{1, 3, 5, 7, 9, 11, 13, 15, 17, 18, 19, 20\}$
   c) $A' = \{9, 10, 11, 12, 13, 14, 15, 16, 17, 18, 19, 20\}$
   d) $C' = \{1, 2, 3, 5, 7, 11, 13, 17, 19, 20\}$

4. a) $\{x: x \in \mathbb{Z}, x \leqslant 0\}$
   b) $\{x: x \in \mathbb{Z}, x \leqslant 10 \text{ or } x \geqslant 20\}$
   c) $\{x: x \in \mathbb{R}, x < 1 \text{ or } x > 2\}$
   d) $\{x: x \in \mathbb{Q}', x < -4\}$ or $\{x: x \in \mathbb{R}, x \geqslant -4\}$

5. a) $n(A') = 44$ b) $n(B') = 21$
   c) $n(C') = 58$ d) $n(D') = 83$

6. Each element is either in a particular subset or it isn't.
   Therefore there are 2 choices for each element and $2^5 = 32$
   possible subsets.

7. Using the same process as in question 6, and knowing that
   there are 25 prime numbers less than 100, we get 2^{25}.

8. a) True   b) False   c) True   d) False

Exercise 8.3

1. a) $\{1, 2, 3, 4, 5, 6\}$ b) $\{3, 4, 5, 6, 7, 8, 12, 16\}$
   c) $\{0, 1, 2, 3, 4, 6, 9, 12, 18, 36\}$ d) $\mathbb{R}$

2. a) $\{2\}$ b) $\{4\}$ c) $\{1, 4, 9\}$ d) $\{\}$

3. a) $\{6, 8, 10\}$ b) $\{0, 2, 4, 6, 8\}$
   c) $\{-1\}$ d) $\{9\}$

4. a) 18   b) 3   c) 18   d) 12

5. a) False   b) True   c) False   d) False

6. a) $\{1, 2, 3, 4, 5, 6, 7\}$ b) $\{-6, -5, -4, -3, -2, -1\}$
   c) $\{\}$ d) 13

7. a) Answers may vary. $A = \{2, 3, 5, 7, 11, 13\}$
   b) (i) $A = \{2, 3, 5, 7\}$ and $B = \{11, 13, 17, 19\}$
      (ii) $A = \{2, 3, 5, 7, 11\}$ and $B = \{2, 3, 5, 7, 11\}$

Exercise 8.4

1. a) 18   b) 20   c) 30   d) 8
2. a) 56   b) 47   c) 22   d) 81
3.

4.

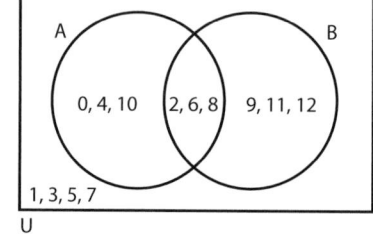

5.

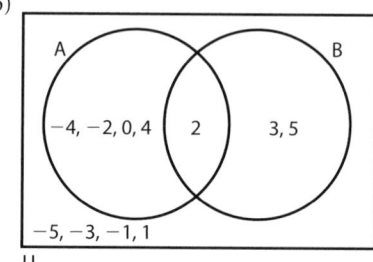

6. a)

b) $n((A \cup B)') = 4$

7. a) $U = \{-5, -4, -3, -2, -1, 0, 1, 2, 3, 4, 5\}$
   $A = \{-4, -2, 0, 2, 4\}$
   $B = \{2, 3, 5\}$
   b)

c) $n(A \cup B) = 7$

8.

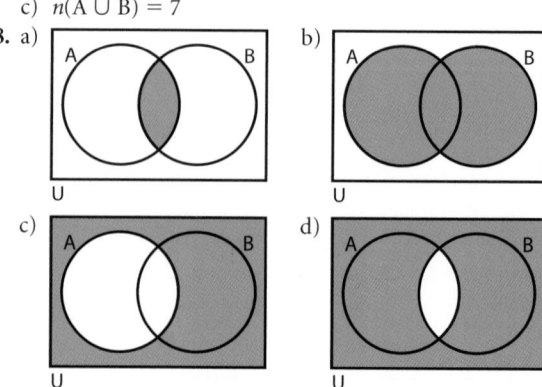

9. a) A   b) $(A \cup B)'$
   c) $(A \cap B)' \cap (A \cup B)$ or $(A \cap B') \cup (A' \cap B)$
   d) $(A \cap B) \cup (A \cup B)'$

10. a) A   b) B

     c) $A \cap B'$ d) $A' \cup B$

11. a) (i) These are all the prime numbers in U – {2, 3, 5, 7, 11, 13, 17, 19, 23}

        (ii) These are the factors of 24 in U – {1, 2, 3, 4, 6, 8, 12, 24}

     b)

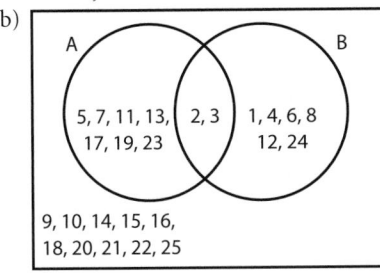

     c) We need to shade the portion where set A overlaps with set B)

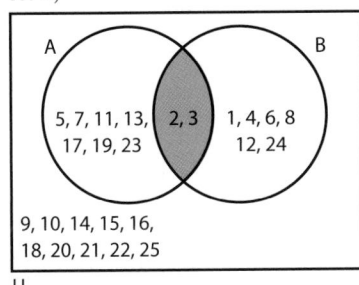

     d) The number of elements in A ∪ B is 15, and the number of elements in U is 25.

        Therefore A ∪ B is $\frac{15}{25} \times 100 = 60\%$ of U.

Exercise 8.5

1. a) 25   b) 25

2. a)

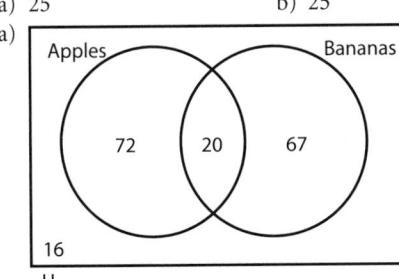

     b) 83

3. a)

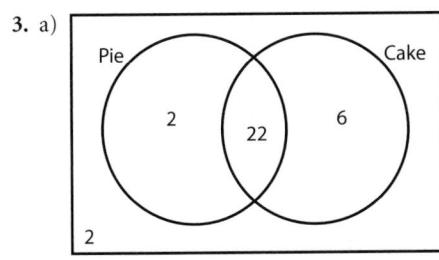

     b) 22

4. a)

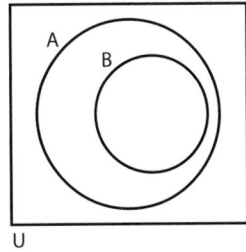

     b) 120

     c) 200

5. a)

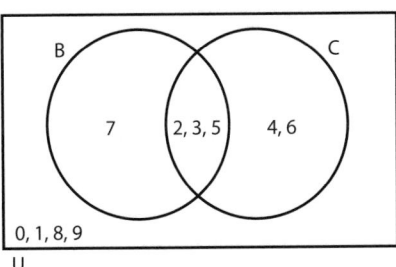

     b) B is a subset of A.

     c) (i)

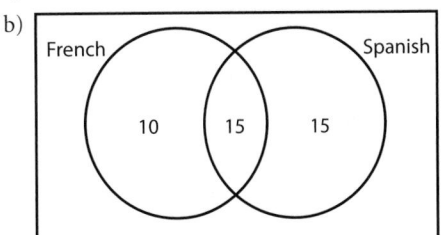

        (ii) The planets in the solar system that are not gas giants.

6. a) (i) {2, 3, 5, 7}

        (ii) {2, 3, 5}

        (iii) {0, 1, 2, 3, 5, 7, 8, 9}

     b)

7. a) 15 students

     b)

c)
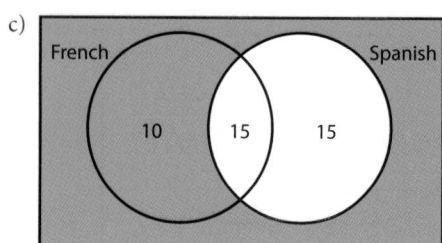
Foreign language students

8. a) (i) {0, 1, 2, 3, 4, 5, 6, 7, 8, 9, 10, 11, 12, 13}
(ii) { 4, 5, 6, 7, 8, 9, 10, 11, 12, 13, 14}

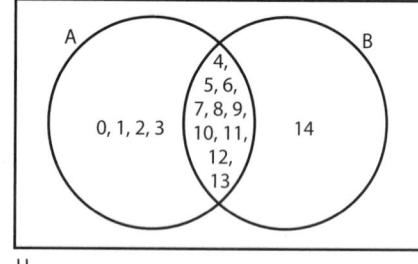

b) ∪
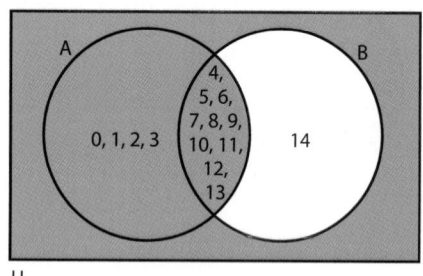

c) See diagram in part b). Shaded portion is $(A \cup B')$.

9. a)
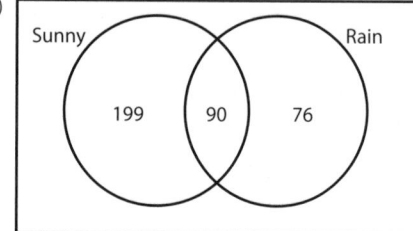
Climate in Vancouver

b) 90 days   c) 54.5%

10. a) $30 - x + 20 - x + 8 + x = 50$, hence 13 movies.

b)
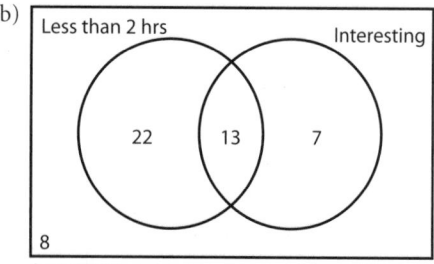
Last 50 movies

c) $20 \times \$100 + 30 \times \$30 = \$2900$, so hence $2900 million.
d) $35 \times \$50 + 15 \times \$70 = \$2800$, so hence $2800 million.

11. a) We use the formula $28 + 30 + x + x = 100$ to solve for the percentage of people who only recycle paper and who only glass. Therefore $x = 21$.

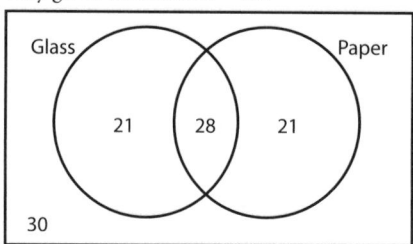
Recycling in the city

b) 21% of 8 million is $\frac{21}{100} \times 8$ million $= 1\,680\,000$ people.
c) $200\,000$ is 2.5% of the city's population, hence $28 + 2.5 = 30.5\%$.
d) The percent increase is always given as
$$\frac{\text{New} - \text{Original}}{\text{Original}} \times 100 = \text{percent increase}.$$
Hence $\frac{72.5 - 70}{70} \times 100 = 3.6\%$.

e) Since the error is $\pm 1\%$ in any category, the minimum percentage of people who are *not* recycling is 29%.
Hence $\frac{71}{100} \times 8$ million $= 5.68 \times 10^6$ people.

Exercise 8.6

1. a) 35   b) 60   c) 8   d) 44
2. a) 55   b) 20   c) 90   d) 35
3. a) 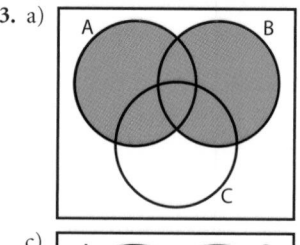   b)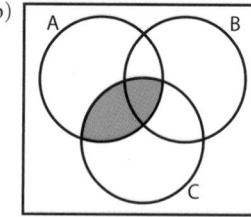

c) 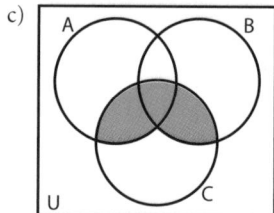   d)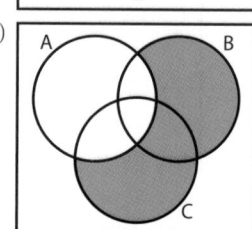

4. a) 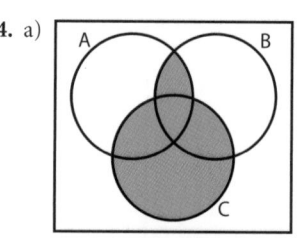   b)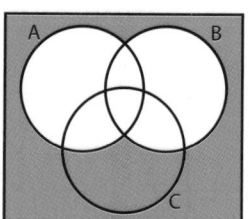

c) 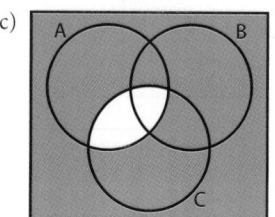   d)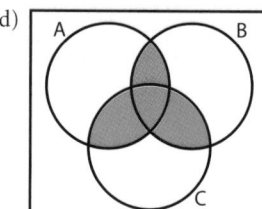

5. a)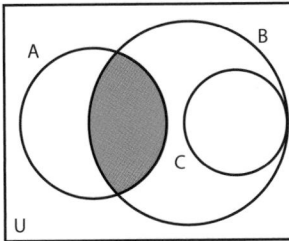

b)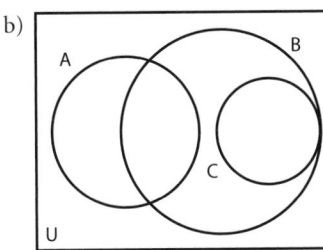

6. a) $A \cap B$
b) $(A \cup B)'$
c) $(A' \cap B)$
d) $(A \cup B) \cap C$

7. a)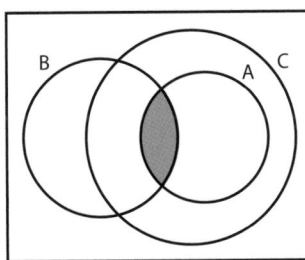

b)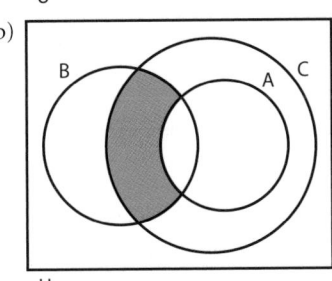

c) Set A is a proper subset of set C)

8. a) 18%   b) 48   c) 1008

9. a)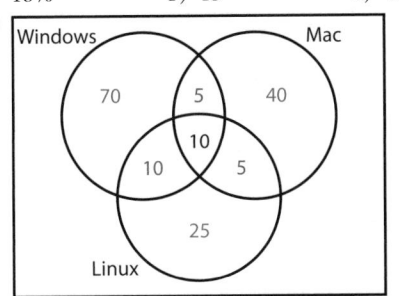

b) 135 students   c) 20 students

10. a)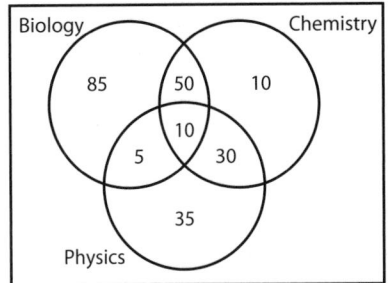

b) 10   c) 125   d) 35

11. a)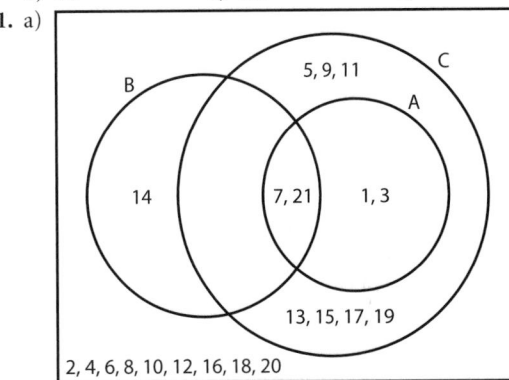

2, 4, 6, 8, 10, 12, 16, 18, 20

b) $\{1, 3, 7, 21\}$
c) (i) $\{1, 3, 7, 14, 21\}$ (ii) $\{14\}$
(iii) $\{1, 3, 5, 9, 11, 13, 15, 17, 19\}$

12. a)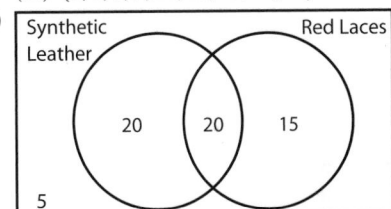

60 models of shoes

b) From our diagram the company has 20 models of shoes with both synthetic leather and red laces.

c)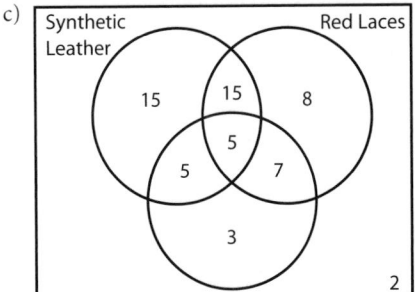

Gel packed soles

d)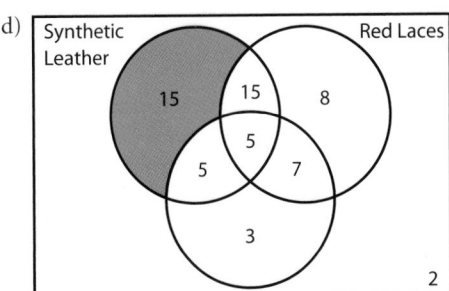

Gel packed soles

e) From above we can see that this is 5 out of 60 shoes, which is $\frac{5}{60} \times 100$ which is 8.3% when rounded to the nearest tenth.

Chapter 9

Exercise 9.1

1. a) Opinion   b) Proposition
   c) Opinion   d) Proposition
2. a) P or Q.   b) Not P and not Q.
   c) If and only if P then Q.
3. a) If the clock has stopped then I will be late.
   b) The clock has stopped and I will be late.
   c) The clock has stopped or I will be late.
   d) If the clock has not stopped then I will be late.
4. a) $\{x : x \in \mathbb{R}, x \geqslant 7\}$
   b) $\{x : x \in \mathbb{R}, x \geqslant 2\}$
   c) $(A \cap B) = \{x : x \in \mathbb{R}, x \geqslant 7\}$
5. a) The slope is wet, Jane will slip.
   b) The car can drive fast, The car is red.
   c) Stars are hot, Stars are far away.
6. a) If $x = 3$ then $2x^2 + 2 = 20$.
   b) $x = 3$ and $2x^2 + 2 = 20$.
   c) $x = 3$ or $2x^2 + 2 = 20$.
   d) Not $2x^2 + 2 = 20$ and not $x = 3$.
7. a) If the piano is in tune the choir is in key.
   b) If the piano is not in tune then the choir is not in key.
   c) Either the piano is not in tune or the choir is not in key.
   d) If and only if the piano is in tune then the choir is in key.

Exercise 9.2

1. a) True   b) False   c) True   d) True
2. a) True   b) False   c) False   d) True
3. a) False   b) True   c) True   d) True
4. a) If x is a factor of 6 then x is a factor of 24.
   b) If x is a factor of 24 then x is a factor 6.
   c) If and only if x is a factor of 6 then x is a factor of 24.
   d) If and only if x is not a factor of 6 then x is not a factor of 24.
5. a) $x = 3$
   b) If x is a prime number then x is a multiple of 3.
   c) P is false, and Q is true, so hence Q $\Rightarrow$ P is false)
   d) P is true, but Q is false, so hence P $\Leftrightarrow$ Q is false)
6. a) P $\Rightarrow$ Q   b) P $\Leftrightarrow$ Q   c) Q $\Rightarrow$ P   d) Q $\Leftrightarrow$ P
   P: the wind is strong
   Q: the waves are large
7. a) If the temperature is less than 0 degrees Celsius then the water is frozen.
   b) If the water is frozen then the temperature is less than 0 degrees Celsius.
   c) If and only if the temperature is less than 0 degrees Celsius then the water is frozen.
   d) If and only if the temperature is not less than 0 degrees Celsius then the water is not frozen.

Exercise 9.3

1. a) True   b) False   c) False   d) True
2. a) False   b) True   c) True   d) False
3. a) x is an even number and x is a composite number.

b) x is not an even number.
c) x is not a composite number.
d) x is not an even number and x is a composite number.
4. a) $P \wedge Q$
   b) $\neg P$
   c) $\neg Q \wedge \neg P$
   d) $(\neg Q \wedge P) \Rightarrow \neg Q$
5. a) If the speed limit is 50 km/h then the car is not travelling at 60 km/h.
   b) The speed limit is 50 km/h and the car is not travelling at 60 km/h
   c) If the speed limit is not 50 km/h then the car is speeding.
   d) If the speed limit is 50 km/h and the car is travelling at 60 km/h then the car is speeding.
6. a) If birds do not lay eggs then fish do not swim.
   b) Birds do not lay eggs and fish do not swim.
   c) If birds lay eggs and fish swim, then birds lay eggs.
   d) If birds lay eggs and fish swim, then fish do not swim.
7. a) (i) $\{x : x \in \mathbb{R}, x \leqslant 15\}$
      (ii) $\{x : x \in \mathbb{R}, x > 3\}$
   b) (i) $\{x : x \in \mathbb{R}, x > 15\}$
      (ii) $\{x : x \in \mathbb{R}, x \leqslant 3\}$
   c) $\frac{x}{3} - 4 \leqslant 1$ and $3x > 9$.
   d) $\{x : x \in \mathbb{R}, 3 < x \leqslant 15\}$

Exercise 9.4

1. a) x is a multiple of 5 or x is a factor of 100.
   b) Either x is a multiple of 5 or x is a factor of 100.
   c) x is not a multiple of 5 or x is a factor of 100.
   d) Either x is a multiple of 5 or x is not a factor of 100.
2. a) False   b) False   c) True   d) False
3. a) True   b) False   c) True   d) True
4. a) It is raining and it is not snowing
   b) Either it is raining or it is snowing
   c) It is not raining or it is not snowing
   d) It is raining and it is snowing
5. a) False   b) False   c) False   d) True
6. a) (i) A square is a type of rhombus and a rhombus has all four sides equal.
      (ii) If and only if a square is a type of rhombus or a rhombus has all four sides equal, then a parallelogram has opposite sides equal.
   b) (i) True
      (ii) False
7. a) (i) Q $\Rightarrow$ P   (ii) Q $\veebar$ $\neg$R
   b) (i) The sun rises in the east and the Earth does not rotate on its axis.
      (ii) If the sun rises in the east or the Earth rotates on its axis, then the Moon orbits the Earth.
   c) (i) False   (ii) True

Exercise 9.5

1.

P	Q	$\neg P$	$\neg P \wedge Q$
T	T	F	F
T	F	F	F
F	T	T	T
F	F	T	F

2.

P	Q	¬P	¬Q	¬P ⇒ ¬Q
T	T	**F**	**F**	**T**
T	F	**F**	**T**	**T**
F	T	**T**	**F**	**F**
F	F	**T**	**T**	**T**

3.

P	Q	¬P	¬Q	(Q ⇒ P)	(¬P ⇒ ¬Q)	(Q ⇒ P) ⇔ (¬P ⇒ ¬Q)
T	T	**F**	**F**	**T**	**T**	**T**
T	F	**F**	**T**	**T**	**T**	**T**
F	T	**T**	**F**	**F**	**F**	**T**
F	F	**T**	**T**	**T**	**T**	**T**

4.

P	Q	R	(P ⊻ Q)	(P ⊻ Q) ∧ R
T	T	T	**F**	**F**
T	T	F	**F**	**F**
T	F	T	**T**	**T**
T	F	F	**T**	**F**
F	T	T	**T**	**T**
F	T	F	**T**	**F**
F	F	T	**F**	**F**
F	F	F	**F**	**F**

5.

P	Q	¬P	¬Q	(P ∧ Q)	¬(P ∧ Q)	¬P ∨ ¬Q
T	T	**F**	**F**	**T**	**F**	**F**
T	F	**F**	**T**	**F**	**T**	**T**
F	T	**T**	**F**	**F**	**T**	**T**
F	F	**T**	**T**	**F**	**T**	**T**

6.

P	Q	R	(P ∧ Q)	(P ∧ Q) ⇒ R
T	T	T	T	T
T	T	F	T	F
T	F	T	F	T
T	F	F	F	T
F	T	T	F	T
F	T	F	F	T
F	F	T	F	T
F	F	F	F	T

(P ∧ Q) ⇒ R is false when both P and Q are true, but R is false)

7. a) x is a multiple of four or x is a square number and x is not a factor of 36.

b)

P	Q	R	¬Q	P ∨ R	(P ∨ R) ∧ ¬Q
T	T	T	F	T	F
T	T	F	F	T	F
T	F	T	T	T	T
T	F	F	T	T	T
F	T	T	F	T	F
F	T	F	F	F	F
F	F	T	T	T	T
F	F	F	T	F	F

c) $x = 8$
Answers will vary.

8.

P	Q	¬Q	(P ∧ ¬Q)	(P ∨ Q)	(P ∧ ¬Q) ⇒ (P ∨ Q)
T	T	F	F	T	**T**
T	F	T	T	**T**	**T**
F	T	F	**F**	T	**T**
F	F	T	F	F	**T**

9.

P	Q	P ⇔ Q	(P ⇔ Q) ∧ P	[(P ⇔ Q) ∧ P] ⇔ Q
T	T	**T**	**T**	**T**
T	F	**F**	**F**	**T**
F	T	**F**	**F**	**F**
F	F	**T**	**F**	**T**

10. a) If the train arrives on time then I am not late for school.
b) $\neg p \wedge \neg q$
c)

p	q	¬p	¬q	$p \Rightarrow \neg q$	$\neg p \wedge \neg q$
T	T	F	F	F	F
T	F	F	T	T	**F**
F	T	T	F	**T**	**F**
F	F	T	T	T	T

d) They are not always the same

Exercise 9.6

1. Inverse: 'If the basketball is not flat, then the basketball will bounce.'
   Converse: 'If the basketball will not bounce, then the basketball is flat.'
   Contrapositive: 'If the basketball will bounce, then the basketball is not flat.'
2. Inverse: 'If there is not a drought, then the crops will not fail.'
   Converse: 'If the crops will fail, then there is a drought.'
   Contrapositive: 'If the crops will not fail, then there is not a drought.'
3. a) Inverse: P ⇒ Q
      Converse: ¬Q ⇒ ¬P

Contrapositive: Q ⇒ P

b) Inverse: P ⇒ ¬Q
Converse: Q ⇒ ¬P
Contrapositive: ¬Q ⇒ P

c) Inverse: ¬(P ∧ Q) ⇒ ¬Q
Converse: Q ⇒ (P ∧ Q)
Contrapositive: ¬Q ⇒ ¬(P ∧ Q)

d) Inverse: ¬P ⇒ ¬(P ∨ Q)
Converse: (P ∨ Q) ⇒ P
Contrapositive: ¬(P ∨ Q) ⇒ ¬P

4. a) (i) If the number ends in zero then the number is divisible by 5.
(ii) If the number is divisible by 5 then the number ends in zero.

b) (i) ¬P ⇒ ¬Q
(ii) ¬Q ⇒ ¬P

5. a) (i) If it is a rhombus then a figure is a square.
(ii) If a figure is not a square then it is not a rhombus.
(iii) If it is not a rhombus then a figure is not a square.

b) Statement iii. must be true, since the inverse and converse have the same truth value.

6. a) (i) If British Columbia is in North America, then Canada is in North America.
(ii) If Canada is not in North America, then British Columbia is not in North America.
(iii) If British Columbia is not in North America, then Canada is not in North America.

b) Statement iii since the inverse and converse of an implication have the same truth value.

7. a) If the grass is not watered regularly, then the grass is green.

b)

P	Q	¬Q	¬Q ⇒ P
T	T	**F**	**T**
T	F	**T**	**T**
F	T	**F**	**T**
F	F	**T**	**F**

c) ¬P ⇒ Q

8. a) If dolphins are not porpoises, then porpoises are not mammals.

b)

P	Q	¬P	¬Q	¬P ⇒ ¬Q
T	T	**F**	**F**	**T**
T	F	**F**	**T**	**T**
F	T	**T**	**F**	**F**
F	F	**T**	**T**	**T**

c) P ⇒ Q

9. a) If Sarah eats lots of carrots then Sarah can see well in the dark.

b) Sarah does not eat lots of carrots and Sarah can see well in the dark.

c) ¬Q ⇒ ¬P

d) Contrapositive

10. a) If Peter eats his vegetables then Peter is healthy.

b) (i) P ⇒ ¬Q
(ii) Q ⇒ ¬P

c)

P	Q	¬P	P ⇒ Q	¬P ∨ Q
T	T	F	T	T
T	F	F	F	F
F	T	T	T	T
F	F	T	T	T

Exercise 9.7

1.

P	¬P	P ∧ ¬P
T	F	F
F	T	F

Since P ∧ ¬P is always false, it is a contradiction.

2.

P	P ⇒ P
T	T
F	T

Since P ⇒ P is always true, it is a tautology.

3.

P	Q	¬P	¬Q	(P ∧ Q)	(¬P ∧ ¬Q)	(P ∧ Q) ∧ (¬P ∧ ¬Q)
T	T	F	F	T	F	F
T	F	F	T	F	F	F
F	T	T	F	F	F	F
F	F	T	T	F	T	F

Since (P ∧ Q) ∧ (¬P ∧ ¬Q) is always false, it is a contradiction.

4.

P	Q	¬P	¬Q	(P ∨ Q)	(¬P ∨ ¬Q)	(P ∨ Q) ∨ (¬P ∨ ¬Q)
T	T	F	F	T	F	T
T	F	F	T	T	T	T
F	T	T	F	T	T	T
F	F	T	T	F	T	T

Since (P ∧ Q) ∧ (¬P ∧ ¬Q) is always true, it is a tautology.

5. a) If the book is short or the book is not short then the book is easy to read and the book is not easy to read.

b)

P	Q	¬P	¬Q	(P ∨ ¬P)	(Q ∧ ¬Q)	(P ∨ ¬P) ⇒ (Q ∧ ¬Q)
T	T	F	F	**T**	F	**F**
T	F	F	T	**T**	F	**F**
F	T	T	F	**T**	F	**F**
F	F	T	T	**T**	F	**F**

c) Contradiction

6. a) (i) x = 3
(ii) x = 3

b) 3

c) No, since they are true for the same value of x.

d)

P	Q	¬Q	P ∨ ¬Q
T	T	**F**	**T**
T	**T**	**F**	**T**
F	**F**	**T**	**T**
F	**F**	**T**	**T**

e) Tautology

7. a)

1	2	3	4	5	6
P	Q	P ∧ Q	¬P	P ∨ Q	(P ∧ Q) ∧ (¬P ∧ ¬Q)
T	T	T	F	T	F
T	F	F	F		F
F	T	F	T	T	F
F	F	**T**	**F**		F

b)

1	2	3	4	5	6
P	Q	P ∧ Q	¬P	P ∨ Q	(P ∧ Q) ∧ (¬P ∧ ¬Q)
T	T	T	F	T	F
T	F	F	F	**T**	F
F	T	F	T	T	F
F	F	T	F	**F**	F

c) Contradiction

8.

P	Q	¬P	¬Q	¬P ⇒ ¬Q	Q ⇒ P
T	T	F	F	T	T
T	F	F	T	T	T
F	T	T	F	F	F
F	F	T	T	T	T

Since ¬P ⇒ ¬Q and Q ⇒ P have the same truth value, they are logically equivalent.

9.

P	Q	¬P	¬Q	P ⇒ Q	¬Q ⇒ ¬P
T	T	F	F	T	T
T	F	F	T	F	F
F	T	T	F	T	T
F	F	T	T	T	T

Since P ⇒ Q and ¬Q ⇒ ¬P have the same truth value, they are logically equivalent.

10. a) (i) ¬Q ⇒ ¬P
(ii) Q ⇒ (P ⊻ R)
b) If Alex does not play the flute then Alex is neither a scientist nor from Uruguay.

c)

P	Q	R	¬R	Q ∨ P	¬(Q ∨ P)	¬R ⇒ ¬(Q ∨ P)
T	T	T	F	T	F	T
T	T	F	T	T	F	F
T	F	T	F	T	F	T
T	F	F	T	T	F	F
F	T	T	F	T	F	T
F	T	F	T	T	F	F
F	F	T	F	F	T	T
F	F	F	T	F	T	T

The argument ¬R ⇒ ¬(Q ∨ P) is not a tautology, so the argument is not logically valid.

11. a) Good music students go to good universities.
Good mathematics students get good jobs.
b) (i) Good music students are not good mathematics students .
(ii) Good mathematics students go to good universities and students who go to good universities get good jobs.
c) Not a valid argument. A student who goes to a good university and gets a good job will not necessarily be a good mathematics student.

12. a) (i) P ⇒ Q
(ii) If Matthew does not cook dinner, then Jill does not wash the dishes.
b) (i)

P	Q	R	P ⇒ Q	Q ⇒ R	¬R	(P ⇒ Q) ∧ (Q ⇒ R) ∧ ¬R	¬P	[(P ⇒ Q) ∧ (Q ⇒ R) ∧ ¬R] ⇒ ¬P
T	T	T	**T**	**T**	F	**F**	**F**	T
T	T	F	**T**	**F**	T	**F**	**F**	T
T	F	T	**F**	**T**	F	**F**	**F**	T
T	F	F	**F**	**T**	T	**F**	**F**	T
F	T	T	**T**	**T**	F	**F**	**T**	T
F	T	F	**T**	**F**	T	**F**	**T**	T
F	F	T	**T**	**T**	F	**F**	**T**	T
F	F	F	**T**	**T**	T	**T**	**T**	T

(ii) [(P ⇒ Q) ∧ (Q ⇒ R) ∧ ¬R] ⇒ ¬P is a tautology.

13. a) (i) $p \Rightarrow q$ (ii) $\neg q \Rightarrow \neg p$
b) $q \Rightarrow p$
c) No, as Ali might arrive at school on time but not come by metro.

14.

P	Q	P ⇒ Q	(P ⇒ Q) ∧ P	[(P ⇒ Q) ∧ P] ⇒ Q
T	T	T	T	T
T	F	F	F	T
F	T	T	F	T
F	F	T	F	T

[(P ⇒ Q) ∧ P] ⇒ Q is a valid argument as all the entries in the last column are T(rue).

Chapter 10

Exercise 10.1

1. SS = A = {−3, −2, −1, 0, 1, 2, 3, 4, 5, 6, 7, 8, 9, 10}

2.

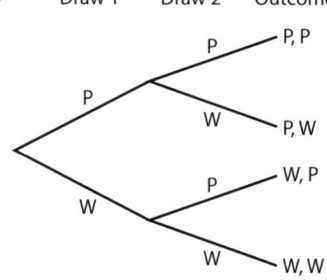

Draw 1 Draw 2 Outcomes

P, P
P, W
W, P
W, W

3.

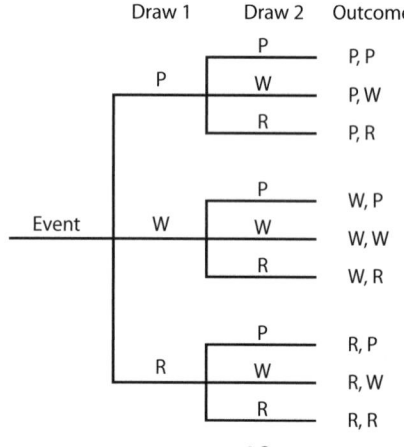

Draw 1 Draw 2 Outcomes

P, P
P, W
P, R
W, P
W, W
W, R
R, P
R, W
R, R

4.

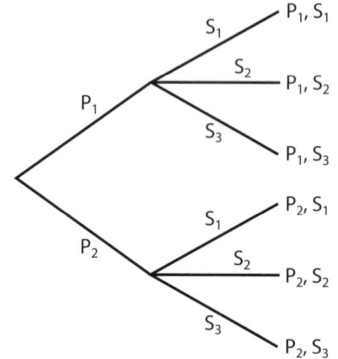

6 Outcomes

P₁, S₁
P₁, S₂
P₁, S₃
P₂, S₁
P₂, S₂
P₂, S₃

5.

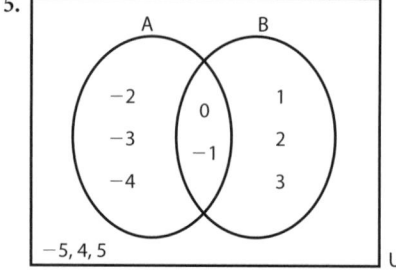

6.

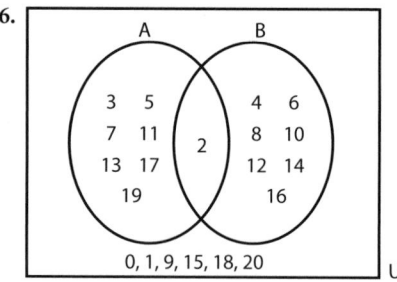

7.

Die 1	Die 2					
	1	2	3	4	5	6
1	1 1	1 2	1 3	1 4	1 5	1 6
2	2 1	2 2	2 3	2 4	2 5	2 6
3	3 1	3 2	3 3	3 4	3 5	3 6
4	4 1	4 2	4 3	4 4	4 5	4 6
5	5 1	5 2	5 3	5 4	5 5	5 6
6	6 1	6 2	6 3	6 4	6 5	6 6

8.

A♦ K♦ Q♦ J♦ 10♦ 9♦ 8♦ 7♦ 6♦ 5♦ 4♦ 3♦ 2♦

A♥ K♥ Q♥ J♥ 10♥ 9♥ 8♥ 7♥ 6♥ 5♥ 4♥ 3♥ 2♥

A♣ K♣ Q♣ J♣ 10♣ 9♣ 8♣ 7♣ 6♣ 5♣ 4♣ 3♣ 2♣

A♠ K♠ Q♠ J♠ 10♠ 9♠ 8♠ 7♠ 6♠ 5♠ 4♠ 3♠ 2♠

9. a) 216
   b) 6
   c) 10

10. a)

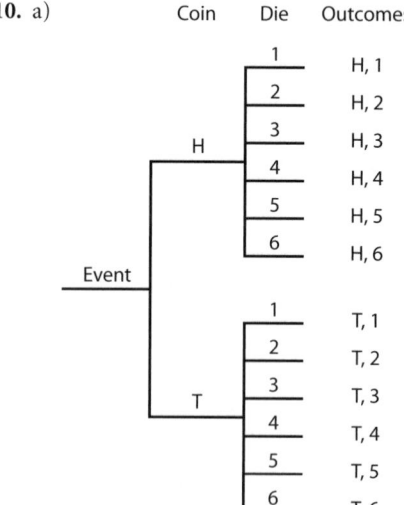

Coin Die Outcomes

H, 1
H, 2
H, 3
H, 4
H, 5
H, 6
T, 1
T, 2
T, 3
T, 4
T, 5
T, 6

b) {H1, H2, H3, H4, H5, H6, T1, T2, T3, T4, T5, T6}

11.

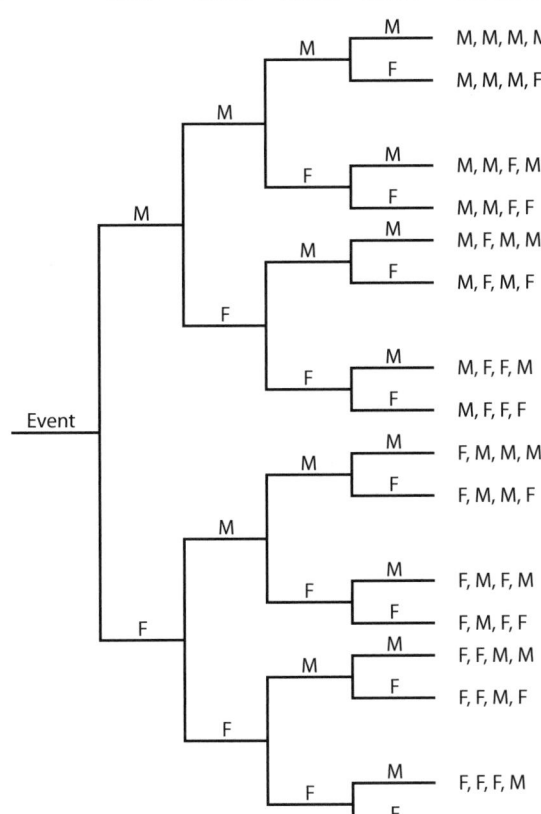

| | Child 1 | Child 2 | Child 3 | Child 4 | Outcomes |

(tree diagram with outcomes:)

M, M, M, M
M, M, M, F
M, M, F, M
M, M, F, F
M, F, M, M
M, F, M, F
M, F, F, M
M, F, F, F
F, M, M, M
F, M, M, F
F, M, F, M
F, M, F, F
F, F, M, M
F, F, M, F
F, F, F, M
F, F, F, F

12. Diamond face card

	K	Q	J
1	1 K	1 Q	1 J
2	2 K	2 Q	2 J
3	3 K	3 Q	3 J
4	4 K	4 Q	4 J
5	5 K	5 Q	5 J
6	6 K	6 Q	6 J

Six-sided die

13.

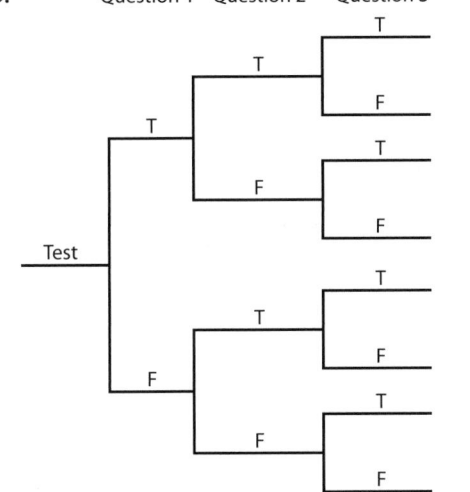

| Question 1 | Question 2 | Question 3 | Outcomes |

T, T, T
T, T, F
T, F, T
T, F, F
F, T, T
F, T, F
F, F, T
F, F, F

Test

14.

Alright

Men (80)
More than 6 ft (24)
Less than 6 ft (56)

Women (120)
More than 6 ft (12)
Less than 6 ft (108)

15.

C1	C2	C3	C4
H	H	H	H
H	H	H	T
H	H	T	H
H	H	T	T
H	T	H	H
H	T	H	T
H	T	T	H
H	T	T	T
T	H	H	H
T	H	H	T
T	H	T	H
T	H	T	T
T	T	H	H
T	T	H	T
T	T	T	H
T	T	T	T

Exercise 10.2

1. $\frac{1}{2}$ **2.** $\frac{1}{2}$

3. a) $\frac{7}{13}$ b) $\frac{6}{13}$

4. $\frac{1}{4}$

5. a) $\frac{1}{216}$ b) $\frac{215}{216}$

6. a) $\frac{1}{8}$ b) $\frac{7}{8}$

7. a) $\frac{1}{6}$ b) $\frac{1}{6}$ c) 0 d) 1 e) $\frac{11}{36}$

8. a) $\frac{1}{2}$ b) 0 c) $\frac{1}{4}$ d) $\frac{3}{13}$ e) 0 f) $\frac{1}{13}$

9. 40%

10. $\frac{2}{5}$ **11.** $\frac{1}{18}$

12. a) $\frac{1}{16}$ b) $\frac{3}{8}$ **13.** $\frac{15}{16}$

14. a) $\frac{1}{26}$ b) $\frac{2}{13}$ c) $\frac{4}{13}$ d) $\frac{7}{13}$

15. a) $\frac{1}{12}$ b) 0 c) $\frac{1}{6}$

16. a) $\frac{3}{7}$ b) $\frac{25}{49}$

Exercise 10.3

1. $A \cap B = \varnothing$

2. Yes. One can belong to one party or another, but cannot belong to both.

3. $P(A \cup B) = P(A) + P(B)$

4.

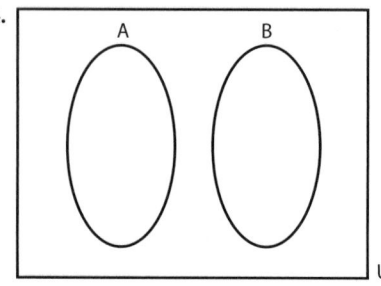

5. a) Is mutually exclusive
b) Not ME. Some 7's are red cards. (7 of diamonds, 7 of hearts)
c) Not ME. The student could play both sports.
d) ME
e) Not ME. All squares are rectangles.
f) ME

6. a) $\frac{2}{13}$ b) $\frac{1}{2}$ c) $\frac{4}{13}$ d) 1

7. a) $\frac{13}{18}$ b) $\frac{1}{2}$

8. a) $\frac{280}{527}$ b) $\frac{412}{527}$

9. a) $\frac{1}{8}$ b) $\frac{1}{4}$

10. 60%

11. a) $\frac{1}{6}$ b) $\frac{7}{18}$

12. a) $\frac{1}{8}$ b) $\frac{3}{8}$

13. a) $\frac{17}{25}$ b) $\frac{37}{50}$ c) $\frac{3}{10}$

Exercise 10.4

1. $n(A \cup B) = n(A) + n(B) - n(A \cap B)$
2. $P(A \cup B) = P(A) + P(B) - P(A \cap B)$
3.

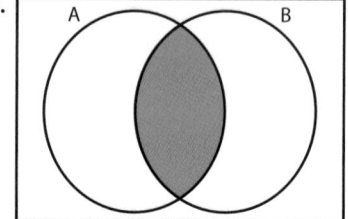

4. a) Not ME   b) Not ME   c) Not ME
d) Not ME   e) ME   f) Not ME

5. a) $\frac{1}{2}$ b) $\frac{1}{13}$ c) $\frac{1}{26}$ d) $\frac{7}{13}$

6. a) $\frac{1}{6}$ b) $\frac{1}{12}$ c) $\frac{1}{36}$ d) $\frac{2}{9}$

7. $0.90 = 90\%$

8. $0.65 = 65\%$

9. $\frac{31}{183}$

10. $\frac{2}{365}$

11. a) 0.79   b) 0.63   c) 0.45   d) 0.24

12. a) 0.16   b) 0.25   c) 0.71   d) 0.13

13. a) $\frac{5}{16}$ b) $\frac{15}{16}$ c) $\frac{11}{16}$

14. a)

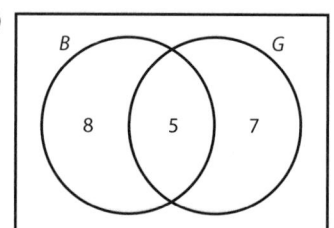

b) 0.86
c) 0.83
d) 0.26
e) 0.09
f) 0.74

15. a) $\frac{1}{2}$ b) $\frac{43}{50}$ c) $\frac{3}{50}$

16. a) $\frac{8}{17}$ b) $\frac{11}{17}$ c) $\frac{3}{17}$

Exercise 10.5

1. Two events are independent if the probability of the first does not influence the probability of the second.

2. Draw a sample space such as a Venn diagram.

3. a, c, e

4. 0.0156

5. $\frac{1}{12}$

6. a) $\frac{6}{25}$ b) $\frac{12}{25}$ c) $\frac{4}{25}$ d) $\frac{16}{25}$

7. a) $\frac{5}{144}$ b) $\frac{5}{24}$ c) $\frac{1}{16}$ d) $\frac{1385}{1728}$

8. a) $\frac{1}{16}$ b) $\frac{3}{52}$ c) $\frac{1}{169}$ d) $\frac{2}{169}$

9. a) 0.00287 to 3 s.f.
b) 0.0115 to 3 s.f.
c) 0.0975 to 3 s.f.
d) 0.000506 to 3 s.f.
e) 0.522 to 3 s.f.

10. a)

B G
8 5 7
U

b) (i) $\frac{3}{5}$ (ii) $\frac{7}{20}$ (iii) $\frac{1}{4}$ (iv) 1
c) $P(B \cap G) = \frac{5}{20} = \frac{1}{4}$ and $P(B) \cdot P(G) = \frac{13}{20} \cdot \frac{12}{20} = \frac{39}{100}$. Therefore, since $P(B \cap G) \neq P(B) \cdot P(G)$, the events are not independent.
d) Since $P(B \cap G) \neq 0$, the events are not mutually exclusive.

11. $\frac{1}{1024}$

12. $\frac{1}{12}$

13. a) $\frac{1}{216}$ b) $\frac{1}{108}$ c) $\frac{1}{36}$

14. a) (i) 0.55   (ii) 0.5   (iii) 0.7
b) No; $0.35 \neq (0.55)(0.5)$
c) $P(A \cap B) \neq 0$, therefore events A and B are not mutually exclusive.

15. a)

Toss 1	Toss 2	Toss 3
H	H	H
H	H	T
H	T	H
H	T	T
T	H	H
T	H	T
T	T	H
T	T	T

b) (i) $\frac{1}{8}$

   (ii) $\frac{4}{8} = \frac{1}{2}$

Exercise 10.6

1. If the probability of the first event influences the probability of the second then the events are dependent.

2. a, b, d

3. a) $\frac{2}{15}$ b) $\frac{4}{15}$ c) $\frac{8}{15}$

4. a) $\frac{5}{408}$ b) $\frac{5}{204}$ c) $\frac{5}{68}$

   d) $\frac{55}{272}$ e) $\frac{5}{408}$ f) $\frac{265}{408}$

5. a) $\frac{1}{221}$ b) $\frac{25}{102}$ c) $\frac{4}{663}$ d) $\frac{8}{663}$

6. a) $\frac{1}{8}$ b) $\frac{3}{8}$ c) $\frac{7}{8}$ d) $\frac{7}{8}$

7. a) $\frac{8}{609}$ b) $\frac{22}{203}$ c) $\frac{32}{1305}$

8. a) $\frac{256}{625}$ b) $\frac{1}{625}$ c) $\frac{16}{625}$

   d) $\frac{16}{625}$ e) $\frac{96}{625}$

9. a)

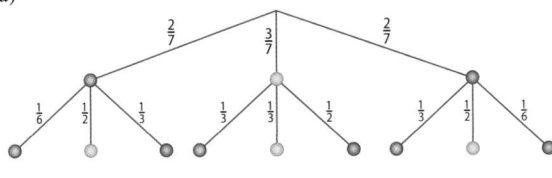

b) (i) $\frac{4}{7}$ (ii) $\frac{2}{7}$ c) $\frac{5}{28}$ d) $\frac{3}{84} = \frac{1}{28}$

Exercise 10.7

1. Answers vary: $P(B|G) = \dfrac{P(B \cap G)}{P(G)}$, $P(X|Y) = \dfrac{P(X \cap Y)}{P(Y)}$,

   $P(C|D) = \dfrac{P(C \cap D)}{P(D)}$, $P(M|N) = \dfrac{P(M \cap N)}{P(N)}$,

   $P(R|S) = \dfrac{P(R \cap S)}{P(S)}$

2. Draw a sample space and reduce the sample space to the given information.

3. $\frac{3}{4}$

4. $\frac{3}{5}$

5. a)

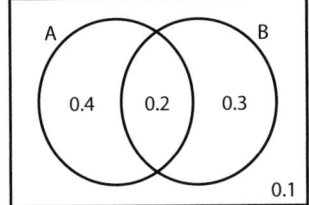

b) 0.9
c) 0.1
d) $\frac{2}{5}$
e) $\frac{1}{3}$

6. a)

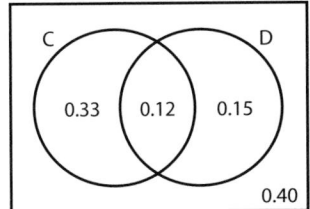

b) 0.60
c) 0.40
d) $\frac{4}{9}$
e) $\frac{4}{15}$

7. a)

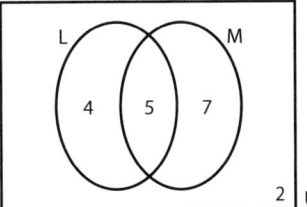

b) $\frac{8}{9}$
c) $\frac{1}{9}$
d) $\frac{5}{9}$
e) $\frac{5}{12}$

8. a)

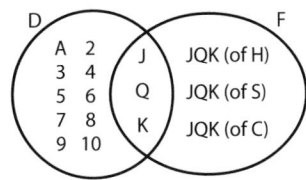

A 2 3 4 5 6 7 8 9 10 (of Hearts)
A 2 3 4 5 6 7 8 9 10 (of Spades)
A 2 3 4 5 6 7 8 9 10 (of Clubs)

b) $\frac{11}{26}$
c) $\frac{15}{26}$
d) $\frac{3}{13}$
e) $\frac{1}{4}$

9. a) $\frac{8}{15}$ b) $\frac{3}{5}$ c) $\frac{7}{30}$ d) $\frac{7}{16}$ e) $\frac{1}{4}$

10. a) $\frac{33}{80}$ b) $\frac{3}{8}$ c) $\frac{1}{10}$ d) $\frac{4}{15}$ e) $\frac{1}{4}$

11. a) $\frac{16}{43}$ b) $\frac{9}{40}$ c) $\frac{57}{112}$ d) $\frac{34}{45}$

Exercise 10.8

1. $\frac{7}{8}$ 2. $\frac{15}{16}$ 3. $\frac{14}{15}$

4. $\frac{115}{143}$ 5. $\frac{31}{32}$ 6. 38.6% to 3 s.f.

Exercise 10.9

1. a) (i) $P(\text{red}) = \frac{7}{12}$ (ii) $P(\text{not red}) = \frac{5}{12}$

   b) (i) $\frac{6}{11}$ (ii) $\frac{5}{33}$ (iii) $\frac{35}{66}$

   c) (i) $\frac{7}{44}$ (ii) $\frac{7}{22}$ (iii) $\frac{21}{22}$

2. a)

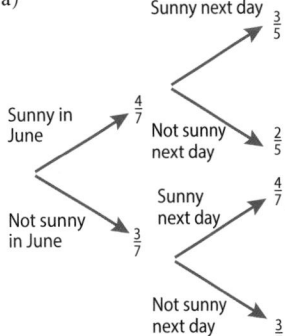

   b) $\frac{116}{245}$
   c) $\frac{3}{5}$

3. a) $\frac{1}{2}$
   b) $\frac{2}{3}$

4. $\frac{2}{3}$

Exercise 10.10

1. Answers vary

2. Answers vary.

	Independent	Dependent
Mutually Exclusive	Perform event B and then D $P(B) = \frac{5}{26}$ $P(D) = \frac{21}{26}$ $P(B \cap D) = 0 \Rightarrow$ Mut. Excl. $P(D)$ is not affected by $A \Rightarrow$ independent	$P(A \cap C) = 0 \Rightarrow$ Mut. Exclusive $P(A) \cdot P(C) \neq P(A \cap C) \Rightarrow$ depend.
Not Mutually Exclusive	Perform event B and then E. $P(B) = \frac{5}{26}$ $P(E) = \frac{4}{26}$ $P(B \cap E) = \frac{1}{26} \Rightarrow$ Not ME Since $P(E)$ is not affected by event $B \Rightarrow$ independent.	Perform event A and then E. $P(A) = \frac{5}{26}$ $P(E) = \frac{4}{26}$ $P(B \cap E) = \frac{1}{26} \Rightarrow$ Not ME. $P(E|A) \neq P(E) \Rightarrow$ dependent

Chapter 11

Exercise 11.1

1. Nominal, Ordinal, Interval
2. Nominal
3. Ordinal, Interval
4. Discrete, Continuous
5.

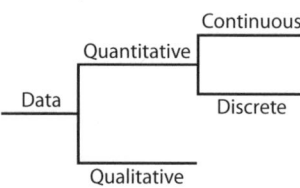

6.

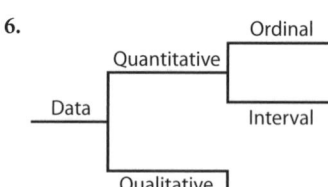

7. a) Qualitative   b) Quantitative
   c) Qualitative   d) Quantitative
   e) Quantitative
8. a) Interval   b) Nominal
   c) Nominal   d) Ordinal
   e) Interval
9. a) Discrete   b) Continuous
   c) Continuous   d) Discrete
   e) Continuous

Exercise 11.2

1. Answers vary. Tally marks are a convenient way to record data in a concise way. Tally marks can be used to quickly record data)
2. a)

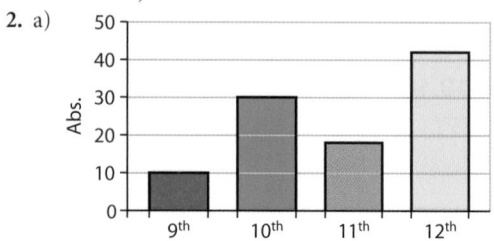

b)

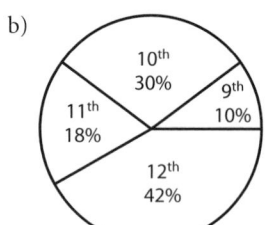

c)  = 5 absences

   9th grade absences: (2 figures)

   10th grade absences: (6 figures)

   11th grade absences: (3½ figures)

   12th grade absences: (8½ figures)

3. a)

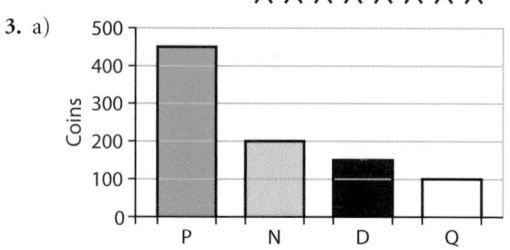

b)

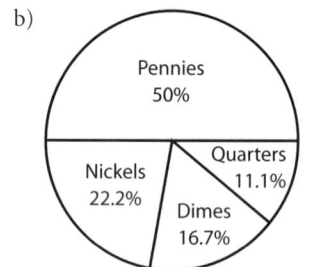

c)

○ = 50 coins

Pennies: ○ ○ ○ ○ ○ ○ ○ ○ ○

Nickels: ○ ○ ○ ○

Dimes: ○ ○ ○

Quarters: ○ ○

4.

Score	Frequency
3	1
4	0
5	1
6	1
7	2
8	1
9	1
10	2
11	3
12	1
13	1
14	1
15	2
16	0
17	4
18	2
19	2
20	1

5.

Age	Frequency
22	12
23	9
24	13
25	8
26	9
27	2
28	9
29	3
30	2
31	4
32	2
33	1
34	0
35	1

6. $a = 6$

Exercise 11.3

1. a) Discrete   b) Continuous
   c) Continuous   d) Continuous
   e) Discrete

2. a) When the data is countable and the range of the data is large ($>$ 15 or so).
   b) When the data is measurable and the range of the data is large ($>$ 15 or so).

3. a) When the data is countable and the range of the data is small ($<$ 15 or so).
   b) When the data is measurable and the range of the data is small ($<$ 15 or so).

4. a)

Class	Class boundaries
5–8	**4.5–8.5**
9–12	**8.5–12.5**
13–16	**12.5–16.5**

b)

Class	Class boundaries
2.50–3.00	**2.45–3.05**
3.10–3.60	**3.05–3.65**
3.70–4.20	**3.65–4.25**

c)

Class	Class boundaries
0–7	**−0.5–7.5**
8–15	**7.5–15.5**
16–23	**15.5–23.5**

d)

Class	Class boundaries
5.40–5.80	**5.35–5.85**
5.90–6.30	**5.85–6.35**
6.40–6.80	**6.35–6.85**

e)

Class	Class boundaries
3–7	**2.5–7.5**
8–12	**7.5–12.5**
13–17	**12.5–17.5**

5. a) $99.5 \leqslant L < 100.5$ b) $24.5 \leqslant L < 25.5$
   c) $6.5 \leqslant L < 7.5$

6. a)

Number of text messages	Frequency
13–14	12
15–16	9
17–18	6
19–20	14
21–22	5
23–24	10
25–26	6
27–28	14
29–30	4

b)
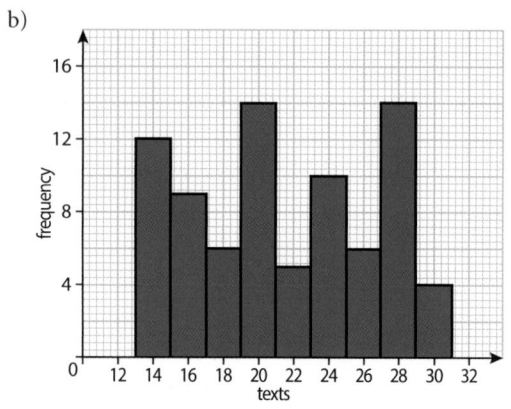

7. a) Lower class boundary = 40, and upper boundary = 50

b)
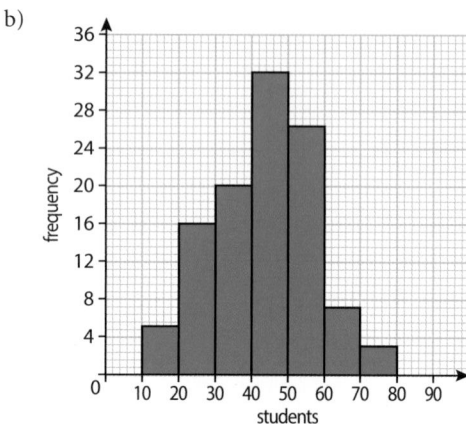

8.
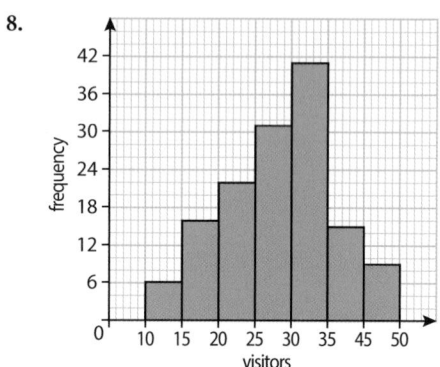

9.
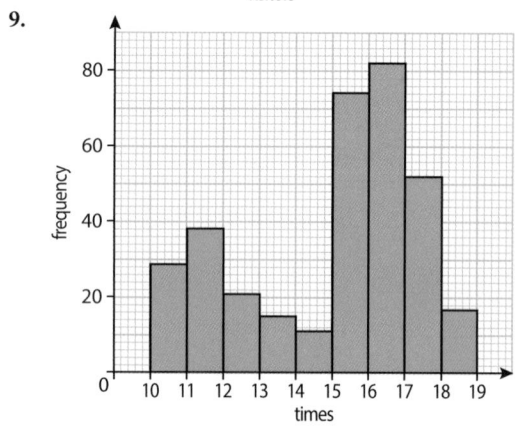

Exercise 11.4

1. mean, median, mode
2. a) 5   b) 29
3. a) 8.08 to 3 s.f.   b) 6.86 to 3 s.f.
4. a) 17.6118 = 17.6 to 3 s.f.   b) 12.0824 = 12.1 to 3 s.f.
5. a) median = 19, mode = 26
   b) median = 12.1, mode = 12.5
6. a) 12.2   b) 11 − 15   c) 11 − 15
7. a) 28.2   b) 21 − 29   c) 12 − 20
8. a) 11   b) 9.5 − 12.5
   c) 6.5 − 9.5 and 15.5 − 18.5 (bimodal)
9. a) 35.6 to 3 s.f.   b) 34.3 − 37.3   c) 37.3 − 40.3

Exercise 11.5

1. a)

Class	Class boundaries	Tally	Freq.	Mid-Intervals	cf
2–4	1.5–4.5	||	2	3	2
5–7	4.5–7.5	||||	4	6	6
8–10	7.5–10.5	⦀⦀ ||	7	9	13
11–13	10.5–13.5	⦀⦀	5	12	18
14–16	13.5–16.5	||	2	15	20
			20		

b)(i) Histogram
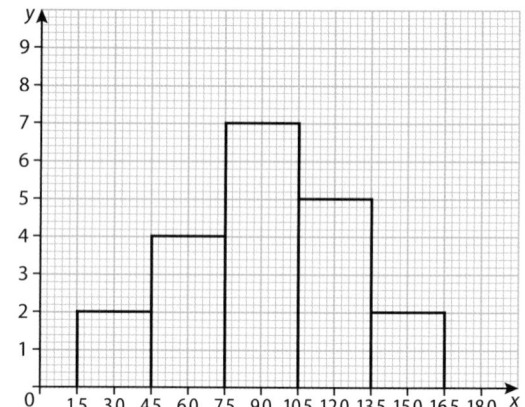

(ii) Cumulative frequency graph
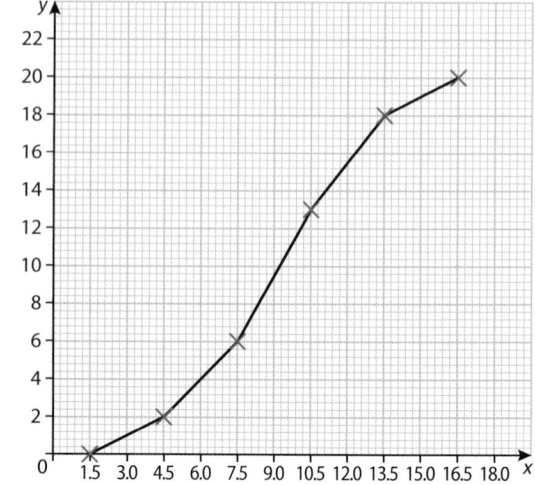

c) The median is approximately 90

2.

Class	Class boundaries	Tally	Freq.	Mid-Intervals	cf
10.5–10.9	**10.45–10.95**	| |	2	10.7	2
11.0–11.4	**10.95–11.45**	₮ℍℓ	5	11.2	7
11.5–11.9	**11.45–11.95**	₮ℍℓ ₮ℍℓ |	11	11.7	18
12.0–12.4	**11.95–12.45**	₮ℍℓ ₮ℍℓ ₮ℍℓ |	16	12.2	34
12.5–12.9	**12.45–12.95**	₮ℍℓ ||||	9	12.7	43
13.0–13.4	**12.95–13.45**	₮ℍℓ ||	7	13.2	50
			50		

b)(i) Histogram

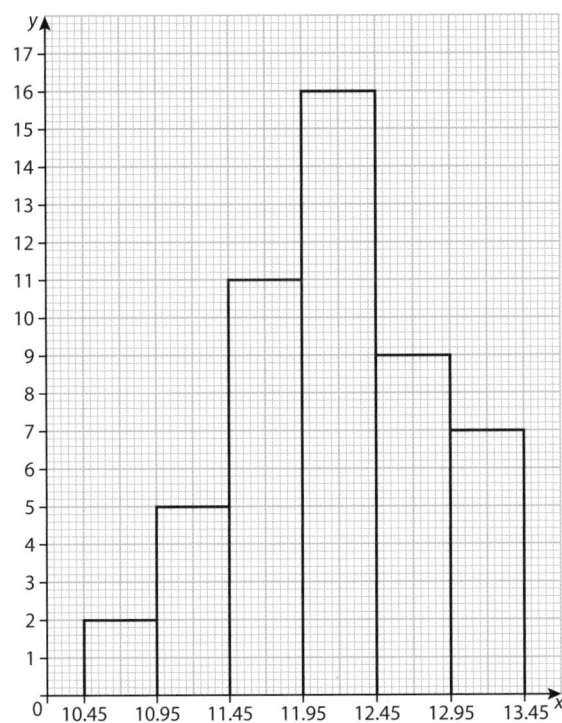

(ii) Cumulative frequency graph.

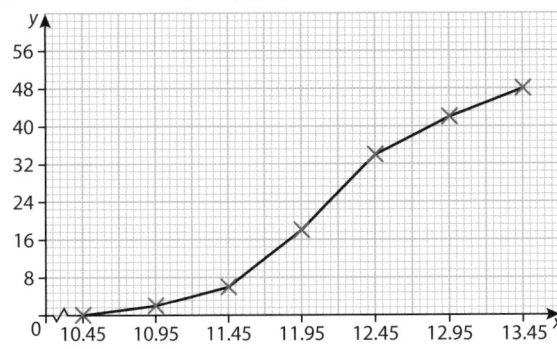

3. a)

Time taken (minutes)	Number of students	Mid interval	Cumulative freq.
$10 \leqslant t < 20$	5	15	5
$20 \leqslant t < 30$	6	25	11
$30 \leqslant t < 40$	10	35	21
$40 \leqslant t < 50$	12	45	33
$50 \leqslant t < 60$	6	55	39
$60 \leqslant t < 70$	5	65	44
$70 \leqslant t < 80$	6	75	50

b) (i)

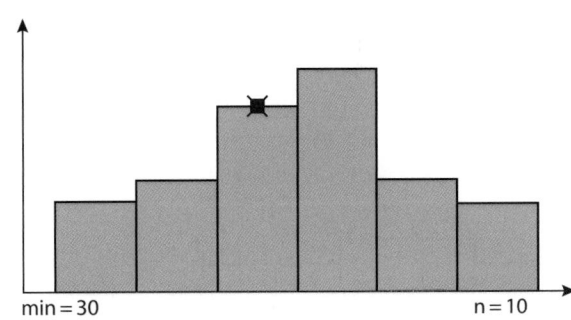

min = 30
max < 40   n = 10

(ii)

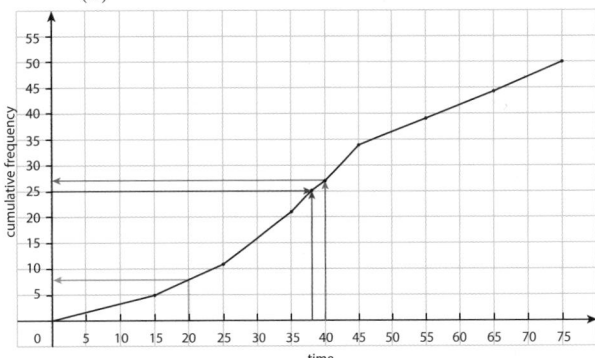

c) (i) 27 (red line)
(ii) 16% (8 students, green line)
(iii) 38 minutes (blue line)

4. a) $a = 38$, $b = 128$
b) 128
c) 38
d) 12.5%

5. a) (i) $a = 52$

(ii)

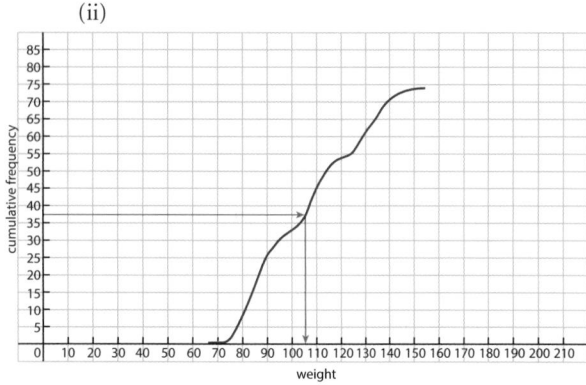

b) Red line indicates median = 106 kg.
c) 42 players

Exercise 11.6

1. a) (i) Minimum = 2
      $Q_1 = 17$
      Median = 28
      $Q_3 = 45.5$
      Maximum = 75
   (ii) IQR = 45.5 − 17 = 28.5

   b)
   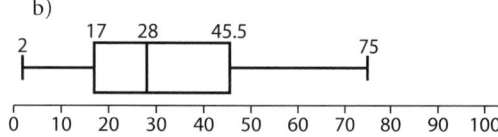

2. a) (i) Minimum = 1, Maximum = 40,
      Median = $\dfrac{18 + 31}{2}$ = 24.5, $Q_1 = \dfrac{16 + 17}{2}$ = 16.5,
      and Q_3 = 37.
   (ii) IQR = $Q_3 - Q1$ = 37 − 16.5 = 24.5

   b)
   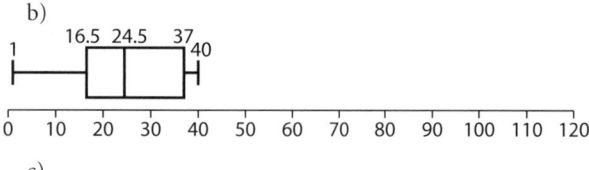

   c)
   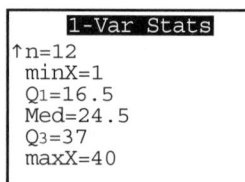
   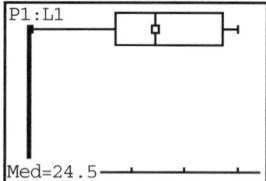

3. a) Minimum = 6, Maximum = 44, Median = 27, $Q_1 = 17$
      and Q_3 = 34.

   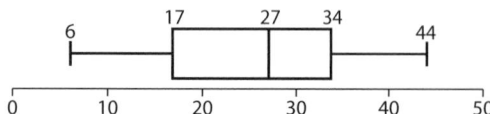

b) Use a GDC to check your answers.

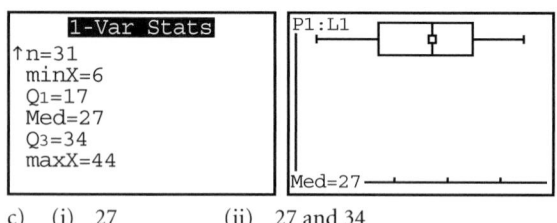

c) (i) 27   (ii) 27 and 34
   (iii) 38   (iv) 17
   (iv) 17   (v) 34

4. a)

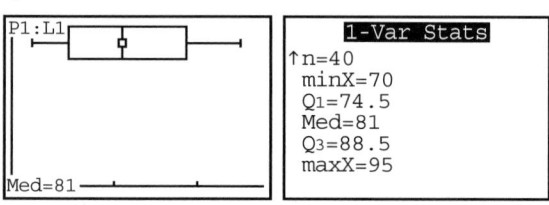

   (i) Median = 81.
   (ii) Mode = 81
   (iii) Range = 95 − 70 = 25
   (iv) IQR = $Q_3 - Q_1$ = 88.5 − 74.5 = 14
   (v) Q1 = 74.5
   (vi) Q3 = 88.5

5. Minimum = 1
   $Q_1 = 3$
   Median = 4
   $Q_3 = 5$
   Maximum = 5
   IQR = 5 − 3 = 2

6. Minimum = 78
   $Q_1 = 80$
   Median = 81
   $Q_3 = 82$
   Maximum = 87
   IQR = 82 − 80 = 2

7. a) 92   b) 91   c) 85   d) 65
   e) 83.5   f) 104   g) 20.5

Exercise 11.7

1. a) mean of the sample
   b) mean of the population
   c) variance of the sample
   d) standard deviation of the sample
   e) variance of the population
   f) standard deviation of the population
2. It is a measure of how the data is spread around the mean.
3. variance
4. variance, standard deviation, range
5. The data is spread out more in set A than it is in set B.
6. a) σ_x = 5.1768 = 5.18 to 3 s.f.
   b) σ_x = 4.7434 = 4.74 to 3 s.f.
7. a) σ_x = 10.6 to 3 s.f.   b) σ_x = 3.77 to 3 s.f.
8. a) (i) 36.1 to 3 s.f.   (ii) 42   (iii) 12.7 to 3 s.f.
   b) (i) 34.9 to 3 s.f.   (ii) 33   (iii) 10.4 to 3 s.f.
   c) Answers vary. The mean score was higher for the
      Mathematics test than for the English test, but the scores
      for the English test were grouped closer to the mean.

There wasn't as much deviation among the students' scores on the English test.

9. a) 10 b) 16.3 c) 22% d) 16 e) 1.53

Chapter 12

Exercise 12.1

1. a) $P(X < 175) = 0.252$ to 3 s.f.

b) $P(X > 190) = 0.369$ to 3 s.f.

c) $P(175 < X < 198) = 0.554$ to 3 s.f.

2. a)

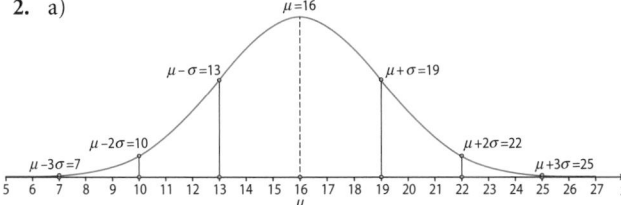

b) As 19 is 1 standard deviation above the mean, then $P(X < 19) = 0.5 + 0.34 = 0.84$

c) As 10 is 2 standard deviations below the mean, then $P(X > 10) = 1 - P(X < 10) = 1 - 0.025 = 0.975$

d) As 10 is 2 standard deviations below the mean, then $P(10 < X < 19) = 0.475 + 0.34 = 0.815$

3. a)

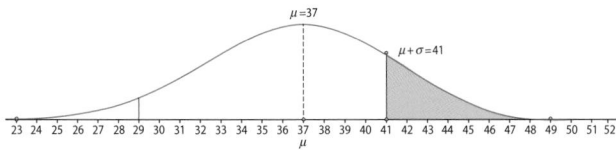

b) As 41 is 1 standard deviation from the mean, then $P(X > 41) = 0.16$

c)

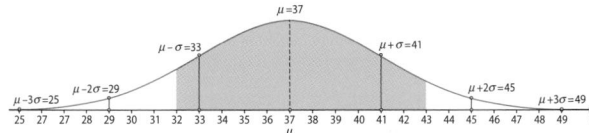

d) $P(32 < X < 43) = 0.746$ to 3 s.f.

4. a)

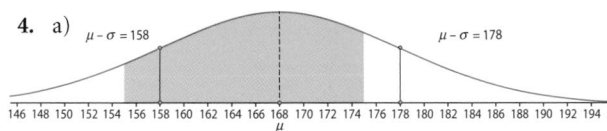

b) 66.1% of females are between height of 155 and 175 cm.

c) 48 students will be greater than 175 cm.

5. a) 15.4%

b) $n = 9530$ hours (3 s.f.)

6. a)

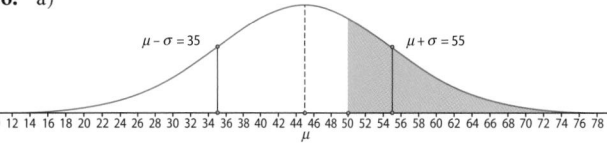

b) 30.9% (3 s.f.) of all vehicles are speeding.

c) 10% of vehicles are trvelling at less than 32.2 (3 s.f.) kmh⁻¹.

7. a) P(Bag weighs < 1 kg) = 0.106 or 10.6%

b) P(Bag weighs < 1 kg) = 0.006 or 0.6%

c) Expected value 6 bags

8. a) and b)

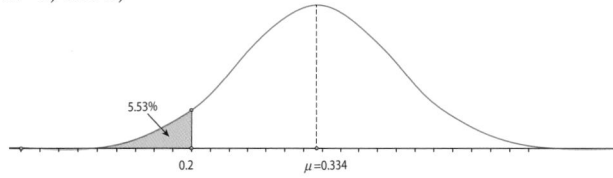

c) 5.53% of the students surveyed have the potential to be successful at computer games.

9. a) (i) Weight < 490 g = 0.252 (3 s.f.) or 25.2%

(ii) Weight > 520 g on GDC = 0.092 (3 s.f.) or 9.2%

(iii) 490 g < weight < 520 g = 0.656 (3 s.f.) or 65.6%

b) $w = 475$ g

10. a) (i) Distance < 2100 m = 0.0223 (3 s.f.) or 2.23%

(ii) Distance > 2300 m = 0.841 (3 s.f.) or 84.1%

(iii) 2000 m < distance < 3000 m = 0.988 (3 s.f.) or 98.8%

b) 90% of our pens last longer than 2240 m.

c) Expected number of pens lasting less than 2000 m = 6.21 i.e. approximately 6 pens.

Exercise 12.2

1. The r-value tells you how scattered, how spread out, the ordered pairs are in the scatter plot. It gives you information about the strength of the correlation between the variables.

2. a) $-1 \leqslant r \leqslant 1$ b) $[-1, 1]$

3. The mean point: $(\bar{x}, \bar{y})$

4. No. Yes.

5. There will not be a linear correlation between the variables. Hence, it serves no purpose to draw a regression line.

6. No. The r-value only tells you the strength of the correlation.

7. $(\bar{x}, \bar{y})$ and one other point that lies on the regression line.

8. Answers vary. Example: Height and the salary of a beginning high school teacher.

9. Answers vary. Example: IQ score and SAT score.

10. Answers vary. Example: Age and strength

11. Answers may vary slightly.

a) There is a strong positive correlation between the variables.

b) There is no correlation between the variables.

c) There is a moderately strong negative correlation between the variables.

d) There is a moderately weak to moderate positive correlation between the variables.

e) There is a strong negative correlation between the variables.

12. Answers vary.

a)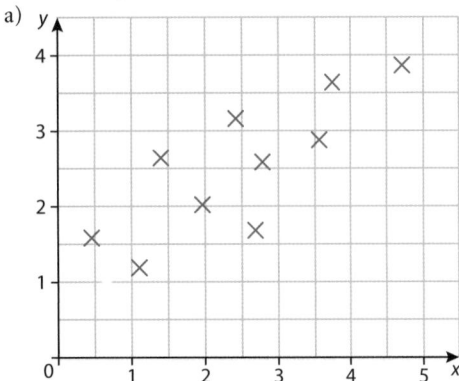

b)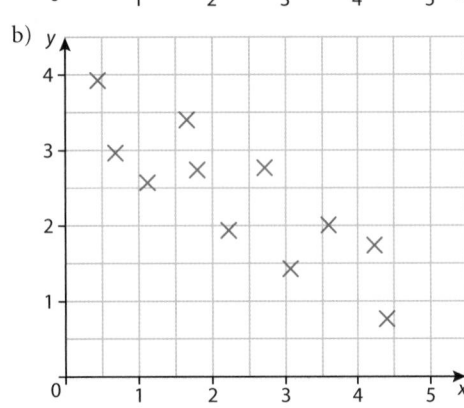

c)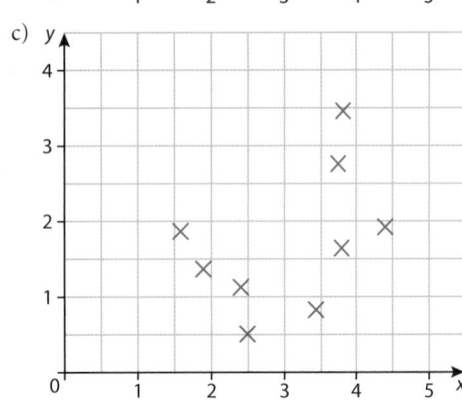

d)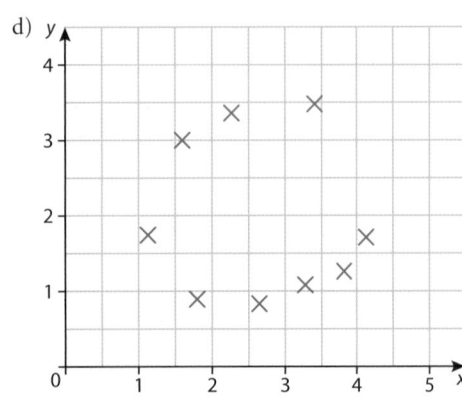

13. Answers vary.

a) *m* is the gradient of the line that contains all points, where *a* represents the gradient of the regression line that best represents the sample data (the scatter points).

b) The *y*-intercept, *b*, for $y = mx + b$, represents where the functional value for the set of ordered pairs will intersect the y-axis, where *y*-intercept, *b*, for $y = ax + b$, represents the value for the line that best represents the data points collected.

c) In general $y = mx + b$ represents the set of ordered pairs that lie on the graph where $y = ax + b$ represents the best representation possible for the data values collected)

14. a) $y = 2.5x - 7.5$ b) $y = -5.43x + 194.862$
c) $y = 8.825x - 79.61875$ d) $y = -0.623x + 25.607$

15. a)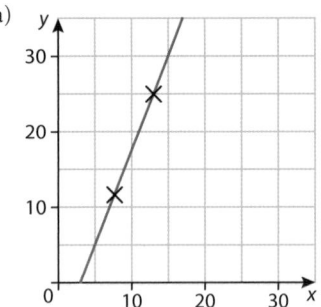

b)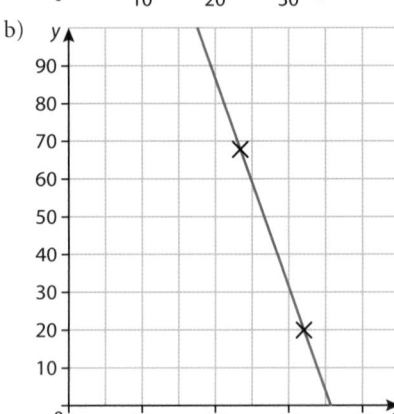

c)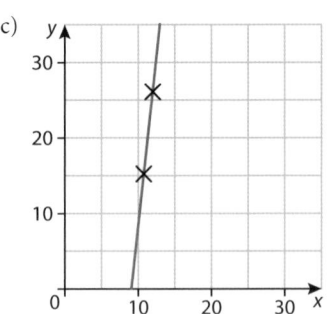

d)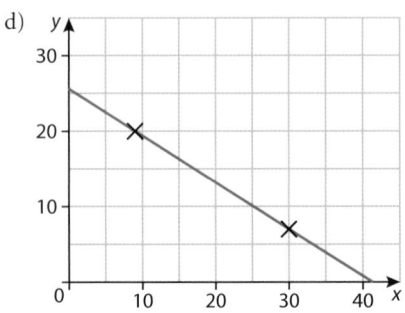

16. a) $y = 5.48$ to 3 s.f. via the TI Calc

b) $x = 8.59$ to 3 s.f. via the TI Calc

17. a), f)

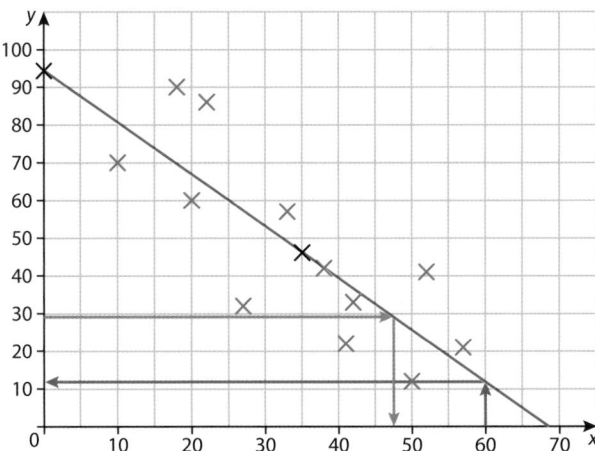

b) $r = -0.802$

c) There is a strong negative correlation between the variables.

d) $y = -1.37x + 94.0$

e) $(\bar{x}, \bar{y}) = (34.2, 47.2)$

g) $y = 11.8$ to 3 s.f. via the TI Calc; $y = 11$ via the graph.

h) $x = 48.2$ to 3 s.f. via the TI Calc; $x = 47.5$ via the graph.

i) Since $x = 60$ falls outside the data range we are less confident about our prediction.

Since $y = 28$ falls inside the data range we are more confident about our prediction.

18. a), b)

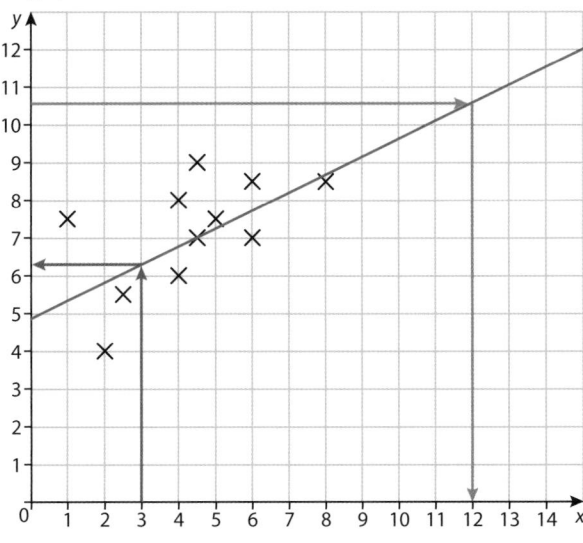

c) $r \approx 0.5$

d) (i) $y \approx 6.4$ (ii) $x \approx 12$

e) Since $x = 3$ falls inside the data range we are more confident about our prediction.

Since $y = 11$ falls outside the data range we are less confident about our prediction.

19. a), b)

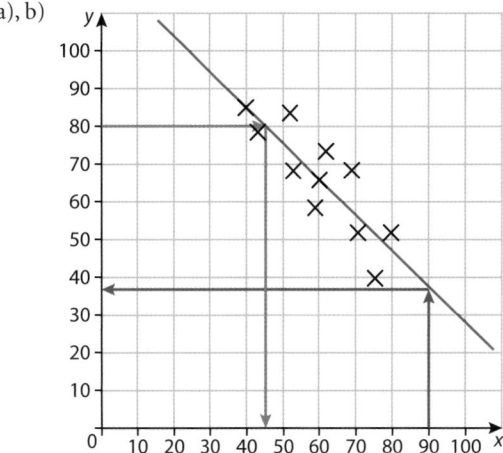

c) $r \approx -0.8$

d) (i) $y \approx 37$ (ii) $x \approx 45$

e) Since $x = 90$ falls outside the data range we are less confident about our prediction.

Since $y = 80$ falls inside the data range we are more confident about our prediction.

20. a)

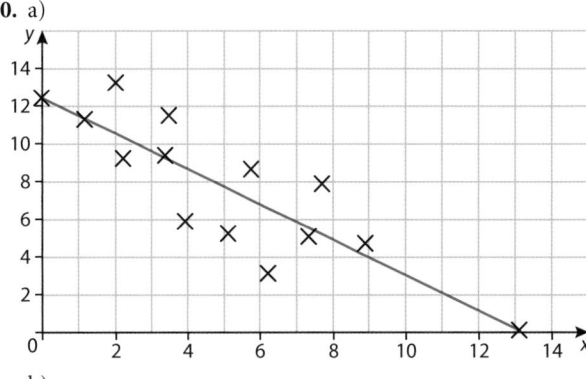

b)

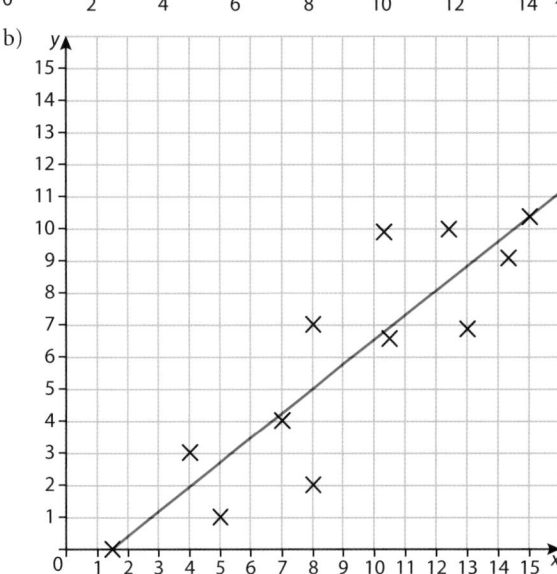

21. a)

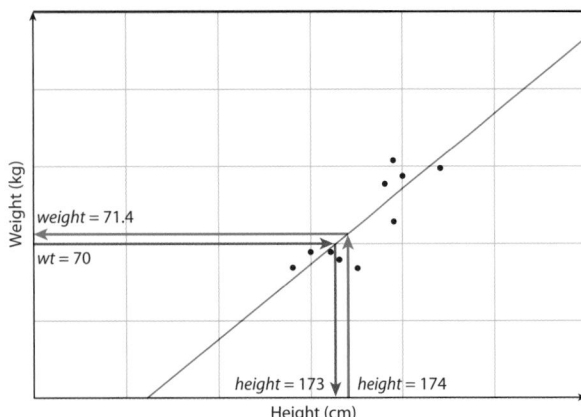

weight = 71.4

wt = 70

height = 173   *height = 174*

Height (cm)

(y-axis: Weight (kg))

b) $(\bar{x}, \bar{y}) = (175.8, 73.1)$

c) $r = 0.8501$, strong positive correlation

d) $y = 0.981x - 99.3$ (3 s.f.)

e) On graph above

f) (i) 70 kilogram athlete will be 173 cm tall. (Blue line)

(ii) A 174 cm tall athlete will weigh 71.4 kg. (Green line)

22. a)

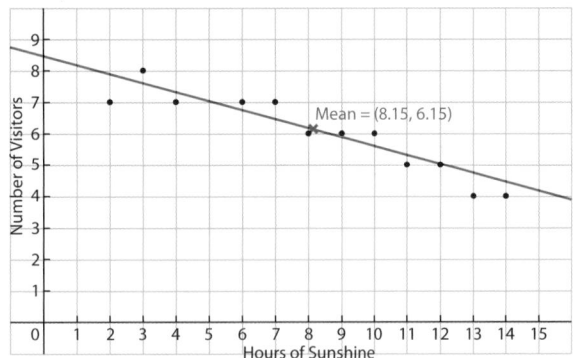

Mean = (8.15, 6.15)

Hours of Sunshine

(y-axis: Number of Visitors)

b) $r = 0.890$ 3 s.f.

c) There is a strong negative correlation between hours of sunlight and number of visitors.

d) $(\bar{x}, \bar{y}) = (8.154, 6.154)$

e) (i) regression equation visitors $= -0.2839$ hours $+ 8.4686$

(ii) See red line on graph above

(iii) 5

Exercise 12.3

1. Step 1. $H_0: \mu = 1300$
$H_1: \mu \neq 1300$

Step 2. p-value $= 0.0935$

Step 3. a) $\alpha = 0.01$
b) $\alpha = 0.05$
c) $\alpha = 0.10$

a) Step 4. $0.0935 > 0.01$

Step 5. Do not reject H_0. There is not enough evidence against the null hypothesis to suggest the average price is significantly different from 1300 €.

b) Step 4. $0.0935 > 0.05$

Step 5. Do not reject H_0. There is not enough evidence against the null hypothesis to suggest the average price is significantly different from 1300 €.

c) Step 4. $0.0935 < 0.10$

Step 5. Reject H_0. There is enough evidence against the null hypothesis to suggest the average price is significantly different from 1300 €.

2. Step 1. $H_0: \mu = 83.85$
$H_1: \mu \neq 83.85$

Step 2. p-value $= 0.0439$

Step 3. a) $\alpha = 0.01$
b) $\alpha = 0.05$
c) $\alpha = 0.10$

a) Step 4. $0.0439 > 0.01$

Step 5. Do not reject H_0. There is not enough evidence against the null hypothesis to suggest the average cost is significantly different from \$ 83.85.

b) Step 4. $0.0439 < 0.05$

Step 5. Reject H_0. There is enough evidence against the null hypothesis to suggest the average cost is significantly different from \$ 83.85.

c) Step 4. $0.0439 < 0.10$

Step 5. Reject H_0. There is enough evidence against the null hypothesis to suggest the average cost is significantly different from\$ 83.85

3. Step 1. $H_0: \mu_1 = \mu_2$
$H_1: \mu_1 \neq \mu_2$

Step 2. p-value $= 0.0272$

Step 3. a) $\alpha = 0.01$
b) $\alpha = 0.05$
c) $\alpha = 0.10$

a) Step 4. $0.0272 > 0.01$

Step 5. Do not reject H_0. There is not enough evidence against the null hypothesis to suggest that the average number of minutes used is significantly different for males and females.

b) Step 4. $0.0272 < 0.05$

Step 5. Reject H_0. There is enough evidence against the null hypothesis to suggest that the average number of minutes used is significantly different for males andfemales.

c) Step 4. $0.0272 < 0.10$

Step 5. Reject H_0. There is enough evidence against the null hypothesis to suggest that the average number of minutes is used significantly different for males and females.

4. Step 1. $H_0: \mu_1 = \mu_2$
$H_1: \mu_1 \neq \mu_2$

Step 2. p-value $= 0.0763$

Step 3. a) $\alpha = 0.01$
b) $\alpha = 0.05$
c) $\alpha = 0.10$

a) Step 4. $0.0763 > 0.01$

Step 5. Do not reject H_0. There is not enough evidence against the null hypothesis to suggest that the average number of gallons used is significantly different between those reside in single family homes and those who reside in apartments.

b) Step 4. $0.0763 > 0.05$

Step 5. Do not reject H_0. There is not enough evidence

against the null hypothesis to suggest that the average number of gallons used is significantly different between those reside in single family homes and those who reside in apartments.

c) Step 4. $0.0763 < 0.10$

Step 5. Reject H_0. There is enough evidence against the null hypothesis to suggest that the average number of gallons used is significantly different between those reside in single family homes and those who reside in apartments.

5. Step 1. H_0: $p = 0.70$
H_1: $p \neq 0.70$

Step 2. p-value $= 0.0147$

Step 3. a) $\alpha = 0.01$
b) $\alpha = 0.05$
c) $\alpha = 0.10$

a) Step 4. $0.0147 > 0.01$

Step 5. Do not reject H_0. There is not enough evidence against the null hypothesis to suggest the number of greenware pieces that are poured from ceramic molds are dinnerware pieces is significantly different from 70%.

b) Step 4. $0.0147 < 0.05$

Step 5. Reject H_0. There is enough evidence against the null hypothesis to suggest the number of greenware pieces that are poured from ceramic molds are dinnerware pieces is significantly different from 70%.

c) Step 4. $0.0147 < 0.10$

Step 5. Reject H_0. There is enough evidence against the null hypothesis to suggest the number of greenware pieces that are poured from ceramic molds are dinnerware pieces is significantly different from 70%.

6. H_0: There no correlation between the variables
H_1: There is a correlation between the variables
The p-value $= 0.0450$ to 3 s.f.
Therefore in order to reject H_0, and thus accept that there is a correlation, $\alpha > 0.0450$.

Exercise 12.4

1. Step 1. H_0: A complete stop at a stop sign is independent of gender.
H_1: A complete stop at a stop sign is dependent of gender.

Step 2. p-value $= 0.0349$ to 3 s.f.

Step 3. $\alpha = 0.01, 0.05, 0.10$

a) Step 4. $0.0349 > 0.01$

Step 5. Do not reject H_0. There is not enough evidence against H_0 to suggest a complete stop at a stop sign is dependent of gender.

b) Step 4. $0.0349 < 0.05$

Step 5. Reject H_0. There is enough evidence against H_0 to suggest a complete stop at a stop sign is dependent of gender.

c) Step 4. $0.0349 < 0.10$

Step 5. Reject H_0. There is enough evidence against H_0 to suggest a complete stop at a stop sign is dependent of gender.

2. Step 1. H_0: The number of cups of coffee is not related to age
H_1: The number of cups of coffee is related to age.

Step 2. p-value $= 0.0710$

Step 3. $\alpha = 0.01, 0.05, 0.10$

a) Step 4. $0.0710 > 0.01$

Step 5. Do not reject H_0. There is not enough evidence against H_0 to suggest the number of cups of coffee is related to age.

b) Step 4. $0.0710 > 0.05$

Step 5. Do not reject H_0. There is not enough evidence against H_0 to suggest the number of cups of coffee is related to age.

c) Step 4. $0.0710 < 0.10$

Step 5. Reject H_0. There is enough evidence against H_0 to suggest the number of cups of coffee is related to age.

3. Step 1. H_0: Extracurricular activities are independent on movie genre.
H_1: Extracurricular activities are dependent on movie genre.

Step 2. p-value $= 1.54 \times 10^{-6}$

Step 3. $\alpha = 0.01, 0.05, 0.10$

Step 4. $1.54 \times 10^{-6} < 0.01, 0.05,$ and 0.10

Step 5. Reject H_0. There is enough evidence against H_0, at each α-level, to suggest the extracurricular activities are dependent on movie genre.

4. Step 1. H_0: The number of hours exercised are independent on gender.
H_1: The number of hours exercised are dependent on gender.

Step 2. p-value $= 0.0959$

Step 3. $\alpha = 0.01, 0.05, 0.10$

a) Step 4. $0.0959 > 0.01$

Step 5. Do not reject H_0. There is not enough evidence against H_0 to suggest the number of hours exercised are dependent on gender.

b) Step 4. $0.0959 > 0.05$

Step 5. Do not reject H_0. There is not enough evidence against H_0 to suggest the number of hours exercised are dependent on gender.

c) Step 4. $0.0959 < 0.10$

Step 5. Reject H_0. There is enough evidence against H_0 to suggest the number of hours exercised are dependent on gender.

Exercise 12.5

1. The null hypothesis says that there is no significant difference between two numbers.
2. The alternative hypothesis says that there is a significant difference between two numbers.
3. Type I
4. I, H_0, true
5. area, critical
6. critical, reject the null hypothesis, accept the null hypothesis
7. percent, against, null
8. 6
9. CV, TV or p-value, α-level
10. Do not reject

11. (i) One variable is independent of the other
    (ii) One variable is not related to the other
12. (i) One variable is dependent of the other
    (ii) One variable is related to the other
13. a) Null Hypothesis. Two numbers are not different.
    b) Alternative Hypothesis. Two numbers are different.
    c) The mean of the population
    d) Degrees of freedom
    e) Critical Value. It separates the "reject the H_0" region from the "accept the H_0" region.
    f) Chi-square value
    g) The percent of evidence against the H_0.
    h) The percent chance of making a Type I error.
14. number
15. p-values, α-levels
16. Descriptive statistics counts all data in the event. Inferential statistics selects a sample, counts that data, and then infers about the population based on that information.
17. There is always a chance a mistake has been made.
18. a) 2.706   b) 7.815   c) 6.251
    d) 13.277   e) 12.592   f) 14.684
19. a) accept   b) accept   c) reject
    d) reject   e) reject   f) accept
    g) accept   h) reject
20. a)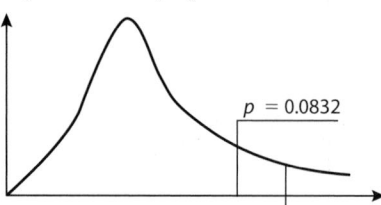

    $p = 0.0832$

    $\alpha = 0.05$

    b)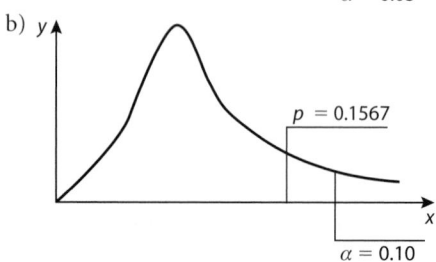

    $p = 0.1567$

    $\alpha = 0.10$

    c)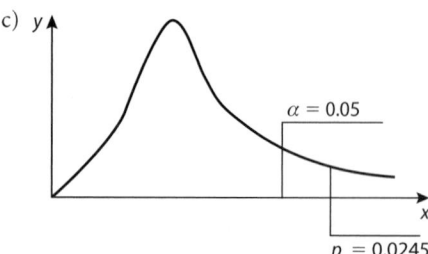

    $\alpha = 0.05$

    $p = 0.0245$

    d)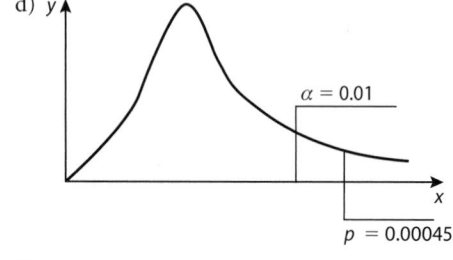

    $\alpha = 0.01$

    $p = 0.000453$

21. e)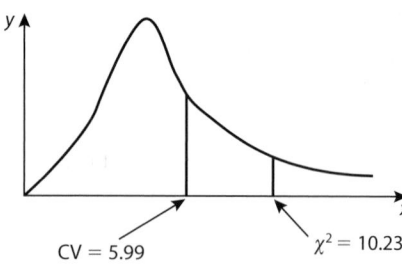

    CV = 5.99 $\chi^2 = 10.23$

    f)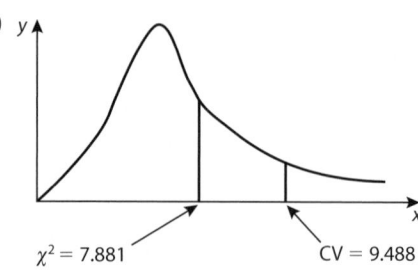

    $\chi^2 = 7.881$ CV = 9.488

    g)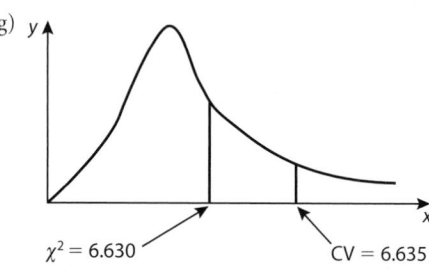

    $\chi^2 = 6.630$ CV = 6.635

    h)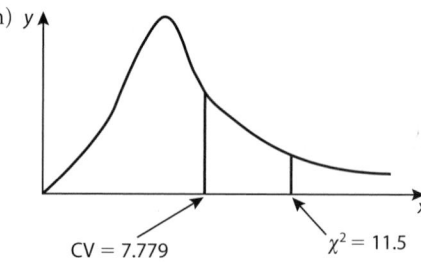

    CV = 7.779 $\chi^2 = 11.5$

22. $\chi^2 = 5.18$
    $p = 0.0751$
    df = 2
    a) Critical value at 5% level is 5.991. Therefore, do not reject H_0 (accept H_0).
    b) No
    c) Critical value at 1% level is 4.605. Therefore, reject H_0.
    d) Yes
23. $\chi^2 = 5.04$
    $p = 0.0248$
    df = 1
    a) p-value = 0.0248 > significance level (0.01). Therefore, do not reject H_0 (accept H_0).
    b) Yes
    c) p-value = 0.0248 < significance level (0.01). Therefore, reject H_0.
    d) No
24. a) (i) $\dfrac{82}{225} \times \dfrac{103}{225} \times 225 = 37.54 = 37.5$
    (ii) $q = 27$, $r = 44.5$, $s = 32$
    b) (i) Null hypothesis H_0: favourite colour is independent of gender

(ii) Alternative hypothesis H_1: favourite colour is
    dependent on gender

c) (i) df = 3

    (ii) $\chi^2 = 4.896$

d) $\chi^2_{calc} = 4.896 < 7.815$. Therefore, do not reject H_0 (accept H_0).

Chapter 13

Exercise 13.1

1. a) 3   b) 2.5

    c) 2.1   d) 2.01

2. a)

x	$f(x) = \frac{1}{3}x^3$	$B(x, f(x))$	Slope of secant line between A and B
1.5	1.1250	(1.5, 1.1250)	1.5833
1.1	0.4437	(1.1, 0.4437)	1.1033
1.01	0.3434	(1.01, 0.3434)	1.0100
1.001	0.3343	(1.001, 0.3343)	1.0010
1.0001	0.3334	(1.0001, 0.3334)	1.0001

b) Estimate is 1.

3. a)

x	$f(x) = \frac{4x-3}{2}$	$B(x, f(x))$	Slope of secant line between A and B
0.5	−0.5000	(0.5, −0.5000)	2
0.1	−1.3000	(0.1, −1.3000)	2
0.01	−1.4800	(0.01, −1.4800)	2
0.001	−1.4980	(0.001, −1.4980)	2
0.0001	−1.4998	(0.0001, −1.4998)	2

b) Since the slope of the secant line is always 2, this is a good estimate for the slope)

4. $f'(1) = 3$

5. $f'(-1) = -4$

6. a) $y = 2x - 2$

b)

x	$f(x) = x^2 - x$	$B(x, f(x))$	Slope of secant line between A and B
2	2	(2, 2)	2
1.5	0.75	(1.5, 0.75)	1.5
1.1	0.11	(1.1, 0.11)	1.1
1.01	0.0101	(1.01, 0.0101)	1.01
1.001	0.001 001	(1.001, 0.001 001)	1.001

c) $f'(1) = 1$

7. a) $y = 7x - 6$

b)

x	$f(x) = x^3$	$B(x, f(x))$	Slope of secant line between A and B
2	8.000	(2, 8.000)	7
1.5	3.375	(1.5, 3.375)	4.75
1.1	1.331	(1.1, 1.331)	3.31
1.01	1.030	(1.01, 1.030)	3.030
1.001	1.003	(1.001, 1.003)	3.003

c) $f'(1) = 3$

8. a)

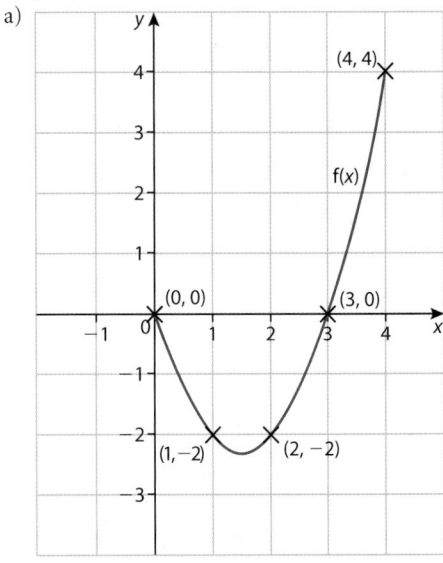

b) (i) $m = \frac{3}{2}$

     $y = \frac{3}{2}x - \frac{9}{2}$

(ii)

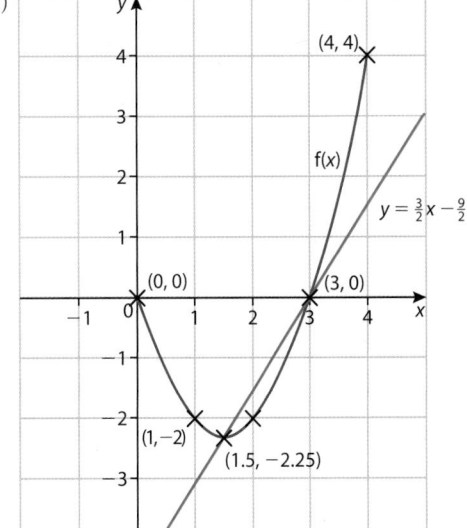

c) (i) $f'(3) = 3$

(ii)

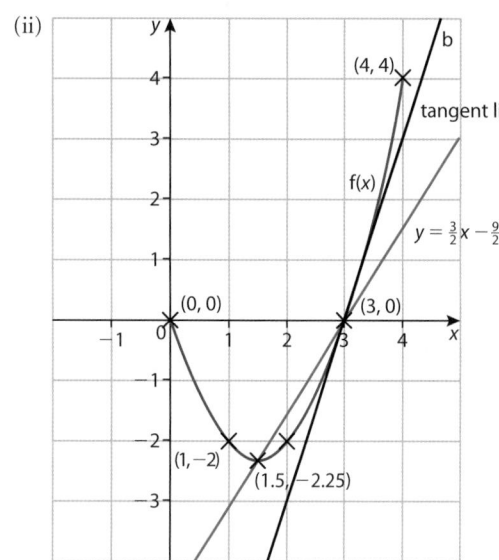

Exercise 13.2

1. a) $f'(x) = 12x - 5$ b) $f'(x) = 12x^2 + 10$
   c) $f'(x) = 2x^2 + 9x$
   d) $f'(x) = 21x^6 - 20x^3 - 6x$

2. a) $f'(x) = -6x^{-4} + \frac{3}{4}$
   b) $f'(x) = -16x^{-5} - 2x^{-2}$
   c) $f'(x) = -6x^{-4} - 2x^{-3}$
   d) $f'(x) = -3x^{-5} + \frac{1}{2}x^{-3}$

3. a) $f'(1) = 9$ b) $f'(1) = 0$
   c) $f'(1) = -7.6$ d) $f'(1) = -18$

4. a) $f'(-1) = -2$ b) $f'(-1) = -1$
   c) $f'(-1) = 22$ d) $f'(-1) = 20\,200$

5. a) $\dfrac{d^2y}{dx^2} = 2$ b) $f''(x) = 42x^5 + 20x^3$
   c) $D''(t) = -20$ d) $\dfrac{d^2y}{dx^2} = 0$

6. a) $f''(1) = 18$
   b) $f''(1) = 18$
   c) $H''(1) = 24$
   d) $g''(1) = 8$

7. a) $B(11) = 21.1$ million balloons.
   b) (i) $B'(x) = 0.052x^4 - 1.2232x^3 + 9.297x^2 - 24.598x^2 + 16.983$
      (ii) $B'(11) = 4.5948$, hence 4.6 million balloons.

8. a) (i)

x	0	2	4	6	8	10	12
$f(x)$	0	11.9	15.7	14.9	13.1	13.8	20.6

   (ii)

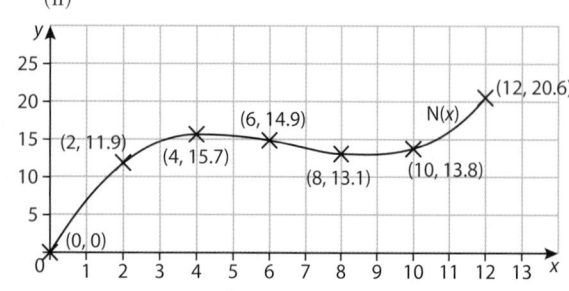

   b) (i) $N'(x) = 0.222x^2 - 2.92x + 8.58$
      (ii) $N'(2) = 3.63$ thousand users per month.

9. a) 100 m
   b) ≈ 4.47 seconds
   c) (i) $D'(t) = -10t$
      (ii) ≈ -44.7 ms^{-1}
      (iii) Since the sign is negative, the ball is travelling down.
   d) -10 ms^{-2}

10. a) (i) $f'(x) = 6x^2 + 6x - 12$
       (ii) $f'(-1) = 6(-1)^2 + 6(-1) - 12 = -12$
    b) (i)

x	-3	-2	-1	0	1	2
$f(x)$	9	20	13	0	-7	4

    (ii)

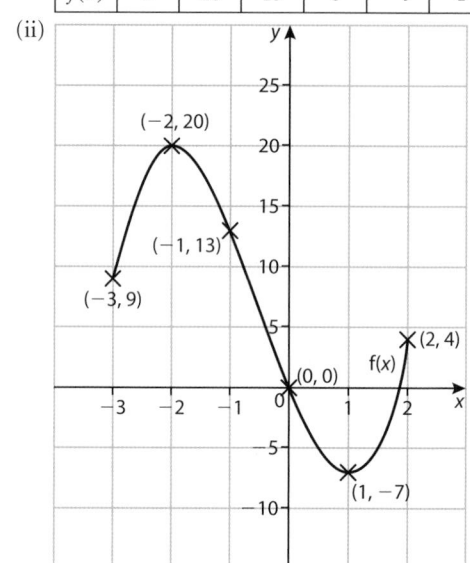

    c)

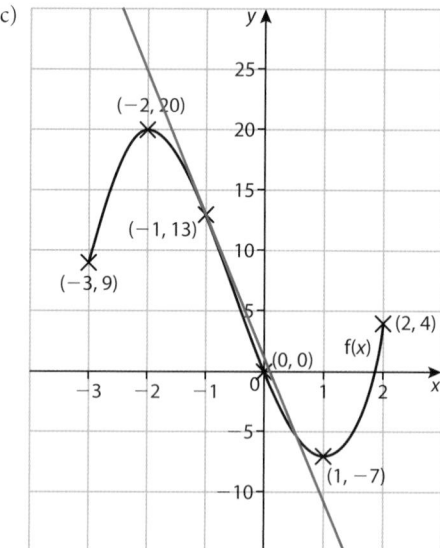

    d) (i) $f''(x) = 12x + 6$
       (ii) $f''(-1) = -6$

Exercise 13.3

1. a) 0 b) 8
   c) 0 d) 400 40
2. a) -4 b) 220
   c) -196 d) -50

3. a) $y = \dfrac{x + 382}{4}$

b) $y = \dfrac{52\,798 - x}{220}$

c) $y = \dfrac{x + 1\,882\,232}{196}$

d) $y = \dfrac{x - 20}{50}$

4. a) $x = 5000$

b) $x = \frac{5}{6}$

c) $x = 3, x = -2$

d) $x = -3, x = 3$

5. a) $x = 3$

b) $x = -3, x = 3$

c) $x = -1, x = 5$

b) All $x \in \mathbb{R}$

6. a) Eqn. of tangent: $y = 3x + 5$, eqn. of normal
$y = -\dfrac{1}{3}x + 15$

b) Eqn. of tangent: $y = 6x - 9$, eqn. of normal
$y = -\dfrac{1}{6}x + \dfrac{19}{2}$ or $6y + x = 57$

c) Eqn. of tangent: $y = -6x + 29$, eqn. of normal
$y = \dfrac{1}{6}x + \dfrac{21}{2}$ or $6y - x = 63$

d) Eqn. of tangent: $y = 21x - 43$, eqn. of normal
$y = \dfrac{-1}{21}x + \dfrac{141}{7}$ or $21y + x = 423$

7. a)

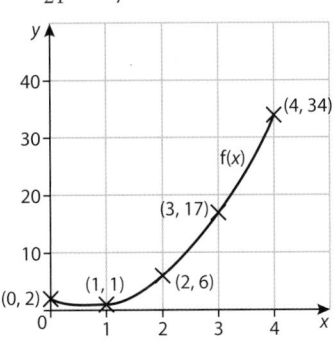

b) (i) $f'(x) = 6x - 4$
(ii) $y = 8x - 10$
(iii)

8. a) $g'(x) = 6x^2 - 30x + 24$
b) (i) $y = -12x + 28$
(ii) $y = \dfrac{1}{12}x + \dfrac{23}{6}$

9. a) $H'(x) = 20x - 2x^2$ b) $x = 0, x = 10$
c) $y = 0, y = \dfrac{1000}{3}$

10. a) $f'(x) = 3x^2 + 14x - 5$ b) $f'(1) = 12$
c) $x = -5$ or $x = \frac{1}{3}$

11. a) $\dfrac{dy}{dx} = 3x^2 + 2x - 3$ b) $\dfrac{dy}{dx} = 13$

c) Increasing d) $y = 13x - 16$
e) $x = -1.79$ or $x = 1.12$ (to 3 s.f.)

f) Equation of normal $y = -\dfrac{(x + 132)}{13}$

12. a) (i) $f'(x) = 2x - 8$
(ii) $x = 4$
b) (i) $f(2) = (2)^2 - 8(2) = 4 - 16 = -12$
(ii) -4
(iii) $x = 2$
(iv) $y = -4x - 4$

13. a) (i) $f'(x) = 2ax - 4$
(ii) $a = 1$
(iii)

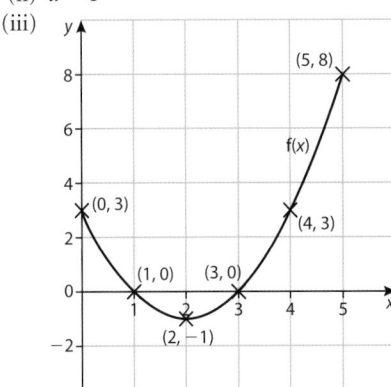

b) (i) $y = 2x - 5$
(ii) $y = 2x - 6$
(iii)

Exercise 13.4

1. a) $[-\frac{3}{2}, \infty)$ b) $[3, \infty)$
c) $(-\infty, -4]$ and $[-2, \infty)$ d) $(-\infty, 4]$

2. a) $(-\infty, \infty)$ b) $[5, \infty)$
c) $[2, \infty)$ d) $[3, 5]$

3. a) $(3, -6)$ b) $(\frac{1}{3}, \frac{1}{2})$ or $(3, 9)$
c) $(1, 5)$ or $(2, 4)$ d) $(0, 0)$ or $(2, -16)$

4. a) $[0, \infty)$ b) $(-\infty, \infty)$
c) $(-\infty, \infty)$ d) $[-1, 1]$ and $[3, \infty)$

5. a) $(-\infty, \infty)$ b) $[0, \infty)$
c) $[-2, 2]$ d) $(-\infty, -1]$ and $[1, 3]$

6. a) 2000, 2001, 2002, 2003, 2007
b) 2004, 2006
c) 2004

7. a) Answers will vary
b) Answers may vary but must include $x = 3$ and $x = 5$.
8. a)

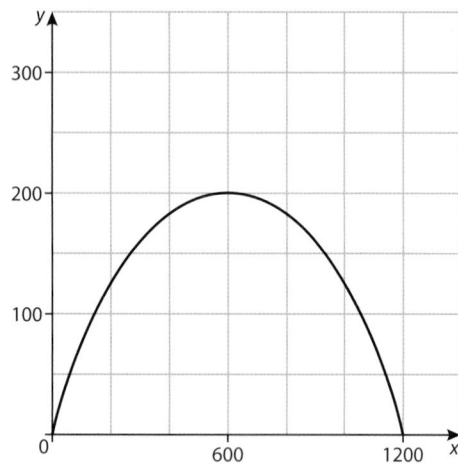

b) (i) $f(0) = 0$ and $f(1200) = 0$.
(ii) $x = 600$
c) (i) $f'(x) = 1200a - 2ax$
(ii) $a = \frac{1}{1800}$
9. a) B, C, D, E
b) From A to B, C to D, and E to F
c) (i) True
(ii) False
(iii) True
10. Answers may vary.
11. a) $f'(x) = 3x^2 - 6x + 3$
b)

x	-1	0	1	2	3
$f(x)$	-7	0	1	2	9
$f'(x)$	12	3	0	3	12

c)

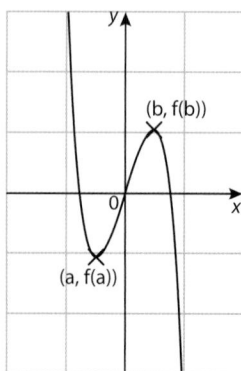

d) 12.
e) (i) From the graph $f(x)$ is increasing for all x so our

interval is $(-\infty, \infty)$
(ii) From the graph $f(x)$ is never decreasing.
12. a) (i) B is a maximum point
(ii) C is a minimum point
b) The function decreases from B to C, where it increases from C to D.
c) Since the gradient of the function is positive in these intervals, it is increasing there:
A to B, C to E and on
d) $m = \dfrac{f(a + 4) - f(a)}{4}$

Exercise 13.5

1. a) Local minimum at $(1, -7)$
b) Local maximum at $(40, 1600)$
c) Local maximum at $(-1, 21.5)$, local minimum at $(6, -150)$
d) Local minimum at $(-1, -3)$, local maximum at $(0, 0)$, local minimum at $(1, -3)$
2. a) Local minimum at $(2, -4)$
b) Local maximum at $(1, 1)$
c) Local maximum at $(-1, \frac{2}{3})$, local minimum at $(1, -\frac{2}{3})$
d) Local maximum at $(-3, -4.5)$, local minimum at $(-2, -4\frac{2}{3})$
3. a) $(5, 6.75)$
b) $(3, 4.5)$
c) $(4, 12)$
d) $(-3, 108)$
4. a) Absolute minimum at $(1.5, -2.25)$, absolute maximum at $(4, 4)$.
b) Absolute minimum at $(0, 0)$ and $(4, 0)$, absolute maximum at $(2, 4)$.
c) Absolute minimum at $(0, 0)$ and $(3, 0)$, absolute maximum at $(1, 4)$ and $(4, 4)$.
d) Absolute minimum at $(-2, -10)$, absolute maximum at $(2, 18.5)$.
5. a) $x = 2$ (double root)
b) (i) $f'(x) = 2x - 4$
(ii) $x = 2$
c) $(5, 9)$
6. a) $P'(x) = 200 - 2x$
b) $x = 100$
c) 10 million dollars.
7. a) $L = 1.2 - 2x$
b) $V = x(1.2 - 2x)^2$
c) Maximum volume will be 0.128 m^3.
8. a) *Possible* curve below.

b) (i) False (ii) True (iii) False
(iv) True (v) False

9. a) $g'(x) = 2px + q$ b) $p = 1, q = 6$
c) (i) $x = -3$ (ii) $c = -3$

10. a) $a = 2, b = 20, c = 9, d = 8, e = 32$
b) $A(x) = x(12 - x)$
c) The length and width are both 6 m.

11. a) $L = 2x + y$.
b) We substitute our value of 2500 metres for L into
$L = 2x + y$.
$2500 = 2x + y$
c) (i) Area of a rectangle is length times width, hence:
$A = L \times W$
$A(x) = x(2500 - 2x)$
$A(x) = 2500x - 2x^2$
(ii) $A'(x) = 2500 - 4x$
(iii) $x = 625$
(iv) $781\,250\text{ m}^2$

12. a) (i) $v = 1\text{ ms}^{-1}$
(ii) $v = 1.125\text{ ms}^{-1}$
b) $a = 0.375\ b = 3$
c) (i) $\frac{dv}{dt} = 3t^2 - 8t + 4t = \frac{2}{3}$ and $t = 2$.
$t = \frac{2}{3}$ corresponds to a local maximum and $t = 2$ corresponds to a local minimum.
(ii) The function increases up to its maximum, then decreases. At the maximum point the gradient of the function is 0.

d)
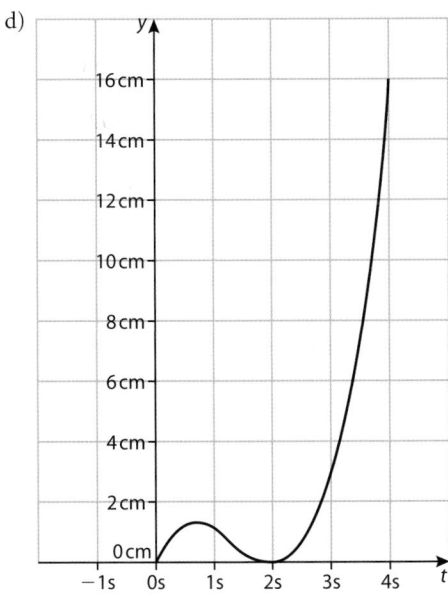

e) In the interval $[0, \frac{2}{3}]$ v is increasing, in the interval $[\frac{2}{3}, 2]$ v is decreasing, and in the interval $[2, 4]$ v is increasing.

Chapter 14

1. a) 14 b) -4 c) $\frac{32}{21}$
d) $6.5 = \frac{13}{2}$ e) $\frac{25}{99}$ f) 5

2. 2,3,5,7,11,13,17,19,23,29
3. a) $2^3 \cdot 3^2$ b) $2^2 \cdot 61$
4. a) 1,2,4,5,8,10,16,20,40,80 b) 1,2,4,17,34,68
5. a) 7,14,21,28,35 b) 26,52,78,104,130
6. a) GCF $= 4$
LCM $= 4392$
b) GCF $= 8$
LCM $= 720$
7. 250
8. $322\,581
9. 4.8 m
10. a) $-2x + 7$
b) $x^2 + 2x - 15$
c) $a^2 + 10a + 25$
d) $4x^3 - 24x^2 + 21x + 49$
e) $3x^2 - 48$
f) $y^3 + 6y^2 + 12y + 8$
11. a) $3(x - 2y + 5)$
b) $(x - 8)(x + 7)$
c) cannot be factorized
d) $(x + 6)(x - 6)$
e) $3(x + 3)(x - 3)$
f) $(2x + 1)(x - 5)$
12. a) $y = \dfrac{-ax - d}{b} = \dfrac{-a}{b}x - \dfrac{d}{b}$
b) $b = \dfrac{360}{P}$
c) $r = \sqrt{\dfrac{A}{4\pi}} = \dfrac{1}{2}\sqrt{\dfrac{A}{\pi}}$
d) $a = \dfrac{2A}{b \sin C}$
13. a) 8π b) 30.6π c) -15
14. a) $\frac{9}{5}$ b) 11 c) 6
15. a) $x = 3, y = 0$ b) $x = -8, y = -15$
16. a) $\frac{1}{2}$ b) $\frac{1}{27}$ c) 1
d) 1024 e) 1024 f) $\frac{1}{343}$
17. a) $x > 4$ b) $x \geqslant -3$
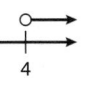 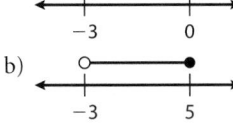
18. a) 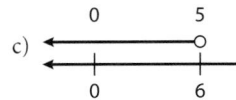 b)
c)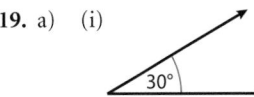
19. a) (i) (ii)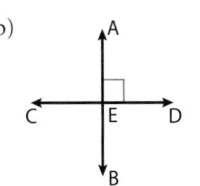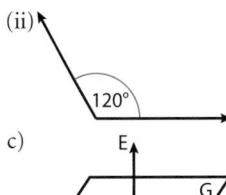
b) c)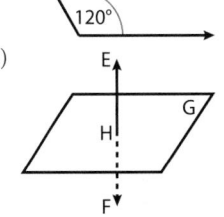

20. a) $y = 0$ b) $x = 0$

c), d)

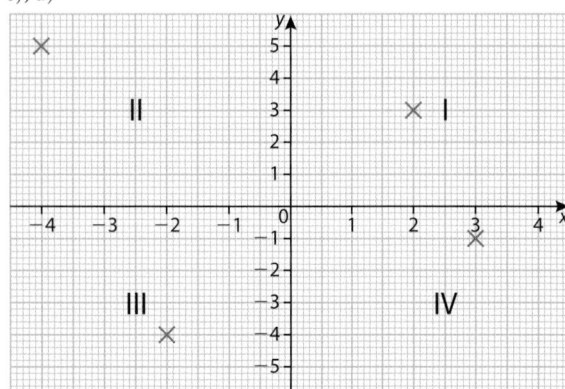

21. a) $(1,1)$ b) $(1.5,1.5,6)$

22. a) 10 b) $\sqrt{146} = 12.1$ to 3 s.f.

23. a)

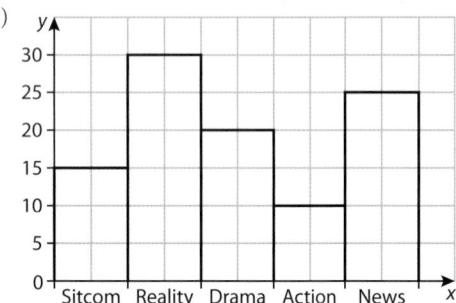

b)

24.

 = 1000 homes

25. a) 1019.1 b) 49 280.50

26. a) $\{0,1,2,3,\ldots\}$

b) $\{0,1,2,3,\ldots\}$

c) $\frac{1}{1},\frac{1}{2},\frac{1}{3},\ldots$
$\frac{2}{1},\frac{2}{2},\frac{2}{3},\ldots$
$\frac{3}{1},\frac{3}{2},\frac{3}{3},\ldots$
⋮ ⋮ ⋮

d) $\{\ldots,-3,-2,-1,0,1,2,3,\ldots\}$

27. a) -1080 b) 2.3×10^3

c) 2310 d) -4946.47

28. a) $a = 8.02$, $b = 10.5$, $\sin C = 0.838$

b) $A = \frac{1}{2}(8.0243)(10.548)(0.838\,47) = 35.5$ to 3 s.f.

c) The student rounded off too early.

29. a) -258 b) 0.402%

30. No. This slope is not steep enough. It appears to be a slope of about $\frac{1}{2}$.

31. a) 1.06 b) 3.78 c) 68.0 d) 124.3

e) 0.350 f) 15.3 g) 23.9

32. a) 1523.5 b) 3281.92

33. a) 30 b) 1970 c) £1007.26

34. $(3.37, 1.42)$

35. $x = \dfrac{-1}{2}$ or $x = 3$

36. a) $x = \dfrac{7 \pm \sqrt{41}}{4}$ b) $x = 3.35$ or $x = 0.149$

37. 152

38. £8880

39. 5050

40. €282 500

41. 2187

42. a) $48\,737.67 b) 22 years

43. 88 573

44. a) 11.8 m b) $188.681\,24 = 189$ m to 3 s.f.

45. a) 5105.17 b) 2805.17

46. a) 3750 b) 21 250 c) 9428.74

47. 14.2 years (accept 14 or 15 years with explanation)

48. a) C b) C c) D d) C

e) D f) D g) C h) C

49. a)

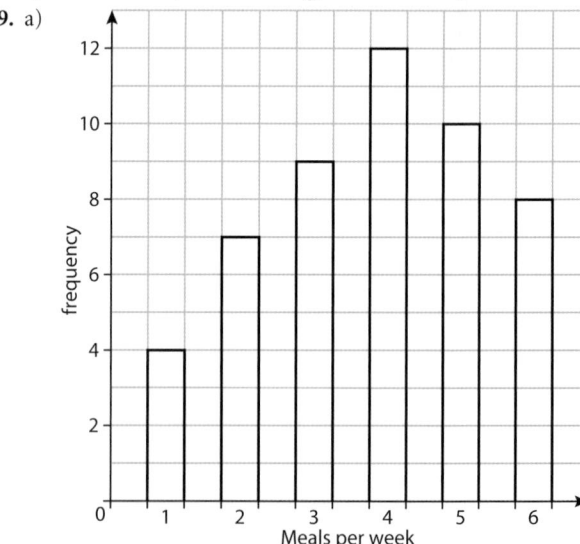

b) 3.82 c) 4

d) 4 e) 1.51 to 3 s.f.

f)

cf
4
11
20
32
42
50

g) 20 couples ate up to 3 meals per week.

50. a)

Pages Read	Frequency f	Mid-Intervals	$f \cdot MI$	Cumulative Frequency
0–10	15	5	75	15
11–21	20	16	320	35
22–32	30	27	810	65
33–43	22	38	836	87
44–54	13	49	637	100
Totals	100		2678	

  b) $26.78 = 27$ pages to the nearest page
  c) $22 - 32$
  d) $22 - 32$
  e) 87 students read up 43 pages per day

51. a) 5.69 to 3 s.f.
  b) $4 - 6$ class
  c) $4 - 6$ class
  d) 2.63 to 3 s.f.

52. a) range = 24
    median = 16
    $Q_1 = 10$
    $Q_3 = 19$
    IQR = 9
    Standard deviation = 6.34 to 3 s.f.
    no mode
  b) range = 18
    median = 6
    $Q_1 = 2.5$
    $Q_3 = 8.5$
    IQR = 6
    Standard deviation = 4.73 to 3 s.f.
    Mode = 3 and 6
  c) range = 193
    median = 255
    $Q_1 = 212$
    $Q_3 = 278$
    IQR = 66
    Standard deviation = 48.5 to 3 s.f.
    no mode

53. a)

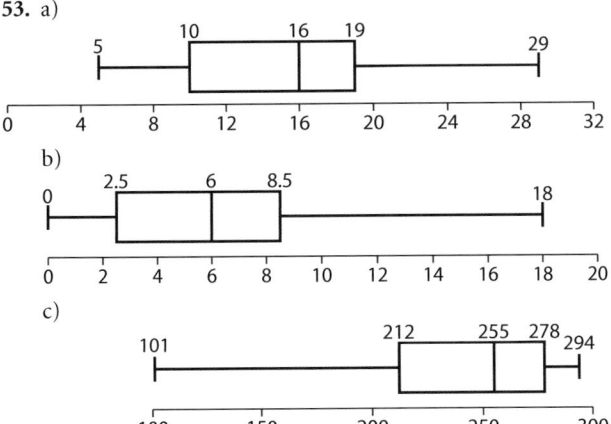

  b)

  c)

54. a) 39
  b) 36
  c) 57
  d) 21

55.

Class limits	Class boundaries	Tallymarks	Frequency	Class mid-interval	Cumulative frequency	Relative c.f.
12–20	11.5–20.5	III	3	16	3	0.0769
21–29	20.5–29.5	THL II	7	25	10	0.256
30–38	29.5–38.5	THL III	8	34	18	0.462
39–47	38.5–47.5	THL THL	10	43	28	0.718
48–56	47.5–56.5	THL	5	52	33	0.846
57–65	56.5–65.5	IIII	4	61	37	0.949
66–74	65.5–74.5	II	2	70	39	1.00

  a)

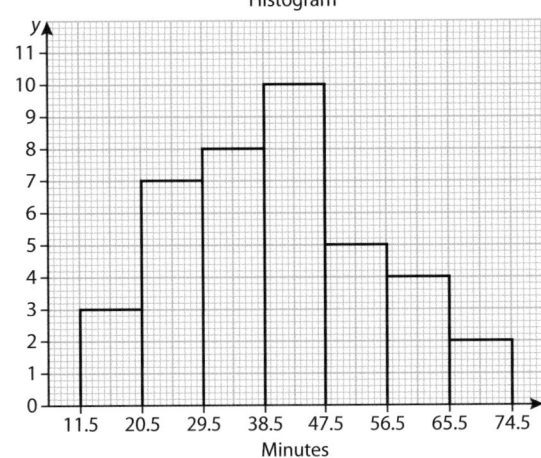

  b)

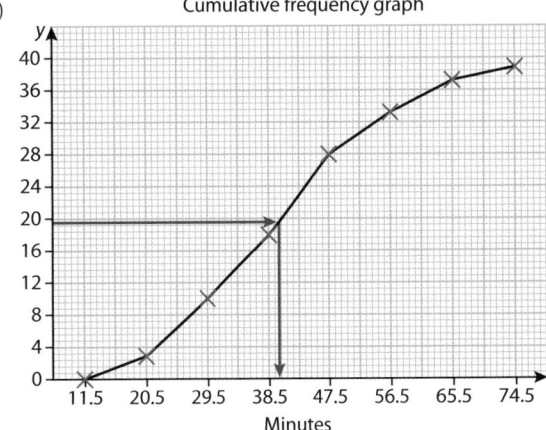

c)

Percentile graph

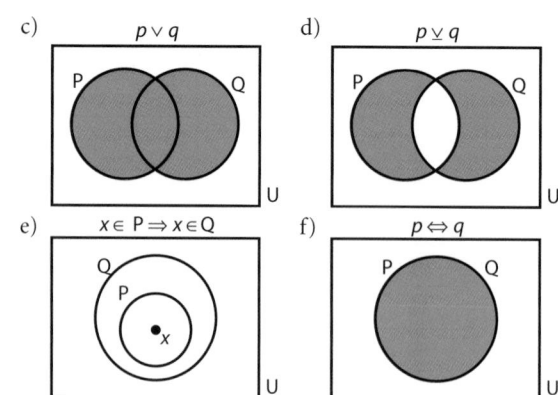

Minutes

d) 9   e) 20.5   f) 65.5

g) Total values below a given number

h) 53

i) 80% (about 31) of the students took 53 minutes or less to get to school.

j) (i) ≈ 39.9   (ii) 29.5   (iii) ≈ 49.8

k) 40.23

l) 40

m) The mid-interval values only approximate the data.

n) ≈ 39.9

o) 39

p) Answers obtained from a graph are only approximate.

56. a) 47.4   b) 24.6 to 3 s.f.

57.

p	q	$\neg p$	$p \wedge q$	$p \vee q$	$p \Rightarrow q$	$p \veebar q$
T	T	F	T	T	T	F
T	F	F	F	T	F	T
F	T	T	F	T	T	T
F	F	T	F	F	T	F

58. ¬; negation; not

∧; conjunction; and

∨; disjunction; or

⇒; implication; implies; if … then

⊻; exclusive disjunction; one or the other, but not both

59.

		(a)	(b)	(c)	(d)
p	q	$\neg p \Rightarrow q$	$(p \Rightarrow q) \Leftrightarrow (\neg q \Rightarrow p)$	$\neg(p \vee q) \Rightarrow (\neg p \wedge \neg q)$	$(p \veebar q) \wedge p$
T	T	T	T	T	F
T	F	T	F	T	T
F	T	T	T	T	F
F	F	F	F	T	F

60. a) ¬p   b) p ∧ q

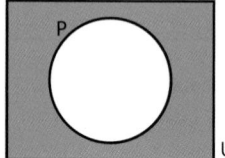   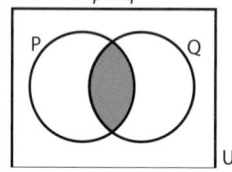

c) $p \vee q$ d) $p \veebar q$

e) $x \in P \Rightarrow x \in Q$ f) $p \Leftrightarrow q$

61. $[(R \Rightarrow W) \wedge W] \Rightarrow R$.

62. The argument is not valid since it is not a tautology.

R	W	$[(R \Rightarrow W) \wedge W] \Rightarrow R$
T	T	T
T	F	T
F	T	F
F	F	T

63. a)

p	q	$\neg p \wedge q$
T	T	F
T	F	F
F	T	T
F	F	F

b) $\neg p \wedge q$ is true when p is false and q is true. (See line 3.)

64. a) M ⇒ P

b) ¬M ⇒ ¬P. If this is not a maths test, then I will not pass it.

c) P ⇒ M. If I pass the test, then it is a maths test.

d) ¬P ⇒ ¬M. If I do not pass the test, then it is not a maths test.

e) d, the contrapositive.

65. a) $p \Rightarrow r$ b) a c) n

d) p e) $\neg a$

66. { }

{a}, {b}, {c}, {d}

{a,b}, {a,c}, {a,d}, {b,c}, {b,d}, {c,d}

{a,b,c}, {a,b,d}, {a,c,d}, {b,c,d}

{a,b,c,d}

67

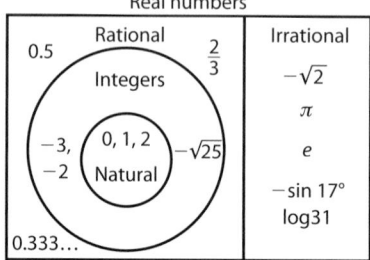

Real numbers

68. a) 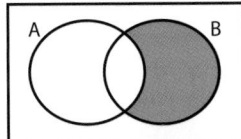 b)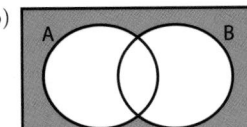

c)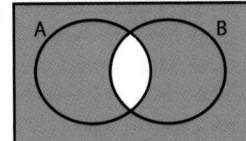

69. a)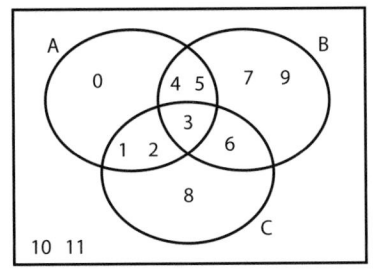

b) (i) {3,4,5}
(ii) {0,1,2,3,4,5,6,8}
(iii) {6,7,8,9,10,11}
(iv) {0,10,11}
(v) {0,3,4,5,6,7,9,10,11}

70. a) $\frac{1}{16}$ b) $\frac{1}{4}$ c) $\frac{15}{16}$
d) $\frac{1}{4}$ e) $\frac{11}{16}$

71. 0.05

72. a) (i) $\frac{40}{323} = 0.124$ to 3 s.f. (ii) $\frac{121}{969} = 0.125$ to 3 s.f.
(iii) $\frac{683}{969} = 0.705$ to 3 s.f. (iv) $\frac{60}{323} = 0.186$ to 3 s.f.
(v) $\frac{65}{323} = 0.201$ to 3 s.f.

b) (i) $\frac{1000}{6859} = 0.146$ to 3 s.f. (ii) $\frac{1027}{6859} = 0.150$ to 3 s.f.
(iii) $\frac{4662}{6859} = 0.680$ to 3 s.f. (iv) $\frac{1080}{6859} = 0.157$ to 3 s.f.
(v) $\frac{1404}{6859} = 0.205$ to 3 s.f.

73.

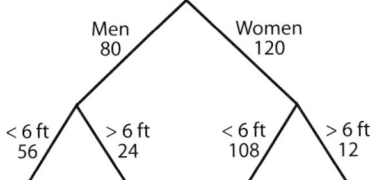

74. a) $\frac{492\,960}{7\,880\,400} = 0.0626$ to 3 s.f.
b) $\frac{1\,685\,040}{7\,880\,400} = 0.214$ to 3 s.f.
c) $\frac{9}{50} = 0.18$
d) $\frac{7}{10} = 0.70$
e) $\frac{2\,275\,200}{7\,880\,400} = 0.289$ to 3 s.f.
f) $\frac{7\,387\,440}{7\,880\,400} = 0.937$ to 3 s.f.
g) $\frac{952}{2189} = 0.435$ (3 s.f.)

75. a) 0.20
b)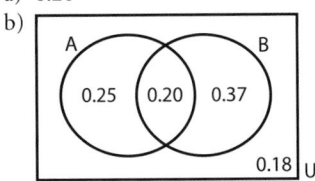

76. a) $\frac{1}{2}$ b) $\frac{7}{13}$ c) $\frac{2}{13}$ d) $\frac{11}{26}$

77. $\frac{1024}{3125} = 0.327\,68$

78. a) $\frac{12}{25}$ b) $\frac{17}{50}$ c) $\frac{37}{50}$ d) $\frac{9}{17}$ e) $\frac{6}{13}$

79. 22.7%

80. a), e)

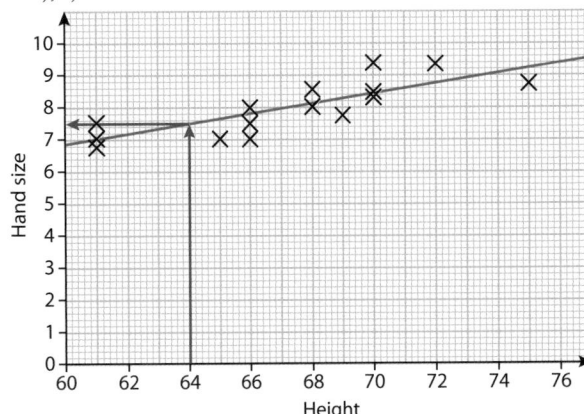

b) (i) $r = 0.819665 = 0.820$ to 3 s.f.
c) There is a strong positive correlation between height and hand size.
d) (i) $y = 0.162\,648x - 2.9966$
$y = 0.163x - 3.00$ to 3 s.f.
f) (i) 7.5 inches to the nearest 0.5 inches
(ii) 58.5 inches to the nearest 0.5 inches
Since 6.5 inches is outside the data range we are less confident in this answer than if we were considering a value inside the data range.
g) (i) 10.0 inches to the nearest 0.5 inches
Since 80 inches is outside the data range we are less confident in this answer than if we were considering a value inside the data range.
(ii) $73.76 = 74.0$ inches to the nearest 0.5 inches

81. a) H_0: Recycling is independent of gender
H_1: Recycling is dependent on gender
b) p-value = 0.0779
c) (i) Do not reject (accept) H_0 since $0.0779 > 0.01$
(ii) Do not reject (accept) H_0 since $0.0779 > 0.05$
(iii) Reject H_0 since $0.0779 < 0.10$.
d) (i), (ii) There is enough evidence to support Nick's conjecture that recycling is independent of gender.
(iii) There is not enough evidence to support Nick's conjecture that recycling is independent of gender.

82. $8x - 5y + 1 = 0$
83. 12 or (0,12)
84. $2x - 3y + 13 = 0$

85.

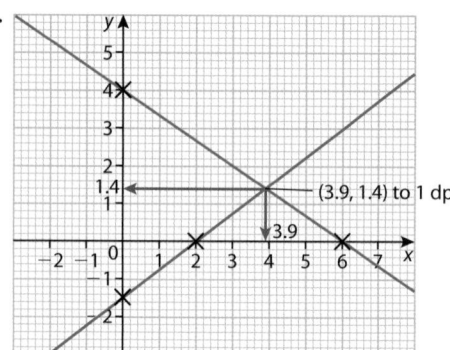

86. a) $(\frac{32}{11}, -\frac{19}{11})$
b) $(2.91, -1.73)$ to 3 s.f.

87. a) $y = \frac{3}{5}x - \frac{11}{5}$ (accept: $y = 0.6x - 2.2$)
b) $3x - 5y - 11 = 0$

88. a) 0.3907   b) 0.08716   c) 2.747

89. a) $\frac{a}{c}$ b) $\frac{b}{c}$ c) $\frac{a}{b}$ d) $\frac{b}{c}$ e) $\frac{a}{c}$ f) $\frac{b}{a}$

90. a) 5   b) 15   c) 16
d) $\sqrt{240} = 4\sqrt{15}$ e) 20

91. a) 65°   b) 8.58   c) 9.46

92. $EF = 5.18$, $\angle F = 37.4°$ to 3 s.f., and $\angle E = 104°$ to 3 s.f.

93. a) 33.7°   b) 13.6°   c) 45

94. a)

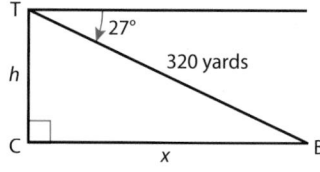

b) 285 yards to the nearest yard
c) 145 yards to the nearest yard
d) 27°

95. 7.71 sq. units

96. a) 10
b) (i) 2.5   (ii) 53.1° to 3 s.f.
c) 5
d) $\sqrt{116} = 10.8$ to 3 s.f.
e) (i) 5   (ii) 26.6° to 3 s.f.   (iii) 38.7° to 3 s.f.
f) 21.8° to 3 s.f.
g) 63.4° to 3 s.f.
h) (i) $\sqrt{104} = 10.2$ to 3 s.f.
   (ii) 11.3 to 3 s.f.

97. a) (i) 8   (ii) 90°   (iii) 53.1° to 3 s.f.
b) (i) 5.29 to 3 s.f.   (ii) 90°
  (iii) 48.6° to 3 s.f.   (iv) 6
  (v) 8.49 to 3 s.f.   (vi) 31.9° to 3 s.f.
c) 336 sq units.

98. a) 0   b) 5   c) 28   d) −9

99. a) Domain = $\{x \mid x \in \mathbb{R}\}$
  Range = $\{y \mid y \geq 1\}$
b) Domain = $\{x \mid x \geq 0\}$
  Range = $\{y \mid y \geq 0\}$
c) Domain = $\{x \mid x \in \mathbb{R}, x \neq 0\}$
  Range = $\{y \mid y \in \mathbb{R}, y \neq 0\}$

100. a)   b)

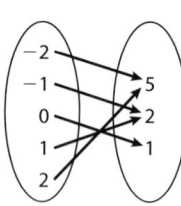
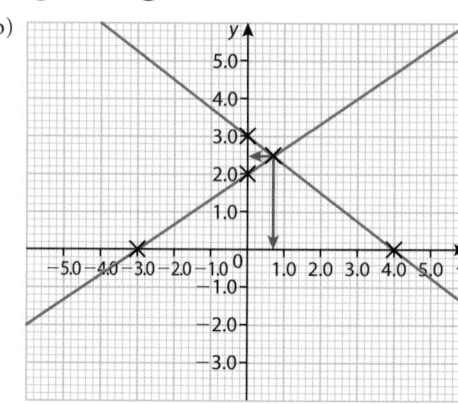

101. a), b)

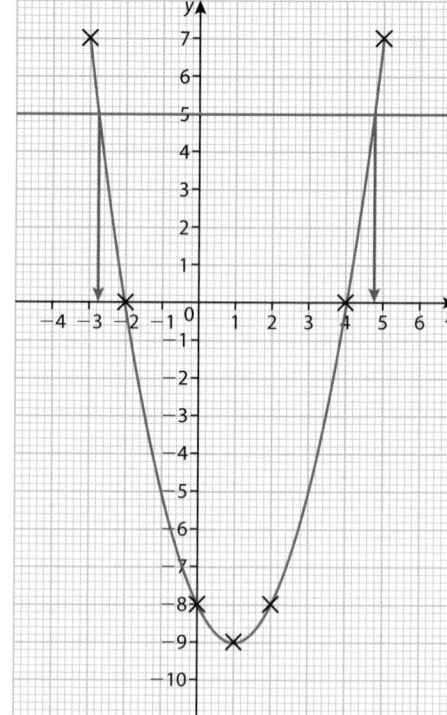

102. $(0.7, 2.5)$ to 1 decimal place

103. a) 4, −2   b) −8
c) $x = 1$ d) $y = (x - 1)^2 - 9$
e) $(1, -9)$
f) See graph

g) $x = -2.7$ or 4.7 to 1 decimal place
h) Domain = $\{x \mid x \in \mathbb{R}\}$
  Range = $\{y \mid y \geq -9\}$

104. a) 350   b) 200.60
c) 4.14 years to 3 sf   d) 150
e) Answers vary. Electronic equipment has little or no value after new technology is introduced.
The model is probably accurate for a short time.

105. a)

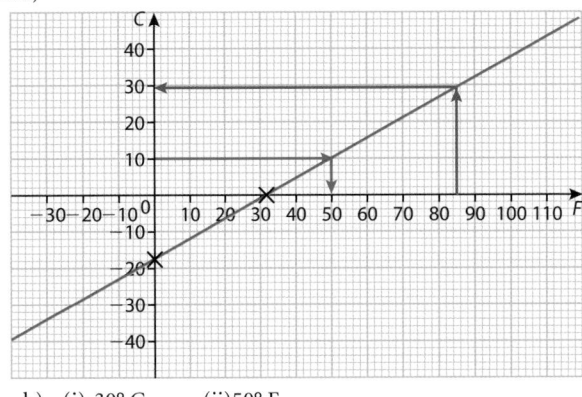

b) (i) 30° C (ii)50° F

106. a)

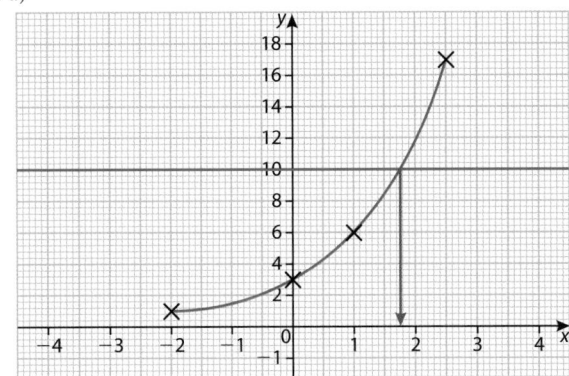

b) $y = 0$

c) Domain = $(-\infty, 2.5]$
   Range = $(0, 16.971])$ y is correct to 5 s.f.

d) 1.7 to 1 dp

107. a) HA: $y = -1$

b) VA: $x = -3$
   HA: $y = 0$

c) VA: $x = 0$
   HA: $y = 2$

d) VA: $x = 5$
   HA: $y = 3$

e) HA: $y = 2$

108. a)

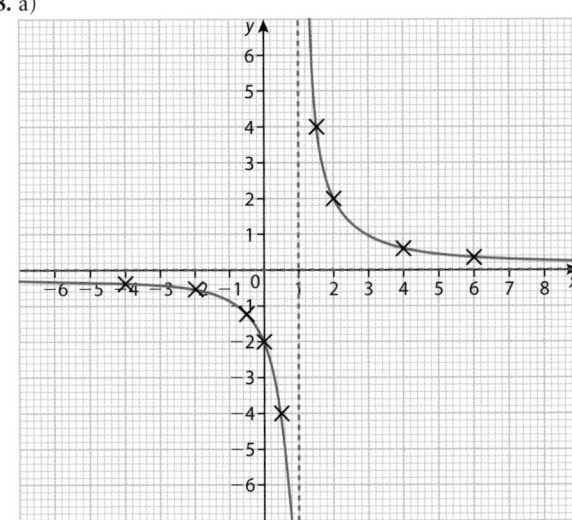

b) 1

c) $x = 1$

d) $y = 0$

e) Domain = $\{x | x \in \mathbb{R}, x \neq 1\}$
   Range = $\{y | y \in \mathbb{R}, y \neq 0\}$

109. a)

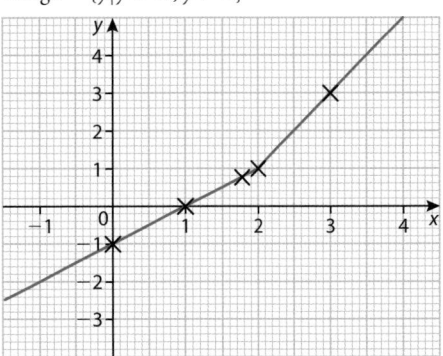

b) Yes. (2,2) exists on both pieces of the graph. You can trace the graph without lifting you pencil. (Note: This is a loose definition of continuity).

110. a) $(-0.281, -1.56)$ and $(1.78, 2.56)$ to 3 s.f.

b) -2.32 to 3 s.f.

c) $(-0.767, 0.588)$ to 3 s.f., $(2, 4)$, and $(4, 16)$

d) 0.269 or 2.17 to 3 s.f.

111. a) $f(1) = 4; P(1, 4)$

b) $f(1.1) = 4.21; Q_1(1.1, 4.21)$

c) $m = 2.1$

d) $f(1.01) = 4.0201; Q_2(1.01, 4.0201)$

e) $m = 2.01$

f) 2

112. For $y = f(x)$, the derivative is given by:
$$f'(x) = \lim_{h \to 0} \frac{f(x + h) - f(x)}{h}$$

113. The gradient of the curve at any given x-value.
The gradient of the tangent line to the curve at any given x-value.
The instantaneous rate of change of y with respect to x.

114. Any three of: $y', \dfrac{dy}{dx}, f'(x), \dfrac{d}{dx}y, D_x y$

115. a) $y' = 6x^2$

b) $f'(x) = -12x^2 + 4x - 1$

c) $y' = -1x^{-2} = \dfrac{-1}{x^2}$

d) $y' = -6x^{-3} = \dfrac{-6}{x^3}$

e) $f'(x) = x^{-2} = \dfrac{1}{x^2}$

f) $f'(x) = \frac{3}{5}x^2 - \frac{8}{3}x + \frac{1}{2}$

g) $f'(x) = \dfrac{-9}{x^4} + \frac{4}{5}x - 3$

h) $f'(x) = 1$

i) $f'(x) = -6x$

116. a) $y'' = 4x + 7$

b) $f''(x) = 6x^{-4} = \dfrac{6}{x^4}$

117. $x = \frac{7}{2}$

118. $y = 5x - 10$

119. a) > b) = c) <

120. a) $(-\infty, -1.55]$ and $[0.215, \infty)$
  b) $[-1.55, 0.215]$
  c) $(-1.55, 3.63)$
  d) $(0.215, 0.887)$

121. a) $x = \frac{1}{4}$ b) $\frac{7}{8}$ c) $\left(\frac{1}{4}, \frac{7}{8}\right)$

122. $\frac{1}{2}$

123. a) Perimeter $P = 2x + y = 270$
  b) Area $A = xy$
  c) $2x + y = 270$
      $y = 270 - 2x$
      Area $A(x) = x(270 - 2x) = 270x - 2x^2$
  d) Area $A'(x) = 270 - 4x$

e) Maximum area $A'(x) = 0 = 270 - 4x$
   $x = \frac{270}{4} = 67.5$ m
   Perimeter $P = 2x + y = 270$ and $x = 67.5$
   $2(67.5) + y = 270$
   $y = 270 - 135 = 135$ m
   Dimensions of the field are 67.5 metres by 135 metres.

f)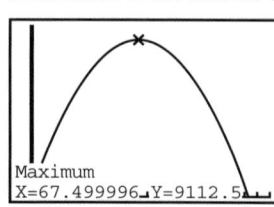
   Maximum area = 9112.5 m^2 (9110 m^2 to 3 s.f.)

Index

Page numbers in italics refer to information boxes and hint boxes.